T0180947

Applied Mathematical Sciences

Volume 184

For further volumes:
http://www.springer.com/series/34

Shangjiang Guo • Jianhong Wu

Bifurcation Theory of Functional Differential Equations

 Springer

Shangjiang Guo
College of Mathematics and Econometrics
Hunan University
Changsha, Hunan
China, People's Republic

Jianhong Wu
Department of Mathematics and Statistics
York University
Toronto, Ontario, Canada

ISSN 0066-5452
ISBN 978-1-4899-8896-6   ISBN 978-1-4614-6992-6 (eBook)
DOI 10.1007/978-1-4614-6992-6
Springer New York Heidelberg Dordrecht London

Mathematics Subject Classification (2010): 34K18, 34C14, 34C15, 34C20, 34C23, 34C25, 35B32, 37C80, 37C85, 37G05, 37G15

Printed on acid-free paper

Springer is part of Springer Science+Business Media (www.springer.com)

Preface

A functional differential equation (FDE) describes the evolution of a dynamical system for which the rate of change of the state variable depends on not only the current but also the historical and even future states of the system. FDEs arise naturally in economics, life sciences, and engineering, and the study of FDEs has been a major source of inspiration for advancement of nonlinear analysis and infinite-dimensional dynamical systems. Therefore, FDEs provide an excellent theoretical platform for developing an interdisciplinary approach to understanding complex nonlinear phenomena via appropriate mathematical techniques.

Unfortunately, the study of FDEs is difficult for newcomers, since a background in nonlinear analysis, ordinary differential equations, and dynamical systems is a prerequisite. On the other hand, the novelty and challenge of fundamental research in the field of FDEs has often been underappreciated. This is especially so in our effort to describe the qualitative behaviors of solutions near equilibria or periodic orbits: these qualitative behaviors can be derived from those of finite-dimensional (ordinary differential) systems obtained through a center and center-unstable manifold reduction process, and hence the (local) bifurcation theory that deals with significant changes in these qualitative behaviors is in principle a consequence of the corresponding theory for finite-dimensional (ordinary differential) systems. The highly nontrivial and often lengthy calculation of center manifold reduction, however, not only leads to enormous duplication of calculation efforts, but also prevents us from discovering simple and key mechanisms behind observed bifurcation phenomena due to the infinite-dimensionality of FDEs. This, in turn, makes it difficult to express bifurcation results explicitly in terms of model parameters and to compare and validate different results. Another challenge is the study of the birth and global continuation of bifurcation of periodic solutions and the coexistence of multiple periodic solutions when the parameters are far from the bifurcation/critical values. There has been substantial progress dedicated to this global bifurcation problem, and remarkably, the presence of a delayed or advanced argument in the nonlinearity can sometimes facilitate the application of topological methods such as equivalent degrees to examine the global continua of branches of periodic solutions, and this has inspired interesting developments in the spectral analysis of circulant matrices.

On the other hand, the study of dynamical systems with symmetries has become well established as a major branch of nonlinear systems theory. The current interest in the field dates mainly to the appearance of the equivariant branching lemma of Vanderbauwhede and Cicogna and the equivariant Hopf bifurcation theorem of Golubitsky and Stewart, both of which are reviewed in the book by Golubitsky, Stewart and Schaeffer. Since then, important new theories have been developed for more complex dynamical phenomena, including the existence, stability, and bifurcations of systems of heteroclinic connections, and the symmetry groups and bifurcations of chaotic attractors.

To a large extent, the phenomenal growth in the subject has been due to its effectiveness in explaining the bifurcations and dynamical phenomena that are seen in a wide range of physical systems including coupled oscillators, reaction–diffusion systems, convecting fluids, and mechanical systems. A local symmetric bifurcation theory for FDEs can be derived from that of but since some special properties of spatiotemporal symmetry of FDEs may be reflected generically in the reduced finite-dimensional systems, one can and should make general observations about the particular bifurcation patterns of symmetric FDEs.

The purpose of this book is to summarize some practical and general approaches and frameworks for the investigation of bifurcation phenomena of FDEs depending on parameters. The book aims to be self-contained, so the reader should find in this book all relevant materials on bifurcation, dynamical systems with symmetry, functional differential equations, normal forms, and center manifold reduction. This material was used in graduate courses on functional differential equations at Hunan University (China) and York University (Canada). We want to thank all students in these courses for their careful reading and some helpful comments. We would like especially to thank Dr. Jing Fang and Dr. Xiang-Sheng Wang for their careful reading of an early version of the manuscript and for their critical comments.

This work was supported in part by the National Natural Science Foundation (China), the Program for New Century Excellent Talents in University of Ministry of Education (China), the Research Fund for the Doctoral Program of Higher Education of China, the Hunan Provincial Natural Science Foundation, the NCE Centre Mathematics for Information Technology and Complex Systems, Mprime, the Canada Research Chairs Program, and the Natural Sciences and Engineering Research Council of Canada.

Changsha, Hunan, China   Shangjiang Guo
Toronto, ON, Canada   Jianhong Wu

Contents

Chapter 1
Introduction to Dynamic Bifurcation Theory

1.1 Introduction

The change in the qualitative behavior of solutions as a control parameter (or control parameters) in a system is varied and is known as a *bifurcation*. When the solutions are restricted to neighborhoods of a given equilibrium, a bifurcation occurs often when the zero solution of the linearization of the system at the equilibrium changes its stability. To illustrate the basic concepts of bifurcation phenomena, we consider the following continuous dynamical system defined by the C^r ($r \geq 1$) vector field f: $\Lambda \times U \to \mathbb{R}^n$:

$$\dot{x} = f(\mu, x), \quad \mu \in \Lambda \subseteq \mathbb{R}^m, \quad x \in U \subseteq \mathbb{R}^n, \tag{1.1}$$

where U and Λ are open sets, x is the state variable, and μ is the (bifurcation) parameter.

Continuously varying μ may change the qualitative behavior of the solutions of (1.1). A value $\mu \in \Lambda$ for which such a change occurs is called a *bifurcation (critical) value*. The set of all bifurcation values is called the *bifurcation set* in the parameter space $\mathbb{R}^m$. We may use a bifurcation diagram to schematically show the considered solutions (equilibria/fixed points, closed orbits/periodic orbits, invariant tori) of a system as a function of a bifurcation parameter in the system. It is normal to represent stable solutions with solid lines and unstable solutions with dashed lines.

Local bifurcations are relevant to the birth or initiation of bifurcations when the bifurcation parameter is close to a bifurcation value. A local bifurcation from a given solution (an equilibrium, a periodic orbit, etc.) can normally be detected from a local stability analysis at the given solution. The global bifurcation thereby concerns the continuation of a local bifurcation when the bifurcation parameter is away from the bifurcation value.

The bifurcation phenomena is linked closely to the concepts of topological equivalence, structural stability, and genericity, which are described in the next section.

S. Guo and J. Wu, *Bifurcation Theory of Functional Differential Equations*,
Applied Mathematical Sciences 184, DOI 10.1007/978-1-4614-6992-6_1,
© Springer Science+Business Media New York 2013

1.2 Topological Equivalence

In the study of dynamical systems, we are interested in not only specific solutions of a specific system, but also classification of solutions of a particular system and classification of systems according to general qualitative behaviors, that is, the number, position, and stability of equilibria, periodic orbits, and other isolated invariant sets.

In what follows, we will not distinguish a flow and a dynamical system. This means that we consider a continuous mapping $\Phi: \mathbb{R} \times U \to U$ over an open set $U \subseteq \mathbb{R}^n$ such that $\Phi(0,x) = x$ and $\Phi(t, \Phi(s,x)) = \Phi(t+s,x)$ for $t, s \in \mathbb{R}$, and $x \in U$. Sometimes, we write it as $\Phi^t := \Phi(t, \cdot): U \to U$ for $t \in \mathbb{R}$.

We consider two dynamical systems to be (locally) equivalent if their (local) phase portraits are similar in a qualitative sense, that is, if they can be locally transformed into each other through a continuous transformation. More precisely, we introduce the following definition.

Definition 1.1. A dynamical system Φ in $\mathbb{R}^n$ is said to be *topologically equivalent* in a region $U \subset \mathbb{R}^n$ to a dynamical system Ψ in a region $V \subset \mathbb{R}^n$ if there exists a homeomorphism $h: U \to V$ that maps the orbits of Φ in U onto the orbits of Ψ in V, preserving the direction of time.

A homeomorphism is an invertible map such that both the map and its inverse are continuous. A homomorphism is called a diffeomorphism if it is C^1-smooth and its inverse is also C^1-smooth. The definition of topological equivalence can be generalized to cover more general cases in which the state space is a complete metric or, in particular, a Banach space. The definition also remains meaningful when the state space is a smooth finite-dimensional manifold in $\mathbb{R}^n$, for example, a two-dimensional torus $\mathbb{T}^2$ or sphere $\mathbb{S}^2$. The phase portraits of topologically equivalent systems are often said to be topologically equivalent.

Example 1.1. Consider the flows Φ^t and Ψ^t associated with the differential equations

$$\dot{x} = -x \quad \text{and} \quad \dot{y} = -3y,$$

respectively. The homeomorphism $h: \mathbb{R} \to \mathbb{R}$ given by $h(x) = x^3$ for $x \in \mathbb{R}$ maps the orbits of Φ onto those of Ψ.

Definition 1.2. Two flows Φ^t (on U) and Ψ^t (on V) are called *topologically conjugate* if there exists a homeomorphism $h: U \to V$ such that

$$\Psi^t = h \circ \Phi^t \circ h^{-1} \quad \text{for} \quad t \in \mathbb{R}.$$

We also use the term *smoothly conjugate* (or *diffeomorphic*) if the involved homeomorphism is a diffeomorphism and the flows are smooth.

For example, for a continuous-time system

$$\dot{x} = f(x), \quad x \in \mathbb{R}^n, \tag{1.2}$$

if h is a diffeomorphism from $\mathbb{R}^n$ to $\mathbb{R}^n$, and $x = h(y)$, then the system

$$\dot{y} = g(y), \quad y \in \mathbb{R}^n \tag{1.3}$$

with $g(y) = [Dh(y)]^{-1} f(h(y))$ for all $y \in \mathbb{R}^n$ is smoothly equivalent (or diffeomor-phic) to system (1.2). In fact, denoting by $\Phi^t(x)$ the flow associated with system (1.2), and letting $\Psi^t(y) = h^{-1}(\Phi^t(h(y)))$, we have

$$Dh(\Psi^t(y))\frac{d}{dt}\Psi^t(y) = f(\Phi^t(h(y))),$$

and so

$$\frac{d}{dt}\Psi^t(y) = [Dh(\Psi^t(y))]^{-1} f(\Phi^t(h(y))) = g(\Psi^t(y)),$$

which implies that $\Psi^t(y)$ is the flow associated with system (1.3). Therefore, systems (1.2) and (1.3) are smoothly equivalent (or diffeomorphic).

In what follows, if the degree of smoothness of h is of interest, we also use the term C^k-equivalent or C^k-diffeomorphic.

Two diffeomorphic systems are practically identical and can be viewed as the same system written using different coordinates. Two diffeomorphic systems have similar qualitative behaviors. For such systems, the eigenvalues of corresponding equilibria are the same: Let x_0 and $y_0 = h(x_0)$ be such equilibria and let $A(x_0)$ and $B(y_0)$ denote corresponding Jacobian matrices. Then we have

$$A(x_0) = M^{-1}(x_0)B(y_0)M(x_0),$$

where $M(x) = Dh(x)$. Therefore, the characteristic polynomials for the matrices $A(x_0)$ and $B(y_0)$ coincide.

It is easy to construct nondiffeomorphic but topologically equivalent flows. To see this, consider a smooth scalar position function $\mu: \mathbb{R}^n \to (0, \infty)$ and assume that the right-hand sides of (1.2) and (1.3) are related by

$$f(x) = \mu(x)g(x) \quad \text{for} \quad x \in \mathbb{R}^n. \tag{1.4}$$

Then systems (1.2) and (1.3) are topologically equivalent since their orbits are identical, and it is the velocity of the motion that makes them different. Thus, the homeomorphism h in Definition 1.1 is the identity map $h(x) = x$. In other words, these two systems are distinguished only by the time parameterization along the orbits. We say that two systems (1.2) and (1.3) satisfying (1.4) for a smooth pos-itive function μ are *orbitally equivalent*. Usually, two orbitally equivalent systems can be nondiffeomorphic, having cycles that look like the same closed curve in the phase space but different periods. For example, the system

$$\dot{r} = r(1-r), \quad \dot{\theta} = 1$$

and the system

$$\dot{\rho} = 2\rho(1-\rho), \quad \dot{\phi} = 2$$

in $\mathbb{R}^2$ using polar coordinates are topologically equivalent, but not topologically conjugate, because their periodic orbits $r = 1$ and $\rho = 1$ have periods 2π and π, respectively.

Let x_0 be an equilibrium of the system (1.2), that is, $f(x_0) = 0$, and let A denote the Jacobian matrix $Df(x)$ evaluated at $x = x_0$. Let n_-, n_0, and n_+ be the numbers of eigenvalues of A (counting multiplicities) with negative, zero, and positive real part, respectively. Recall that an equilibrium is called *hyperbolic* if $n_0 = 0$, that is, if A has no purely imaginary eigenvalues. A hyperbolic equilibrium is called a *hyperbolic saddle* if $n_- n_+ \neq 0$.

Topological equivalence of linear systems is generally easy to determine. If the linearized flow near an equilibrium is asymptotically stable, then the equilibrium is asymptotically stable. Moreover, two asymptotically stable n-dimensional linear flows are topologically equivalent.

Example 1.2. Consider two linear planar systems:

$$\dot{x} = -x, \quad \dot{y} = -y, \tag{1.5}$$

and

$$\dot{x} = -x - y, \quad \dot{y} = x - y. \tag{1.6}$$

Clearly, the origin is a stable equilibrium in both systems. All other trajectories of (1.5) are straight lines, while those of (1.6) are spirals. The equilibrium of the first system is a node, while in the second systems it is a focus. These two systems are neither orbitally nor smoothly equivalent. However, they are topologically equivalent.

We can further claim that *near a hyperbolic equilibrium p, the system behaves essentially like the linearized one.* In other words, Φ^t is topologically equivalent to $e^{Df(p)t}$ in a sufficiently small neighborhood of a hyperbolic equilibrium p (Grobman–Hartman theorem). See Grobman [123], Hartman [161, 162], Hirsch [163], Hale and Kocak [152] for details. As a result, determining topological equivalence near hyperbolic equilibria boils down to counting the dimensions of the local stable and unstable subspaces (manifolds).

Theorem 1.1. *Two systems of differential equations with hyperbolic equilibria are topologically equivalent near these equilibria if and only if their linearizations have the same number n_+ of eigenvalues with positive real parts and the same number n_- of eigenvalues with negative real parts.*

1.3 Structural Stability

There are dynamical systems whose phase portrait (in some domain) does not change qualitatively under all sufficiently small perturbations. For example, suppose that (1.1) has an equilibrium x_0 when $\mu = \mu_0$, that is,

$$f(\mu_0, x_0) = 0. \tag{1.7}$$

It is natural to ask about the stability of this equilibrium and how the stability or instability is affected as μ is varied. Thus, we first linearize (1.1) at (μ_0, x_0) to get

$$\dot{x} = D_x f(\mu_0, x_0)x, \quad x \in \mathbb{R}^n. \tag{1.8}$$

If the eigenvalues of the linearized matrix $D_x f(\mu_0, x_0)$ are all nonzero, then the linearized matrix is invertible, and by an application of the implicit function theorem, there is a curve $\mu \to \beta(\mu)$ in $\mathbb{R}^n$ such that $\beta(\mu_0) = x_0$ and $f(\mu, \beta(\mu)) \equiv 0$ for all sufficiently small $|\mu - \mu_0|$. In other words, for each μ in the domain of β, the point $\beta(\mu) \in \mathbb{R}^n$ corresponds to an equilibrium point for the member of the family (1.1) at the parameter value μ.

If the equilibrium x_0 is hyperbolic, that is, none of eigenvalues of the linearized matrix $D_x f(\mu_0, x_0)$ lie on the imaginary axis, then the linearized matrix of (1.1) at $(\mu, \beta(\mu))$ is $D_x f(\mu, \beta(\mu))$ it depends smoothly on μ and coincides with $D_x f(\mu_0, x_0)$ at $\mu = \mu_0$. Recall that if $D_x f(\mu_0, x_0)$ has no eigenvalues on the imaginary axis, then neither does $D_x f(\mu, \beta(\mu))$ for each μ in a sufficiently small neighborhood of μ_0. In other words, $\beta(\mu)$ is a hyperbolic equilibrium of (1.1) for all μ in a sufficiently small neighborhood of μ_0. Moreover, the numbers n_+ and n_- of the positive and negative eigenvalues of $D_x f(\mu, \beta(\mu))$ are fixed for these values of μ. In view of Theorem 1.1, system (1.1) is locally topologically equivalent to $\dot{x} = f(\mu_0, x)$ near the equilibria. This means that *a hyperbolic equilibrium is structurally stable under smooth perturbations.*

Inspired by the above property, we now can define a structurally stable system, which means that every sufficiently close system is topologically equivalent to the structurally stable one.

Definition 1.3. A flow Φ is said to be *structurally stable* in a region $D \subset \mathbb{R}^n$ if for every flow Ψ that is sufficiently C^1-close to Φ, there exist regions U and V with $D \subset U$ such that Ψ is topologically equivalent in V to Φ in U.

The following theorem results from the previous discussion.

Theorem 1.2. *A flow with a hyperbolic equilibrium is structurally stable in a neighborhood of the equilibrium.*

In Definition 1.3, we require the C^1 metric, instead of C^0, because two C^0 curves may be arbitrarily close to each other but have different numbers of equilibria. Moreover, it would be nice to show that structurally stable systems are generic. The following classical theorem gives necessary and sufficient conditions for a continuous-time system in a plane to be structurally stable.

Theorem 1.3 (Andronov and Pontryagin [16]). *A smooth dynamical system*

$$\dot{x} = f(x), \quad x \in \mathbb{R}^2,$$

is structurally stable in a region $D_0 \subset \mathbb{R}^2$ if and only if

(i) The number of equilibria and periodic orbits is finite and each is hyperbolic;
(ii) There are no orbits connecting saddle points.

Furthermore, for two-dimensional vector fields on compact manifolds, we have the following result due to Peixoto [244].

Theorem 1.4 (Peixoto's theorem [244]). *Let $\mathscr{D}$ be a compact two-dimensional manifold without boundary and let $\mathscr{X}^k(\mathscr{D})$ denote the C^k ($k \geq 1$) vector fields defined on $\mathscr{D}$. Then $f \in \mathscr{X}^k(\mathscr{D})$ is structurally stable on $\mathscr{D}$ if and only if*

(i) The number of equilibria and periodic orbits is finite and each is hyperbolic;
(ii) There are no orbits connecting saddle points;
(iii) The nonwandering set consists of equilibria and periodic orbits.

Moreover, if D is orientable, then the set of such vector fields is open and dense in $\mathscr{X}^k(\mathscr{D})$.

This theorem is useful because it spells out precise conditions for structural stability on the dynamics of a vector field on a compact two-manifold without boundary under which it is structurally stable. Unfortunately, we do not have a similar theorem in higher dimensions. This is in part due to the presence of complicated recurrent motions (e.g., the Smale horseshoe). In light of this theorem, it appears to be practically convenient to ignore more structurally unstable vector fields defined on a compact two-dimensional manifold without boundary, because an arbitrarily small perturbation will usually turn a structurally unstable vector field into a structurally stable one. However, as we shall see, if this vector field depends on a parameter, more complicated dynamics will take place.

1.4 Codimension-One Bifurcations of Equilibria

Let x_0 be a hyperbolic equilibrium point of (1.1) for $\mu = \mu_0$. As we have seen in the previous section, under a small parameter variation, the equilibrium moves slightly but remains hyperbolic. Therefore, we can vary the parameter further and control the equilibrium. It is clear that there are, generically, only two ways in which the hyperbolicity condition can be violated. Either a simple real eigenvalue approaches zero, or a pair of simple complex eigenvalues reaches the imaginary axis for some values of the parameter.

If the equilibrium x_0 of (1.1) is not hyperbolic, that is, $D_x f(\mu_0, x_0)$ has some eigenvalues on the imaginary axis, then the topology of the local phase portrait of the corresponding differential equation (1.1) at this equilibrium point may change under perturbation, that is, a bifurcation occurs. For example, equilibria can be created or destroyed, and time-dependent behavior such as periodic, quasiperiodic, homoclinic, heteroclinic, or even chaotic dynamics can be created. Moreover, the more eigenvalues on the imaginary axis, the more complicated the dynamics will be.

For equilibria of flows, a (generic) codimension-one bifurcation means that the crossing of the stability region (the imaginary axis) is taking place with either one eigenvalue of the linear part going through 0 or one pair of complex conjugate eigenvalues crossing the imaginary axis. This section will be devoted essentially to the proof that a nonhyperbolic equilibrium satisfying one of these two conditions is structurally unstable and to the analysis of the corresponding bifurcations of the local phase portrait under variation of the parameter.

Definition 1.4. The bifurcation associated with the appearance of eigenvalue 0 is called a *fold* (or *tangent*) bifurcation.

This bifurcation is also associated with a lot of other names, including limit point and turning point.

Definition 1.5. The bifurcation corresponding to the presence of a pair of complex purely imaginary eigenvalues is called a *Hopf* (or *Andronov–Hopf*, or *Poincaré–Andronov–Hopf*) bifurcation.

As pointed out repeatedly by Arnold [19], examples of Hopf bifurcation can be found in the work of Poincaré [248]. The first specific study and formulation of a theorem in this area was due to Andronov [14]. However, the work of Poincaré and Andronov was concerned with two-dimensional vector fields. The existence of such a bifurcation was found in the context of general n-dimensional ordinary differential equations (ODEs) by Hopf [167] in 1942. This was before the discovery of the center manifold theorem. For these reasons, we usually refer to this kind of bifurcation as a Poincaré–Andronov–Hopf bifurcation.

In the 1970s, Hsu and Kazarinoff [169], Poore [250], Marsden and McCracken [217], and others discussed in their works the computation of important features of the Hopf bifurcation, especially the direction of bifurcation and dynamical aspects (stability, attractiveness, etc.), both from theoretical and numerical standpoints. A very important new achievement was the proof by Alexander and Yorke [10] of what is known as the global Hopf bifurcation theorem, which, roughly speaking, describes the global continuation of the local branch. The theory was also extended to allow further degeneracies (more than two eigenvalues crossing the imaginary axis, or multiplicity higher than one, etc.), leading notably to the development of the generalized Hopf bifurcation theory (Bernfeld et al. [31, 32], Negrini and Salvadori [228]).

Now, if these phenomena were taking place in a linear system, then there would be just a low-dimensional (1 or 2, respectively) invariant subspace to be affected by the bifurcations. In what follows, we first study these bifurcations in systems of smallest possible dimension for the bifurcations to take place. Here, the effort will be to obtain expressions for these systems that are as simple as possible while still capturing the bifurcations of interest, and at the same time to show that other systems undergoing the same bifurcation are locally topologically equivalent to these simple ones. In subsequent chapters, we shall see that center manifold reduction can transform the bifurcation problem in general functional differential equations (of course, general n-dimensional ODEs) into that of ordinary differential equations on a one- or two-dimensional invariant manifold. Therefore, this part of study is basic and crucial for discussing bifurcations in general functional differential equations (see Chap. 7).

1.4.1 Fold Bifurcation

Consider the following one-parameter scalar ODE:

$$\dot{x} = f(\mu, x), \quad x, \mu \in \mathbb{R}, \tag{1.9}$$

where $f(0,0) = 0$. That is, (1.9) has an equilibrium $x_0 = 0$ when $\mu = \mu_0 = 0$. The condition ensuring a fold bifurcation of (1.9) is that $f_x(0,0) = 0$. Usually, we may encounter three situations, as discussed in this section.

Example 1.3. Consider the family of differential equations

$$\dot{x} = \mu - x^2, \quad x, \mu \in \mathbb{R}.$$

We see that $\mu = 0$ is the bifurcation value. In particular, if $\mu > 0$, then there are two equilibria: an unstable equilibrium $-\sqrt{\mu}$ and a stable one $\sqrt{\mu}$. At the bifurcation value $\mu = 0$, there is only one equilibrium, which is not hyperbolic. If $\mu < 0$, there are no equilibria. The bifurcation diagram is the parabola $\mu = x^2$ labeled as in Fig. 1.1. Notice that the parameter μ is assigned to the horizontal axis, while the stable equilibria are drawn in solid lines and the unstable equilibria in dashed lines.

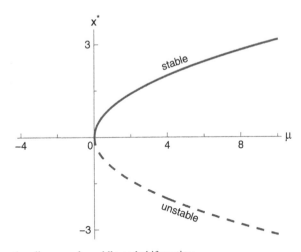

Fig. 1.1 Bifurcation diagram of a saddle-node bifurcation

The type of bifurcation described in Example 1.3—on one side of a parameter value there are no equilibria, and on the other side there are two equilibria—is referred to as a *saddle-node bifurcation*. The next theorem lists sufficient conditions for a saddle-node bifurcation to occur at $(\mu, x) = (0,0)$ in the scalar system (1.9). A more general theorem on saddle-node bifurcation will be formulated and proved later for general delay differential equations.

Theorem 1.5. *Suppose f in (1.9) is sufficiently smooth and satisfies*

$$f(0,0) = 0, \quad f_x(0,0) = 0, \quad f_\mu(0,0)f_{xx}(0,0) \neq 0. \tag{1.10}$$

Then there are smooth invertible local changes of coordinate and parameter that transform the system (1.9) into the following normal form:

$$\dot{y} = \gamma \pm y^2 + O(|y|^3). \tag{1.11}$$

Therefore, if $f_\mu(0,0)f_{xx}(0,0) < 0$ (respectively, > 0), then near the origin, only two equilibria exist for $\mu > 0$ (respectively, < 0), only one equilibrium $x = 0$ exists for $\mu = 0$, and no equilibria exist for $\mu < 0$ (respectively, > 0). In the case that two equilibria exist, one is asymptotically stable and the other is unstable.

Proof. Expanding f with respect to x around $\mu = 0$ yields

$$f(\mu,x) = f_0(\mu) + f_1(\mu)x + f_2(\mu)x^2 + O(x^3),$$

where

$$f_j(\mu) = \frac{1}{j!}\frac{\partial^j f}{\partial x^j}(\mu,0), \quad j = 0,1,2,\ldots.$$

Obviously, $f_0(0) = f(0,0) = 0$ and $f_1(0) = f_x(0,0) = 0$. Set $\xi = x + \delta$, where δ is a constant independent of t. Then (1.9) can be transformed into

$$\dot{\xi} = f_0(\mu) - f_1(\mu)\delta + f_2(\mu)\delta^2 + O(\delta^3)$$
$$+ [f_1(\mu) - 2f_2(\mu)\delta + O(\delta^2)]\xi + [f_2(\mu) + O(\delta)]\xi^2 + O(\xi^3). \tag{1.12}$$

Noting that $f_1(0) = 0$ and $f_2(0) = \frac{1}{2}f_{xx}(0,0) \neq 0$, and using the implicit function theorem, we can find $\delta(\mu)$ for small μ such that $f_1(\mu) - 2f_2(\mu)\delta + O(\delta^2) = 0$. This gives

$$\delta(\mu) = \frac{f_{\mu x}(0,0)}{f_{xx}(0,0)}\mu + O(\mu^2).$$

Using this $\delta(\mu)$, we have

$$\dot{\xi} = \beta(\mu) + [f_2(\mu) + O(\mu)]\xi^2 + O(\xi^3), \tag{1.13}$$

where $\beta(\mu) = f_0'(0)\mu + O(\mu^2)$. Recall that $f_0'(0) = f_\mu(0,0) \neq 0$. Then the function β is invertible near the origin. Hence, we can obtain $\mu(\beta)$ with $\mu(0) = 0$. Thus, (1.13) can be changed into the form

$$\dot{\xi} = \beta \pm c(\beta)\xi^2 + O(\xi^3),$$

where the sign is that of $f_{xx}(0,0)$ and c is a smooth positive function. Take $y = c(\beta)\xi$ and $\gamma = c(\beta)\beta$. Then we obtain (1.11), which is obviously topologically equivalent to $\dot{y} = \gamma \pm y^2$. The rest of the proof follows from Example 1.3. $\square$

Remark 1.1. In the study of bifurcations, we usually have bifurcation conditions and genericity conditions (nondegeneracy conditions). For the saddle-node bifurcation of (1.9), the bifurcation conditions are $f(0,0) = 0$ and $f_x(0,0) = 0$, and the genericity conditions are $f_\mu(0,0) = 0$ and $f_{xx}(0,0) = 0$. The bifurcation conditions will be used to numerically search for bifurcation points, while the genericity conditions will be used to verify whether a bifurcation point is really of the type we are looking for, i.e., to guarantee that locally, nothing more complicated can occur.

1.4.2 Poincaré–Andronov–Hopf Bifurcation

We start with a simple example in which a pair of simple complex conjugate eigenvalues cross the imaginary axis.

Example 1.4. Consider the following planar system:

$$\dot{x} = \mu x - y - x(x^2 + y^2),$$
$$\dot{y} = x + \mu y - y(x^2 + y^2), \tag{1.14}$$

where $x, y, \mu \in \mathbb{R}$. Using the complex and polar coordinates $z = x + iy = re^{i\theta}$, system (1.14) takes the forms

$$\dot{z} = (\mu + i)z - z|z|^2$$

and

$$\dot{r} = r(\mu - r^2), \quad \dot{\theta} = 1,$$

which can be solved for (r, θ):

$$r = \begin{cases} \sqrt{\mu(1 + Ce^{-2\mu t})^{-1}}, & \mu \neq 0, \\ \sqrt{(2t + C)^{-1}}, & \mu = 0, \end{cases} \tag{1.15}$$
$$\theta = t - t_0,$$

where C and t_0 are determined by the initial condition. Variations of the phase portrait of system (1.14) as μ passes through zero can be easily analyzed using the polar form (1.15), since the equations for r and θ are uncoupled. We can see that system (1.14) always has a unique equilibrium at the origin, which is a stable focus for $\mu < 0$ and an unstable focus for $\mu > 0$. This equilibrium is surrounded for $\mu > 0$ by an isolated closed orbit (limit cycle) that is unique and stable. This bifurcation is *supercritical* because the closed orbit (limit cycle) appears after the bifurcation.

The bifurcation diagram for periodic solutions of (1.14) is simply a plot of the solutions of $\mu = r^2$ in the (μ, r)-plane together with the line $r = 0$ (see Fig. 1.2). As usual, stable periodic orbits are indicated by solid curves, and unstable ones with dashed curves.

Similarly, the system

$$\dot{x} = \mu x - y + x(x^2 + y^2),$$
$$\dot{y} = x + \mu y + y(x^2 + y^2), \tag{1.16}$$

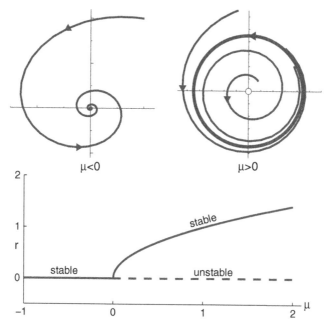

Fig. 1.2 A supercritical Hopf bifurcation

can be rewritten as

$$\dot{z} = (\mu + \mathrm{i})z + z|z|^2$$

or

$$\dot{r} = r(\mu + r^2), \quad \dot{\theta} = 1.$$

This has an unstable periodic solution (limit cycle) for $\mu < 0$. Hence this bifurcation is *subcritical*.

As described in Example 1.4, in a Poincaré–Andronov–Hopf bifurcation, an equilibrium of a system loses stability as a pair of complex conjugate eigenvalues of the linearization around the equilibrium cross the imaginary axis of the complex plane. Under reasonably generic assumptions about the dynamical system, we can expect a small-amplitude limit cycle branching from the fixed point. Either the limit cycle is orbitally stable and the bifurcation is supercritical, or the limit cycle is unstable and the bifurcation is subcritical.

The next theorem lists sufficient conditions for a Poincaré–Andronov–Hopf bifurcation to occur in a planar system.

Theorem 1.6 (Hassard and Wan [159]). *Consider the following system:*

$$\begin{bmatrix} \dot{x} \\ \dot{y} \end{bmatrix} = \begin{bmatrix} \mu & \omega \\ -\omega & \mu \end{bmatrix} \begin{bmatrix} x \\ y \end{bmatrix} + \begin{bmatrix} f^1(x,y) \\ f^2(x,y) \end{bmatrix}, \tag{1.17}$$

where $\omega > 0$ and f^j is three times differentiable, satisfying $f_x^j(0,0) = f_y^j(0,0) = 0$, $j = 1,2$. Then there exists a branch of periodic solutions of (1.17) bifurcating from the trivial solution $x = 0$, and the Poincaré–Andronov–Hopf bifurcation is supercritical (subcritical), i.e., bifurcating periodic solutions exist for $\mu > 0$ (respectively, < 0) if $\Upsilon < 0$ (respectively, > 0), where

$$\Upsilon = f_{xxx}^1 + f_{xyy}^1 + f_{xxy}^2 + f_{yyy}^2$$
$$+ \frac{1}{\omega}[f_{xy}^1(f_{xx}^1 + f_{yy}^1) - f_{xy}^2(f_{xx}^2 + f_{yy}^2) - f_{xx}^1 f_{xx}^2 + f_{yy}^1 f_{yy}^2].$$

More generally, in order to investigate Poincaré–Andronov–Hopf bifurcations in high-dimensional ODEs, even in infinite-dimensional ODEs generated by partial differential equations (PDEs) and functional differential equations (FDEs), we may employ center manifold reduction and normal form theory to obtain the following system:

$$\dot{z} = \lambda(\mu)z + C(\mu)z|z|^2 + O(|z|^5), \quad (\mu, z) \in \mathbb{R} \times \mathbb{C}, \tag{1.18}$$

where $\lambda(0) = i\omega$ and $\omega > 0$. Detailed analysis can be found in Sects. 3.4.1, 4.3.1, and 7.3.2. Also see [54, 55, 74, 152, 200, 257, 282, 302] for more background on Poincaré-Andronov-Hopf bifurcation.

Definition 1.6. The first Lyapunov coefficient of a Hopf bifurcation is defined by $l_1(0) = \text{Re}\{C(0)\}/\omega$.

As stated in Lemma 3.7 of Kuznetsov [200], if $\text{Re}\{\lambda'(0)\}\text{Re}\{C(0)\} \neq 0$, then (1.18) can be transformed by a parameter-dependent linear coordinate transformation, a time rescaling, and a nonlinear-time reparameterization into an equation of the form

$$\dot{z} = (\beta + i)z + sz|z|^2 + O(|z|^5), \quad (\mu, z) \in \mathbb{R} \times \mathbb{C}, \tag{1.19}$$

where $s = \text{sgn}\text{Re}\{C(0)\} = \text{sgn}l_1(0)$ and β is the new parameter. Obviously, the truncated system of (1.19) is equivalent to either (1.14) (in the cases in which $s = -1$) or (1.16) (in the cases in which $s = 1$). Thus, the bifurcation direction and stability of bifurcated periodic solutions are determined by the signs of $\text{Re}\{\lambda'(0)\}$ and $\text{Re}\{C(0)\}$ (or equivalently, $l_1(0)$).

1.5 Transcritical and Pitchfork Bifurcations of Equilibria

In a saddle-node bifurcation, on one side of a parameter value there is no equilibrium, and on the other side there are two equilibria. In some examples, we may meet another type of bifurcation: both equilibria exist before and after the bifurcation value, and there is one unstable equilibrium and one stable one; however, their stability is exchanged when they collide. So the unstable equilibrium becomes stable and vice versa. We refer to this type as a *transcritical bifurcation*, as shown in the following example.

Example 1.5. Consider a vector field

$$\dot{x} = \mu x - x^2, \quad x, \mu \in \mathbb{R}.$$

If $\mu < 0$, there are two equilibria: $x = 0$, which is stable, and $x = \mu$, which is unstable. These two equilibria coalesce at the bifurcation value $\mu = 0$. If $\mu > 0$, there are also two equilibria: $x = 0$ is unstable, while $x = \mu$ is stable. The bifurcation diagram is depicted in Fig. 1.3.

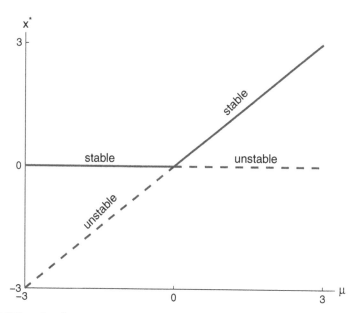

Fig. 1.3 Bifurcation diagram of a transcritical bifurcation

Similarly to the proof of Theorem 1.5, we may list sufficient conditions for a transcritical bifurcation for the scalar system (1.9).

Theorem 1.7. *Suppose f in (1.9) is sufficiently smooth and satisfies*

$$f(\mu,0) = 0, \quad f_x(0,0) = 0, \quad f_{x\mu}(0,0)f_{xx}(0,0) \neq 0. \tag{1.20}$$

Then there are smooth invertible local coordinate and parameter changes that transform the system (1.9) into the following normal form:

$$\dot{y} = \gamma y \pm y^2 + O(|y|^3). \tag{1.21}$$

Therefore, besides the trivial solution, system (1.9) has a nonzero equilibrium, which continuously depends on μ for all sufficiently small $|\mu|$ and is stable for all sufficiently small μ such that $\mu f_{x\mu}(0,0) > 0$.

To illustrate another generic equilibrium bifurcation, we consider the following example.

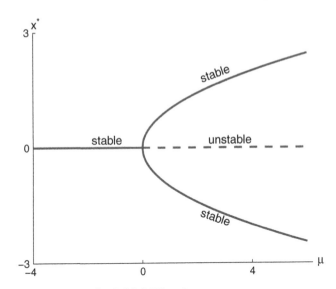

Fig. 1.4 Bifurcation diagram of a pitchfork bifurcation

Example 1.6. The vector field

$$\dot{x} = \mu x - x^3, \qquad x, \mu \in \mathbb{R}$$

has one stable equilibrium $x = 0$ if $\mu < 0$, and it has an unstable equilibrium $x = 0$ and two stable equilibria $x = \pm\sqrt{\mu}$ if $\mu > 0$. See Fig. 1.4.

The bifurcation diagram is shown in Fig. 1.4, and this kind of bifurcation is known as a *pitchfork bifurcation*. Note that $x = 0$ is always an equilibrium. However, as the parameter μ passes through the bifurcation value $\mu = 0$, the equilibrium at the origin loses its stability and two new stable equilibria are bifurcated from the origin. This is also an example of spontaneous symmetry breaking, because the two bifurcated equilibria do not have the symmetry $\mathbb{Z}_2$ possessed by the system. Moreover, this pitchfork bifurcation is called *supercritical* because new equilibria exist for a parameter μ that is greater than the bifurcation value $\mu = 0$. When additional equilibria exist for a parameter μ smaller than the bifurcation value $\mu = 0$, the bifurcation is called *subcritical*. An example of a subcritical pitchfork bifurcation can be seen in the equation $\dot{x} = \mu x + x^3$.

Similarly, we may list sufficient conditions for a pitchfork bifurcation in the scalar system (1.9). A more general theorem on pitchfork bifurcation will be formulated and proved in Sect. 7.2.

Theorem 1.8. *Suppose f in (1.9) is sufficiently smooth and satisfies*

$$f(\mu, -x) = -f(\mu, x), \quad f_x(0,0) = 0, \quad f_{x\mu}(0,0)f_{xxx}(0,0) \neq 0. \qquad (1.22)$$

Then there are smooth invertible local coordinate and parameter changes that transform the system (1.9) into the following normal form:

$$\dot{y} = \gamma y \pm y^3 + o(|y|^3). \tag{1.23}$$

Therefore, if $f_{x\mu}(0,0)f_{xxx}(0,0) < 0$ (respectively, > 0), then two nontrivial equilibria exist for $\mu > 0$ (respectively, < 0), and only the trivial equilibrium continues to exist for $\mu < 0$ (respectively, > 0). Moreover, the two nontrivial equilibria coalesce into zero as μ goes to 0.

Remark 1.2. The codimension of a bifurcation is the number of parameters that must be varied for the bifurcation to occur. It coincides with the number of transversality conditions. This also corresponds to the codimension of the parameter set for which the bifurcation occurs within the full space of parameters. Saddle-node bifurcations and Hopf bifurcations are the only generic local bifurcations that are really of codimension one, while transcritical and pitchfork bifurcations both have a higher codimension. However, transcritical and pitchfork bifurcations are also often thought of as begin of codimension one, because the normal forms (1.21) and (1.23) can be written with only one parameter.

Remark 1.3. In Theorems 1.7 and 1.8, we study the transcritical and pitchfork bifurcations of equilibria in the one-parameter scalar system (1.9). Based on center manifold reduction (Chap. 3) and normal form theory (Chap. 4), we can discuss these bifurcations in high-dimensional systems, even in infinite-dimensional systems such as functional differential equations. See Sect. 7.2 for more details.

1.6 Bifurcations of Closed Orbits

When (1.1) has a periodic orbit Γ_0 when $\mu = \mu_0$, one may also be interested in the qualitative behaviors of solutions of (1.1) in a neighborhood of the periodic orbit Γ_0 for the parameter μ near μ_0.

The so-called Poincaré map is a technical tool for studying the local behaviors of solutions of (1.1) near a periodic orbit. To describe this tool, we consider a local transversal section L_ε to the periodic orbit Γ_0 (see Fig. 1.5). There are $\alpha_0 > 0$ and $\delta > 0$ such that for $0 \le |\mu - \mu_0| < \alpha_0$ and $x_0 \in L_\delta$, there is a first time $T(\mu, x_0) > 0$ such that the solution $x(t; \mu, x_0)$ of (1.1) satisfies $x(T(\mu, x_0); \mu, x_0) \in L_\varepsilon$. Therefore, we define the Poincaré map depending on parameters as $\Pi(\mu, x_0) = x(T(\mu, x_0); \mu, x_0)$ mapping L_δ to L_ε. Periodic orbits near Γ_0 correspond to fixed points of $\Pi(\mu, x_0)$. The periodic orbit through the point $x_0 \in L_\delta$ is said to be hyperbolic if x_0 is a hyperbolic fixed point of the Poincaré map $\Pi(\mu_0, \cdot)$, that is, none of the eigenvalues of the linearized operator $D_x\Pi(\mu_0, x_0)$ (also referred to as Floquet multipliers) lie on the unit circle.

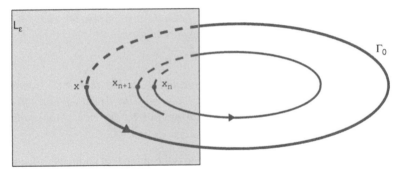

Fig. 1.5 An illustration of the Poincaré map for a periodic orbit, where x_{n+1} is the image of x_n under the Poincaré map

By means of the Poincaré map, we can investigate the behavior of solutions near a periodic solution. If Γ_0 is hyperbolic, then for each μ with $|\mu - \mu_0|$ small, there is a unique periodic orbit Γ_μ near Γ_0, and Γ_μ is also hyperbolic. When Γ_0 is nonhyperbolic, the bifurcations near the periodic orbit Γ_0 can be determined from those of the Poincaré map $\Pi(\mu, x_0)$.

Example 1.7 (Saddle-node bifurcation of periodic orbits). Consider the planar system

$$\dot{x} = \mu x - y + x(x^2 + y^2)(1 - x^2 - y^2), \qquad (1.24)$$
$$\dot{y} = x + \mu y + y(x^2 + y^2)(1 - x^2 - y^2),$$

where $x, y, \mu \in \mathbb{R}$. In polar coordinates $x + iy = re^{i\theta}$, the system (1.24) has the form

$$\dot{r} = r(\mu + r^2 - r^4), \qquad (1.25)$$
$$\dot{\theta} = 1.$$

Since the two equations above are uncoupled, we may investigate directly the local fold bifurcations for the r-equation using the general arguments in Sect. 1.4. However, the r-equation is so special that we can employ the following arguments to depict the global bifurcation explicitly and directly.

Indeed, if $\mu = -0.25$, then the periodic orbit is given by $r = \frac{\sqrt{2}}{2}$, and the transversal section L becomes

$$L = \{(r, \theta) \in \mathbb{R} \times \mathbb{S}^1 : r > 0, \theta = 0\}.$$

So the poincaré map $\Pi(-0.25, r)$ has a fixed point at $r = \frac{\sqrt{2}}{2}$. Moreover, it is easy to see that $D_r \Pi(-0.25, \frac{\sqrt{2}}{2}) = 1$. Consequently, the corresponding periodic orbit is nonhyperbolic. Moreover, since the first equation is independent of θ, it is easy to see that in the radial direction, system (1.25) undergoes a saddle-node bifurcation as the parameter μ passes through -0.25. If $\mu \in (-0.25, 0)$, system (1.25) has two periodic orbits: a stable periodic orbit $r = \sqrt{0.5 + \sqrt{\mu + 0.25}}$ and an unstable periodic orbit $r = \sqrt{0.5 - \sqrt{\mu + 0.25}}$. If $\mu < -0.25$, then system (1.25) has no periodic orbits, because $\dot{r} < 0$ and all the solutions tend to the origin as $t \to \infty$; see

Fig. 1.6. At the bifurcation value $\mu = -0.25$, there is only one semistable periodic orbit $r = \frac{\sqrt{2}}{2}$, which is not hyperbolic and has a Floquet multiplier equal to one.

For a nonhyperbolic periodic orbit of a higher-dimensional continuous dynamical system, there may be some bifurcations of closed orbits, which cannot happen in a planar system. For example, if at $\mu = \mu_0$, the closed orbit has a Floquet multiplier -1 and the modulus of all the remaining Floquet multipliers are not equal to 1, then a period-doubling bifurcation (also referred to as flip or subharmonic bifurcation) of the closed orbit may take place (see Fig. 1.7). Namely, as μ passes through μ_0, the closed orbit Γ_0 becomes another closed orbit Γ_μ with approximately twice the period of Γ_0. See Arnold [19], Newhouse–Palis–Takens [230], Feigenbaum [94] for further information. If at $\mu = \mu_0$ the closed orbit Γ_0 has a pair of complex conjugate Floquet multipliers on the unit circle, then as μ passes through μ_0, this nonhyperbolic closed orbit may bifurcate into a two-dimensional invariant torus Γ_μ (or $\mathbb{T}^2$). This bifurcation has many names. Some call it Neimark–Sacker bifurcation, while others call it the secondary Andronov–Hopf bifurcation due to its similarity to that for flows discussed in the previous section. Detailed analysis of Neimark–Sacker bifurcations can be found in Ruelle and Takens [256], Sacker [258], and Kuznetsov [200]. For details and further results on periodic orbits and their bifurcations, see, for example, [17, 63, 64, 98–102, 121, 122, 143, 151–155, 185, 212–214, 225, 288].

1.7 Homoclinic Bifurcation

A homoclinic orbit of a system is given by the intersection of the stable and unstable manifolds of a saddle-type invariant set (see Andronov and Leontovich [15], Kuznetsov [200]). Recall that the stable manifold is defined as the set of all trajectories that tend to the invariant set in forward time, and the unstable manifold is defined as the set of all trajectories that tend to the invariant set in backward time. Here, the invariant sets that we consider are steady states (equilibria) and/or periodic solutions.

For example, an orbit Γ_0 starting at a point $x \in \mathbb{R}$ is called homoclinic to the equilibrium x_0 of system (1.1) with $\mu = \mu_0$ if the solution $\varphi(t; x, \mu_0)$ tends to x_0 as $t \to \pm\infty$. In particular, if at $\mu = \mu_0$, system (1.1) has a homoclinic loop Γ_0, and the intersection of the stable and unstable manifolds of equilibria or closed orbits of system is not transversal,[1] then system (1.1) is not structurally stable. A slight perturbation of the parameter μ makes the stable and unstable manifolds either non-intersecting or transversally intersecting, and so may change the topological structure of the vector field of (1.1). Thus, closed orbits can be created or destroyed, and time-dependent behaviors such as invariant tori and even chaotic dynamics can be created. Therefore, a homoclinic orbit to a steady state is of codimension one; it may be destroyed by small perturbations to the system parameters.

---

[1] Two smooth manifolds $M, N \in \mathbb{R}^n$ intersect transversally if there exist n linearly independent vectors that are tangent to at least one of these manifolds at every intersection point.

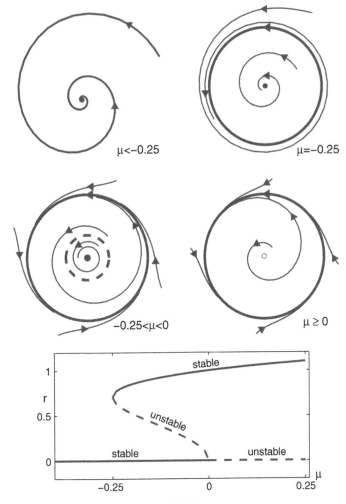

Fig. 1.6 The bifurcation phenomena of system (1.24)

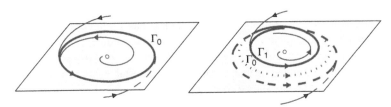

Fig. 1.7 Period-doubling bifurcation of a closed orbit

However, if a homoclinic orbit to a periodic solution is the transversal intersection of the stable and unstable manifolds of the periodic solution (Kuznetsov [200, Sects. 6.1 and 7.2.1]), then it cannot be destroyed by a small perturbation, but it can be destroyed through a codimension-one homoclinic tangency. This occurs when the intersection of the stable and unstable manifolds becomes tangential, and thus a small perturbation can separate the manifolds completely. A transition between a homoclinic orbit of a saddle-focus-type steady state and a homoclinic orbit of a periodic solution occurs at a codimension-two Shil'nikov–Hopf bifurcation (see Hirschberg and Knobloch [165]). At the Shil'nikov–Hopf bifurcation, the homoclinic orbit is *transferred* from the steady state to the periodic solution.

Example 1.8 (Periodic orbit from a homoclinic loop). Consider the planar system

$$\dot{x} = 2y, \tag{1.26}$$
$$\dot{y} = 2x - 3x^2 - y(x^3 - x^2 + y^2 - \mu),$$

where $x, y, \mu \in \mathbb{R}$. For all values of μ, system (1.26) always has two equilibria: one saddle $(0,0)$ and one source $(2/3,0)$ when $\mu > -4/27$. When $\mu = 0$, we can employ Lyapunov functions $V(x,y) = x^3 - x^2 + y^2$ and phase portrait analysis to show that system (1.26) has a homoclinic orbit loop through the origin and attracts from inside, as seen in Fig. 1.8. For $-4/27 < \mu < 0$, using the invariance principle, one can show that there is an orbitally asymptotically stable periodic orbit lying on the curves $x^3 - x^2 + y^2 - \mu = 0$. As μ increases and tends to zero, the periodic orbit grows until it collides with the saddle point. At the bifurcation point $\mu = 0$, the period of the periodic orbit has grown to infinity, and it has become a homoclinic orbit. For $\mu > 0$, the homoclinic loop is broken, and also there is no periodic orbit. This sequence of bifurcations is illustrated in Fig. 1.8. Therefore, there is a homoclinic bifurcation at $\mu = 0$.

A homoclinic bifurcation often occurs when a periodic orbit collides with a saddle point. Homoclinic bifurcations can occur supercritically or subcritically. In three or more dimensions, bifurcations of higher codimension can occur, producing complicated, possibly chaotic, dynamics [297, 298].

1.8 Heteroclinic Bifurcation

An orbit Γ_0 starting at a point $x \in \mathbb{R}$ is called heteroclinic to the equilibria x_1 and $x_2 \neq x_1$ of system (1.1) with $\mu = \mu_0$ if the solution $\varphi(t; x, \mu_0)$ tends to x_1 as $t \to \infty$ and to x_2 as $t \to \infty$. The nontransversal heteroclinic case is somehow trivial, since the disappearance of the connecting orbit is the only essential event in a sufficiently small neighborhood of $\Gamma_0 \cup \{x_1, x_2\}$ (see Example 1.9).

Example 1.9 (Heteroclinic bifurcation). Consider the planar system

$$\dot{x} = x^2 - y^2 - 1, \tag{1.27}$$
$$\dot{y} = \mu + y^2 - xy,$$

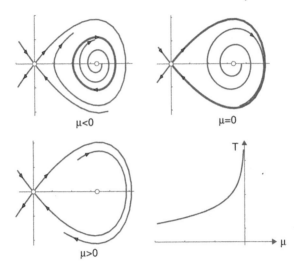

Fig. 1.8 Bifurcation phenomena in (1.26)

where $x, y, \mu \in \mathbb{R}$. When $\mu = 0$, system (1.27) has a heteroclinic trajectory connecting the two saddle points $(1,0)$ and $(-1,0)$. However, there is no heteroclinic trajectory when $\mu \neq 0$. Therefore, there is a heteroclinic bifurcation at $\mu = 0$ (Fig. 1.9).

1.9 Two-Parameter Bifurcations of Equilibria

Here we briefly review the generic bifurcations in two-parameter families of differential equations. We only give a list for them, and refer to Kuznetsov [200, 201], Guckenheimer [126], or Guckenheimer and Holmes [125] for analysis. There are two categories of generic bifurcations in two-parameter families: (1) extra eigenvalues can approach the imaginary axis; (2) some of the genericity con-

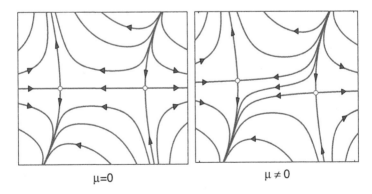

Fig. 1.9 Bifurcation phenomena in (1.27)

ditions for the codimension-one bifurcations can be violated. Thus, we can identify five bifurcation (Bogdanov–Takens bifurcation, cusp bifurcation, Bautin bifurcation, fold-Hopf bifurcation, and Hopf–Hopf bifurcation) points that one can meet in generic two-parameter systems while moving along codimension-one curves. Each of these bifurcations is characterized by two independent conditions. This section is devoted to the study of these bifurcations in the least possible phase-space dimensions.

1.9.1 Bogdanov–Takens Bifurcation

The Bogdanov–Takens bifurcation is a bifurcation of an equilibrium point in a two-parameter family of autonomous ODEs at which the critical equilibrium has a zero eigenvalue of (algebraic) multiplicity two. It is named after Rifkat Bogdanov and Floris Takens, who independently and simultaneously described this bifurcation. The main features of Bogdanov–Takens bifurcation were known to mathematicians of the Andronov school in the late 1960s. However, the complete picture is due to Bogdanov [35], as announced by Arnold [20] and Takens [274]. Their analysis is based on the Pontryagin [249] technique.

The usual normal form of the Bogdanov–Takens bifurcation is

$$\dot{x} = y,$$
$$\dot{y} = \mu_1 + \mu_2 x + x^2 \pm xy + O(\sqrt{(x^2 + y^2)^3}),$$

which was introduced by Bogdanov (see Sect. 7.4.2 for more details), while the normal form derived by Takens is

$$\dot{x} = y + \mu_2 x + x^2 + O(\sqrt{(x^2 + y^2)^3}),$$
$$\dot{y} = \mu_1 \pm x^2 + O(\sqrt{(x^2 + y^2)^3}).$$

These two systems are equivalent, and their detailed analysis can be found, for example, in Guckenheimer and Holmes [125] and Kuznetsov [200]. In the above systems, four associated bifurcation curves meet at the Bogdanov–Takens bifurcation: two branches of the saddle-node bifurcation curve, an Andronov–Hopf bifurcation curve, and a saddle homoclinic bifurcation curve. Moreover, these bifurcations are nondegenerate, and no other bifurcations occur in a small fixed neighborhood of $(x,y) = (0,0)$ for parameter values sufficiently close to $\mu = 0$. In this neighborhood, the system has at most two equilibria and one limit cycle.

If system (1.1) has a fixed equilibrium $x = x_0$ for all parameters μ, and the equilibrium x_0 has a zero eigenvalue of (algebraic) multiplicity two at $\mu = 0$, then the normal form of (1.1) at $(\mu, x) = (0, x_0)$ is not equivalent to the above two systems derived by Bogdanov or Takens. See Sect. 7.4.3 for more details. Therefore, the goal of this subsection is to investigate the following two-parameter system:

$$\dot{x} = y, \tag{1.28}$$
$$\dot{y} = \mu_1 x + \mu_2 y + x^2 + xy,$$

where $(\mu_1,\mu_2) \in \mathbb{R}^2$. At $(\mu_1,\mu_2) = (0,0)$, the linearization of (1.28) at the equilibrium $O = (0,0)$ has exactly one eigenvalue 0 of geometric multiplicity one and algebraic multiplicity two. The critical point $(\mu_1,\mu_2) = (0,0)$ is referred to as a Bogdanov–Takens point.

It is easy to see that system (1.28) always has two equilibria: $O = (0,0)$ and $E = (-\mu_1,0)$. Moreover, the characteristic equation of (1.28) at the equilibria O and E are $\lambda^2 - \mu_2\lambda - \mu_1 = 0$ and $\lambda^2 - (\mu_2 - \mu_1)\lambda + \mu_1 = 0$, respectively. Each of these two equations can have between zero and two real roots. However, the discriminant parabolas $\{(\mu_1,\mu_2) : \mu_2^2 + 4\mu_1 = 0\}$ and $\{(\mu_1,\mu_2) : (\mu_2 - \mu_1)^2 - 4\mu_1 = 0\}$ are not bifurcation curves at which the equilibrium O or E undergoes a node to focus transition. Moreover, it is easy to see that the equilibrium O (respectively, E) is a saddle for all parameters $\mu_1 > 0$ (respectively, $\mu_1 < 0$).

We can check that the equilibria O and E have a pair of purely imaginary eigenvalues on the lines $l_1 = \{(\mu_1,\mu_2) : \mu_1 < 0, \mu_2 = 0\}$ and $l_2 = \{(\mu_1,\mu_2) : \mu_1 = \mu_2 \geq 0\}$, respectively. This implies that the equilibrium O (or E) undergoes a nondegenerate Hopf bifurcation along the line l_1 (respectively, l_2), giving rise to an unstable limit cycle, since the first Lyapunov coefficients are both $1/|\mu_1| > 0$. The cycle exists near l_1 (or l_2) for $\mu_2 < 0$ (respectively, $\mu_2 < \mu_1$). We have the following results on the existence of a homoclinic bifurcation.

Theorem 1.9. *There exist exactly two smooth curves m_1 and m_2 corresponding to saddle homoclinic bifurcations in system (1.28) that originate at $(\mu_1,\mu_2) = (0,0)$ and have the following local representation:*

$$m_1 = \left\{ (\mu_1,\mu_2) : \mu_2 = \frac{1}{7}\mu_1 + o(|\mu_1|), \mu_1 \leq 0 \right\}$$

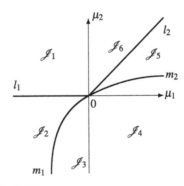

Fig. 1.10 Bifurcation sets for (1.28)

and

$$m_2 = \left\{ (\mu_1,\mu_2) : \mu_2 = \frac{6}{7}\mu_1 + o(|\mu_1|), \mu_1 \geq 0 \right\}.$$

Moreover, for $\|(\mu_1,\mu_2)\|$ small, system (1.28) has a unique and hyperbolic unstable cycle for parameter values inside the region bounded by the Hopf bifurcation curve

l_1 (or l_2) and the homoclinic bifurcation curve m_1 (respectively, m_2), and no cycles outside these regions.

Proof. First, we restrict our attention to the parameter region $\mathscr{I}_2$ in Fig. 1.10. Performing a singular rescaling and introducing a new time

$$u = x/(-\mu_1), \quad v = y/\sqrt{(-\mu_1)^3}, \quad s = t\sqrt{(-\mu_1)},$$

reduces (1.28) to

$$\dot{u} = v, \tag{1.29}$$
$$\dot{v} = u(u-1) - (\gamma_1 v + \gamma_2 uv),$$

where the dots mean derivatives with respect to the new time s and

$$\gamma_1 = -\mu_2/\sqrt{(-\mu_1)}, \quad \gamma_2 = -\sqrt{(-\mu_1)}. \tag{1.30}$$

Clearly, system (1.29) is orbitally equivalent to a system induced by (1.28) with the help of (1.30). Studying the limit cycles and homoclinic orbits of (1.29) for $(\gamma_1, \gamma_2) \neq (0,0)$ provides complete information on these objects in (1.28). As stated in Kuznetsov 1998 [200, Sect. 8.8], there is a unique smooth curve m corresponding to a saddle homoclinic bifurcation in system (1.29) that originates at $(\gamma_1, \gamma_2) = (0,0)$ and has the following local representation:

$$m = \left\{ (\gamma_1, \gamma_2) : \gamma_1 = -\frac{1}{7}\gamma_2 + o(|\gamma_2|), \gamma_2 \le 0 \right\}.$$

This homoclinic curve is mapped by (1.30) into the curve m_1. Using arguments similar to those in Kuznetsov 1998 [200, Sect. 8.8], we see that the cycle in (1.28) is unique and hyperbolic within the region bounded by l_1 and m_1.

In what follows, we focus on the parameter region $\mathscr{I}_5$, where O is a saddle and E is a stable focus. Translate the origin of the coordinate system to the left (antisaddle) equilibrium E of system (1.28):

$$\xi_1 = x + \mu_1, \quad \xi_2 = y.$$

This obviously gives

$$\dot{\xi}_1 = \xi_2, \tag{1.31}$$
$$\dot{\xi}_2 = -\mu_1 \xi_1 + (\mu_2 - \mu_1)\xi_2 + \xi_1^2 + \xi_1 \xi_2.$$

Performing a singular rescaling and introducing a new time

$$u = x/\mu_1, \quad v = y/\sqrt{\mu_1^3}, \quad s = t\sqrt{\mu_1}$$

reduces (1.31)–(1.29) with

$$\gamma_1 = (\mu_1 - \mu_2)/\sqrt{\mu_1}, \quad \gamma_2 = -\sqrt{\mu_1}. \tag{1.32}$$

Thus, the homoclinic curve m is mapped by (1.32) into the curve m_2. Similarly, the cycle in (1.28) is unique and hyperbolic within the region bounded by l_2 and m_2. □

Thus, for $(\mu_1, \mu_2) \in m_2$ (or $(\mu_1, \mu_2) \in m_1$), there is an orbit homoclinic to the equilibrium O (respectively, E). In fact, we can also have nearly explicit expressions for the homoclinic orbits. Scaling system (1.28) by

$$t^* = \varepsilon t, \quad x^* = x/\varepsilon^2, \quad \mu_1^* = \mu_1/\varepsilon^2, \quad \mu_2^* = \mu_2/\varepsilon^2,$$

and then dropping the $*$ gives

$$x'' - \varepsilon[\mu_2 x' + xx'] - (\mu_1 x + x^2) = O(\varepsilon^2). \tag{1.33}$$

Letting $\varepsilon = 0$, the equation has an explicit homoclinic orbit for $\mu_1 > 0$:

$$x = -\frac{2\mu_1}{2} \left[1 - \tanh^2\left(\frac{\sqrt{\mu_1}}{2} t \right) \right].$$

Using the Melnikov method (see, for example, Guckenheimer and Holmes 1983 [125]), we can compute parameter values for which the homoclinic orbit to the equilibrium O persists for ε. Moreover, the nearly explicit expressions for the homoclinic orbit to the equilibrium E can be discussed analogously.

Make a round trip near the Bogdanov–Takens point $(\mu_1, \mu_2) = (0,0)$ (see Fig. 1.10), starting from region $\mathscr{J}_1$, where equilibrium E is a saddle. There is a nonbifurcation curve (not shown in the figure) located in $\mathscr{J}_1$ and passing through the origin at which the equilibrium O undergoes an unstable node to an unstable focus transition. Entering from region $\mathscr{J}_1$ into region $\mathscr{J}_2$ through the Hopf bifurcation boundary l_1, the unstable focus O gains stability, and an unstable limit cycle $\mathscr{O}_1$ is present for sufficiently small parameters $|\mu_1|$ and $|\mu_2|$ satisfying $\mu_1 < 0$ and $\mu_2 < 0$. If we continue the journey counterclockwise, the unstable limit cycle $\mathscr{O}_1$ grows and approaches the saddle, turning into a homoclinic orbit at m_1. There are no cycles in region $\mathscr{J}_3$, where the equilibrium E remains a saddle while the stable focus O becomes a stable node. Entering from region $\mathscr{J}_3$ into region $\mathscr{J}_4$ through the negative μ_2-axis, the two equilibria O and E coalesce into zero and then exchange their properties, i.e., the stable node O becomes a saddle, while the saddle E becomes a stable node. In region $\mathscr{J}_4$, the equilibrium O remains a saddle, while the stable node E becomes a stable focus. Due to Theorem 1.9, system (1.28) has a homoclinic orbit at the curve m_2. As (μ_1, μ_2) continues moving counterclockwise in region $\mathscr{J}_5$, the homoclinic orbit turns into an unstable limit cycle, which shrinks and collides with equilibrium E and then disappears at the curve l_2. In region $\mathscr{J}_6$, the unstable focus E turns into an unstable node, while O remains a saddle.

1.9.2 Cusp Bifurcation

Cusp bifurcation is a bifurcation of equilibria in a two-parameter family of autonomous ODEs at which the critical equilibrium has one zero eigenvalue and the quadratic coefficient for the saddle-node bifurcation vanishes. Let us begin by considering the following example.

Example 1.10. Consider the following two-parameter system

$$\dot{x} = \mu_1 + \mu_2 x - x^3, \quad (x, \mu_1, \mu_2) \in \mathbb{R}^3. \tag{1.34}$$

At $(\mu_1, \mu_2) = (0,0)$, the linearization of (1.34) at the equilibrium 0 has exactly a simple eigenvalue 0. The critical point $(\mu_1, \mu_2) = (0,0)$ is referred to as a cusp point. The local bifurcation diagram of (1.34) is presented in Fig. 1.11. The cusp point $(\mu_1, \mu_2) = (0,0)$ is the origin of two branches of the saddle-node bifurcation curve:

$$LP_{\pm} = \{(\mu_1, \mu_2) : \mu_1 = \mp \frac{2}{3\sqrt{3}} \mu_2^{3/2}, \mu_2 > 0\},$$

which divides the parameter plane into two regions $\mathscr{I}_{1,2}$. Inside the region $\mathscr{I}_1$, there are three equilibria, two stable and one unstable. In the region $\mathscr{I}_2$, there is a single equilibrium, which is stable. A nondegenerate fold bifurcation (with respect to the parameter μ_1) takes place if we cross either LP_+ or LP_- at any point other than the origin. More precisely, if the curve LP_+ is crossed from region $\mathscr{I}_1$ to region $\mathscr{I}_2$, the right stable equilibrium collides with the unstable one, and then both disappear. The same happens to the left stable equilibrium and the unstable equilibrium at the curve LP_-. In the symmetric case $\mu_1 = 0$, one observes a pitchfork bifurcation as μ_2 is reduced, with one stable solution suddenly splitting into two stable solutions and one unstable solution as the physical system passes to $\mu_2 > 0$ through the cusp point $\mu = 0$ (an example of *spontaneous symmetry breaking*). In other words, if we approach the cusp point from inside the region $\mathscr{I}_1$, all three equilibria merge together into a triple root of the right-hand side of (1.34). Away from the cusp point, there is no sudden change in a physical solution being followed: when passing through the curve of saddle-node bifurcations, all that happens is that an alternative second solution becomes available.

In view of the above example, at the cusp bifurcation point, two branches of the saddle-node bifurcation curve meet tangentially, forming a semicubic parabola. For nearby parameter values, the system can have three equilibria that collide and disappear pairwise via the saddle-node bifurcations. The cusp bifurcation implies the presence of a hysteresis phenomenon.

Cusp bifurcation occurs also in infinite-dimensional ODEs generated by PDEs and DDEs, to which the center manifold theorem (see Chap. 3) applies. See Arrowsmith and Place [21] for details. The nomenclature and analysis of cusp bifurcations is based on cusps in singularity theory, where they appear as one of Thom's seven elementary catastrophes [275, 276]. The following theorem lists sufficient conditions for a general one-dimensional ODE.

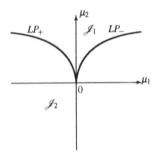

Fig. 1.11 Bifurcation sets for (1.34)

Theorem 1.10. *Suppose the system*

$$\dot{x} = f(\mu,x), \quad x \in \mathbb{R}, \ \mu = (\mu_1,\mu_2) \in \mathbb{R}^2, \tag{1.35}$$

with a smooth function f, has at $\mu = 0$ the equilibrium $x = 0$ for which the cusp bifurcation conditions are satisfied, namely, $f_x(0,0) = f_{xx}(0,0) = 0$. Assume that the following genericity conditions are satisfied:

$$f_{xxx}(0,0) \neq 0, \quad \det \begin{bmatrix} f_{\mu_1} & f_{\mu_2} \\ f_{x\mu_1} & f_{x\mu_2} \end{bmatrix}_{(\mu,x)=(0,0)} \neq 0. \tag{1.36}$$

Then there are smooth invertible coordinate and parameter changes transforming the system into

$$\dot{y} = \beta_1 + \beta_2 y + s y^3 + O(y^4), \tag{1.37}$$

where the $s = \operatorname{sign} f_{xxx}(0,0)$ and $O(y^4)$ terms depend smoothly on (β_1,β_2).

Proof. Expand f with respect to x around $\mu = 0$:

$$f(\mu,x) = f_0(\mu) + f_1(\mu)x + f_2(\mu)x^2 + f_3(\mu)x^3 + O(x^3),$$

where

$$f_j(\mu) = \frac{1}{j!} \frac{\partial^j f}{\partial x^j}(\mu,0), \quad j = 0,1,2,\dots.$$

Obviously, $f_0(0) = f(0,0) = 0$, $f_1(0) = f_x(0,0) = 0$, and $f_2(0) = \frac{1}{2} f_{xx}(0,0) = 0$. Set $\xi = x + \delta(\mu)$, where δ is a constant independent of t. Then (1.35) can be transformed into

$$
\begin{aligned}
\dot{\xi} = {}& [f_0(\mu) - f_1(\mu)\delta + \delta^2\varphi(\mu,\delta)] + [f_1(\mu) - 2f_2(\mu)\delta + \delta^2\phi(\mu,\delta)]\xi \\
& + [f_2(\mu) - 3f_3(\mu)\delta + \delta^2\psi(\mu,\delta)]\xi^2 + [f_3(\mu) + \delta\theta(\mu,\delta)]\xi^3 + O(\xi^4)
\end{aligned}
\tag{1.38}
$$

for some smooth functions φ, ϕ, ψ, and θ. Since $f_2(0) = 0$, we cannot use the implicit function theorem to select a function $\delta(\mu)$ to eliminate the linear terms in ξ in the above equation. However, in view of $f_3(0) = \frac{1}{6}f_{xxx}(0,0) \neq 0$, there is a smooth shift function $\delta(\mu)$, $\delta(0) = 0$, that annihilates the quadratic terms in the equation for all sufficiently small $\|\mu\|$. Indeed, let $F(\mu, \delta) = f_2(\mu) - 3f_3(\mu)\delta + \delta^2\psi(\mu, \delta)$. Then we have $F(0,0) = 0$ and $F_\delta(0,0) = -3f_3(0) \neq 0$. Therefore, the implicit function theorem gives the (local) existence and uniqueness of a smooth scalar function $\delta = \delta(\mu)$ such that $\delta(0) = 0$ and $F(\mu, \delta(\mu)) = 0$ for $\|\mu\|$ small enough. Now with $\delta(\mu)$ as constructed above, (1.38) contains no quadratic terms. Let $\gamma(\mu) = (\gamma_1(\mu), \gamma_2(\mu))$ be defined as

$$\gamma_1(\mu) = f_0(\mu) - f_1(\mu)\delta(\mu) + \delta^2(\mu)\varphi(\mu, \delta(\mu)),$$
$$\gamma_2(\mu) = f_1(\mu) - 2f_2(\mu)\delta(\mu) + \delta^2(\mu)\phi(\mu, \delta(\mu)).$$

Clearly, $\gamma(0) = 0$, and the Jacobian matrix of the map $\gamma = \gamma(\mu)$ is nonsingular at $\mu = 0$:

$$\det\left(\frac{\partial\gamma}{\partial\mu}\right)\bigg|_{\mu=0} = \det\begin{bmatrix} f_{\mu_1} & f_{\mu_2} \\ f_{x\mu_1} & f_{x\mu_2} \end{bmatrix}_{\mu=0} \neq 0. \tag{1.39}$$

Thus, the inverse function theorem implies the local existence and uniqueness of a smooth inverse function $\mu = \mu(\gamma)$ with $\mu(0) = 0$. Therefore, the equation for ξ now reads

$$\dot{\xi} = \gamma_1 + \gamma_2\xi + c(\gamma)\xi^3 + O(\xi^4),$$

where $c(\gamma) = f_3(\mu(\gamma)) + \delta(\mu((\gamma))\theta(\mu(\gamma), \delta(\mu(\gamma)))$ is a smooth function of γ and $c(0) = f_3(0) = \frac{1}{6}f_{xxx}(0,0) \neq 0$. Finally, the above equation can be transformed into (1.37) by performing a linear scaling $y = \xi\sqrt{|c(\gamma)|}$ and introducing new parameters: $\beta_1 = \gamma_1\sqrt{|c(\gamma)|}$, $\beta_2 = \gamma_2$. $\qquad\square$

1.9.3 Fold–Hopf Bifurcation

The fold–Hopf bifurcation is a bifurcation of an equilibrium point in a two-parameter family of autonomous ODEs at which the critical equilibrium has a zero eigenvalue and a pair of purely imaginary eigenvalues. This phenomenon is also called the zero–Hopf bifurcation or Gavrilov–Guckenheimer bifurcation. An early example of this bifurcation in a specific system is provided by the Brusselator reaction–diffusion system in one spatial dimension (Guckenheimer [124], Wittenberg and Holmes [299]).

The usual norm form of the fold–Hopf bifurcation is

$$
\begin{aligned}
\dot{y} &= \mu_1 + b(u^2 + v^2) - y^2 + \text{h.o.t.}, \\
\dot{u} &= \mu_2 u - v + ayu + \text{h.o.t.}, \\
\dot{v} &= u + \mu_2 v + ayv + \text{h.o.t.},
\end{aligned}
\tag{1.40}
$$

where $\mu = (\mu_1, \mu_2)$ and h.o.t. stands for "higher-order terms." System (1.40) has been studied by Broer and Vegter [44], Chow–Li–Wang [66], Dumortier and Ibáñez [84], Gamero–Freire–Rodríguez–Luis [106], Gaspard [107], Gavrilov [108, 109], Guckenheimer [124], Keener [187], Langford [202], Takens [272–274]. The bifurcation point $\mu = 0$ in the μ-parameter plane lies at a tangential intersection of curves of saddle-node bifurcations and Poincaré–Andronov–Hopf bifurcations. Depending on the system, a branch of torus bifurcations can emanate from the fold–Hopf bifurcation point. In such cases, other bifurcations occur for nearby parameter values, including saddle-node bifurcations of periodic orbits on the invariant torus, torus breakdown, and bifurcations of Shil'nikov homoclinic orbits to saddle-foci and heteroclinic orbits connecting equilibria. See Guckenheimer and Holmes [125] for more details.

If system (1.1) has a fixed equilibrium $x = x_0$ for all parameters μ, and the equilibrium x_0 has a zero eigenvalue and a pair of purely imaginary eigenvalues at $\mu = 0$, then the normal form of (1.1) at $(\mu, x) = (0, x_0)$ is not equivalent to system (1.40). Therefore, in this subsection we consider the following two-parameter system:

$$
\begin{aligned}
\dot{y} &= \mu_1 y + y^2 + u^2 + v^2, \\
\dot{u} &= \mu_2 u - v + ayu + y^2 u, \\
\dot{v} &= u + \mu_2 v + ayv + y^2 v,
\end{aligned}
\tag{1.41}
$$

where $0 \neq a \in \mathbb{R}$, $\mu = (\mu_1, \mu_2) \in \mathbb{R}^2$, and $(y, u, v) \in \mathbb{R}^3$. At $\mu = 0$, the linearization of (1.41) at the equilibrium $(0, 0, 0)$ has a zero eigenvalue $\lambda_1 = 0$ and a pair of purely imaginary eigenvalues $\lambda_{2,3} = \pm i$. Let $z = u + iv = \sqrt{\rho} e^{i\theta}$. Then system (1.41) can be rewritten as

$$
\begin{aligned}
\dot{y} &= \mu_1 y + y^2 + |z|^2, \\
\dot{z} &= (\mu_2 + i) z + ayz + y^2 z,
\end{aligned}
\tag{1.42}
$$

and

$$
\begin{aligned}
\dot{y} &= \mu_1 y + y^2 + \rho, \\
\dot{\rho} &= 2\rho(\mu_2 + ay + y^2), \\
\dot{\theta} &= 1.
\end{aligned}
\tag{1.43}
$$

The first two equations of (1.43) are decoupled from the third one. The equation for θ describes a rotation around the y-axis with constant angular velocity $\dot{\theta} = 1$. Thus, to understand the bifurcations in (1.43), we only need to study the planar system for (y, ρ) with $\rho \geq 0$:

$$\dot{y} = \mu_1 y + y^2 + \rho,$$
$$\dot{\rho} = 2\rho(\mu_2 + ay + y^2). \tag{1.44}$$

It is easy to see that system (1.44) always has two equilibria, $E_1 = (0,0)$ and $E_2 = (-\mu_1, 0)$, and that there always exists one orbit connecting E_1 and E_2 due to the symmetry that the y-axis is always invariant. Other equilibria (y, ρ) of (1.44) with $\rho > 0$ satisfy

$$\mu_1 y + y^2 + \rho = 0 \quad \text{and} \quad \mu_2 + ay + y^2 = 0, \tag{1.45}$$

which can have zero, one, or two solutions in the interior of the quadrants with $\rho > 0$. Since we consider the dynamics of (1.44) only with μ sufficiently close to 0, we can require the parameters μ to be in $\mathscr{J} = \{\mu = (\mu_1, \mu_2): |\mu_1| < \frac{1}{2}|a|$ and $|\mu_2| < \frac{1}{4}a^2\}$. Thus, the second equation of (1.45) has two solutions y_1 and y_2 with $y_1 < y_2$. Moreover, $y_1 < y_2 < 0$ if $\mu_2 > 0$ and $a > 0$, while $y_1 < 0 < y_2$ if $\mu_2 < 0$ and $a > 0$. Next, we determine the signs of $\rho_j = -y_j^2 - \mu_1 y_j$, $j = 1,2$, because we consider the equilibrium (y, ρ) of (1.44) only with $\rho \geq 0$.

We first consider the case $a > 0$. We divide the region $\mathscr{J}$ into six parts:

$$\mathscr{J}_{11} = \{\mu \in \mathscr{J} : \mu_1 < 0 \text{ and } \mu_2 > 0\},$$
$$\mathscr{J}_{12} = \{\mu \in \mathscr{J} : \mu_1 < 0 \text{ and } \mu_2 < 0 \text{ and } \mu_1^2 - a\mu_1 + \mu_2 > 0\},$$
$$\mathscr{J}_{13} = \{\mu \in \mathscr{J} : \mu_1 < 0 \text{ and } \mu_1^2 - a\mu_1 + \mu_2 < 0\},$$
$$\mathscr{J}_{14} = \{\mu \in \mathscr{J} : \mu_1 > 0 \text{ and } \mu_1^2 - a\mu_1 + \mu_2 > 0\},$$
$$\mathscr{J}_{15} = \{\mu \in \mathscr{J} : \mu_1 > 0 \text{ and } \mu_2 > 0 \text{ and } \mu_1^2 - a\mu_1 + \mu_2 < 0\},$$
$$\mathscr{J}_{16} = \{\mu \in \mathscr{J} : \mu_1 > 0 \text{ and } \mu_2 < 0\}.$$

These regions are illustrated in Fig. 1.12a, where the bold curve l_4 represents the parabola $\mu_1^2 - \mu_1 a + \mu_2 = 0$.

Lemma 1.1. *Suppose $a > 0$. Then in the interior of the quadrants of the (y, ρ)-plane with $\rho > 0$, system (1.45) has no solution (respectively, one solution (y_2, ρ_2) with $y_2 > 0$, one solution (y_2, ρ_2) with $y_2 < 0$) for parameters μ in $\mathscr{J} \setminus (\mathscr{J}_{12} \cup \mathscr{J}_{15})$ (respectively, $\mathscr{J}_{12}$, $\mathscr{J}_{15}$).*

Proof. We distinguish two cases:

Case 1: $\mu_1 < 0$. Then $y^2 + \mu_1 y$ is negative if $0 < y < -\mu_1$ and positive otherwise. If $\mu \in \mathscr{J}_{11}$, then $y_1 < y_2 < 0$, and hence $\rho_j = -y_j^2 - \mu_1 y_1 < 0$, $j = 1, 2$. If $\mu \in \mathscr{J}_{12}$, then $y_1 < 0 < y_2 < -\mu_1$, and hence $\rho_1 < 0$ and $\rho_2 > 0$. If $\mu \in \mathscr{J}_{13}$, then $y_1 < 0 < -\mu_1 < y_2$, and hence $\rho_j = -y_j^2 - \mu_1 y_1 < 0$, $j = 1, 2$.

Case 2: $\mu_1 > 0$. Then $y^2 + \mu_1 y$ is negative if $-\mu_1 < y < 0$ and positive otherwise. If $\mu \in \mathscr{J}_{14}$, then $y_1 < y_2 < -\mu_1 < 0$, and hence $\rho_j < 0$, $j = 1, 2$. If $\mu \in \mathscr{J}_{15}$, then $y_1 < -\mu_1 < y_2 < 0$, and hence $\rho_1 < 0$ and $\rho_2 > 0$. If $\mu \in \mathscr{J}_{16}$, then $y_1 < -\mu_1 < 0 < y_2$, and hence $\rho_1 < 0$ and $\rho_2 < 0$. The proof is complete. $\qquad\square$

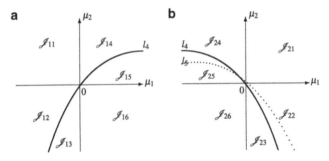

Fig. 1.12 Bifurcation sets for (1.44)

Moreover, for parameters $\mu \in \mathscr{I}_{12} \cup \mathscr{I}_{15}$, the characteristic equation of (1.44) at the equilibrium $E_3 = (y_2, \rho_2)$ is

$$\varsigma^2 - (\mu_1 + 2y_2)\varsigma - 2\rho_2(a + 2y_2) = 0.$$

The two eigenvalues $\varsigma_{1,2}$ satisfy $\varsigma_1 \varsigma_2 = -2\rho_2(a + 2y_2)$. For $\mu \in \mathscr{I}_{12}$, it follows from the proof of Lemma 1.1 that $y_2 > 0$, and so $\varsigma_1 \varsigma_2 < 0$. For $\mu \in \mathscr{I}_{15}$, it follows from the proof of Lemma 1.1 that $-\frac{1}{2}a < -\mu_1 < y_2 < 0$, and hence $\varsigma_1 \varsigma_2 < 0$. Thus, we obtain the following:

Proposition 1.1. *Suppose $a > 0$. Then, in the quadrants of the (y, ρ)-plane with $\rho \geq 0$, we have the following information on the equilibria of system (1.44):*

(i) *There are two equilibria E_1 and E_2 for $\mu \in \mathscr{I}_{11}$, where E_1 is a saddle and E_2 is a source.*

(ii) *There are three equilibria E_1, E_2, and E_3 for $\mu \in \mathscr{I}_{12}$, where E_1 is a sink, E_2 is a source, $E_3 = (y_2, \rho_2)$ satisfying $y_2 > 0$ and $\rho_2 > 0$ is a saddle.*

(iii) *There are two equilibria E_1 and E_2 for $\mu \in \mathscr{I}_{13}$, where E_1 is a sink and E_2 is a saddle.*

(iv) *There are two equilibria E_1 and E_2 for $\mu \in \mathscr{I}_{14}$, where E_1 is a source and E_2 is a saddle.*

(v) *There are three equilibria E_1, E_2, and E_3 for $\mu \in \mathscr{I}_{15}$, where E_1 is a source, E_2 is a sink, $E_3 = (y_2, \rho_2)$ satisfying $y_2 < 0$ and $\rho_2 > 0$ is a saddle.*

(vi) *There are two equilibria E_1 and E_2 for $\mu \in \mathscr{I}_{16}$, where E_1 is a saddle and E_2 is a sink.*

The following theorem follows immediately from the equivalence mentioned before.

Theorem 1.11. *Suppose $a > 0$. Then a semistable limit cycle of (1.41) appears as μ crosses the negative μ_1-axis from $\mathscr{I}_{11}$ to $\mathscr{I}_{12}$, which is always present for $\mu \in \mathscr{I}_{12}$, and then disappears as μ crosses the parabola l_4 from $\mathscr{I}_{12}$ to $\mathscr{I}_{13}$. Similarly, a semistable limit cycle of (1.41) appears as μ crosses the parabola l_4 from $\mathscr{I}_{14}$ to $\mathscr{I}_{15}$, which is always present for $\mu \in \mathscr{I}_{15}$, and then disappears as μ crosses the positive μ_1-axis from $\mathscr{I}_{15}$ to $\mathscr{I}_{16}$.*

Now we consider the case $a < 0$. Again, we divide the region $\mathscr{J}$ into six parts:

$$\mathscr{J}_{21} = \{\mu \in \mathscr{J} : \mu_1 > 0 \text{ and } \mu_2 > 0\},$$
$$\mathscr{J}_{22} = \{\mu \in \mathscr{J} : \mu_1 > 0 \text{ and } \mu_2 < 0 \text{ and } \mu_1^2 - \mu_1 a + \mu_2 > 0\},$$
$$\mathscr{J}_{23} = \{\mu \in \mathscr{J} : \mu_1 > 0 \text{ and } \mu_1^2 - \mu_1 a + \mu_2 < 0\},$$
$$\mathscr{J}_{24} = \{\mu \in \mathscr{J} : \mu_1 < 0 \text{ and } \mu_1^2 - \mu_1 a + \mu_2 > 0\},$$
$$\mathscr{J}_{25} = \{\mu \in \mathscr{J} : \mu_1 < 0 \text{ and } \mu_2 > 0 \text{ and } \mu_1^2 - \mu_1 a + \mu_2 < 0\},$$
$$\mathscr{J}_{26} = \{\mu \in \mathscr{J} : \mu_1 < 0 \text{ and } \mu_2 < 0\}.$$

These regions are illustrated in Fig. 1.12b, where the bold curve l_4 and the dotted curve l_5 represent the parabolas $\mu_1^2 - \mu_1 a + \mu_2 = 0$ and $\mu_1^2 - 2\mu_1 a + 4\mu_2 = 0$, respectively. Similarly, we have the following result.

Lemma 1.2. *Suppose $a < 0$. Then, in the interior of the quadrants of the (y,ρ)-plane with $\rho > 0$, system (1.45) has no solution (respectively, one solution (y_1,ρ_1) with $y_1 < 0$, one solution (y_1,ρ_1) with $y_1 > 0$) for parameters μ in $\mathscr{J} \setminus (\mathscr{J}_{22} \cup \mathscr{J}_{25})$ (respectively, $\mathscr{J}_{22}$, $\mathscr{J}_{25}$).*

Moreover, for parameters $\mu \in \mathscr{J}_{22} \cup \mathscr{J}_{25}$, the characteristic polynomial of (1.44) at the equilibrium $E_4 = (y_1,\rho_1)$ is

$$\varsigma^2 - (\mu_1 + 2y_1)\varsigma - 2\rho_1(a + 2y_1) = 0.$$

The two eigenvalues $\varsigma_{1,2}$ satisfy $\varsigma_1 \varsigma_2 = -2\rho_1(a + 2y_1)$, which can be shown to be positive. Then, we need to consider the sign of $\varsigma_1 + \varsigma_2$ in order to discuss the stability of the equilibrium E_4. In fact,

$$\varsigma_1 + \varsigma_2 = \mu_1 + 2y_1 = \mu_1 - a - \sqrt{a^2 - 4\mu_2}.$$

It follows from $2|\mu_1| < |a|$ and $a < 0$ that $\mu_1 - a > 0$, and hence

$$\text{sign}(\varsigma_1 + \varsigma_2) = \text{sign}\{(\mu_1 - a)^2 - a^2 + 4\mu_2\}$$
$$= \text{sign}\{\mu_1^2 - 2\mu_1 a + 4\mu_2\}.$$

Let

$$\mathscr{J}^+ = \{\mu : \mu_1^2 - 2\mu_1 a + 4\mu_2 > 0\}$$

and

$$\mathscr{J}^- = \{\mu : \mu_1^2 - 2\mu_1 a + 4\mu_2 < 0\}.$$

Then we have the following:

Lemma 1.3. *Suppose $a < 0$. For parameters $\mu \in \mathscr{J}_{22} \cup \mathscr{J}_{25}$, besides equilibria E_1 and E_2, system (1.45) has a third equilibrium E_4, which is a sink if $\mu \in \mathscr{J}^- \cap (\mathscr{J}_{22} \cup \mathscr{J}_{25})$ and a source if $\mu \in \mathscr{J}^+ \cap (\mathscr{J}_{22} \cup \mathscr{J}_{25})$.*

Proposition 1.2. *Suppose $a < 0$. Then, in the quadrants of the (y, ρ)-plane with $\rho \geq 0$, we have the following information about equilibria of system (1.44):*

(i) *There are two equilibria E_1 and E_2 for $\mu \in \mathscr{J}_{21}$, where E_1 is a source and E_2 is a saddle.*

(ii) *There are three equilibria E_1, E_2, and E_4 for $\mu \in \mathscr{J}_{22}$, where E_1 and E_2 are saddles, and $E_4 = (y_1, \rho_1)$ satisfying $y_1 < 0$ and $\rho_1 > 0$ is a sink if $\mu_1^2 - 2\mu_1 a + 4\mu_2 < 0$ and a source otherwise. Namely, in the region $\mathscr{J}_{22}$, as μ crosses the parabola l_5 from the region $\mathscr{J}_{22} \cap \mathscr{J}^+$ to the region $\mathscr{J}_{22} \cap \mathscr{J}^-$, the equilibrium E_4 gains stability, and hence system (1.44) undergoes a Hopf bifurcation, and a stable limit cycle appears; as μ varies further, this limit cycle can approach a heteroclinic cycle formed by the separatrices of the two saddles E_1 and E_2, i.e., its period tends to infinity and the cycle disappears.*

(iii) *There are two equilibria E_1 and E_2 for $\mu \in \mathscr{J}_{23}$, where E_1 is a saddle and E_2 is a sink.*

(iv) *There are two equilibria E_1 and E_2 for $\mu \in \mathscr{J}_{24}$, where E_1 is a saddle and E_2 is a source.*

(v) *There are three equilibria E_1, E_2, and E_4 for $\mu \in \mathscr{J}_{25}$, where E_1 and E_2 are saddles, and $E_4 = (y_1, \rho_1)$ satisfying $y_1 > 0$ and $\rho_1 > 0$ is a sink if $\mu_1^2 - 2\mu_1 a + 4\mu_2 < 0$ and a source otherwise. Namely, in the region $\mathscr{J}_{25}$, as μ crosses the parabola l_5 from the region $\mathscr{J}_{25} \cap \mathscr{J}^+$ to the region $\mathscr{J}_{25} \cap \mathscr{J}^-$, equilibrium E_4 gains stability, and hence system (1.44) undergoes a Hopf bifurcation, and a stable limit cycle appears; as μ varies further, this limit cycle can approach a heteroclinic cycle formed by the separatrices of the two saddles E_1 and E_2, i.e., its period tends to infinity, and the cycle disappears.*

(vi) *There are two equilibria E_1 and E_2 for $\mu \in \mathscr{J}_{26}$, where E_1 is a sink and E_2 is a saddle.*

Theorem 1.12. *Suppose that $a < 0$. Then the following statements are true:*

(i) *An unstable limit cycle $\mathscr{O}_1$ of (1.41) appears as μ crosses the positive μ_1-axis from $\mathscr{J}_{21}$ to $\mathscr{J}_{22}$. As μ crosses the parabola l_5 from $\mathscr{J}_{22} \cap \mathscr{J}^+$ to $\mathscr{J}_{22} \cap \mathscr{J}^-$, this limit cycle $\mathscr{O}_1$ becomes stable and generates an unstable torus $\mathscr{T}_1$. Under further variation of the parameter μ in $\mathscr{J}_{22} \cap \mathscr{J}^-$, this torus $\mathscr{T}_1$ degenerates to a sphere-like surface $\mathscr{S}_1$ and then disappears. As μ crosses the parabola l_4 from $\mathscr{J}_{22} \cap \mathscr{J}^-$ to $\mathscr{J}_{23}$, the stable limit circle $\mathscr{O}_1$ disappears.*

(ii) *An unstable limit cycle $\mathscr{O}_2$ of (1.41) appears as μ crosses the parabola l_4 from $\mathscr{J}_{24}$ to $\mathscr{J}_{25}$. As μ crosses the parabola l_5 from $\mathscr{J}_{25} \cap \mathscr{J}^+$ to $\mathscr{J}_{25} \cap \mathscr{J}^-$, this limit cycle $\mathscr{O}_2$ becomes stable and generates an unstable torus $\mathscr{T}_2$. Under further variation of the parameter μ in $\mathscr{J}_{25} \cap \mathscr{J}^-$, this torus $\mathscr{T}_2$ degenerates to a sphere-like surface $\mathscr{S}_2$ and then disappears. As μ crosses the negative μ_1-axis from $\mathscr{J}_{25} \cap \mathscr{J}^-$ to $\mathscr{J}_{26}$, the stable limit circle $\mathscr{O}_2$ disappears.*

1.9.4 Bautin Bifurcation

Consider the following two-parameter system:

$$\dot{x}_1 = \mu_1 x_1 - x_2 + \mu_2 x_1 (x_1^2 + x_2^2) + \sigma x_1 (x_1^2 + x_2^2)^2, \qquad (1.46)$$
$$\dot{x}_2 = x_1 + \mu_1 x_2 + \mu_2 x_2 (x_1^2 + x_2^2) + \sigma x_2 (x_1^2 + x_2^2)^2,$$

where $\sigma = \pm 1$, $\mu = (\mu_1, \mu_2) \in \mathbb{R}^2$, and $x = (x_1, x_2) \in \mathbb{R}^2$. At $\mu = 0$, the linearization of (1.46) at the equilibrium $(0,0)$ has a pair of purely imaginary eigenvalues $\pm i$. Let $z = u + iv = \sqrt{\rho} e^{i\theta}$. Then system (1.46) can be rewritten as

$$\dot{z} = (\mu_1 + i) z + \mu_2 z |z|^2 + \sigma z |z|^4, \quad z \in \mathbb{C} \qquad (1.47)$$

and

$$\dot{\rho} = 2\rho (\mu_1 + \mu_2 \rho + \sigma \rho^2), \qquad (1.48)$$
$$\dot{\theta} = 1.$$

The first equation in (1.48) is uncoupled from the second one. Thus, to understand the bifurcations in (1.48), it suffices to study the scalar equation for ρ, that is,

$$\dot{\rho} = 2\rho (\mu_1 + \mu_2 \rho + \sigma \rho^2). \qquad (1.49)$$

It follows that the trivial equilibrium $\rho = 0$ of (1.49) corresponds to the equilibrium $x = 0$ of (1.46), and the existence and stability of positive equilibria of (1.49) determine the existence and stability of periodic solutions of (1.47) and hence of the original system (1.46). In the remaining part of this subsection, we depict the complete bifurcation diagrams of (1.49) on the μ-parameter plane.

We first consider the case $\sigma = -1$. Positive equilibria of (1.49) satisfy $\mu_1 + \mu_2 \rho - \rho^2 = 0$, which can have zero, one, or two positive solutions. These solutions branch from the trivial one along the line l_1 on the μ-parameter plane and collide and disappear at the half-parabola l_2 (see Fig. 1.13a), where

$$l_1 : \mu_1 = 0 \quad \text{and} \quad l_2 : \mu_2^2 + 4\mu_1 = 0 \text{ with } \mu_2 > 0.$$

The details are summarized below.

1. In the region $\mathscr{D}_{11} = \{\mu : \mu_2^2 + 4\mu_1 < 0 \text{ or } \mu_1 < 0 \text{ and } \mu_2 < 0\}$, (1.49) has no positive equilibria. Thus, the equilibrium $\rho = 0$ is globally asymptotically stable, which means that system (1.46) has no periodic solutions in a sufficiently small neighborhood of the stable equilibrium $z = 0$.
2. In the region $\mathscr{D}_{12} = \{\mu : \mu_1 > 0\}$, (1.49) has only one positive equilibrium, which is stable. This means that system (1.46) has exactly one stable periodic solution in a sufficiently small neighborhood of the unstable equilibrium $x = 0$.
3. In the region $\mathscr{D}_{13} = \{\mu : \mu_1 < 0, \mu_2 > 0, \text{ and } \mu_2^2 + 4\mu_1 > 0\}$, (1.49) has two positive equilibria, one stable and the other unstable. This means that system (1.46) has one stable periodic solution and one unstable periodic solution in a sufficiently small neighborhood of the stable equilibrium $x = 0$.

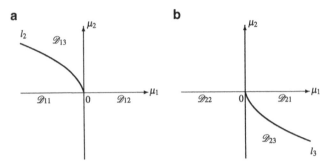

Fig. 1.13 Bifurcation sets for (1.49)

Therefore, on the μ-parameter plane, the line l_1 and the half-parabola l_2 are bifurcation curves. The bifurcation scenario is explained below.

(a) On the μ-parameter plane, if the point μ crosses the line l_1 from region $\mathscr{D}_{11}$ to region $\mathscr{D}_{12}$, then (1.46) undergoes a Hopf bifurcation, and a stable limit cycle bifurcates from $x = 0$.
(b) On the μ-parameter plane, if the point μ crosses the line l_1 from region $\mathscr{D}_{12}$ to region $\mathscr{D}_{13}$, then (1.46) undergoes a Hopf bifurcation, and an unstable limit cycle bifurcates from $x = 0$.
(c) On the μ-parameter plane, if the point μ crosses the line l_2 from region $\mathscr{D}_{13}$ to region $\mathscr{D}_{11}$, then limit cycles of (1.46) undergo a fold bifurcation, i.e., the two limit cycles collide and then disappear.

Now we come to the complete bifurcation diagram of (1.49) with $\sigma = 1$. Positive equilibria of (1.49) satisfy $\mu_1 + \mu_2\rho + \rho^2 = 0$, which can have zero, one, or two positive solutions. These solutions branch from the trivial one along the line l_1 on the μ-parameter plane and collide and disappear at the half-parabola $l_3 : \mu_2^2 - 4\mu_1 = 0$ and $\mu_2 < 0$ (see Fig. 1.13b). We have the following conclusions:

1. In the region $\mathscr{D}_{21} = \{\mu : \mu_2^2 - 4\mu_1 < 0 \text{ or } \mu_1 > 0 \text{ and } \mu_2 > 0\}$, (1.49) has no positive equilibria. Thus, the equilibrium $\rho = 0$ is unstable. This means that system (1.46) has no periodic solutions in a sufficiently small neighborhood of the unstable equilibrium $z = 0$.
2. In the region $\mathscr{D}_{22} = \{\mu : \mu_1 < 0\}$, (1.49) has only one positive equilibrium, which is unstable. This means that system (1.46) has exactly one unstable periodic solution in a sufficiently small neighborhood of the stable equilibrium $x = 0$.
3. In the region $\mathscr{D}_{23} = \{\mu : \mu_1 > 0, \mu_2 < 0, \text{ and } \mu_2^2 - 4\mu_1 > 0\}$, (1.49) has two positive equilibria, one stable and the other unstable. This means that system (1.46) has one stable periodic solution and one unstable periodic solution in a sufficiently small neighborhood of the unstable equilibrium $x = 0$.

Therefore, on the parameter plane μ, the line l_1 and the half-parabola l_3 are bifurcation curves. More specifically, we have the following:

(a) On the μ-parameter plane, if the point μ crosses the line l_1 from region $\mathscr{D}_{21}$ to region $\mathscr{D}_{22}$, then (1.46) undergoes a Hopf bifurcation, and an unstable limit cycle bifurcates from $x = 0$.

(b) On the μ-parameter plane, if the point μ crosses the line l_1 from region $\mathscr{D}_{22}$ to region $\mathscr{D}_{23}$, then (1.46) undergoes a Hopf bifurcation, and a stable limit cycle bifurcates from $x = 0$.

(c) On the μ-parameter plane, if the point μ crosses the line l_3 from region $\mathscr{D}_{23}$ to region $\mathscr{D}_{21}$, then limit cycles of (1.46) undergo a fold bifurcation, i.e., the two limit cycles collide and then disappear.

1.9.5 Hopf–Hopf Bifurcation

The Hopf–Hopf bifurcation is a bifurcation of an equilibrium point in a two-parameter family of autonomous ODEs at which the critical equilibrium has two pairs of purely imaginary eigenvalues. This phenomenon is also called the double Hopf bifurcation. For example, consider the following two-parameter system:

$$\dot{x}_1 = \mu_1 x_1 - \omega_1 x_2 + (A_{11}x_1 - B_{11}x_2)(x_1^2 + x_2^2) + (A_{12}x_1 - B_{12}x_2)(x_3^2 + x_4^2),$$
$$\dot{x}_2 = \omega_1 x_1 - \mu_1 x_2 + (A_{11}x_2 + B_{11}x_1)(x_1^2 + x_2^2) + (A_{12}x_2 + B_{12}x_1)(x_3^2 + x_4^2),$$
$$\dot{x}_3 = \mu_2 x_3 - \omega_2 x_4 + (A_{21}x_3 - B_{21}x_4)(x_1^2 + x_2^2) + (A_{22}x_3 - B_{22}x_4)(x_3^2 + x_4^2),$$
$$\dot{x}_4 = \omega_2 x_3 - \mu_2 x_4 + (A_{21}x_4 + B_{21}x_3)(x_1^2 + x_2^2) + (A_{22}x_4 + B_{22}x_3)(x_3^2 + x_4^2),$$
$$(1.50)$$

where $\sigma = \pm 1$, $\mu = (\mu_1, \mu_2) \in \mathbb{R}^2$, and $x = (x_1, x_2, x_3, x_4) \in \mathbb{R}^4$. At $\mu = 0$, the linearization of (1.50) at the equilibrium $(0,0,0,0)$ has two pairs of purely imaginary eigenvalues $\pm i\omega_1$ and $\pm i\omega_2$. Let $x_1 + ix_2 = \rho_1 e^{i\theta_1}$ and $x_3 + ix_4 = \rho_2 e^{i\theta_2}$. Then system (1.50) can be rewritten as

$$\dot{\rho}_1 = \rho_1(\mu_1 + A_{11}\rho_1^2 + A_{12}\rho_2^2),$$
$$\dot{\rho}_2 = \rho_2(\mu_2 + A_{21}\rho_1^2 + A_{22}\rho_2^2),$$
$$\dot{\theta}_1 = \omega_1 + B_{11}\rho_1^2 + B_{12}\rho_2^2, \qquad\qquad (1.51)$$
$$\dot{\theta}_2 = \omega_2 + B_{21}\rho_1^2 + B_{22}\rho_2^2.$$

Note that the amplitude and phase variables of (1.51) decouple. As a result, the bifurcation and asymptotic behavior of solutions of (1.50) can be studied via the following two-dimensional amplitude equations alone:

$$\dot{\rho}_1 = \rho_1(\mu_1 + A_{11}\rho_1^2 + A_{12}\rho_2^2), \qquad\qquad (1.52)$$
$$\dot{\rho}_2 = \rho_2(\mu_2 + A_{21}\rho_1^2 + A_{22}\rho_2^2).$$

The relation between equilibria of (1.52) and bifurcations of (1.50) is as follows:

(a) If (1.52) has an asymptotically stable (respectively, unstable) equilibrium $(r,0)$ (respectively, $(0,r)$) on either axis, then (1.50) has an asymptotically stable (respectively, unstable) periodic orbit of frequency close to ω_1 (respectively, ω_2).
(b) If (1.52) has an asymptotically stable (respectively, unstable) equilibrium (r_1,r_2) in the interior of the positive quadrant, then (1.50) has an asymptotically stable (respectively, unstable) two-dimensional invariant torus, i.e., (1.50) has a quasiperiodic solution in a neighborhood of the origin.
(c) If (1.52) has an asymptotically stable (respectively, unstable) limit cycle in the interior of the positive quadrant, then (1.50) has an asymptotically stable (respectively, unstable) three-dimensional invariant torus in a neighborhood of the origin.

From the above, we see that sufficiently close to the Hopf–Hopf bifurcation point $\mu = 0$, system (1.50) will exhibit either periodic or quasiperiodic motions. Thus, if we can find combinations of parameters μ_i and A_{ij} $(i, j = 1, 2)$ that yield stable equilibria (r_1, r_2) with $r_1 r_2 \neq 0$, we can conclude that the stable quasiperiodic motions should occur for the corresponding parameter values of system (1.50). Therefore, from now on, we concentrate on describing the behavior of the coupled amplitude equation (1.52) in the μ-parameter plane. The mode interaction equations (1.52) have been investigated by many researchers. See, for example, Guckenheimer and Holmes [125, Sect. 7.5]. Here, for the sake of completeness, we shall employ some techniques from the above-mentioned classical work of Guckenheimer and Holmes (including rescaling in time and variables) to investigate the qualitative behavior of the mode interaction equations (1.52) in the parameter ranges of interest. We discuss these case by case.

First, we consider the case that $A_{11} < 0$ and $A_{22} < 0$. Introducing new phase variables according to

$$r_1 = \sqrt{|A_{11}|}\rho_1, \quad r_2 = \sqrt{|A_{22}|}\rho_2, \tag{1.53}$$

yields

$$\begin{aligned}
\dot{r}_1 &= \mu_1 r_1 - r_1^3 - \theta r_1 r_2^2, \\
\dot{r}_2 &= \mu_2 r_2 - r_2^3 - \Delta r_2 r_1^2,
\end{aligned} \tag{1.54}$$

where $\theta = A_{12}/A_{22}$ and $\Delta = A_{21}/A_{11}$. Notice that the r_1- and r_2-axes are invariant lines for the flow of (1.54). Simple linear analysis reveals the following results about equilibria of (1.54):

(a) $(r_1, r_2) = (0, 0)$ is always an equilibrium. It is a stable sink if $\max\{\mu_1, \mu_2\} < 0$, a saddle if $\mu_1 \mu_2 < 0$, and an unstable source if $\min\{\mu_1, \mu_2\} > 0$.
(b) $(r_1, r_2) = (\sqrt{\mu_1}, 0)$ is an equilibrium if $\mu_1 > 0$. If, in addition, $\Delta \mu_1 > \mu_2$, then it is a sink; otherwise, it is a saddle.
(c) $(r_1, r_2) = (0, \sqrt{\mu_2})$ is an equilibrium if $\mu_2 > 0$. If, in addition, $\theta \mu_2 > \mu_1$, then it is a sink; otherwise, it is a saddle.

(d) $(r_1, r_2) = (\sqrt{[\mu_1 - \theta\mu_2]/[1 - \theta\Delta]}, \sqrt{[\mu_2 - \Delta\mu_1]/[1 - \theta\Delta]})$ is an equilibrium if both radicands are positive. It is a saddle if $\theta\Delta > 1$ and a sink if $\theta\Delta < 1$.

Therefore, we deduce that bifurcations to the pure modes $(\sqrt{\mu_1}, 0)$ and $(0, \sqrt{\mu_2})$ occur on the lines $\mu_1 = 0$ and $\mu_2 = 0$, whereas bifurcations to the mixed mode occur on the lines $\mu_1 = \theta\mu_2$ and $\mu_2 = \Delta\mu_1$ if they exist. In addition, we need check that no closed orbits (or limit cycles) can occur. Since the r_1- and r_2-axes are invariant, any such closed orbit would have to lie in the interior of the positive quadrant and must enclose at least one equilibrium with Poincaré index equal to 1.

If $\theta\Delta > 1$ and $\mu_1 - \theta\mu_2 < 0$ and $\mu_2 - \Delta\mu_1 < 0$, then system (1.54) has an equilibrium $(\tilde{r}_1, \tilde{r}_2)$ with $\tilde{r}_1\tilde{r}_2 \neq 0$. Recall that $(\tilde{r}_1, \tilde{r}_2)$ is a saddle with Poincaré index equal to -1. We immediately see that no closed orbit can occur around $(\tilde{r}_1, \tilde{r}_2)$. If $\theta\Delta < 1$ and $\mu_1 - \theta\mu_2 > 0$ and $\mu_2 - \Delta\mu_1 > 0$, then system (1.54) has an equilibrium $(\tilde{r}_1, \tilde{r}_2)$ with $\tilde{r}_1\tilde{r}_2 \neq 0$, which is a sink. In what follows, we distinguish several cases to conclude that no closed orbits can occur around the sink $(\tilde{r}_1, \tilde{r}_2)$ when $\theta\Delta < 1$ and $\mu \in \mathscr{E} = \{\mu: \mu_1 - \theta\mu_2 > 0 \text{ and } \mu_2 - \Delta\mu_1 > 0\}$.

Case 1: $\theta > 0$ and $\Delta > 0$. We follow a directional arc $\overrightarrow{l_1}$ crossing the line $\mu_1 = \theta\mu_2 > 0$ and then passing through the sector $\mathscr{E}$ and finally crossing the line $\mu_2 = \Delta\mu_1 > 0$. When $\mu \in \overrightarrow{l_1}$ crosses the line $\mu_1 = \theta\mu_2 > 0$, the sink $(0, \sqrt{\mu_2})$ becomes a saddle, a sink $(\tilde{r}_1, \tilde{r}_2)$ bifurcates from $(0, \sqrt{\mu_2})$, and the unstable separatrix of the saddle $(0, \sqrt{\mu_2})$ limits this bifurcated sink $(\tilde{r}_1, \tilde{r}_2)$. Thus, after bifurcation there is no closed orbit around this sink. The only way whereby the closed orbit can appear in the positive quadrant is by Hopf bifurcation from $(\tilde{r}_1, \tilde{r}_2)$. But this is impossible, because $(\tilde{r}_1, \tilde{r}_2)$ remains stable for all $\mu \in \mathscr{E}$.

Case 2: $\theta > 0 > \Delta$. Similar arguments as those in Case 1 show that there is no closed orbit in the positive quadrant when μ is in the sector $0 < \mu_2 < \mu_1/\theta$. In order to rule out the existence of closed orbits in the positive quadrant when μ is in the sector $\Delta\mu_1 < \mu_2 < 0$, we follow another directional arc $\overrightarrow{l_2}$ crossing the line $\mu_2 = \Delta\mu_1$ and then passing through the sector $\Delta\mu_1 < \mu_2 < 0$. When $\mu \in \overrightarrow{l_2}$ crosses the line $\mu_2 = \Delta\mu_1$, the sink $(\sqrt{\mu_1}, 0)$ becomes a saddle, a sink $(\tilde{r}_1, \tilde{r}_2)$ bifurcates from $(\sqrt{\mu_1}, 0)$, and the unstable separatrix of the saddle $(\sqrt{\mu_1}, 0)$ limits this bifurcated sink $(\tilde{r}_1, \tilde{r}_2)$. Thus, after bifurcation there is no closed orbit around this sink. Similarly, no Hopf bifurcation can occur from $(\tilde{r}_1, \tilde{r}_2)$, since it remains stable for all $\mu \in \mathscr{E}$.

Case 3: $\theta < 0 < \Delta$. Similar arguments as those in Case 1 tell us that there is no closed orbit in the positive quadrant when μ is in the sector $\theta\mu_2 < \mu_1 < 0$, while arguments like those in Case 2 yield that there is no closed orbit in the positive quadrant when μ is in the sector $0 < \mu_1 < \mu_2/\Delta$.

Case 4: $\theta < 0$ and $\Delta < 0$. The discussion is similar to that in Case 1 and hence is omitted.

In summary, we have proved the following theorem.

Theorem 1.13. *No closed orbit of system (1.54) can occur around the mixed mode* $(\tilde{r}_1, \tilde{r}_2)$.

Second, for the case that $A_{11} > 0$ and $A_{22} > 0$, we introduce new phase variables and rescale time in (1.52) according to

$$r_1 = \sqrt{|A_{11}|}\rho_1, \quad r_2 = \sqrt{|A_{22}|}\rho_2, \quad t^* = -t. \tag{1.55}$$

After dropping $*$, we obtain

$$\dot{r}_1 = -\mu_1 r_1 - r_1^3 - \theta r_1 r_2^2, \tag{1.56}$$
$$\dot{r}_2 = -\mu_2 r_2 - r_2^3 - \Delta r_2 r_1^2,$$

where θ and Δ are the same as before. System (1.56) is quite similar to (1.54), and hence similar arguments can be employed. We omit the details here.

Third, for the case that $A_{11} > 0$ and $A_{22} < 0$, we introduce new phase variables and rescale time in (1.52) as (1.53). After dropping $*$, we obtain

$$\dot{r}_1 = \mu_1 r_1 + r_1^3 - \theta r_1 r_2^2, \tag{1.57}$$
$$\dot{r}_2 = \mu_2 r_2 - r_2^3 + \Delta r_2 r_1^2,$$

where θ and Δ are the same as before. Simple linear analysis produces the following results:

(a) $(r_1, r_2) = (0, 0)$ is always an equilibrium. It is a stable sink if $\max\{\mu_1, \mu_2\} < 0$, a saddle if $\mu_1 \mu_2 < 0$, and an unstable source if $\min\{\mu_1, \mu_2\} > 0$.
(b) $(r_1, r_2) = (\sqrt{-\mu_1}, 0)$ is an equilibrium if $\mu_1 < 0$. If, in addition, $\Delta \mu_1 < \mu_2$, then it is a source; otherwise, it is a saddle.
(c) $(r_1, r_2) = (0, \sqrt{\mu_2})$ is an equilibrium if $\mu_2 > 0$. If, in addition, $\theta \mu_2 > \mu_1$, then it is a sink; otherwise, it is a saddle.
(d) $(r_1, r_2) = (\sqrt{[\mu_1 - \theta\mu_2]/[\theta\Delta - 1]}, \sqrt{[\Delta\mu_1 - \mu_2]/[\theta\Delta - 1]})$ is an equilibrium if both radicands are positive. If $\theta\Delta < 1$, then it is a saddle; if $\theta\Delta > 1$ and $\tilde{r}_1 > \tilde{r}_2$, then it is a source; if $\theta\Delta > 1$ and $\tilde{r}_1 < \tilde{r}_2$, then it is a sink.

It follows from the above results that bifurcations to the pure modes $(\sqrt{-\mu_1}, 0)$ and $(0, \sqrt{\mu_2})$ occur on the lines $\mu_1 = 0$ and $\mu_2 = 0$, whereas bifurcations to the mixed modes occur on the lines $\mu_1 = \theta\mu_2$ and $\mu_2 = \Delta\mu_1$ if they exist. Since the r_1- and r_2-axes are invariant, any such closed orbit would have to lie in the interior of the positive quadrant and must enclose at least one equilibrium with Poincaré index equal to 1. If $\theta\Delta < 1$, $\mu_1 - \theta\mu_2 < 0$, and $\mu_2 - \Delta\mu_1 > 0$, then system (1.57) has an equilibrium $(\tilde{r}_1, \tilde{r}_2)$ with $\tilde{r}_1\tilde{r}_2 \neq 0$, which is a saddle with Poincaré index equal to -1. We immediately conclude the following result.

Theorem 1.14. *Assume that $\theta\Delta < 1$, $\mu_1 - \theta\mu_2 < 0$, and $\mu_2 - \Delta\mu_1 > 0$. Then no closed orbit of system (1.57) can occur around $(\tilde{r}_1, \tilde{r}_2)$.*

If $\theta\Delta > 1$, (μ_1, μ_2) is in the sector $\mathscr{I} = \{\mu : \mu_1 - \theta\mu_2 > 0, \text{ and } \mu_2 - \Delta\mu_1 < 0\}$, then system (1.57) has an equilibrium $(\tilde{r}_1, \tilde{r}_2)$ with $\tilde{r}_1\tilde{r}_2 \neq 0$. It follows from the expressions for $\tilde{r}_1$ and $\tilde{r}_2$ that $\text{sign}(\tilde{r}_1 - \tilde{r}_2) = \text{sign}(1 - \theta)\text{sign}\{\mu_2 - \chi\mu_1\}$, where $\chi = (1 - \Delta)/(\theta - 1)$. Furthermore, if $\theta > 1$, then $\chi < 1/\theta$ and $\chi < \Delta$; if $0 < \theta < 1$, then $\chi > 1/\theta$ and $\chi > \Delta$; if $\theta < 0$, then $\Delta < \chi < \frac{1}{\theta}$. Therefore, we have the following observations:

Lemma 1.4. *If $\Delta > 1/\theta > 0$ and $(\mu_1, \mu_2) \in \mathscr{I}$, then system (1.57) has a mixed mode $(\tilde{r}_1, \tilde{r}_2)$. Moreover, it is a sink (respectively, source) if μ is in the sector $\mathscr{I}_1$ (respectively, $\mathscr{I}_2$), where*

$$\mathscr{I}_1 = \begin{cases} \{\mu: \chi\mu_1 < \mu_2 < \mu_1/\theta\} & \text{if } \theta > 1, \\ \{\mu: \mu_2 < \chi\mu_1 \text{ and } \mu_2 < \mu_1/\theta\} & \text{if } \theta < 1, \end{cases}$$

$$\mathscr{I}_2 = \begin{cases} \{\mu: \mu_2 < \chi\mu_1 \text{ and } \mu_2 < \Delta\mu_1\} & \text{if } \theta > 1, \\ \{\mu: \chi\mu_1 < \mu_2 < \Delta\mu_1\} & \text{if } \theta < 1. \end{cases}$$

Lemma 1.5. *If $\Delta < 1/\theta < 0$ and $(\mu_1, \mu_2) \in \mathscr{I}$, then system (1.57) has a mixed mode $(\tilde{r}_1, \tilde{r}_2)$. Moreover, it is a sink (respectively, source) if μ is in the sector $\mathscr{I}_3$ (respectively, $\mathscr{I}_4$), where*

$$\mathscr{I}_3 = \{\mu: \mu_1/\theta < \mu_2 < \chi\mu_1\},$$
$$\mathscr{I}_4 = \{\mu: \chi\mu_1 < \mu_2 < \Delta\mu_1\}.$$

The following result describes the phase portrait of (1.57).

Theorem 1.15. *Assume $\theta\Delta > 1$. Then for some points $\mu \in \mathscr{I}$, system (1.57) has closed orbits surrounding the mixed mode $(\tilde{r}_1, \tilde{r}_2)$.*

Proof. Here, we consider only the case $\theta > 1 > \Delta > 1/\theta > 0$, because other cases can be handled similarly. If $\theta\Delta > 1$ and $\theta > 1$, then $\mu \in \mathscr{I}$ and system (1.57) has a mixed mode $(\tilde{r}_1, \tilde{r}_2)$. We follow a directional arc in the μ-parameter plane that starts from a point in the sector $\mu_1/\theta < \mu_2 < \Delta\mu_1$, then crosses the line $\mu_1 = \theta\mu_2 > 0$ into the sector $\mathscr{I}_1$, and finally successively crosses the line $\mu_2 = \chi\mu_1 > 0$ and the positive μ_1-axis. When the point μ is in the sector $\mu_1/\theta < \mu_2 < \Delta\mu_1$, system (1.57) has a source $(0,0)$ and a sink $(0, \sqrt{\mu_2})$. As μ crosses the line $\mu_1 = \theta\mu_2 > 0$, a mixed mode $(\tilde{r}_1, \tilde{r}_2)$ (which is a sink) bifurcates from $(0, \sqrt{\mu_2})$, and the unstable separatrix of the saddle $(0, \sqrt{\mu_2})$ limits the newly bifurcated mixed mode. Thus, immediately after bifurcation, no closed orbit can surround the mixed mode. However, as μ crosses the line $\mu_2 = \chi\mu_1 > 0$, the mixed mode $(\tilde{r}_1, \tilde{r}_2)$ loses its stability, and hence system (1.57) undergoes a Hopf bifurcation, i.e., a stable closed orbit appears in the positive quadrant. Moreover, as μ crosses the positive μ_1-axis, the pure mode $(0, \sqrt{\mu_2})$ collides with $(0,0)$ and disappears. □

Theorem 1.15 implies that crossing the line $\mu_2 = \chi\mu_1$ in the sector $\mathscr{I}$ results in the branching of a three-dimensional torus from the two-dimensional torus of system (1.52).

Finally, for the case $A_{11} < 0$ and $A_{22} > 0$, we can obtain the reparameterized equation of the form (1.57) by reversing time, and hence the details are omitted.

1.10 Some Other Bifurcations

1. *Nontransversal homoclinic orbit to a hyperbolic cycle.* Consider a three-dimensional system (1.1) with a hyperbolic limit cycle Γ_μ. Its stable and unstable two-dimensional invariant manifolds, $W^s(\Gamma_\mu)$ and $W^u(\Gamma_\mu)$, can intersect along homoclinic orbits, tending to Γ_μ as $t \to \pm\infty$. Generically, such an intersection is transversal. It implies the presence of an infinite number of saddle limit cycles near the homoclinic orbit. However, at a certain parameter value, say $\mu = \mu_0$, the manifolds can become tangent to each other and then no longer intersect. At $\mu = \mu_0$, there is a homoclinic orbit to Γ_0 along which the manifolds $W^s(\Gamma_\mu)$ and $W^u(\Gamma_\mu)$ generically have a quadratic tangency. It has been proved that an infinite number of limit cycles can exist for sufficiently small $|\mu - \mu_0|$, even if the manifolds do not intersect. Passing the critical parameter value is accompanied by an infinite number of period-doubling and fold bifurcations of limit cycles. See, for example, Poincaré [246], Birkhoff [34], Smale [268], Neimark [229], and Shil'nikov [263], Gavrilov and Shilnikov [110], Palis and Takens [243].

2. *Homoclinic orbits to a nonhyperbolic limit cycle.* Suppose a three-dimensional system (1.1) has at $\mu = \mu_0$ a nonhyperbolic limit cycle Γ_0 with a simple multiplier $\lambda_1 = 1$, while the second multiplier satisfies $|\lambda_2| < 1$. Under generic perturbations, this cycle Γ_0 will either disappear or split into two hyperbolic cycles (i.e., via fold bifurcation for cycles). However, the locally unstable manifold $W^u(\Gamma_0)$ of the cycle can *return* to the cycle Γ_0 at the critical parameter value $\mu = \mu_0$, forming a set composed of homoclinic orbits that approach Γ_0 as $t \to \pm\infty$. Thus, at the critical parameter value, there may exist a smooth invariant torus or a *strange* attracting invariant set that contains an infinite number of saddle and stable limit cycles, or a *blue-sky* catastrophe. See, for example, Afraimovich and Shil'nikov [4], Palis and Pugh [242], Medvedev [218], Turaev and Shil'nikov [279].

3. *Bifurcations on invariant tori.* Continuous-time dynamical systems with phase-space dimension $n > 2$ can have invariant tori. For example, a stable cycle in $\mathbb{R}^3$ can lose stability when a pair of complex-conjugate multipliers crosses the unit circle. It will be much more interesting to discuss changes of the orbit structure on an invariant 2-torus under variation of the parameters of the system. These bifurcations are responsible for such phenomena as frequency and phase locking. See, for example, Arnold [19], Fenichel [95, 96], Kuznetsov [200].

Chapter 2
Introduction to Functional Differential Equations

There are different types of functional differential equations (FDEs) arising from important applications: delay differential equations (DDEs) (also referred to as retarded FDEs [RFDEs]), neutral FDEs (NFDEs), and mixed FDEs (MFDEs). The classification depends on how the current change rate of the system state depends on the history (the historical status of the state only or the historical change rate and the historical status) or whether the current change rate of the system state depends on the future expectation of the system. Later we will also see that the delay involved may also depend on the system state, leading to DDEs with state-dependent delay.

2.1 Infinite Dynamical Systems Generated by Time Lags

In Newtonian mechanics, the system's state variable changes over time, and the law that governs the change of the system's state is normally described by an ordinary differential equation (ODE). Assuming that the function involved in this ODE is sufficiently smooth (locally Lipschitz, for example), the corresponding Cauchy initial value problem is well posed, and thus knowing the current status, one is able to reconstruct the history and predict the future of the system.

In many applications, a close look at the physical or biological background of the modeling system shows that the change rate of the system's current status often depends not only on the current state but also on the history of the system, see, for example, [50, 76, 198, 199]. This usually leads to so-called DDEs with the following prototype:

$$\dot{x}(t) = f(x(t), x(t - \tau)), \tag{2.1}$$

where $x(t)$ is the system's state at time t, $f \colon \mathbb{R}^n \times \mathbb{R}^n \to \mathbb{R}^n$ is a given mapping, and the time lag $\tau > 0$ is a constant.

Such an equation arises naturally, for example, from the population dynamics of a single-species structured population. In such an example, if $x(t)$ denotes the pop-

S. Guo and J. Wu, *Bifurcation Theory of Functional Differential Equations*,
Applied Mathematical Sciences 184, DOI 10.1007/978-1-4614-6992-6_2,
© Springer Science+Business Media New York 2013

ulation density of the mature/reproductive population, and if the maturation period is assumed to be a constant, then we have

$$f(x(t), x(t-\tau)) = -d_m x(t) + e^{-d_i \tau} b(x(t-\tau)), \tag{2.2}$$

where d_m and d_i are the death rates of the mature and immature populations, respectively, and $b \colon \mathbb{R} \to \mathbb{R}$ is the birth rate. Death is instantaneous, so the term $-d_m x(t)$ is without delay. However, the rate into the mature population is the maturation rate (not the birth rate), that is, the birth rate at time τ, multiplied by the survival probability $e^{-d_i \tau}$ during the maturation process.

Clearly, to specify a function $x(t)$ of $t \geq 0$ that satisfies (2.1) (called a solution of (2.1)), we must prescribe the history of x on $[-\tau, 0]$. On the other hand, once the initial value data

$$\varphi : [-\tau, 0] \to \mathbb{R}^n \tag{2.3}$$

is given as a continuous function and if $f \colon \mathbb{R}^n \times \mathbb{R}^n \ni (x, y) \to f(x, y) \in \mathbb{R}^n$ is continuous and locally Lipschitz with respect to the first state variable $x \in \mathbb{R}^n$, then (2.1) on $[0, \tau]$ becomes an ODE for which the initial value problem

$$\dot{x}(t) = f(x(t), \varphi(t-\tau)), \quad t \in [0, \tau], \ x(0) = \varphi(0), \tag{2.4}$$

is solvable. If such a solution exists on $[0, \tau]$, we can repeat the argument to the initial value problem

$$\begin{cases} \dot{x}(t) = f(x(t), \underbrace{x(t-\tau)}_{\text{given}}), \quad t \in [\tau, 2\tau], \\ x(\tau) \text{ is given in the previous step,} \end{cases} \tag{2.5}$$

to obtain a solution on $[\tau, 2\tau]$. This process may be continued to yield a solution of (2.1) subject to $x|_{[-\tau, 0]} = \varphi$ given in (2.3).

Let $C_{n,\tau} = C([-\tau, 0]; \mathbb{R}^n)$ be the Banach space of continuous mappings from $[-\tau, 0]$ to $\mathbb{R}^n$ equipped with the supremum norm

$$\|\phi\| = \sup_{-\tau \leq \theta \leq 0} |\phi(\theta)| \text{ for } \phi \in C_{n,\tau},$$

and if we define $x_t \colon C_{n,\tau} \to C_{n,\tau}$ by the segment of x on the interval $[t-\tau, t]$ translated back to the initial interval $[-\tau, 0]$, namely,

$$x_t(\theta) = x(t+\theta), \quad \theta \in [-\tau, 0], \tag{2.6}$$

then (2.1) subject to $x_0 = \varphi \in C_{n,\tau}$ gives a semiflow $[0, \infty] \ni t \mapsto x_t \in C_{n,\tau}$. This clearly shows that an appropriate state space of a DDE is $C_{n,\tau}$ and that a DDE gives an infinite-dimensional dynamical system on this phase space.

Many applications call for the study of asymptotic behaviors (as $t \to \infty$) of solutions of (2.1), and such a study seems to be very difficult due to the infinite-dimensionality of the phase space and the generated semiflow, even for a scalar

DDE (2.1) (that is, when $n = 1$). Even to restrict the study of the asymptotic behaviors of solutions near a specified solution is highly nontrivial. Take a steady state as an example. A vector $x^* \in \mathbb{R}^n$ is called an equilibrium of (2.1) if

$$f(x^*, x^*) = 0. \tag{2.7}$$

This vector gives a state $\hat{x}^* \in C_{n,\tau}$, which is a constant mapping on $[-\tau, 0]$ with the constant value $x^* \in \mathbb{R}^n$, and a solution of (2.1) with the initial value $\hat{x}^*$ is a constant function $x: [0, \infty) \to \mathbb{R}^n$ with the constant value x^*. Behaviors of solutions of (2.1) in a neighborhood of $\hat{x}^*$ may be determined by the zero solution of the linearization

$$\dot{x}(t) = D_x f(x^*, x^*) x(t) + D_y f(x^*, x^*) x(t - \tau) \tag{2.8}$$

with

$$D_x f(x^*, x^*) \stackrel{\text{def}}{=} \frac{\partial}{\partial x} f(x, y) \bigg|_{x=x^*, y=x^*},$$

$$D_y f(x^*, x^*) \stackrel{\text{def}}{=} \frac{\partial}{\partial y} f(x, y) \bigg|_{x=x^*, y=x^*}.$$

In the case $\tau > 0$, even when $n = 1$, the behaviors of solutions of (2.8) can be more complicated than any given linear system of ODEs, since (2.8) even when $n = 1$ may have infinitely many linearly independent solutions $e^{\lambda t}$ with λ being given by the so-called characteristic equation

$$\lambda = D_x f(x^*, x^*) + D_y f(x^*, x^*) e^{-\lambda \tau}. \tag{2.9}$$

In particular, the infinite-dimensionality of the problem (2.1) leads to a transcendental equation (rather than a polynomial), which can have multiple zeros on the imaginary axis, giving rise to complicated critical cases.

On the other hand, some special features (specially the eventual compactness of the solution semiflow) of DDEs ensure that the sequence of zeros of the characteristic equation on the imaginary axis (counting multiplicity, either algebraically or geometrically, as will be specified later) must be finite. This gives a finite-dimensional center manifold of system (2.1) in a neighborhood of the equilibrium state $\hat{x}^*$, so that the asymptotic behaviors of solutions of (2.1) in a neighborhood of $\hat{x}^*$ can be captured by the reduced system on the center manifold, and this reduced system is an ODE system even though its dimension can be high.

We aim to introduce systematically the approach that enables us to derive the specific form of the reduced ODE system on the center manifold, explicitly in terms of the original system (2.1). Some forms of system (2.1) from application problems come with a parameter, and since the asymptotic behaviors of solutions near a given equilibrium may change qualitatively when the parameter varies (the so-called bifurcation), our focus will be on how the center manifold and the reduced ODE system on the center manifold change when the parameter is varied.

We should mention the step-by-step method in solving (2.1) on $[0,\tau]$, $[\tau,2\tau]$, ... inductively, which, though effectively numerically, may not give useful qualitative information about asymptotic behaviors of solutions. This method is also not useful in solving the kind of DDE with distributed delay such as

$$\dot{x}(t) = \int_{-\tau}^{0} f(x(t), x(t+\theta))d\theta$$

or

$$\dot{x}(t) = f\left(x(t), \int_{-\tau}^{0} g(x(t+\theta))d\theta\right)$$

with $g: \mathbb{R}^n \to \mathbb{R}^n$. One should also mention that in case the change rate of $x(t)$ depends on the historical value of $\dot{x}(t+\theta)$ with $\theta \in [-\tau,0]$, such as

$$\dot{x}(t) = c\dot{x}(t-\tau) + f(x(t), x(t-\tau)),$$

we encounter additional difficulties, which shall be discussed later.

2.2 The Framework for DDEs

2.2.1 Definitions

Assume that $\mathbb{R}^n$ is equipped with the Euclidean norm $|\cdot|$. For a given constant $\tau \geq 0$, $C_{n,\tau} \overset{\text{def}}{=} C([-\tau,0],\mathbb{R}^n)$ denotes the Banach space of continuous mappings from $[-\tau,0]$ into $\mathbb{R}^n$ equipped with the supremum norm $\|\phi\| = \sup_{-\tau \leq \theta \leq 0} |\phi(\theta)|$ for $\phi \in C_{n,\tau}$. Moreover, if $t_0 \in \mathbb{R}$, $A \geq 0$, and $x: [t_0-\tau, t_0+A] \to \mathbb{R}^n$ is a continuous mapping, then for every $t \in [t_0, t_0+A]$, $x_t \in C_{n,\tau}$ is defined by $x_t(\theta) = x(t+\theta)$ for $\theta \in [-\tau,0]$.

If $f: C_{n,\tau} \to \mathbb{R}^n$ is a mapping, we say that the equation

$$\dot{x} = f(x_t) \tag{2.10}$$

is a retarded functional differential equation (RFDE), or a delay differential equation (DDE). A function x is said to be a solution of (2.10) on $[t_0, t_0+A]$ if there are $t_0 \in \mathbb{R}$ and $A > 0$ such that $x \in C([t_0-\tau, t_0+A), \mathbb{R}^n)$, and $x(t)$ is differentiable and satisfies (2.10) for all $t \in [t_0, t_0+A]$. If f is locally Lipschitz (i.e., for every $\varphi \in C_{n,\tau}$ there exist a neighborhood $U \subseteq C_{n,\tau}$ of φ and a constant L such that $\|f(\phi) - f(\psi)\| \leq L\|\phi - \psi\|$ for all $\phi, \psi \in U$), then for each given initial condition $(t_0, \varphi) \in \mathbb{R} \times C_{n,\tau}$, system (2.10) has a unique mapping $x^{\varphi}: [t_0-\tau, \beta) \to \mathbb{R}^n$ such that $x^{\varphi}|_{[t_0-\tau,t_0]} = \varphi$, x^{φ} is continuous for all $t \geq t_0 - \tau$, is differentiable, and satisfies (2.10) for $t \in (t_0, \beta)$, the maximal interval of existence of the solution x^{φ}. Furthermore, if $\beta < \infty$, then there exists a sequence $t_k \to \beta^-$ such that $|x^{\varphi}(t_k)| \to \infty$ as $k \to \infty$. For further results on existence, uniqueness, continuation, and continuous dependence of solutions for DDEs, see, for example, [18, 30, 51, 70, 120, 144–147, 154, 206, 208, 300, 302].

System (2.10) includes the following DDE with distributed delay

$$\dot{x}(t) = \int_{-\tau}^{0} g(\theta, x(t+\theta))d\theta, \tag{2.11}$$

and the following DDE with discrete delay

$$\dot{x}(t) = h(x(t), x(t-\tau_1), \dots, x(t-\tau_k)), \tag{2.12}$$

where $\tau = \max\{\tau_1, \dots, \tau_k\}$, $g: [-\tau, 0] \times \mathbb{R}^n \to \mathbb{R}^n$, and $h: \mathbb{R}^n \times \cdots \times \mathbb{R}^n (= \mathbb{R}^{n(k+1)}) \to \mathbb{R}^n$ are continuous. In these cases, for $\varphi \in C_{n,\tau}$,

$$f(\varphi) = \int_{-\tau}^{0} g(\theta, \varphi(\theta))d\theta$$

and

$$f(\varphi) = h(\varphi(0), \varphi(-\tau_1), \dots, \varphi(-\tau_k)),$$

respectively. It can be shown that if h is locally Lipschitz (in (2.12)), then so is f. Similarly, if for every $x \in \mathbb{R}^n$ there exist a neighborhood U of $x \in \mathbb{R}^n$ and a constant $L > 0$ such that $|g(\theta, z) - g(\theta, y)| \le L|z - y|$ for all $\theta \in [-\tau, 0]$ and $z, y \in U$, then the corresponding f is locally Lipschitz.

2.2.2 An Operator Equation

Throughout this chapter, we always assume that $f: C_{n,\tau} \to \mathbb{R}^n$ is continuously differentiable. Without loss of generality, assume that $f(0) = 0$, that is, 0 is an equilibrium point of (2.10). Let L be the linearized operator of f at this equilibrium point. Then the linearization of system (2.10) at this equilibrium point is

$$\dot{x}(t) = Lx_t. \tag{2.13}$$

We will consider the above linear system with a general linear operator L : $C_{n,\tau} \to \mathbb{R}^n$. Such an operator is clearly locally Lipschitz. For $\varphi \in C_{n,\tau}$, let $x = x^\varphi$ be the unique solution of (2.13) satisfying $x_0^\varphi = \varphi$. Then we have $|x(t)| \le |\varphi(0)| + \int_0^t |L| \|x_s\| ds$ for all $t \ge 0$, from which it follows that $\|x_t\| \le \|\varphi\| + \int_0^t |L| \|x_s\| ds$ for $t \ge 0$ and hence $\|x_t\| \le \|\varphi\| e^{|L|t}$ for $t \ge 0$. This implies that the solution is defined for all $t \ge 0$. Here we use $|L|$ to denote the operator norm of the bounded operator L.

Define the solution operators $T(t) : C_{n,\tau} \to C_{n,\tau}$ by the relation

$$(T(t)\varphi)(\theta) = x_t^\varphi(\theta) = x(t+\theta) \tag{2.14}$$

for $\varphi \in C_{n,\tau}$, $\theta \in [-\tau, 0]$, $t \ge 0$. Then (2.13) can be thought of as maps from $C_{n,\tau}$ to $C_{n,\tau}$. Moreover,

(i) $T(t)$ is bounded and linear for $t \ge 0$;
(ii) $T(0)\varphi = \varphi$ or $T(0) = \text{Id}$;

(iii) $\lim_{t \to t_0^+} \| T(t)\varphi - T(t_0)\varphi \| = 0$ for $\varphi \in C_{n,\tau}$.

Note that the inverse of $T(t), t \geq 0$, does not necessarily exist. Therefore, $T(t), t \geq 0$, is a *strongly continuous semigroup*.

An *infinitesimal generator* of a semigroup $T(t)$ is defined by

$$\mathscr{A}\varphi = \lim_{t \to 0^+} \frac{T(t)\varphi - \varphi}{t} \quad \text{for} \quad \varphi \in C_{n,\tau}.$$

In the case of the linear system (2.13), the infinitesimal generator can be constructed as

$$(\mathscr{A}\varphi)(\theta) = \begin{cases} d\varphi/d\theta, & \text{if} \quad \theta \in [-\tau, 0), \\ L\varphi, & \text{if} \quad \theta = 0. \end{cases} \tag{2.15}$$

We can show that the domain of $\mathscr{A}$ is given by

$$\mathrm{dom}(\mathscr{A}) = \{\varphi : \phi \in C_{n,\tau}^1, \varphi'(0) = L\varphi\}.$$

Then $T(t)\varphi$ satisfies

$$\frac{\mathrm{d}}{\mathrm{d}t} T(t)\varphi = \mathscr{A}T(t)\varphi,$$

where

$$\frac{\mathrm{d}}{\mathrm{d}t} T(t)\varphi = \lim_{h \to 0} \frac{T(t+h)\varphi - T(t)\varphi}{h}.$$

We may enlarge the phase space $C_{n,\tau}$ in such a way that (2.10) can be written as an abstract ODE in a Banach space. To accomplish this, for a positive integer n, let BC_n be the set of all functions from $[-\tau, 0]$ to $\mathbb{R}^n$ that are uniformly continuous on $[-\tau, 0)$ and may have a possible jump discontinuity at 0. We also introduce $X_0 : [-\tau, 0] \to BL(\mathbb{R}^n)$ defined by

$$X_0(\theta) = \begin{cases} \mathrm{Id}_n, & \theta = 0 \\ 0, & \theta \in [-\tau, 0). \end{cases} \tag{2.16}$$

Then every $\psi \in BC_n$ can be expressed as $\psi = \varphi + X_0\xi$ with $\varphi \in C_{n,\tau}$ and $\xi \in \mathbb{R}^n$. Thus BC_n can be identified with $C_{n,\tau} \times \mathbb{R}^n$. Equipped with the norm $|\varphi + X_0\xi|_{BC_n} = \|\varphi\| + |\xi|$, BC_n is a Banach space. In BC_n, we consider an extension of the infinitesimal generator of $\{T(t)\}_{t \geq 0}$, still denoted by $\mathscr{A}$,

$$\mathscr{A} : C_{n,\tau}^1 \ni \psi \mapsto \dot{\psi} + X_0[L\psi - \dot{\psi}(0)] \in BC_n,$$

where $\dot{\psi} = \frac{\mathrm{d}}{\mathrm{d}\theta}\psi$. Thus, the abstract ODE in BC_n associated with (2.10) can be rewritten in the form

$$\frac{\mathrm{d}}{\mathrm{d}t} x_t = \mathscr{A}x_t + X_0 F(x_t), \tag{2.17}$$

where $F(x_t) = f(x_t) - Lx_t$. For $\theta \in [-\tau, 0)$, (2.17) is just the trivial equation $du_t/dt = du_t/d\theta$; for $\theta = 0$, it is (2.10).

2.2.3 Spectrum of the Generator

If the linear operator $L: C_{n,\tau} \to \mathbb{R}^n$ defined in (2.13) is continuous, then by the Riesz representation theorem, there exists an $n \times n$ matrix-valued function $\eta : [-\tau, 0] \to \mathbb{R}^{n^2}$ whose elements are of bounded variation such that (see, for example, Hale and Verduyn Lunel [154] for more details)

$$L\varphi = \int_{-\tau}^{0} d\eta(\theta)\varphi(\theta), \qquad \varphi \in C_{n,\tau}. \tag{2.18}$$

For example, consider $x'(t) = -x(t) + bx(t-1)$. Let $\eta : [-1,0] \to \mathbb{R}$ be given such that $\eta(\theta) = 0$ for all $\eta \in (-1,0)$ and $\eta(0) = -1$ and $\eta(-1) = -b$. Then $\int_{-1}^{0} d\eta(\theta)\varphi(\theta) = -\varphi(0) + b\varphi(-1)$ for $\varphi \in C_{1,1}$.

In general, the spectrum of an operator may consist of three different types of points, namely, the residual spectrum, the continuous spectrum, and the point spectrum. Moreover, points of the point spectrum are called eigenvalues of this operator. It is interesting to see that the spectrum $\sigma(\mathscr{A})$ of $\mathscr{A}$ consists of only the point spectrum. This implies that $\sigma(\mathscr{A})$ consists of eigenvalues of $\mathscr{A}$ and that λ is in $\sigma(\mathscr{A})$ if and only if λ satisfies the characteristic equation

$$\det \Delta(\lambda) = 0, \tag{2.19}$$

where $\Delta(\lambda)$ is the characteristic matrix of (2.13) and is given by

$$\Delta(\lambda) = \lambda \operatorname{Id}_n - \int_{-\tau}^{0} e^{\lambda \theta} d\eta(\theta). \tag{2.20}$$

Here and in what follows, Id_n is the $n \times n$ identity matrix. We will not use the subscript n if that does not cause confusion.

For any $\lambda \in \sigma(\mathscr{A})$, the generalized eigenspace $\mathscr{M}_\lambda(\mathscr{A})$ is finite-dimensional, and there exists an integer k such that $\mathscr{M}_\lambda(\mathscr{A}) = \operatorname{Ker}((\lambda \operatorname{Id} - \mathscr{A})^k)$ and we have the direct sum decomposition

$$C_{n,\tau} = \operatorname{Ker}((\lambda \operatorname{Id} - \mathscr{A})^k) \oplus \operatorname{Ran}((\lambda \operatorname{Id} - \mathscr{A})^k),$$

where $\operatorname{Ker}((\lambda \operatorname{Id} - \mathscr{A})^k)$ and $\operatorname{Ran}((\lambda \operatorname{Id} - \mathscr{A})^k)$ represent the kernel and image of $(\lambda \operatorname{Id} - \mathscr{A})^k$, respectively. Clearly, $\mathscr{A}\mathscr{M}_\lambda(\mathscr{A}) \subseteq \mathscr{M}_\lambda(\mathscr{A})$.

The dimension $\mathscr{M}_\lambda(\mathscr{A})$ is the same as the order of zero for $\det\Delta(\lambda) = 0$.

Let $d = \dim \mathcal{M}_\lambda(\mathscr{A})$, let $\varphi_1, \ldots, \varphi_d$ be a basis for $\mathcal{M}_\lambda(\mathscr{A})$, and let $\Phi_\lambda = (\varphi_1, \ldots, \varphi_d)$. Then there exists a $d \times d$ constant matrix B_λ such that $\mathscr{A}\Phi_\lambda = \Phi_\lambda B_\lambda$. Moreover, we have the following properties:

(i) the only eigenvalue of B_λ is λ;
(ii) $\Phi_\lambda(\theta) = \Phi_\lambda(0)e^{B_\lambda \theta}$;
(iii) $T(t)\Phi_\lambda = \Phi_\lambda e^{B_\lambda t}$.

Therefore, we have the following result.

Theorem 2.1 (Hale and Verduyn Lunel [154]). *Suppose Λ is a finite set $\{\lambda_1, \ldots, \lambda_p\}$ of eigenvalues of (2.13), and let $\Phi_\Lambda = (\Phi_{\lambda_1}, \ldots, \Phi_{\lambda_p})$ and $B_\Lambda = \mathrm{diag}(B_{\lambda_1}, \ldots, B_{\lambda_p})$, where Φ_{λ_j} is a basis for the generalized space of $\mathscr{A}$ associated with λ_j and B_{λ_j} is the matrix defined by $\mathscr{A}\Phi_{\lambda_j} = \Phi_{\lambda_j} B_{\lambda_j}$, $j = 1, 2, \ldots, p$. Then the only eigenvalue of B_{λ_j} is λ_j, and for every vector v of the same dimension as the space P_Λ spanned by Φ_Λ, the solution $T(t)\Phi_\Lambda v$ with initial value $\Phi_\Lambda v$ at $t = 0$ may be defined on $(-\infty, \infty)$ by the relations*

$$T(t)\Phi_\Lambda v = \Phi_\Lambda e^{B_\Lambda t} v$$

and

$$\Phi_\Lambda(\theta) = \Phi_\Lambda(0)e^{B_\Lambda \theta}, \quad -\tau \le \theta \le 0.$$

Furthermore, there exists a subspace Q_Λ of $C_{n,\tau}$ such that $T(t)Q_\Lambda \subseteq Q_\Lambda$ for all $t \ge 0$ and

$$C_{n,\tau} = P_\Lambda \oplus Q_\Lambda. \tag{2.21}$$

2.2.4 An Adjoint Operator

We now describe a formal adjoint operator associated with (2.15). Let $C_{n,\tau}^* = C([0,\tau]; \mathbb{R}^{n*})$ be the space of continuous functions from $[0,\tau]$ to $\mathbb{R}^{n*}$ with

$$\|\psi\| = \sup_{t \in [0,\tau]} |\psi(t)|$$

for $\psi \in C_{n,\tau}^*$, where $\mathbb{R}^{n*}$ is the space of n-dimensional real row vectors. The formal adjoint equation associated with the linear RFDE (2.13) is given by

$$\dot{y} = -\int_{-\tau}^{0} y(t-\theta)d\eta(\theta). \tag{2.22}$$

For $\psi \in C_{n,\tau}^*$, let y^ψ be the unique solution of (2.22) satisfying $y_0^\psi = \psi$ (in this subsection, $y_t \in C_{n,\tau}^*$ is defined as $y_t(s) = y(t+s)$ for $s \in [0,\tau]$).
    If we define

$$(T^*(t)\psi)(\theta) = y_t^\psi(\theta) = y(t+\theta) \tag{2.23}$$

for $\psi \in C_{n,\tau}^*$, $\theta \in [0,\tau]$, $t \leq 0$, then (2.23) defines a strongly continuous semigroup with the infinitesimal generator

$$(\mathscr{A}^*\psi)(\xi) = \begin{cases} -d\psi(\xi)/d\xi, & \text{if } \xi \in (0,\tau], \\ \int_{-\tau}^0 \psi(-\theta)d\eta(\theta), & \text{if } \xi = 0. \end{cases} \tag{2.24}$$

Note that although the formal infinitesimal generator for (2.23) is defined as

$$A^*\psi = \lim_{t \to 0^-} \frac{T(t)\psi - \psi}{t} \quad \text{for } \varphi \in C_{n,\tau},$$

Hale [144], for convenience, takes $\mathscr{A}^* = -A^*$ in (2.24) as the formal adjoint to (2.15). This family of operators (2.23) satisfies

$$\frac{d}{dt}T^*(t)\psi = -\mathscr{A}^*T^*(t)\psi.$$

In addition, it is easy to obtain the following results.

Theorem 2.2. *The following hold:*

(i) λ *is an eigenvalue of* $\mathscr{A}$ *if and only if* $\overline{\lambda}$ *is an eigenvalue of* $\mathscr{A}^*$.
(ii) *The dimensions of the eigenspaces of* $\mathscr{A}$ *and* $\mathscr{A}^*$ *are finite and equal.*
(iii) *The dimensions of the generalized eigenspaces of* $\mathscr{A}$ *and* $\mathscr{A}^*$ *are finite and equal.*

2.2.5 A Bilinear Form

In contrast to $\mathbb{R}^n$, the space $C_{n,\tau}$ does not have a natural inner product associated with its norm. However, following Hale [144], one can introduce a substitute device that acts like an inner product in $C_{n,\tau}$. This is an approach that is often taken when a function space does not have a natural inner product associated with its norm. Throughout, we will be assuming the complexification of the spaces so that we can work with complex eigenvalues and eigenvectors.

Define two operators $\Pi: C^1(\mathbb{R};\mathbb{R}^n) \to C(\mathbb{R};\mathbb{R}^n)$ and $\Omega: C^1(\mathbb{R};\mathbb{R}^{n*}) \to C(\mathbb{R};\mathbb{R}^{n*})$ as follows:

$$(\Pi x)(t) = \dot{x}(t) - \int_{-\tau}^0 d\eta(\theta)x(t+\theta)$$

and

$$(\Omega y)(t) = \dot{y}(t) + \int_{-\tau}^0 y(t-\theta)d\eta(\theta).$$

Then we have

$$\overline{y}(t)(\Pi x)(t) + (\overline{\Omega y})(t)x(t) = \frac{d}{dt}\langle y, x\rangle(t),$$

where

$$\langle y, x \rangle (t) = \overline{y}(t) x(t) - \int_{-\tau}^{0} \int_{0}^{\theta} \overline{y}(t + \xi - \theta) d\eta(\theta) x(t + \xi) d\xi. \tag{2.25}$$

Thus, if $x \in C^1(\mathbb{R}; \mathbb{R}^n)$ and $y \in C^1(\mathbb{R}; \mathbb{R}^{n*})$ satisfy $\Pi x = 0$ and $\Omega y = 0$, then $\langle y, x \rangle (t)$ is constant, and one can set $t = 0$ in (2.25) to define the bilinear form

$$\langle \psi, \varphi \rangle = \overline{\psi}(0)\varphi(0) - \int_{-\tau}^{0} \int_{0}^{\theta} \overline{\psi}(\xi - \theta) d\eta(\theta) \varphi(\xi) d\xi, \quad \psi \in C_{n,\tau}^*, \varphi \in C_{n,\tau}. \tag{2.26}$$

In terms of (2.15) and (2.24), we see that

$$\langle \psi, \mathscr{A} \varphi \rangle = \langle \mathscr{A}^* \psi, \varphi \rangle$$

for $\varphi \in C_{n,\tau}$ and $\psi \in C_{n,\tau}^*$.

Let Λ be a set of some eigenvalues of $\mathscr{A}$ satisfying $\overline{\lambda} \in \Lambda$ if $\lambda \in \Lambda$. Denote by P and P^* the generalized eigenspaces of $\mathscr{A}$ and $\mathscr{A}^*$ associated with Λ, respectively. It follows from Theorem 2.2 that $\dim P = \dim P^*$. If $\varphi_1, \varphi_2, \ldots, \varphi_m$ is a basis for P and $\psi_1, \psi_2, \ldots, \psi_m$ is a basis for P^*, then construct the matrices $\Phi = (\varphi_1, \varphi_2, \ldots, \varphi_m)$ and $\Psi = (\psi_1, \psi_2, \ldots, \psi_m)^T$. Define the bilinear form between Ψ and Φ by

$$\langle \Psi, \Phi \rangle = \begin{bmatrix} \langle \psi_1, \varphi_1 \rangle & \cdots & \langle \psi_1, \varphi_m \rangle \\ \vdots & \ddots & \vdots \\ \langle \psi_m, \varphi_1 \rangle & \cdots & \langle \psi_m, \varphi_m \rangle \end{bmatrix}.$$

This matrix is nonsingular and can be chosen so that $\langle \Psi, \Phi \rangle = \mathrm{Id}_m$. In fact, if $\langle \Psi, \Phi \rangle$ is not the identity, then a change of coordinates can be performed by setting $K = \langle \Psi, \Phi \rangle^{-1}$ and $\tilde{\Psi} = K\Psi$. Then $\langle \tilde{\Psi}, \Phi \rangle = \langle K\Psi, \Phi \rangle = K \langle \Psi, \Phi \rangle = \mathrm{Id}_m$. The decomposition (2.21) of $C_{n,\tau}$ given by Theorem 2.21 may be written explicitly as

$$\varphi = \varphi_p + \varphi_q,$$

where $\varphi_p = \Phi_\Lambda \langle \Psi_\Lambda, \varphi \rangle \in P_\Lambda$, $\varphi_q \in Q_\Lambda = \{\phi : \langle \Psi_\Lambda, \phi \rangle = 0\}$.

Remark 2.1. The bilinear form in $C_{n,\tau}^* \times C_{n,\tau}$ given by (2.26) can be extended in a natural way to $C_{n,\tau}^* \times BC_n$ by setting $\langle \psi, X_0 \rangle = \overline{\psi}(0)$. We defer to Sect. 2.3 for a discussion how this extended bilinear form allows us to cast a functional differential equation to a system defined on the spaces P and Q_Λ.

2.2.6 Neural Networks with Delay: A Case Study on Characteristic Equations

In this section, we provide a detailed case study for the characteristic equation of the linearization at the trivial equilibrium of a coupled network of neurons with delayed feedback. Such a network with feedback with different interneuron and intraneuron time lags arises naturally in biological neural populations and their hardware implementation, and such a network also provides a simple-looking delay differential system that can exhibit complicated dynamics due to the existence of multiple eigenvalues of the infinitesimal generator of the linearized system at a given equilibrium when the synaptic connections and signal transmission delays are in certain ranges.

2.2.6.1 General Additive Neural Networks with Delay

We first describe an artificial neural network consisting of electronic neurons (amplifiers) interconnected through a matrix of resistors. Here an electronic neuron, the building block of the network, consists of a nonlinear amplifier that transforms an input signal u_i into the output signal v_i, and the input impedance of the amplifier unit is described by the combination of a resistor ρ_i and a capacitor C_i. We assume that the input–output relation is completely characterized by a voltage amplification function $v_i = f_i(u_i)$. The synaptic connections of the network are represented by resistors R_{ij} that connect the output terminal of the amplifier j with the input part of the neuron i. In order for the network to function properly, the resistances R_{ij} must be able to take on negative values. This can be realized by supplying each amplifier with an inverting output line that produces the signal $-v_j$. The number of rows in the resistor matrix is doubled, and whenever a negative value of R_{ij} is needed, this is realized using an ordinary resistor that is connected to the inverting output line.

The time evolution of the signals of the network is described by the Kirchhoff's law. Namely, the strengths of the incoming and outgoing current at the amplifier input port must balance. Consequently, we arrive at

$$C_i \frac{du_i}{dt} + \frac{u_i}{\rho_i} = \sum_{j=1}^{n} \frac{1}{R_{ij}}(v_j - u_i).$$

Let

$$\frac{1}{R_i} = \frac{1}{\rho_i} + \sum_{j=1}^{n} \frac{1}{R_{ij}}.$$

We get

$$C_i R_i \frac{du_i}{dt} + u_i = \sum_{j=1}^{n} \frac{R_i}{R_{ij}} v_j.$$

In the above derivation of the model equation for an artificial neural network, we implicitly assumed that the neurons communicate and respond instantaneously. Consideration of the finite switching speed of amplifiers requires that the input–output relation be replaced by $v_i = f_i(u_i(t - \tau_i))$ with a positive constant $\tau_i > 0$, and thus we obtain the following system of delay differential equations (see also [168, 209, 252, 267, 278]):

$$C_i R_i \frac{du_i(t)}{dt} = -u_i(t) + \sum_{j=1}^{n} \frac{R_i}{R_{ij}} f_j(u_j(t - \tau_j)), \quad 1 \leq i \leq n.$$

In what follows, for the sake of simplicity, we assume that

$$C_i = C, \quad R_i = R, \quad 1 \le i \le n,$$

and thus all local relaxation times $C_i R_i = CR$ are the same. Rescaling the time delay with respect to the network's relaxation time and rescaling the synaptic connection by

$$x_i(t) = u_i(CRt), \quad r_j = \frac{\tau_j}{RC}, \quad w_{ij} = \frac{R}{R_{ij}},$$

we get

$$x_i'(t) = -x_i(t) + \sum_{j=1}^{n} w_{ij} f_j(x_j(t - r_j)).$$

It is now easy to observe that it is the relative size of the delay r_j that determines the dynamics and the computational performance of the network, and designing a network to operate more quickly will increase this relative size of the delay.

It is therefore important to examine the effect of signal delays on the network dynamics. An important issue that has been addressed in the literature is how signal delays change the stability of equilibria, causing nonlinear oscillations and inducing periodic solutions. It will be shown that increasing the delay is among many mechanisms to create a network that exhibits periodic oscillations. Obviously, whether delay can generate oscillation also depends on the network connection topology. We refer to the monographs [224, 304] and a book chapter [52] for discussions about the relevance of this type of artificial neural network for the study of biological neural populations. In particular, we emphasize the importance of temporal delays in the coupling between cells, since in many chemical and biological oscillators (cells coupled via membrane transport of ions), the time needed for transport or processing of chemical components or signals may be of considerable length.

2.2.6.2 Special Case: Two Neurons

We now consider the following system of two neurons:

$$\begin{cases} \dot{x}_1(t) = -x_1(t) + \beta f(x_1(t-\tau)) + a_{12} f(x_2(t-\tau_1)), \\ \dot{x}_2(t) = -x_2(t) + \beta f(x_2(t-\tau)) + a_{21} f(x_1(t-\tau_2)), \end{cases} \tag{2.27}$$

where $x_1(t)$ and $x_2(t)$ denote the activations of the two neurons, $\tau_i (i = 1, 2)$ and τ denote the synaptic transmission delays, a_{12} and a_{21} are the synaptic coupling weights, $f : \mathbb{R} \to \mathbb{R}$ is the activation function. Throughout this subsection, we always assume that $\tau_1 + \tau_2 = 2\tau > 0$ and $f : \mathbb{R} \to \mathbb{R}$ is a C^1-smooth function with $f(0)=0$. Without loss of generality, we also assume that $\tau_1 \ge \tau_2$ and $f'(0) = 1$. Letting $x(t) = (x_1(t), x_2(t))^T$ and $x_t(\theta) = x(t + \theta)$ for $\theta \in [-\tau_1, 0]$, we can rewrite (2.27) as

$$\dot{x}(t) = Lx_t + F(x_t)$$

with

$$L\varphi = -\varphi(0) + B_1\varphi(-\tau_1) + B_2\varphi(-\tau_2) + B\varphi(-\tau)$$

and

$$F(\varphi) = \frac{f''(0)}{2}\begin{bmatrix} a_{11}\varphi_1^2(-\tau) + a_{12}\varphi_2^2(-\tau_1) \\ a_{21}\varphi_1^2(-\tau_2) + a_{22}\varphi_2^2(-\tau) \end{bmatrix}$$

$$+ \frac{f'''(0)}{6}\begin{bmatrix} a_{11}\varphi_1^3(-\tau) + a_{12}\varphi_2^3(-\tau_1) \\ a_{21}\varphi_1^3(-\tau_2) + a_{22}\varphi_2^3(-\tau) \end{bmatrix} + o(\|\varphi\|^3)$$

for $\varphi = (\varphi_1, \varphi_2)^T \in C_{2,\tau_1}$, where

$$B_1 = \begin{bmatrix} 0 & a_{12} \\ 0 & 0 \end{bmatrix}, \quad B_2 = \begin{bmatrix} 0 & 0 \\ a_{21} & 0 \end{bmatrix}, \quad B = \begin{bmatrix} \beta & 0 \\ 0 & \beta \end{bmatrix}.$$

The linearized system of (2.27) can be written as

$$\dot{x} = Lx_t = \int_{-\tau_1}^{0} d\eta(\theta)x(t+\theta), \tag{2.28}$$

where the matrix function $\eta(\theta)$ is given by

$$\eta(\theta, \mu) = \begin{cases} B_1 + B + B_2 - \mathrm{Id}_n, & \theta = 0, \\ B_1 + B + B_2, & \theta \in [-\tau_2, 0), \\ B_1 + B, & \theta \in [-\tau, -\tau_2), \\ B_1, & \theta \in (-\tau_1, -\tau), \\ 0, & \theta = -\tau_1, \end{cases}$$

and $\delta(\theta)$ is the Dirac delta function. The formal adjoint equation associated with (2.28) is given by

$$\dot{y}(t) = y(t) - y(t+\tau_1)B_1 - y(t+\tau_2)B_2 - y(t+\tau)B.$$

The bilinear form is

$$\langle \psi, \varphi \rangle = \overline{\psi}(0)\varphi(0) + \int_{-\tau_1}^{0} \overline{\psi}(s+\tau_1)B_1\varphi(s)ds$$

$$+ \int_{-\tau_2}^{0} \overline{\psi}(s+\tau_2)B_2\varphi(s)ds + \int_{-\tau}^{0} \overline{\psi}(s+\tau)B\varphi(s)ds. \tag{2.29}$$

The operators $\mathscr{A}$ and $\mathscr{A}^*$ are given by

$$(\mathscr{A}\varphi)(\theta) = \begin{cases} \frac{d\varphi(\theta)}{d\theta}, & \text{if } \theta \in [-\tau_1, 0), \\ -\varphi(0) + B_1\varphi(-\tau_1) + B_2\varphi(-\tau_2) + B\varphi(-\tau), & \text{if } \theta = 0, \end{cases}$$

and

$$(\mathscr{A}^*\psi)(\theta) = \begin{cases} -\frac{d\psi(\xi)}{d\xi}, & \text{if } \xi \in (0, \tau_1], \\ -\psi(0) + \psi(\tau_1)B_1 + \psi(\tau_2)B_2 + \psi(\tau)B, & \text{if } \xi = 0. \end{cases}$$

Moreover, φ is in $\mathrm{Ker}(\lambda \mathrm{Id} - \mathscr{A})$ if and only if $\varphi(\theta) = e^{\lambda \theta} v$, $-\tau_1 \leq \theta \leq 0$, where v is a vector in $\mathbb{R}^2$ such that $\Delta(\lambda)v = 0$ and the characteristic matrix $\Delta(\lambda)$ is

$$\Delta(\lambda) = \begin{bmatrix} \lambda + 1 - \beta e^{-\lambda \tau} & -a_{12} e^{-\lambda \tau_1} \\ -a_{21} e^{-\lambda \tau_2} & \lambda + 1 - \beta e^{-\lambda \tau} \end{bmatrix}.$$

Thus, the characteristic equation is

$$\det \Delta(\lambda) = [\lambda + 1 - \beta e^{-\lambda \tau}]^2 - a_{12} a_{21} e^{-2\lambda \tau} = 0. \tag{2.30}$$

Also, ψ is in $\mathrm{Ker}(\lambda \mathrm{Id} - \mathscr{A}^*)$ if and only if $\psi(\xi) = e^{\lambda \xi} u$, $0 \leq \xi \leq \tau_1$, where u is a vector in $\mathbb{R}^{2*}$ such that $u\Delta(-\lambda) = 0$.

Let $\gamma_{\pm} = \beta \pm \sqrt{a_{12} a_{21}}$, where $\sqrt{a_{12} a_{21}}$ is a real if $a_{12} a_{21} > 0$ and purely imaginary otherwise. Then, $\det \Delta(\lambda)$ can be decomposed as

$$\det \Delta(\lambda) = [\lambda + 1 - \gamma_+ e^{-\lambda \tau}][\lambda + 1 - \gamma_- e^{-\lambda \tau}].$$

Thus, in order to investigate the distribution of zeros of $\det \Delta(\lambda)$, we first consider the distribution of zeros of the following function:

$$P_z(\lambda) = \lambda + 1 - z e^{-\lambda \tau}, \tag{2.31}$$

where $z \in \mathbb{C}$. Define a parametric curve Σ with the parametric equations

$$\begin{cases} u(t) = \cos \tau t - t \sin \tau t, \\ v(t) = t \cos \tau t + \sin \tau t, \end{cases} \quad t \in \mathbb{R}. \tag{2.32}$$

It is easy to see that the curve Σ is symmetric about the u-axis. Let $\theta(t) = v(t)/u(t)$. Then $\theta'(t) = u^{-2}(t)[1 + \tau + \tau t^2] > 0$ for all $t \in \mathbb{R}$ such that $u(t) \neq 0$. This implies that as t increases, the corresponding point $(u(t), v(t))$ on the curve Σ moves counterclockwise about the origin. Moreover, it follows from $u^2(t) + v^2(t) = 1 + t^2$ that $\Sigma^+ = \{(u(t), v(t)) : t \in \mathbb{R}^+\}$ is simple, i.e., it cannot intersect itself. Let $\{t_n\}_{n=0}^{+\infty}$ be the monotonic increasing sequence of the nonnegative zeros of $v(t)$, and $c_n = u(t_n)$ for all $n \in \mathbb{N}_0 := \{0, 1, 2, \ldots\}$. Obviously, we have $t_0 = 0$ and $t_n \in ((2n-1)\pi/(2\tau), n\pi/\tau)$ for all $n \in \mathbb{N}$. Therefore, the curve Σ intersects with the u-axis at $(c_n, 0)$, $n \in \mathbb{N}_0$. It follows from the counterclockwise property of the curve Σ that $(-1)^n c_n > 0$ for all $n \in \mathbb{N}_0$. In addition, we have $|c_n| = \sqrt{1 + t_n^2}$, which implies that $c_n = (-1)^n \sqrt{1 + t_n^2}$ for $n \in \mathbb{N}_0$ and $\{|c_n|\}_{n \in \mathbb{N}_0}$ is an increasing sequence. In particular, $c_0 = 1$ and $c_1 = \sec \tau t_1 < -1$. Moreover, we claim that

$$(-1)^n v'(t_n) > 0 \quad \text{and} \quad (-1)^n u'(t_n) \geq 0 \quad \text{for } n \in \mathbb{N}_0. \tag{2.33}$$

Equality in the second formula of (2.33) holds if and only if $n = 0$. In fact, we can check that $v'(t_n) \neq 0$ when $v(t_n) = 0$. This, combined with the counterclockwise property of the curve Σ, gives the first inequality in (2.33). From $u^2(t) + v^2(t) = 1 + t^2$, we have $u'(t)u(t) + v'(t)v(t) = t$ for $t \in \mathbb{R}^+$. Particularly, $u'(t_n)c_n = t_n$ for all $n \in \mathbb{N}_0$. This, combined with $(-1)^n c_n > 0$ for $n \in \mathbb{N}_0$, immediately implies the second inequality in (2.33). This proves the claim. Finally, $u^2(t) + v^2(t) = 1 + t^2 \geq 1$ also implies that the curve is not inside the unit circle and it has only one intersection point $(1,0)$ with the unit circle.

For each $n \in \mathbb{N}_0$, let $\Sigma_n = \{(u(t), v(t)) : t \in [-t_{n+1}, -t_n] \cup [t_n, t_{n+1}]\}$, which is a closed curve with $(0,0)$ inside. The curve Σ is schematically illustrated in Fig. 2.1. In the sequel, we will identify Σ with $\{u(t) + iv(t) : t \in \mathbb{R}\} \subset \mathbb{C}$. The following

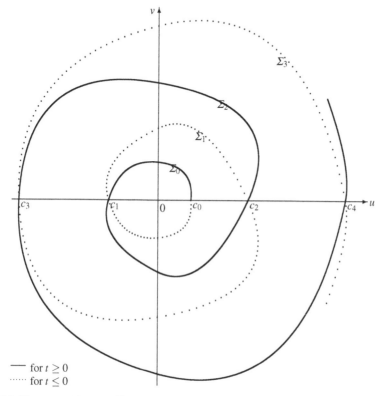

Fig. 2.1 The parametric curve Σ

lemma will play an important role in analyzing the distributions of the roots of (2.1).

Lemma 2.1. *Consider* $P_z(\lambda)$ *defined in (2.31) with* $z \in \mathbb{C}$. *Then the following statements are true:*

(i) $P_z(\lambda)$ *has a purely imaginary zero if and only if* $z \in \Sigma$. *Moreover, if* $z = u(\theta) + iv(\theta)$, *then the purely imaginary zero is* $i\theta$ *except that there is a pair of conjugate purely imaginary zeros* $\pm it_n$ *if* $z = c_n$ *for* $n \in \mathbb{N}$.

(ii) *For each fixed* $z_0 = u(\theta_0) + iv(\theta_0) \in \Sigma$, *there exist an open* δ-*neighborhood of* z_0 *in the complex plane, denoted by* $B(z_0, \delta)$, *and an analytic function* $\lambda :$ $B(z_0, \delta) \to \mathbb{C}$ *such that* $\lambda(z_0) = i\theta_0$ *and* $\lambda(z)$ *is a zero of* $P_z(\lambda)$ *for all* $z \in$

$B(z_0, \delta)$. Moreover, along the outward-pointing normal vector to the curve Σ at z_0, the directional derivative of $\mathrm{Re}\{\lambda(z)\}$ at z_0 is positive.

(iii) $P_z(\lambda)$ has only zeros with strictly negative real parts if and only if z is inside the curve Σ_0, exactly $j \in \mathbb{N}$ zeros with positive real parts if z is between Σ_{j-1} and Σ_j. In particular, if $z \in \Sigma_0$, $P_z(\lambda)$ has either a simple real zero 0 (if $z = 1$) or a simple purely imaginary zero (if $\mathrm{Im}(z) \neq 0$), or a pair of simple purely imaginary zeros (if $z = c_1$), and all other zeros has strictly negative real parts.

Proof. (i) $P_z(\lambda)$ has a purely imaginary zero, say $\lambda = i\theta$, if and only if $e^{i\tau\theta}(1 + i\theta) = z$, which is equivalent to $z \in \Sigma$ by separating the real and imaginary parts of $e^{i\tau\theta}(1 + i\theta)$.

(ii) Note that $P_{z_0}(i\theta_0) = 0$ and $i\theta_0$ is a simple zero of $P_{z_0}(\lambda)$. The existence of δ and the mapping λ follow from the implicit function theorem. Moreover, $\lambda(z)$ is analytic with respect to z. Thus,

$$\lambda'(z) = \frac{\partial}{\partial a}\mathrm{Re}\{\lambda(z)\} + i\frac{\partial}{\partial a}\mathrm{Im}\{\lambda(z)\} = \frac{\partial}{\partial b}\mathrm{Im}\{\lambda(z)\} - i\frac{\partial}{\partial b}\mathrm{Re}\{\lambda(z)\},$$

where $a = \mathrm{Re}(z)$ and $b = \mathrm{Im}(z)$. On the other hand, differentiating $P_z(\lambda) = 0$ with respect to z and using $P_{z_0}(i\theta_0) = 0$, we have

$$\lambda'(z_0) = \varepsilon_1 \left[u_0\varepsilon_2 + \theta_0 v_0 + i(\theta_0 u_0 - v_0\varepsilon_2)\right],$$

where $\varepsilon_1 = [(1 + \tau)^2 + (\tau\theta_0)^2]^{-1}(1 + \theta_0^2)^{-1}$ and $\varepsilon_2 = 1 + \tau + \tau\theta_0^2$. It follows that

$$\nabla\mathrm{Re}\{\lambda(z_0)\} = \left(\frac{\partial}{\partial a}\mathrm{Re}\{\lambda(z_0)\}, \frac{\partial}{\partial b}\mathrm{Re}\{\lambda(z_0)\}\right)^T$$

$$= \varepsilon_1 \left(u_0\varepsilon_2 + \theta_0 v_0, v_0\varepsilon_2 - \theta_0 u_0\right)^T.$$

Let $\vartheta(\xi) = (v'(\theta_0), -u'(\theta_0))M(\xi)$, where $\xi \in (-\pi/2, \pi/2)$ and

$$M(\xi) = \begin{bmatrix} \cos\xi & \sin\xi \\ -\sin\xi & \cos\xi \end{bmatrix}.$$

Obviously, for each fixed $\xi \in (-\pi/2, \pi/2)$, $\vartheta(\xi)$ is an outward-pointing vector to the curve Σ at z_0, Thus, the directional derivative along the vector $\vartheta(\xi)$ at z_0 is

$$\frac{d}{d\vartheta(\xi)}\mathrm{Re}\{\lambda(z_0)\} = \varepsilon_3(v'(\theta_0), -u'(\theta_0))M(\xi)\nabla\mathrm{Re}\{\lambda(z_0)\}$$

$$= \varepsilon_1\varepsilon_3(\varepsilon_2^2 + \theta_0^2)\cos\xi > 0,$$

where $\varepsilon_3 = 1/\sqrt{(1 + \tau)^2 + \tau^2\theta_0^2}$.

(iii) Note that $P_0(\lambda)$ has exactly one zero -1, which obviously has a negative real part. Since zeros of $P_z(\lambda)$ depend continuously on z, there exists a region Ω_0

containing $z = 0$ such that for $z \in \Omega_0$, all zeros of $P_z(\lambda)$ have negative real parts. Moreover, as z varies and passes through the boundary $\partial \Omega_0$, only one (or two if z is real) zero point of $P_z(\lambda)$ varies from a complex number with a negative real part to a purely imaginary number and then to a complex number with a positive real part. By (i), $\partial \Omega_0 = \Sigma_0$. Therefore, $P_z(\lambda)$ has only zeros with negative real parts if z in inside the curve Σ_0. If z is a real number between Σ_{j-1} and Σ_j, then one can easily show that $P_z(\lambda)$ has exactly j zeros with positive real parts (see, for example, the discussion in Chen and Wu [59]). This, combined with (i) and the continuous dependence of zeros of $P_z(\lambda)$ on z, completes the proof.   □

In view of Lemma 2.1, we have the following conclusions:

(1) All zeros of $\det \Delta(\lambda)$ have negative real parts if and only if both of $\gamma_\pm$ are inside the curve Σ.
(2) If and only if $1 \neq \gamma_+ \in \Sigma$ or $1 \neq \gamma_- \in \Sigma$, $\det \Delta(\lambda)$ has a pair of simple conjugate purely imaginary zeros $\pm i\omega$, where $\omega > 0$ satisfies either $u(\omega) + iv(\omega) = \gamma_+$ or $u(\omega) + iv(\omega) = \gamma_-$. In particular, $\omega = t_n$ if either γ_+ or γ_- is equal to c_n for some $n \in \mathbb{N}$.
(3) If and only if only one of γ_+ and γ_- is equal to 1, $\det \Delta(\lambda)$ has a simple zero $\lambda = 0$. Moreover, if $c_1 < \gamma_- < \gamma_+ = 1$, then all zeros but $\lambda = 0$ of $\det \Delta(\lambda)$ have strictly negative real parts.

If $a_{12}a_{21} > 0$ and only one of $\gamma_\pm$ lies on the curve Σ, or $a_{12}a_{21} < 0$ and $\gamma_\pm \in \Sigma$, then on the imaginary axis, the infinitesimal generator $\mathscr{A}$ has only one pair of simple purely imaginary eigenvalues $\pm i\omega$. Let $\Phi = (\varphi_1, \varphi_2)$ and $\Psi = (\psi_1, \psi_2)^T$ be bases for the generalized eigenspaces $P_{\pm i\omega}$ and $P^*_{\pm i\omega}$ of $\mathscr{A}$ and $\mathscr{A}^*$ associated with eigenvalues $\pm i\omega$, respectively. In fact, we can choose

$$\varphi_1(\theta) = \overline{\varphi}_2(\theta) = (1,d)^T e^{i\omega\theta}, \quad \theta \in [-\tau_1, 0],$$
$$\psi_1(\xi) = \overline{\psi}_2(\xi) = \overline{D}(\overline{d}, 1) e^{i\omega\xi}, \quad \xi \in [0, \tau_1],$$

and

$$d = (1 + i\omega - \beta e^{-i\omega\tau}) e^{i\omega\tau_1} / a_{12},$$
$$D = \{2d[1 + \tau(1 + i\omega)]\}^{-1}.$$

Moreover, $\langle \psi_j, \varphi_k \rangle = \delta_{jk}$, $j, k = 1, 2$, where $\langle \cdot, \cdot \rangle$ is defined in (2.29) and

$$\delta_{jk} = \begin{cases} 1, & \text{if } j = k, \\ 0, & \text{if } j \neq k. \end{cases}$$

Assume that $a_{12}a_{21} > 0$. If $\gamma_+ = 1$ and $\gamma_- = c_n$ or $\gamma_- = 1$ and $\gamma_+ = c_n$ for some $n \in \mathbb{N}$, then on the imaginary axis, the infinitesimal generator $\mathscr{A}$ has only simple eigenvalues 0, it_n, and $-it_n$. Here, we consider only the first case. Namely, assume that $a_{12}a_{21} > 0$ and $\gamma_+ = 1$ and $\gamma_- = c_n$ for some $n \in \mathbb{N}$. Let $\Phi = (q_0, q_1, \overline{q}_1)$, and $\Psi = (p_0, p_1, \overline{p}_1)^T$ be bases for the generalized eigenspaces P_Λ and P^*_Λ of $\mathscr{A}$ and $\mathscr{A}^*$ associated with $\Lambda = \{0, it_n, -it_n\}$. In fact, we can choose

$$q_0(\theta) = (1,d_0)^T, \quad q_1(\theta) = (1,d_1)^T e^{it_n\theta}, \quad \theta \in [-\tau_1, 0],$$

and

$$p_0(\xi) = D_0(d_0,1), \quad p_1(\xi) = \overline{D}_1\left(\overline{d}_1,1\right)e^{it_n\xi}, \quad \xi \in [0,\tau_1],$$

where $d_0 = (1-\beta)/a_{12}$, $d_1 = (1 + it_n - \beta e^{-it_n\tau})e^{it_n\tau_1}/a_{12}$, $D_0 = [2d_0(1+\tau)]^{-1}$, and $D_1 = \{2d_1[1+\tau(1+it_n)]\}^{-1}$. Moreover, $\langle p_j, q_k \rangle = \delta_{jk}$ and $\langle p_j, \bar{q}_k \rangle = 0$, $j,k = 0,1$.

Assume that $a_{12}a_{21} > 0$. If $\gamma_+ = c_n$ and $\gamma_- = c_m$ for $n,m \in \mathbb{N}$ such that $c_n > c_m$, then on the imaginary axis, the infinitesimal generator $\mathscr{A}$ has only two pairs of simple purely imaginary eigenvalues $\pm i\omega_1$ and $\pm i\omega_2$, where $\omega_1 = t_n$ and $\omega_2 = t_m$. Let $\Phi = (q_1, \bar{q}_1, q_2, \bar{q}_2)$, and $\Psi = (p_1, \bar{p}_1, p_2, \bar{p}_2)^T$ be bases for the generalized eigenspaces P_Λ and P_Λ^* of $\mathscr{A}$ and $\mathscr{A}^*$ associated with $\Lambda = \{i\omega_1, -i\omega_1, i\omega_2, -i\omega_2\}$. In fact, we can choose

$$q_j(\theta) = (1,d_j)^T e^{i\omega_j\theta}, \quad \theta \in [-\tau_1, 0], \quad j = 1,2,$$

and

$$p_j(\xi) = \overline{D}_j\left(\overline{d}_j,1\right)e^{i\omega_j\xi}, \quad \xi \in [0,\tau_1], \quad j = 1,2,$$

where $d_1 = (1 + i\omega_j - \beta e^{-i\omega_j\tau})e^{i\omega_j\tau_1}/a_{12}$ and $D_j = \{2d_j[1+\tau(1+i\omega_j)]\}^{-1}$. Moreover, $\langle p_j, q_k \rangle = \delta_{jk}$ and $\langle p_j, \bar{q}_k \rangle = 0$, $j,k = 1,2$.

2.3 General Framework of NFDEs

Suppose that $f, h: C_{n,\tau} \to \mathbb{R}^n$ are given continuous mappings. The relation

$$\frac{\mathrm{d}}{\mathrm{d}t}h(x_t) = f(x_t) \tag{2.34}$$

is called a neutral functional differential equation (NFDE). The mapping h will be called the difference operator for NFDE (2.34). If $h(\varphi) = \varphi(0)$ for all φ, then (2.34) becomes (2.10). Consequently, DDEs are special cases of NFDEs.

A function x is said to be a solution of (2.34) on $[t_0, t_0 + A)$ for some $t_0 \in \mathbb{R}$ and $A > 0$ if $x \in C([t_0 - \tau, t_0 + A), \mathbb{R}^n)$, $x_t \in C_{n,\tau}$ for all $t \in [t_0, t_0 + A)$, $h(x_t)$ is continuously differentiable, and $x(t)$ satisfies (2.34) for all $t \in [t_0, t_0 + A)$.

Let $D, L: C_{n,\tau} \to \mathbb{R}^n$ be the two linearized operators of h and f at some equilibrium point, respectively. Without loss of generality, we assume that there exist two $n \times n$ matrix-valued functions $\mu, \eta : [-\tau, 0] \to \mathbb{R}^{n^2}$ whose components each have bounded variation and such that for $\varphi \in C_{n,\tau}$,

$$D\varphi = \varphi(0) - \int_{-\tau}^0 d\mu(\theta)\varphi(\theta), \quad L\varphi = \int_{-\tau}^0 d\eta(\theta)\varphi(\theta).$$

Moreover, we assume that D is atomic at zero, that is, $\mathrm{Var}_{[s,0]}\mu(\theta) \to 0$ as $s \to 0$ (see Hale and Verduyn Lunel [154] for more details). The linear system

$$\frac{\mathrm{d}}{\mathrm{d}t}Dx_t = Lx_t \qquad (2.35)$$

generates a strongly continuous semigroup of linear operators with infinitesimal generator $\mathscr{A}$. The spectrum of $\mathscr{A}$, denoted by $\sigma(\mathscr{A})$, is the point spectrum. Moreover, λ is an eigenvalue of $\mathscr{A}$, i.e., $\lambda \in \sigma(\mathscr{A})$, if and only if λ satisfies $\det \Delta(\lambda) = 0$, where the characteristic matrix $\Delta(\lambda)$ is given by

$$\Delta(\lambda) = \lambda D(e^{\lambda(\cdot)}\mathrm{Id}) - L(e^{\lambda(\cdot)}\mathrm{Id}).$$

It is well known that $\phi \in C_{n,\tau}$ is an eigenvector of $\mathscr{A}$ associated with the eigenvalue λ if and only if $\phi(\theta) = e^{\lambda\theta}b$ for $\theta \in [-\tau,0]$ and some vector $b \in \mathbb{C}^n$ such that $\Delta(\lambda)b = 0$. Here and in the sequel, for the sake of convenience, we shall also allow functions with range in $\mathbb{C}^n$.

Let Λ be a set of some eigenvalues of $\mathscr{A}$, and denote by E_Λ the generalized eigenspace of $\mathscr{A}$ associated with Λ. It is known that $\dim E_\Lambda = m$, where m is the number of zeros of $\det \Delta(\lambda)$ in Λ, counting multiplicities. As we did earlier for DDEs, we define a bilinear form

$$\langle \psi, \varphi \rangle = \overline{\psi}(0)\varphi(0) - \int_{-\tau}^0 \left[\frac{\mathrm{d}}{\mathrm{d}s} \int_0^s \overline{\psi}(\xi - s)d\mu(\theta)\varphi(\xi)d\xi \right]_{s=\theta} \qquad (2.36)$$
$$- \int_{-\tau}^0 \int_0^\theta \overline{\psi}(\xi - \theta)d\eta(\theta)\varphi(\xi)d\xi$$

for $\psi \in C_{n,\tau}^*$ and $\varphi \in C_{n,\tau}$. Let Φ be a basis for E_Λ and Ψ the basis for the dual space E_Λ^* in C_n^* such that $\langle \Psi, \Phi \rangle = \mathrm{Id}_m$. The phase space $C_{n,\tau}$ is decomposed by Λ as $C_{n,\tau} = E_\Lambda \oplus Q_\Lambda$, where $Q_\Lambda = \{\phi \in C_{n,\tau} : \langle \Psi, \phi \rangle = 0\}$. Moreover, there exists an $m \times m$ constant matrix B with $\sigma(B) = \Lambda$ such that

$$\dot{\Phi} = \Phi B \quad \text{and} \quad \dot{\Psi} = -B\Psi.$$

Similarly to the previous sections for DDEs, we may enlarge the phase space $C_{n,\tau}$ such that (2.34) can be written as an abstract ODE in the Banach space BC_n. First, in BC_n, we consider an extension of the infinitesimal generator $\mathscr{A}$, still denoted by $\mathscr{A}$,

$$\mathscr{A} : BC_n \supset C_{n,\tau}^1 \ni \varphi \mapsto \dot{\varphi} + X_0[L\varphi - D\dot{\varphi}] \in BC_n,$$

where $\mathrm{Dom}(\mathscr{A}) = C_{n,\tau}^1 \overset{\text{def}}{=} \{\varphi \in C_{n,\tau} : \dot{\varphi} \in C_{n,\tau}\}$. The bilinear form in $C_{n,\tau}^* \times C_{n,\tau}$ given by (2.36) is extended in a natural way to $C_{n,\tau}^* \times BC_n$ by setting $\langle \psi, X_0 \rangle = \overline{\psi}(0)$. Thus, the abstract ODE in BC_n associated with (2.34) can be rewritten in the form

$$\frac{\mathrm{d}}{\mathrm{d}t}u = \mathscr{A}u + X_0 G(u), \qquad (2.37)$$

where

$$G(u) = f(u) - Lu - \frac{d}{dt}[h(u) - Du].$$

(2.38)

Consider the projection $\pi : BC_n \to E_\Lambda$ given by

$$\pi(\varphi + X_0 \xi) = \Phi[\langle \Psi, \varphi \rangle + \overline{\Psi}(0)\xi].$$

(2.39)

Obviously, π is a continuous projection onto E_Λ, which commutes with $\mathscr{A}$ in $C^1_{n,\tau}$. This allows us to decompose BC_n as a topological direct sum, $BC_n = E_\Lambda \oplus \operatorname{Ker} \pi$, where $Q_\Lambda \subset \operatorname{Ker} \pi$.

Due to the decomposition of BC_n, we can decompose u in (2.37) in the form $u = \Phi x + y$, with $x \in \mathbb{R}^m$, $y \in Q \overset{\text{def}}{=} \operatorname{Ker} \pi \cap C^1_{n,\tau}$. Then (2.37) is equivalent to the system

$$\begin{aligned} \dot{x} &= Bx + \overline{\Psi}(0)G(\Phi x + y), \\ \tfrac{dy}{dt} &= \mathscr{A}_Q y + (I - \pi)X_0 G(\Phi x + y), \end{aligned}$$

(2.40)

where $\mathscr{A}_Q$ is the restriction of $\mathscr{A}$ to Q interpreted as an operator acting in the Banach space $\operatorname{Ker} \pi$. The spectrum of $\mathscr{A}_Q$ will be very important for the construction of normal forms. Similarly, $\mathscr{A}_Q$ has only a point spectrum. Moreover, $\sigma(\mathscr{A}_Q) = \sigma(\mathscr{A}) \setminus \Lambda$.

Chapter 3
Center Manifold Reduction

A center manifold at a given nonhyperbolic equilibrium is an invariant manifold of a given differential equation that is tangent at the equilibrium point to the (generalized) eigenspace of the neutrally stable eigenvalues. Since the local dynamic behavior *transverse* to the center manifold is relatively simple, the potentially complicated asymptotic behaviors of the full system are captured by the flows restricted to the center manifolds.

Center manifold theory plays an important role in the study of the stability of dynamical systems when the equilibrium point is not hyperbolic. The combination of center manifold reduction with the normal form approach has been used extensively to study bifurcations of parameterized dynamical systems. The center manifold theorem provides the theoretical foundation for systematically reducing the dimension of the state spaces.

The classical center manifold theory of equilibria, since first introduced by Pliss [245] and Kelley [189] in the 1960s, has been well developed and treated by Carr [53], Hirsch et al. [164], Sijbrand [266], Hassard et al. [159, 160], Ruelle et al. [255, 256], Ait Babram et al. [5–7], Guo and Man [138], Guckenheimer and Holmes [125], Vanderbauwhede [285], and others [91–93, 147, 183, 184, 211, 286].

3.1 Some Examples of Ordinary Differential Equations

To illustrate the concept and importance of invariant manifolds, we examine some examples of ordinary differential equations (ODEs).

Example 3.1. Consider the system of ODEs

$$\dot{x} = x, \quad \dot{y} = -y + x^2, \quad (x,y) \in \mathbb{R}^2, \tag{3.1}$$

which has a hyperbolic fixed point at $(x,y) = (0,0)$. The associated linearized system is given by

$$\dot{x} = x, \quad \dot{y} = -y,$$

S. Guo and J. Wu, *Bifurcation Theory of Functional Differential Equations*,
Applied Mathematical Sciences 184, DOI 10.1007/978-1-4614-6992-6_3,
© Springer Science+Business Media New York 2013

with stable and unstable subspaces given by

$$E^s = \{(x,y) \in \mathbb{R}^2 : x = 0\}, \quad E^u = \{(x,y) \in \mathbb{R}^2 : y = 0\}.$$

For the nonlinear vector field (3.1), the solution can be obtained explicitly as follows. Eliminating time as the independent variable gives

$$\frac{dy}{dx} = \frac{x^2 - y}{x} \quad \text{if } x \neq 0,$$

which can be solved to obtain

$$y(x) = \frac{x^2}{3} + \frac{c}{x} \quad \text{if } x \neq 0,$$

where c is some constant.

The local unstable manifold of (3.1), denoted by $W^u_{\text{loc}}(0,0)$, is a one-dimensional manifold that is tangent to the unstable subspace at the origin such that solutions starting from this manifold will stay in the manifold for $t \geq 0$ and converge to the origin as $t \to -\infty$. This unstable manifold can be represented by a graph over the x variable, that is, $y = h(x)$ with $h(0) = h'(0) = 0$. The unstable manifold is given by letting $c = 0$:

$$W^u_{\text{loc}}(0,0) = \left\{ (x,y) \in \mathbb{R}^2 : y = \frac{x^2}{3} \right\}.$$

Note that every solution starting in $W^u_{\text{loc}}(0,0)$ remains in $W^u_{\text{loc}}(0,0)$, so this is also the global unstable manifold of the origin.

We can similarly define stable manifolds. Note that if we have initial conditions on the y-axis, then the solution stays on the y-axis and approaches $(0,0)$ as $t \to \infty$. Thus, the local stable manifold and its global extension (global stable manifold) are both $W^s_{\text{loc}}(0,0) = W^s(0,0) = \{(x,y) \in \mathbb{R}^2 : x = 0\}$ (see Fig. 3.1).

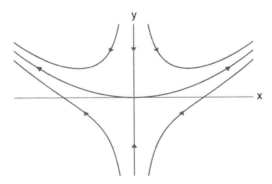

Fig. 3.1 Stable and unstable manifolds of $(x,y) = (0,0)$ in system (3.1)

In Chap. 1, we studied bifurcations of equilibria in one-parameter dynamical systems having the minimum possible phase dimensions. Indeed, the systems we analyzed were either one- or two-dimensional. The corresponding bifurcations occur in essentially the same way for higher-dimensional systems. As we shall see, there are certain parameter-dependent one- or two-dimensional invariant manifolds on which the system exhibits the corresponding bifurcations, while the behavior off the manifolds is somehow *trivial*. Therefore, such manifolds (called center manifolds) are very important in describing dynamic behavior of high-dimensional systems.

The following example shows that the center manifold need not be unique.

Example 3.2. Consider the following ODE system:

$$\dot{x} = x^2, \quad \dot{y} = -y, \quad (x,y) \in \mathbb{R}^2. \tag{3.2}$$

The trajectory through point (x_0, y_0) is (see Fig. 3.2) given by

$$x(t) = \frac{x_0}{1 - tx_0}, \quad y(t) = y_0 e^{-t}.$$

After eliminating time t, we obtain

$$y = (y_0 e^{-1/x_0}) e^{1/x}.$$

It is easy to see that all trajectories starting from the left half of the (x,y)-plane $(x < 0)$ tend to $(0,0)$ as $x \to 0$. The center space E^c of the linearized system (3.2) is the x-axis. System (3.2) possesses a family of one-dimensional manifolds:

$$W_\beta^c = \{(x,y) \in \mathbb{R}^2 : y = \psi_\beta(x)\},$$

where

$$\psi_\beta(x) = \begin{cases} \beta e^{1/x}, & \text{if } x < 0, \\ 0, & \text{if } x \geq 0. \end{cases}$$

Obviously, W_β^c is tangent to E^c at the origin, and it is a family of C^∞ manifolds. According to the definition to be introduced in the next section, these are all center manifolds.

We conclude this section with an example to show how to reduce the dimension and simplify the corresponding bifurcation problem.

Example 3.3. Consider the following two-dimensional system:

$$\begin{aligned} \dot{x} &= \mu x + 2\mu y + x^2 + y^2, \\ \dot{y} &= -y + \mu x + x^2 + xy. \end{aligned} \tag{3.3}$$

By the center manifold theorem, there should be an invariant manifold of the form

$$y = h(x, \mu) = a(\mu)x + b(\mu)x^2 + c(\mu)x^3 + O(|x|^4). \tag{3.4}$$

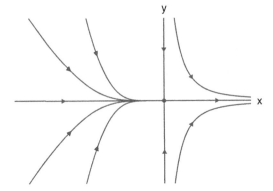

Fig. 3.2 Center manifolds of $(x,y) = (0,0)$ in system (3.2)

To compute the coefficients, we simply differentiate the equation and obtain

$$\dot{y} = a(\mu)\dot{x} + 2b(\mu)x\dot{x} + 3c(\mu)x^2\dot{x} + O(|x|^3)\dot{x}.$$

Now we use the differential equation (3.3) to replace $\dot{y}$ and $\dot{x}$. This yields

$$-y + \mu x + x^2 + xy = [a(\mu) + 2b(\mu)x + 3c(\mu)x^2 + O(|x|^3)][\mu x + 2\mu y + x^2 + y^2].$$

Next, we use (3.4) to substitute for y, and we compare terms of order x, x^2, and x^3 to obtain

$$\begin{aligned}
-a(\mu) + \mu &= a(\mu)\mu + 2\mu[a(\mu)]^2, \\
-b(\mu) + 1 + a(\mu) &= 2b(\mu)\mu + 6\mu a(\mu)b(\mu) + a(\mu) + [a(\mu)]^3, \\
-c(\mu) + b(\mu) &= 8\mu a(\mu)c(\mu) + 4[a(\mu)]^2 b(\mu) + 4\mu[b(\mu)]^2 + 2b(\mu) + 3\mu c(\mu).
\end{aligned}$$

It follows that $a(\mu) = \mu + O(\mu^2)$, $b(\mu) = 1 + O(\mu)$. Finally, we can insert (3.4) back into the first equation of (3.3) and obtain

$$\dot{x} = xf(\mu,x),$$

where

$$f(\mu,x) = \mu + 2\mu a(\mu) + [2\mu b(\mu) + 1 + (a(\mu))^2]x + [2\mu c(\mu) + 2a(\mu)b(\mu)]x^2 + O(x^3).$$

Notice that $f_x(0,0) = 1$ and $f_\mu(0,0) = 1$, so that according to the implicit function theorem, we can solve uniquely for either variable in a neighborhood of $(0,0)$: $x = -\mu + O(\mu^2)$ or $\mu = -x + O(x^2)$. We have thus established the existence of a bifurcating branch of steady solutions.

3.2 Invariant Manifolds of RFDEs

We now state the invariant manifold theory for the DDE

$$\dot{x} = f(x_t), \tag{3.5}$$

where $f: C_{n,\tau} \to \mathbb{R}$ is sufficiently smooth and satisfies $f(0) = 0$. The linearization of system (3.5) at this equilibrium point $x = 0$ takes the form

$$\dot{x}(t) = Lx_t, \tag{3.6}$$

which generates a strongly continuous semigroup with the infinitesimal generator $\mathscr{A} : C_{n,\tau} \to C_{n,\tau}$. Let σ^s, σ^c, and σ^u be the sets of eigenvalues of the infinitesimal generator $\mathscr{A}$ with negative, zero, and positive real parts, respectively. Obviously, $\sigma = \sigma^s \cup \sigma^c \cup \sigma^u$. Let E^s, E^c, and E^u be the generalized eigenspaces corresponding to σ^s, σ^c, and σ^u, respectively. Obviously, E^s, E^c, and E^u are $\mathscr{A}$-invariant subspaces of $C_{n,\tau}$, and are referred to as the stable, unstable, and center subspaces, respectively. Moreover, solutions starting in E^s approach $x = 0$ asymptotically as $t \to \infty$, and solutions starting in E^u approach $x = 0$ asymptotically as $t \to -\infty$.

The space $\mathscr{M} = E^s \oplus E^c \oplus E^u$ is precisely the space of initial functions such that the corresponding solution of (3.6) has a finite expansion in characteristic solutions. In fact, the solution through $\phi \in \mathscr{M}$ exists for all time, and the semigroup $T(t)$ extends to a flow on $\mathscr{M}$. In general, the closure of $\mathscr{M}$ is (possibly properly) contained in $C_{n,\tau}$, and the space $C_{n,\tau}$ can be decomposed as

$$C_{n,\tau} = S \oplus E^c \oplus E^u, \tag{3.7}$$

where S contains E^s and is possibly infinite-dimensional.

Definition 3.1. For a given neighborhood V, the local strongly stable set (or manifold) $M_{\mathrm{loc}}^{ss}(0)$ and the local strongly unstable set (or manifold) $M_{\mathrm{loc}}^{su}(0)$ of the equilibrium point 0 of system (3.5) are defined as

$$M_{\mathrm{loc}}^{ss}(0) \overset{\mathrm{def}}{=} \{\varphi \in C_{n,\tau}: x_t(\cdot, \varphi) \in V \text{ for } t \geq 0 \text{ and approaches } 0 \text{ exponentially as } t \to \infty\},$$

$$M_{\mathrm{loc}}^{su}(0) \overset{\mathrm{def}}{=} \{\varphi \in C_{n,\tau}: x_t(\cdot, \varphi) \in V \text{ for } t \leq 0 \text{ and approaches } 0 \text{ exponentially as } t \to -\infty\}.$$

Definition 3.2. For a given neighborhood V, a local center manifold $M_{\mathrm{loc}}^{c}(0)$ of the equilibrium point 0 of system (3.5) is a C^1 submanifold that is a graph over $V \cap E^c$ in $C_{n,\tau}$, tangent to E^c at 0, and locally invariant under the flow defined by (3.5). Namely,

$$M_{\mathrm{loc}}^{c}(0) \cap V = \{\varphi \in C_{n,\tau}: \varphi = \phi + h(\phi), \phi \in V \cap E^c\},$$

where $h: E^c \to S \oplus E^u$ is a C^1 mapping with $h(0) = 0, D_\phi h(0) = 0$. Moreover, every orbit that begins on $M_{\mathrm{loc}}^{c}(0)$ remains in this set as long as it stays in V.

Definition 3.3. For a given neighborhood V, a local center-stable manifold $M_{\mathrm{loc}}^{cs}(0)$ of the equilibrium point 0 of system (3.5) is a set in $C_{n,\tau}$ such that $M_{\mathrm{loc}}^{cs}(0) \cap V$ is a

C^1 submanifold that is a graph over $V \cap (E^c \oplus S)$, tangent to $E^c \oplus S$ at 0, and locally invariant under the semiflow defined by (3.5). Namely,

$$M_{\text{loc}}^{cs}(0) \cap V = \{\varphi \in C_{n,\tau} : \varphi = \phi + h(\phi), \phi \in V \cap (E^c \oplus S)\},$$

where $h : E^c \oplus S \to E^u$ is a C^1 mapping with $h(0) = 0$, $D_\phi h(0) = 0$. Moreover, every orbit that begins on $M_{\text{loc}}^{cs}(0)$ remains in this set as long as it stays in V. Furthermore, every orbit that stays in V for all $t \geq 0$ must belong to $M_{\text{loc}}^{cs}(0)$. In the same way, we define the local center-unstable manifold $M_{\text{loc}}^{cu}(0)$ of the equilibrium point 0 of (3.5) by replacing $t \geq 0$ by $t \leq 0$, the set $E^c \oplus S$ by $E^c \oplus E^u$, and the set E^u by S.

We have the following basic result on the existence of invariant manifolds.

Theorem 3.1. *If f in (3.5) is a C^k function such that $f(0) = 0$, then there is a neighborhood V of $0 \in C_{n,\tau}$ such that the sets $M_{\text{loc}}^{ss}(0)$, $M_{\text{loc}}^{su}(0)$, $M_{\text{loc}}^{c}(0)$, $M_{\text{loc}}^{cu}(0)$, and $M_{\text{loc}}^{cs}(0)$ exist and are C^k submanifolds of $C_{n,\tau}$. The manifolds $M_{\text{loc}}^{ss}(0)$ and $M_{\text{loc}}^{su}(0)$ are uniquely defined, whereas the manifolds $M_{\text{loc}}^{c}(0)$, $M_{\text{loc}}^{cu}(0)$, and $M_{\text{loc}}^{cs}(0)$ might not be.*

Local stable and unstable manifolds can be extended to global stable and unstable manifolds. In addition, the existence of invariant manifolds for partial functional differential equations was established by Memory [219–221].

Very often, one talks about *stable manifolds*, *unstable manifolds*, or *center manifolds* without specific reference to a particular equilibrium. However, this is a misuse of the term, because the addition of *of an equilibrium* is key: one must speak about the stable, unstable, or center manifolds *of something* in order for the terminology to make sense. The *something* studied thus far has been an equilibrium.

3.3 Center Manifold Theorem

Let Φ be a basis for E^c and Ψ the basis for the dual space E^{c*} in $C_{n,\tau}^*$ such that $\langle \Psi, \Phi \rangle = \text{Id}_m$, where $m = \dim E^c$. There exists an $m \times m$ matrix B such that $\dot{\Phi} = \Phi B$. It follows from (3.7) that we have $BC_n = E^c \oplus \text{Ker}\,\pi$ and $E^s \oplus E^u \subset \text{Ker}\,\pi$, where $\pi : BC_n \to E^c$ is the projection defined by

$$\pi(\varphi + X_0 \xi) = \Phi[\langle \Psi, \varphi \rangle + \overline{\Psi}(0)\xi] \tag{3.8}$$

for $\varphi \in C_{n,\tau}$ and $\xi \in \mathbb{R}^n$.

As we know, the abstract ODE in BC_n associated with (3.5) can be rewritten in the form

$$\frac{d}{dt}x_t = \mathscr{A}x_t + X_0 F(x_t), \tag{3.9}$$

where $F(x_t) = f(x_t) - Lx_t$. Due to the decomposition of BC_n, we can decompose x_t in (3.9) in the form $x_t = \Phi z + y$, with $z \in \mathbb{R}^m$, $y \in Q \overset{\text{def}}{=} \text{Ker}\,\pi \cap C_{n,\tau}^1$. Then (3.9) is equivalent to the system

$$\dot{z} = Bz + \overline{\Psi}(0)F(\Phi z + y),$$
$$\frac{dy}{dt} = \mathscr{A}_Q y + (I - \pi)X_0 F(\Phi z + y), \tag{3.10}$$

where $\mathscr{A}_Q$ is the operator from Q to $\mathrm{Ker}\,\pi$, i.e.,

$$\mathscr{A}_Q \psi = \dot{\psi} + X_0[L\psi - \dot{\psi}(0)] \qquad \text{for } \psi \in Q.$$

In view of Theorem 3.1, we have the following result.

Theorem 3.2 (Center manifold theorem). *There exist* $W \in C^k(\mathbb{R}^m, \mathrm{Ker}\,\pi)$ *with* $W(0) = 0$ *and* $D_z W(0) = 0$ *and a neighborhood* V *of* $u = 0$ *in* $C_{n,\tau}$ *such that the center manifold*

$$M^c_{\mathrm{loc}}(0) \cap V = \{\Phi z + W(z) : z \in \mathbb{R}^m\}$$

has the following properties:

(i) $M^c_{\mathrm{loc}}(0)$ *is locally invariant with respect to (3.5). More precisely, if* $\varphi \in M^c_{\mathrm{loc}}(0) \cap V$ *and* $u_t(\varphi) \in V$ *for* $t \in I$, *then* $u_t(\varphi) \in M^c_{\mathrm{loc}}(0)$ *for* $t \in I$, *where* $I = I(\varphi)$ *is an interval containing* $t = 0$.

(ii) $M^c_{\mathrm{loc}}(0)$ *contains all solutions of (3.5) remaining in* V *for all* $t \in \mathbb{R}$. *That is, if* $\varphi \in V$ *and* $u_t(\varphi) \in V$ *for* $t \in \mathbb{R}$, *then* $u_t(\varphi) \in M^c_{\mathrm{loc}}(0)$.

(iii) *If* $\sigma^u = \emptyset$, *then* $M^c_{\mathrm{loc}}(0)$ *is locally attractive. More precisely, all solutions of (3.5) remaining in* V *for all* $t > 0$ *tend exponentially to some solution of (3.5) on* $M^c_{\mathrm{loc}}(0)$. *That is, if* $\varphi \in V$ *and* $u_t(\varphi) \in V$ *for* $t \geq 0$, *then there exist* $\tilde{t} \in \mathbb{R}$, $\tilde{\varphi} \in M^c_{\mathrm{loc}}(0) \cap V$, *and* $\gamma > 0$ *such that*

$$u(t; \varphi) = u(t - \tilde{t}; \tilde{\varphi}) + O(e^{-\gamma t}) \quad (t \to \infty).$$

Remark 3.1. The domain of definition of $M^c_{\mathrm{loc}}(0)$ may depend on the degree of smoothness k of f in (3.5). Moreover, for analytic (or C^∞) f, the center manifold is not necessarily analytic (C^∞).

In Theorem 3.2, $M^c_{\mathrm{loc}}(0)$ is a C^k manifold of (3.5) parameterized by $z \in \mathbb{R}^m$. Hence $M^c_{\mathrm{loc}}(0)$ has the same dimension as E^c; $M^c_{\mathrm{loc}}(0)$ passes through $u = 0$ and is tangent to E^c at $u = 0$.

Next, we consider the reduced equation of (3.5) on the center manifold $M^c_{\mathrm{loc}}(0)$. If $\varphi \in M^c_{\mathrm{loc}}(0) \cap V$, then $u_t(\varphi) \in M^c_{\mathrm{loc}}(0)$ for t close to $t = 0$. Defining

$$z_0 = \langle \Psi, \varphi \rangle, \quad z(t; z_0) = \langle \Psi, u_t(\varphi) \rangle,$$

we can write

$$u_t(\varphi) = \Phi z(t; z_0) + W(z(t; z_0)).$$

Using (3.5), we obtain the following characterizations:

(i) $z(\cdot; z_0)$ satisfies the m-dimensional nonlinear differential equation

$$\dot{z} = Bz + G(z), \tag{3.11}$$

where $G(z) = \overline{\Psi}(0)F(W(z) + \Phi z)$.

(ii) The map W satisfies

$$\frac{\mathrm{d}}{\mathrm{d}t}W = \mathscr{A}_Q W + H(z), \qquad (3.12)$$

where $H(z) = [X_0 - \Phi\overline{\Psi}(0)]F(W(z) + \Phi z)$.

The reduced system (3.11) governs the flow of (3.5) on the center manifold $M_{\mathrm{loc}}^c(0)$. The next theorem shows that (3.11) contains all the necessary information needed to determine the asymptotic behavior of small solutions of (3.5).

Theorem 3.3. (i) *Suppose that* $\sigma^u = \emptyset$ *and the zero solution of (3.11) is stable (respectively, asymptotically stable, unstable). Then the zero solution of (3.5) is stable (respectively, asymptotically stable, unstable).*

(ii) *Suppose that the zero solution of (3.5) is asymptotically stable. Let* $u(t;\varphi)$ *be a solution of (3.5) with the initial value* φ *sufficiently close to the origin of the phase space. Then there exists a solution* $z(t)$ *of (3.11) such that*

$$u_t(\varphi) = \Phi z(t) + W(z(t)) + O(\mathrm{e}^{-\gamma t})$$

as $t \to \infty$, *where* γ *is a positive constant.*

3.4 Calculation of Center Manifolds

In order to calculate W in (3.12), we expand G and W in their Taylor series around $z = 0$. To avoid awkward formulas, we use a concise representation of Taylor series throughout this book. That is, we write the Taylor series of G as

$$G(z) = \sum_{j=2}^k \frac{1}{j!} G_j(z) + o(\|z\|^k). \qquad (3.13)$$

In this representation, G_j is a symmetric j-linear map from $\mathbb{R}^m \times \cdots \times \mathbb{R}^m$ (j times) into $\mathbb{R}^m$, and $z^{(j)}$ stands for the repetition of j identical arguments of z. Namely,

$$G_j(z) = \sum_{j_1 + \cdots + j_m = j} \frac{j!}{j_1! \cdots j_m!} \cdot \frac{\partial^j G(0)}{\partial z_1^{j_1} \cdots \partial z_m^{j_m}} z_1^{j_1} \cdots z_m^{j_m}.$$

The m components of $G_j(z)$ are real-valued homogeneous polynomials of degree j in the m components of z. The analogous expansions for W and H read

$$W(z) = \sum_{j=2}^k \frac{1}{j!} W_j(z) + o(|z|^k), \quad H(z) = \sum_{j=2}^k \frac{1}{j!} H_j(z) + o(|z|^k). \qquad (3.14)$$

Using (3.12), we have $\frac{\mathrm{d}}{\mathrm{d}t}W = D_z W(z)\dot{z} = D_z W(z)(Bz + G(z))$, and so

$$\left[\sum_{j=2}^{k} \frac{1}{j!} DW_j(z)\right] \left[Bz + \sum_{j=2}^{k} \frac{1}{j!} G_j(z)\right] = \sum_{j=2}^{k} \frac{1}{j!} \mathscr{A} W_j(z) + \sum_{j=2}^{k} \frac{1}{j!} H_j(z) + o(|z|^k).$$

There are no terms of order 1. At order 2, we obtain

$$D_z W_2(z) Bz - \mathscr{A} W_2(z) = H_2(z).$$

At order 3, we have

$$2 D_z W_3(z) Bz - 2 \mathscr{A} W_3(z) = 2 H_2(z) - 3 D_z W_2(z) G_2(z).$$

At order $s \leq k$, we have

$$D_z W_s(z) Bz - \mathscr{A} W_s(z) = H_s(z) - \sum_{l=1}^{s-2} \frac{s!}{(l+1)!(s-l)!} DW_{l+1}(z) G_{s-l}(z).$$

The equations have to be solved step by step with increasing s, starting with $s = 2$.

In the subsequent subsections, we consider the restriction of (3.5) to the center manifold at the critical parameter values when the dimension of the center space is not too high. Assuming sufficient smoothness of f of (3.5), we write

$$f(\varphi) = L\varphi + \frac{1}{2} \mathscr{F}^2(\varphi, \varphi) + \frac{1}{6} \mathscr{F}^3(\varphi, \varphi, \varphi) + \frac{1}{24} \mathscr{F}^4(\varphi, \varphi, \varphi, \varphi)$$

$$+ \frac{1}{120} \mathscr{F}^5(\varphi, \varphi, \varphi, \varphi, \varphi) + \cdots,$$

where

$$\mathscr{F}^j(v_1, v_2, \cdots, v_j) = \frac{\partial^j}{\partial t_1 \partial t_2 \cdots \partial t_j} f\left(\sum_{s=1}^{j} t_s v_s\right)\Bigg|_{t_1 = t_2 = \cdots = t_j = 0}$$

for $j \in \mathbb{N}$.

3.4.1 The Hopf Case

We first recall that the eigenvalues of $\mathscr{A}$ are solutions of the characteristic equation $\det \Delta(\lambda) = 0$, where

$$\Delta(\lambda) = \lambda \mathrm{Id} - \int_{-\tau}^{0} e^{\lambda \theta} d\eta(\theta), \tag{3.15}$$

and $\eta : [-\tau, 0] \to \mathbb{R}^{n^2}$ is an $n \times n$ matrix-valued function whose elements are of bounded variation such that

$$L\varphi = \int_{-\tau}^{0} d\eta(\theta) \varphi(\theta), \qquad \varphi \in C_{n,\tau}. \tag{3.16}$$

Throughout this subsection, we always assume that on the imaginary axis, the infinitesimal generator $\mathscr{A}$ has only one pair of simple purely imaginary eigenvalues $\pm i\omega$. Thus, the center space E^c is spanned by q and $\bar{q}$, and its dual space E^{c*} is spanned by p and $\bar{p}$, where p and q satisfy

$$\bar{p}(0)\Delta(i\omega) = 0, \quad \Delta(i\omega)q(0) = 0,$$

and

$$\langle p,q \rangle = 1, \quad \langle p,\bar{q} \rangle = 0.$$

For a solution u_t of (3.9), we define $z(t) = \langle p, u_t \rangle$ and then define $w(t,\theta) = u_t(\theta) - 2\mathrm{Re}\{z(t)q(\theta)\}$. In fact, z and $\bar{z}$ are local coordinates for M^c_{loc} in the directions of p and $\bar{p}$. Note that w is real if u_t is, which allows us to deal with real solutions only.

It is easy to see that $\langle p, w \rangle = 0$. Note that $\langle p, \dot{u}_t \rangle = \langle p, \mathscr{A}u_t \rangle + \langle p, X_0 F(u_t) \rangle$. Then we have

$$\dot{z} = i\omega z + g(z), \quad z \in \mathbb{C}, \tag{3.17}$$

where $g(z) = \bar{p}(0)F(w(z) + 2\mathrm{Re}\{zq\})$, and

$$w(z) = \sum_{j+k\geq 2} \frac{1}{j!k!} w_{jk} z^j \bar{z}^k$$

satisfies

$$\frac{\mathrm{d}}{\mathrm{d}t}w = \mathscr{A}w + H(z), \quad \theta \in [-\tau, 0], \tag{3.18}$$

with $H(z) = -2\mathrm{Re}\{g(z)q\} + X_0 F(w(z) + 2\mathrm{Re}\{zq\})$. Let

$$g(z) = \sum_{j+k\geq 2} \frac{1}{j!k!} g_{jk} z^j \bar{z}^k.$$

Then we have

$$g_{20} = \bar{p}(0)\mathscr{F}^2(q,q), \quad g_{11} = \bar{p}(0)\mathscr{F}^2(q,\bar{q}), \quad g_{02} = \bar{p}(0)\mathscr{F}^2(\bar{q},\bar{q}),$$

and

$$g_{21} = \bar{p}(0)\mathscr{F}^3(q,q,\bar{q}) + \bar{p}(0)\mathscr{F}^2(w_{20},\bar{q}) + 2\bar{p}(0)\mathscr{F}^2(w_{11},q),$$

$$g_{30} = \bar{p}(0)\mathscr{F}^3(q,q,q) + 3\bar{p}(0)\mathscr{F}^2(w_{20},q),$$

$$g_{32} = \bar{p}(0)\mathscr{F}^5(q,q,q,\bar{q},\bar{q}) + \bar{p}(0)\mathscr{F}^4(q,q,q,\bar{w}_{20}) + 3\bar{p}(0)\mathscr{F}^4(q,\bar{q},\bar{q},w_{20})$$

$$+6\bar{p}(0)\mathscr{F}^4(q,q,\bar{q},w_{11}) + \bar{p}(0)\mathscr{F}^3(\bar{q},\bar{q},w_{30}) + 3\bar{p}(0)\mathscr{F}^3(q,q,\bar{w}_{21})$$

$$+6\bar{p}(0)\mathscr{F}^3(q,\bar{q},w_{21}) + 3\bar{p}(0)\mathscr{F}^3(q,\bar{w}_{20},w_{20}) + 6\bar{p}(0)\mathscr{F}^3(q,\bar{w}_{11},w_{11})$$

$$+6\bar{p}(0)\mathscr{F}^3(\bar{q},w_{20},w_{11}) + 2\bar{p}(0)\mathscr{F}^2(\bar{q},w_{31}) + 3\bar{p}(0)\mathscr{F}^2(q,w_{22})$$

$$+\bar{p}(0)\mathscr{F}^2(\bar{w}_{20},w_{30}) + 3\bar{p}(0)\mathscr{F}^2(\bar{w}_{21},w_{20}) + 6\bar{p}(0)\mathscr{F}^2(w_{11},w_{21}).$$

Let

$$H(z) = \sum_{j+k \geq 2} \frac{1}{j!k!} H_{jk} z^j \bar{z}^k.$$

Then we have

$$H_{20} = -g_{20}q - \bar{g}_{02}\bar{q} + X_0 \mathscr{F}^2(q,q),$$
$$H_{11} = -g_{11}q - \bar{g}_{11}\bar{q} + X_0 \mathscr{F}^2(q,\bar{q}),$$
$$H_{30} = -g_{30}q - \bar{g}_{03}\bar{q} + X_0 \mathscr{F}^3(q,q,q) + 3X_0 \mathscr{F}^2(q,w_{20}),$$
$$H_{21} = -g_{21}q - \bar{g}_{12}\bar{q} + X_0 \mathscr{F}^3(q,q,\bar{q}) + X_0 \mathscr{F}^2(\bar{q},w_{20}) + 2X_0 \mathscr{F}^2(q,w_{11}),$$
$$H_{31} = -g_{31}q - \bar{g}_{13}\bar{q} + X_0 \mathscr{F}^4(q,q,q,\bar{q}) + 3X_0 \mathscr{F}^3(q,q,w_{11}) + 3X_0 \mathscr{F}^3(q,\bar{q},w_{20})$$
$$\qquad + 3X_0 \mathscr{F}^2(w_{20},w_{11}) + X_0 \mathscr{F}^2(\bar{q},w_{30}) + 3X_0 \mathscr{F}^2(q,w_{21}),$$
$$H_{22} = -g_{22}q - \bar{g}_{22}\bar{q} + X_0 \mathscr{F}^4(q,q,\bar{q},\bar{q}) + 4X_0 \mathscr{F}^3(q,\bar{q},w_{11}) + X_0 \mathscr{F}^3(\bar{q},\bar{q},w_{20})$$
$$\qquad + X_0 \mathscr{F}^3(q,q,\overline{w}_{20}) + 2X_0 \mathscr{F}^2(w_{11},w_{11}) + 2X_0 \mathscr{F}^2(q,\overline{w}_{21}) + 2X_0 \mathscr{F}^2(\bar{q},w_{21})$$
$$\qquad + X_0 \mathscr{F}^2(\overline{w}_{20},w_{20}).$$

Comparing the coefficients of (3.18), we have

$$(\mathscr{A} - 2i\omega)w_{20} + H_{20} = 0,$$
$$\mathscr{A}w_{11} + H_{11} = 0,$$
$$(\mathscr{A} - 3i\omega)w_{30} + H_{30} = 3w_{20}g_{20} + 3w_{11}\bar{g}_{02},$$
$$(\mathscr{A} - i\omega)w_{21} + H_{21} = 2w_{20}g_{11} + w_{11}g_{20} + 2w_{11}\bar{g}_{11} + \overline{w}_{20}\bar{g}_{02},$$
$$(\mathscr{A} - 2i\omega)w_{31} + H_{31} = 3w_{20}g_{21} + w_{11}(g_{30} + 3\bar{g}_{12})$$
$$\qquad + 3w_{30}g_{11} + 3w_{21}(g_{20} + g_{11}) + \overline{w}_{20}\bar{g}_{03} + 3w_{12}\bar{g}_{02},$$
$$\mathscr{A}w_{22} + H_{22} = 2(w_{20}g_{12} + \overline{w}_{20}\bar{g}_{12}) + 2w_{11}(g_{21} + \bar{g}_{21}) + w_{30}g_{02} + \overline{w}_{30}\bar{g}_{02}$$
$$\qquad + w_{21}(4g_{11} + \bar{g}_{20}) + \overline{w}_{21}(4\bar{g}_{11} + g_{20}),$$

from which it follows that

$$\dot{w}_{20}(\theta) = 2i\omega w_{20}(\theta) + g_{20}q(\theta) + \bar{g}_{02}\bar{q}(\theta),$$
$$\dot{w}_{11}(\theta) = g_{11}q(\theta) + \bar{g}_{11}\bar{q}(\theta).$$

Solving for $w_{20}(\theta)$ and $w_{11}(\theta)$, we obtain

$$w_{20}(\theta) = -\frac{g_{20}q(\theta)}{i\omega} - \frac{\bar{g}_{02}\bar{q}(\theta)}{3i\omega} + h_{20}e^{2i\omega\theta},$$

$$w_{11}(\theta) = \frac{g_{11}q(\theta)}{i\omega} - \frac{\bar{g}_{11}\bar{q}(\theta)}{i\omega} + h_{11}, \qquad (3.19)$$

where h_{20} and h_{11} are both n-dimensional vectors that can be determined by setting $\theta = 0$ in $H(z,\bar{z},\theta)$. In view of $[(\mathscr{A} - 2i\omega)w_{20}](0) = -H_{20}(0)$ and $[(\mathscr{A} - 2i\omega)w_{11}](0) = -H_{11}(0)$, we have

$$h_{20} = \Delta^{-1}(2i\omega)\mathscr{F}^2(q,q)$$

and

$$h_{11} = \Delta^{-1}(0)\mathscr{F}^2(q,\bar{q}).$$

Some other terms of w_{30}, w_{21}, w_{31}, and $w_{w_{22}}$ can be calculated similarly. Note that not every term will be needed to apply the Hopf bifurcation theory, to be introduced in the following chapters.

3.4.2 The Fold–Hopf Case

Throughout this subsection, we always assume that on the imaginary axis, the infinitesimal generator $\mathscr{A}$ has only simple eigenvalues 0, $i\omega$, and $-i\omega$. Thus, the center space E^c is spanned by q_0, q_1, and $\bar{q}_1$, and its dual space E^{c*} is spanned by p_0, p_1, and $\bar{p}_1$, where p_0, p_1, q_0, and q_1 satisfy

$$\bar{p}_0(0)\Delta(0) = 0, \quad \bar{p}_1(0)\Delta(i\omega) = 0, \quad \Delta(0)q_0(0) = 0, \quad \Delta(i\omega)q_1(0) = 0,$$

and

$$\langle p_j, q_k \rangle = \delta_{jk}, \quad \langle p_j, \bar{q}_k \rangle = 0,$$

for $j,k = 0,1$. Let $\Phi = (q_0,q_1,\bar{q}_1)$, and $\Psi = (p_0,p_1,\bar{p}_1)^T$. Then $\langle \Psi, \Phi \rangle = \mathrm{Id}_3$.

For a solution u_t of (3.9), we define $x(t) = \langle p_0, u_t \rangle$, $z(t) = \langle p_1, u_t \rangle$, and $w(x,z,\bar{z}) = u_t - x(t)q_0 - 2\mathrm{Re}\{z(t)q_1\}$. In fact, x, z, and $\bar{z}$ are local coordinates for M_{loc}^c in the directions of p_0, p_1, and $\bar{p}_1$. It is easy to see that $\langle p_0, w \rangle = 0$ and $\langle p_1, w \rangle = 0$. Then on the center manifold M_{loc}^c, we have

$$\begin{aligned} \dot{x}(t) &= g^0(x,z,\bar{z}), \\ \dot{z}(t) &= i\omega z(t) + g^1(x,z,\bar{z}), \end{aligned} \tag{3.20}$$

where $g^j(x,z,\bar{z}) = \bar{p}_j(0)F(w(x,z,\bar{z}) + xq_0 + 2\mathrm{Re}\{zq_1\})$ for $j = 0,1$ and

$$w(x,z,\bar{z}) = \sum_{j+s+k\geq2} \frac{1}{j!s!k!} w_{jsk} x^j z^s \bar{z}^k$$

satisfies

$$\frac{d}{dt}w = \mathscr{A}w + (X_0 - \Phi\bar{\Psi}(0))F(w(x,z,\bar{z}) + xq_0 + 2\mathrm{Re}\{zq_1\}). \tag{3.21}$$

Let

$$g^j(x,z,\bar{z}) = \sum_{r+s+k\geq2} \frac{1}{r!s!k!} g_{rsk}^j x^r z^s \bar{z}^k, \quad j = 0,1.$$

Since g^0 must be real, we have $g_{rsk}^0 = \bar{g}_{rks}^0$. Therefore, g_{rsk}^0 is real for $s = k$. We have

$$g_{200}^j = \bar{p}_j(0)\mathscr{F}^2(q_0,q_0), \quad g_{110}^j = \bar{p}_j(0)\mathscr{F}^2(q_0,q_1), \quad g_{011}^j = \bar{p}_j(0)\mathscr{F}^2(q_1,\bar{q}_1),$$

and

$$g_{300}^j = \overline{p}_j(0)\mathscr{F}^3(q_0,q_0,q_0) + 3\overline{p}_j(0)\mathscr{F}^2(q_0,w_{200}),$$

$$g_{111}^j = \overline{p}_j(0)\mathscr{F}^3(q_0,q_1,\overline{q}_1) + \overline{p}_j(0)\mathscr{F}^2(q_0,w_{011}) + \overline{p}_j(0)\mathscr{F}^2(\overline{q}_1,w_{110})$$
$$+\overline{p}_j(0)\mathscr{F}^2(q_1,\overline{w}_{110}),$$

$$g_{210}^j = \overline{p}_j(0)\mathscr{F}^3(q_0,q_0,q_1) + 2\overline{p}_j(0)\mathscr{F}^2(q_0,w_{110}) + \overline{p}_j(0)\mathscr{F}^2(q_1,w_{200}),$$

$$g_{021}^j = \overline{p}_j(0)\mathscr{F}^3(q_1,q_1,\overline{q}_1) + 2\overline{p}_j(0)\mathscr{F}^2(q_1,w_{011}) + \overline{p}_j(0)\mathscr{F}^2(\overline{q}_1,w_{020}).$$

Obviously, in order to determine the cubic terms, we need to calculate w_{200}, w_{110}, w_{011}, and w_{020}. In fact, it follows from (3.21) that

$$\mathscr{A}w_{200} + (X_0 - \Phi\overline{\Psi}(0))\mathscr{F}^2(q_0,q_0) = 0, \qquad (3.22)$$

$$(\mathscr{A} - i\omega)w_{110} + (X_0 - \Phi\overline{\Psi}(0))\mathscr{F}^2(q_0,q_1) = 0, \qquad (3.23)$$

$$\mathscr{A}w_{011} + (X_0 - \Phi\overline{\Psi}(0))\mathscr{F}^2(q_1,\overline{q}_1) = 0, \qquad (3.24)$$

$$(\mathscr{A} - 2i\omega)w_{020} + (X_0 - \Phi\overline{\Psi}(0))\mathscr{F}^2(q_1,q_1) = 0. \qquad (3.25)$$

It follows from (3.22) that

$$\dot{w}_{200} = g_{200}^0 q_0 + 2\text{Re}\{g_{200}^1 q_1\} \qquad (3.26)$$

and

$$Lw_{200} = g_{200}^0 q_0 + 2\text{Re}\{g_{200}^1 q_1\} - \mathscr{F}^2(q_0,q_0). \qquad (3.27)$$

Solving (3.26) for w_{200}, we have

$$w_{200}(\theta) = g_{200}^0 q_0 \theta + 2\text{Re}\{\frac{g_{200}^1}{i\omega}q_1(\theta)\} + E_{200},$$

where E_{200} is a constant vector in $\mathbb{R}^n$. Substituting the above equation into (3.27) yields

$$\Delta(0)E_{200} = \mathscr{F}^2(q_0,q_0) - g_{200}^0 q_0 + g_{200}^0 \int_{-\tau}^0 \theta d\eta(\theta)q_0. \qquad (3.28)$$

Since $\Delta(0)q_0 = 0$ and $p_0\Delta(0) = 0$, we see that $\Delta(0)$ is a singular matrix, and so (3.28) cannot be solved easily. By Keller [188], the unique solution E_{200} to (3.28) satisfying $p_0E_{200} = 0$ can be obtained by solving the following nonsingular $(n+1)$-dimensional bordered system:

$$\begin{bmatrix} \Delta(0) & q_0 \\ p_0 & 0 \end{bmatrix} \begin{bmatrix} x \\ y \end{bmatrix} = \begin{bmatrix} \mathscr{F}^2(q_0,q_0) - g_{200}^0 q_0 + g_{200}^0 \int_{-\tau}^0 \theta d\eta(\theta)q_0 \\ 0 \end{bmatrix}. \qquad (3.29)$$

See also Govaerts and Pryce [119] for further generalizations. We write

$$E_{200} = [\Delta(0)]^{\text{inv}}[\mathscr{F}^2(q_0,q_0) - g_{200}^0 q_0 + g_{200}^0 \int_{-\tau}^0 \theta d\eta(\theta)q_0].$$

Similarly, it follows from (3.23), (3.24), and (3.25) that

$$w_{110}(\theta) = -\frac{g_{110}^0 q_0}{i\omega} + g_{110}^1 q_1(\theta)\theta - \frac{\bar{g}_{101}^1 \bar{q}_1(\theta)}{2i\omega} + E_{110}e^{i\omega\theta},$$

$$w_{011}(\theta) = g_{011}^0 q_0\theta + \frac{g_{011}^1 q_1(\theta)}{i\omega} - \frac{\bar{g}_{011}^1 \bar{q}_1(\theta)}{i\omega} + E_{011},$$

$$w_{020}(\theta) = -\frac{g_{020}^0 q_0}{2i\omega} - \frac{g_{020}^1 q_1(\theta)}{i\omega} - \frac{\bar{g}_{002}^1 \bar{q}_1(\theta)}{3i\omega} + E_{020}e^{2i\omega\theta},$$

where E_{110}, E_{011}, and E_{020} are n-dimensional constant vectors given by

$$E_{110} = [\Delta(i\omega)]^{\text{inv}} \left[\mathscr{F}^2(q_0, q_1) - g_{110}^1 q_1(0) + g_{110}^1 \int_{-\tau}^0 \theta d\eta(\theta) q_1(\theta) \right],$$

$$E_{011} = [\Delta(0)]^{\text{inv}} \left[\mathscr{F}^2(q_1, \bar{q}_1) - g_{011}^0 q_0 + g_{011}^0 \int_{-\tau}^0 \theta d\eta(\theta) q_0 \right],$$

$$E_{020} = [\Delta(2i\omega)]^{-1} \mathscr{F}^2(q_1, q_1).$$

3.4.3 The Double Hopf Case

Throughout this subsection, we always assume that on the imaginary axis, the infinitesimal generator $\mathscr{A}$ has only two pairs of simple purely imaginary eigenvalues $\pm i\omega_1$ and $\pm i\omega_2$, where $\omega_1 > \omega_2 > 0$ and $k\omega_1 \neq l\omega_2$ for all integers $k, l > 0$ satisfying $k + l \leq 5$. Since the eigenvalues are simple, there are two complex eigenvectors $q_1, q_2 \in C_{n,\tau}$ corresponding to these eigenvalues,

$$\mathscr{A} q_j = i\omega_j q_j, \quad j = 1, 2.$$

Introduce the adjoint eigenvectors $p_1, p_2 \in C_{n,\tau}^*$ by

$$\mathscr{A}^* p_j = -i\omega_j p_j, \quad j = 1, 2.$$

These eigenvectors can be normalized using the bilinear norm $\langle \cdot, \cdot \rangle \colon C_{n,\tau}^* \times C_{n,\tau} \to \mathbb{R}$, so that $\langle p_j, q_k \rangle = \delta_{jk}$ and $\langle p_j, \bar{q}_k \rangle = 0$, $j, k = 1, 2$. Let $\Phi = (q_1, \bar{q}_1, q_2, \bar{q}_2)$ and $\Psi = (p_1, \bar{p}_1, p_2, \bar{p}_2)^T$. Then $\langle \Psi, \Phi \rangle = \mathrm{Id}_4$.

For a solution u_t of (3.9), we define $z_j(t) = \langle p_j, u_t \rangle$, $j = 1, 2$, and $w(z) = u_t - 2\mathrm{Re}\{z_1(t)q_1 + z_2(t)q_2\}$, where $z = (z_1, z_2) \in \mathbb{C}^2$. In fact, z_j and $\bar{z}_j$ are local coordinates for the center manifold M_{loc}^c in the directions of p_j and $\bar{p}_j$, $j = 1, 2$. It is easy to see that $\langle p_j, w \rangle = 0$. For solutions $u_t \in M_{\text{loc}}^c$ of (3.9), $\langle p_j, \dot{u}_t \rangle = \langle p_j, \mathscr{A}u_t + X_0 F(u_t) \rangle$, $j = 1, 2$. Then on the center manifold M_{loc}^c, we have

$$\dot{z}_j(t) = i\omega_j z_j(t) + g^j(z), \quad j = 1, 2, \tag{3.30}$$

where $g^j(z) = \overline{p}_j(0)F(w(z) + 2\text{Re}\{z_1q_1 + z_2q_2\})$ for $z = (z_1, z_2) \in \mathbb{C}^2$, and $w(z)$ satisfies

$$\frac{d}{dt}w = \mathscr{A}w + (X_0 - \Phi\overline{\Psi}(0))F(w(z) + 2\text{Re}\{z_1q_1 + z_2q_2\}). \qquad (3.31)$$

Let

$$g^j(z) = \sum_{l+s+r+k\geq 2} \frac{1}{l!s!r!k!}g^j_{lsrk}z_1^l\overline{z}_1^s z_2^r\overline{z}_2^k, \quad w(z) = \sum_{l+s+r+k\geq 2} \frac{1}{l!s!r!k!}w_{lsrk}z_1^l\overline{z}_1^s z_2^r\overline{z}_2^k.$$

Then we have

$$g^j_{2000} = \overline{p}_j(0)\mathscr{F}^2(q_1,q_1), \qquad g^j_{0200} = \overline{p}_j(0)\mathscr{F}^2(\overline{q}_1,\overline{q}_1),$$
$$g^j_{0020} = \overline{p}_j(0)\mathscr{F}^2(q_2,q_2), \qquad g^j_{0002} = \overline{p}_j(0)\mathscr{F}^2(\overline{q}_2,\overline{q}_2),$$
$$g^j_{1100} = \overline{p}_j(0)\mathscr{F}^2(q_1,\overline{q}_1), \qquad g^j_{0011} = \overline{p}_j(0)\mathscr{F}^2(q_2,\overline{q}_2),$$
$$g^j_{1010} = \overline{p}_j(0)\mathscr{F}^2(q_1,q_2), \qquad g^j_{0101} = \overline{p}_j(0)\mathscr{F}^2(\overline{q}_1,\overline{q}_2),$$
$$g^j_{1001} = \overline{p}_j(0)\mathscr{F}^2(q_1,\overline{q}_2), \qquad g^j_{0110} = \overline{p}_j(0)\mathscr{F}^2(q_2,\overline{q}_1),$$

and

$$g^j_{2100} = \overline{p}_j(0)\mathscr{F}^3(q_1,q_1,\overline{q}_1) + 2\overline{p}_j(0)\mathscr{F}^2(q_1,w_{1100}) + \overline{p}_j(0)\mathscr{F}^2(\overline{q}_1,w_{2000}),$$
$$g^j_{1011} = \overline{p}_j(0)\mathscr{F}^3(q_1,q_2,\overline{q}_2) + \overline{p}_j(0)\mathscr{F}^2(q_1,w_{0011})$$
$$\qquad + \overline{p}_j(0)\mathscr{F}^2(q_2,w_{1001}) + \overline{p}_j(0)\mathscr{F}^2(\overline{q}_2,w_{1010}),$$
$$g^j_{1110} = \overline{p}_j(0)\mathscr{F}^3(q_1,\overline{q}_1,q_2) + \overline{p}_j(0)\mathscr{F}^2(q_1,\overline{w}_{1001})$$
$$\qquad + \overline{p}_j(0)\mathscr{F}^2(q_2,w_{1100}) + \overline{p}_j(0)\mathscr{F}^2(\overline{q}_1,w_{1010}),$$
$$g^j_{0021} = \overline{p}_j(0)\mathscr{F}^3(q_2,q_2,\overline{q}_2) + 2\overline{p}_j(0)\mathscr{F}^2(q_2,w_{0011}) + \overline{p}_j(0)\mathscr{F}^2(\overline{q}_2,w_{0020}).$$

In what follows, we need to determine w_{1100}, w_{2000}, w_{1010}, w_{1001}, w_{0002}, and w_{0011}. In view of (3.31), we have

$$(\mathscr{A} - 2i\omega_1)w_{2000} + (X_0 - \Phi\overline{\Psi}(0))\mathscr{F}^2(q_1,q_1) = 0.$$

This is equivalent to the following system:

$$\dot{w}_{2000}(\theta) = 2i\omega_1 w_{2000} + g^1_{2000}q_1(\theta) + \overline{g}^1_{0200}\overline{q}_1(\theta) \qquad (3.32)$$
$$\qquad + g^2_{2000}q_2(\theta) + \overline{g}^2_{0200}\overline{q}_2(\theta),$$
$$Lw_{2000} = 2i\omega_1 w_{2000}(0) + g^1_{2000}q_1(0) + \overline{g}^1_{0200}\overline{q}_1(0) \qquad (3.33)$$
$$\qquad + g^2_{2000}q_2(0) + \overline{g}^2_{0200}\overline{q}_2(0) - \mathscr{F}^2(q_1,q_1).$$

From (3.33), we have

$$w_{2000}(\theta) = -\frac{g^1_{2000}q_1(\theta)}{i\omega_1} - \frac{\overline{g}^1_{0200}\overline{q}_1(\theta)}{3i\omega_1} + \frac{g^2_{2000}q_2(\theta)}{i(\omega_2 - 2\omega_1)} - \frac{\overline{g}^2_{0200}\overline{q}_2(\theta)}{i(\omega_2 + 2\omega_1)} + E_{2000}e^{2i\omega_1\theta}, \qquad (3.34)$$

with $E_{2000} \in \mathbb{R}^n$. Substituting (3.34) into (3.34), we have

$$w_{2000} = [\Delta(2i\omega_1)]^{-1}\mathscr{F}^2(q_1,q_1).$$

Similarly, we have

$$w_{1100}(\theta) = \frac{g^1_{1100}q_1(\theta)}{i\omega_1} - \frac{\overline{g}^1_{1100}\overline{q}_1(\theta)}{i\omega_1} + \frac{g^2_{1100}q_2(\theta)}{i\omega_2} - \frac{\overline{g}^2_{1100}\overline{q}_2(\theta)}{i\omega_2} + E_{1100},$$

$$w_{1010}(\theta) = -\frac{g^1_{1010}q_1(\theta)}{i\omega_2} - \frac{\overline{g}^1_{0101}\overline{q}_1(\theta)}{i(2\omega_1+\omega_2)} - \frac{g^2_{1010}q_2(\theta)}{i\omega_1} - \frac{\overline{g}^2_{0101}\overline{q}_2(\theta)}{i(\omega_1+2\omega_2)} + E_{1010}e^{i(\omega_1+\omega_2)\theta},$$

$$w_{1001}(\theta) = \frac{g^1_{1001}q_1(\theta)}{i\omega_2} + \frac{\overline{g}^1_{0110}\overline{q}_1(\theta)}{i(\omega_2-2\omega_1)} + \frac{g^2_{1001}q_2(\theta)}{i(2\omega_2-\omega_1)} - \frac{\overline{g}^2_{0110}\overline{q}_2(\theta)}{i\omega_1} + E_{1001}e^{i(\omega_1-\omega_2)\theta},$$

$$w_{0020}(\theta) = \frac{g^1_{0020}q_1(\theta)}{i(\omega_1-2\omega_2)} - \frac{\overline{g}^1_{0002}\overline{q}_1(\theta)}{i(\omega_1+2\omega_1)} - \frac{g^2_{0020}q_2(\theta)}{i\omega_2} - \frac{\overline{g}^2_{0002}\overline{q}_2(\theta)}{3i\omega_2} + E_{0020}e^{2i\omega_2\theta},$$

$$w_{0011}(\theta) = \frac{g^1_{0011}q_1(\theta)}{i\omega_1} - \frac{\overline{g}^1_{0011}\overline{q}_1(\theta)}{i\omega_1} + \frac{g^2_{0011}q_2(\theta)}{i\omega_2} - \frac{\overline{g}^2_{0011}\overline{q}_2(\theta)}{i\omega_2} + E_{0011},$$

where

$$E_{1100} = [\Delta(0)]^{-1}\mathscr{F}^2(q_1,\overline{q}_1),$$
$$E_{2000} = [\Delta(2i\omega_1)]^{-1}\mathscr{F}^2(q_1,q_1),$$
$$E_{1010} = [\Delta(i\omega_1+i\omega_2)]^{-1}\mathscr{F}^2(q_1,q_2),$$
$$E_{1001} = [\Delta(i\omega_1-i\omega_2)]^{-1}\mathscr{F}^2(q_1,\overline{q}_2),$$
$$E_{0020} = [\Delta(2i\omega_2)]^{-1}\mathscr{F}^2(q_2,q_2),$$
$$E_{0011} = [\Delta(0)]^{-1}\mathscr{F}^2(q_2,\overline{q}_2).$$

3.5 Center Manifolds with Parameters

Consider the following parameterized DDE:

$$\dot{x}(t) = f(\alpha,x_t), \tag{3.35}$$

where $\alpha \in \mathbb{R}^r$ is the parameter, $f \in C^k(\mathbb{R}^r \times C_{n,\tau}, \mathbb{R}^n)$ for a large enough integer k, $f(0,0) = 0$. In most of the literature [27–29, 49, 58–61, 91–93, 113, 114, 128–141, 232–238, 252, 253, 262, 293–296], it is assumed that $f(\alpha,0) = 0$ for all α, i.e., the equilibrium point is always fixed at the origin. In fact, this is not true in a general physical system or an engineering problem. Here, we introduce the work [138] to provide a general framework to obtain the reduced equation on the center manifold in the case that $f(0,0) = 0$ and there is no assumption that $f(\alpha,0) = 0$ for all α.

Let L be the linearized operator of f with respect to x_t at $(\alpha,x_t) = (0,0)$. By the Riesz representation theorem, there exists an $n \times n$ matrix-valued function η : $[-\tau,0] \to \mathbb{R}^{n^2}$ whose elements are of bounded variation such that

$$L\varphi = \int_{-\tau}^{0} d\eta(\theta)\varphi(\theta), \qquad \varphi \in C_{n,\tau}. \tag{3.36}$$

Thus, we can define a bilinear form

$$\langle \psi, \varphi \rangle = \overline{\psi}(0)\varphi(0) - \int_{-\tau}^{0} \int_{0}^{\theta} \overline{\psi}(\xi - \theta)d\eta(\theta)\varphi(\xi)d\xi \tag{3.37}$$

for $\psi \in C_{n,\tau}^*$ and $\varphi \in C_{n,\tau}$. Let $\mathscr{A}$ be the infinitesimal generator associated with the linear system $\dot{x}(t) = Lx_t$, and let $\Delta(\lambda)$ be the characteristic matrix of the operator $\mathscr{A}$, i.e.,

$$\Delta(\lambda) = \lambda \operatorname{Id}_n - \int_{-\tau}^{0} e^{\lambda \theta} d\eta(\theta).$$

If $0 \notin \sigma(\mathscr{A})$, then $\det \Delta(0) = \det[-L(\hat{1})] \neq 0$. Here and in the sequel, for every $s \in \mathbb{N}$ and $\zeta \in \mathbb{R}^s$, $\hat{\zeta} \in C_{s,\tau}$ is a constant map with the value ζ. In fact, $L(\hat{1})$ is the Jacobian matrix of $f(\alpha, \hat{x})$ with respect to $\hat{x}$ at $(\alpha, x) = (0, 0)$. Thus, according to the implicit function theorem, there exists, for small $|\alpha|$, a unique $x(\alpha)$ such that $x(0) = 0$ and $f(\alpha, \hat{x}(\alpha)) = 0$ for all small $|\alpha|$. This means that (3.35) has a unique equilibrium $\hat{x}(\alpha)$ in some neighborhood of the origin for all sufficiently small $|\alpha|$. Thus, we can perform a coordinate shift to place this equilibrium at the origin. Therefore, in this case, we may assume without loss of generality that $\hat{x} = 0$ is the equilibrium point of (3.35) for all sufficiently small $|\alpha|$. Unfortunately, it is not always true that (3.35) has a unique equilibrium $\hat{x}(\alpha)$ for small $|\alpha|$. In fact, if $0 \in \sigma(\mathscr{A})$, then difficulties arise here. In what follows, we will not assume $0 \notin \sigma(\mathscr{A})$.

Let Φ be a basis for the center space E^c of $\mathscr{A}$, and let Ψ be the basis for its dual space E^{c*} in $C_{n,\tau}^*$ such that $\langle \Psi, \Phi \rangle = \operatorname{Id}_m$, where $m = \dim E^c$. Then there exists an $m \times m$ constant matrix such that $\dot{\Phi} = \Phi B$ and $\dot{\Psi} = -B\Psi$. It follows from (3.7) that we have $BC_n = E^c \oplus \operatorname{Ker} \pi$, where $\pi : BC_n \to E^c$ is the projection defined by

$$\pi(\varphi + X_0\xi) = \Phi[\langle \Psi, \varphi \rangle + \overline{\Psi}(0)\xi] \tag{3.38}$$

for $\varphi \in C_{n,\tau}$ and $\xi \in \mathbb{R}^n$. For convenience, let $\pi_h = I - \pi$. It is easy to see that

$$\langle \Psi, \dot{\varphi} \rangle = \overline{\Psi}(0)[\dot{\varphi}(0) - L\varphi] + B\langle \Psi, \varphi \rangle$$

for all $\varphi \in C_{n,\tau}^1$. Thus, we have

$$\pi_h\dot{\varphi} = \dot{\varphi}_h + \Phi\overline{\Psi}(0)[L\varphi - \dot{\varphi}(0)] \tag{3.39}$$

for all $\varphi \in C_{n,\tau}^1$, where $\varphi_h = \pi_h\varphi$.

One way to consider the center manifold for system (3.35) for small $|\alpha|$ is to study the following DDE without parameters:

$$\begin{aligned} \dot{\alpha}(t) &= 0, \\ \dot{x}(t) &= f(\alpha, x_t). \end{aligned} \tag{3.40}$$

Define

$$u(t) = \begin{bmatrix} \alpha(t) \\ x(t) \end{bmatrix} \in \mathbb{R}^{n+r} \quad \text{and} \quad \tilde{f}(u_t) = \begin{bmatrix} 0 \\ f(\alpha(t), x_t) \end{bmatrix} \in \mathbb{R}^{n+r}.$$

Then (3.40) becomes

$$\dot{u}(t) = \tilde{f}(u_t), \tag{3.41}$$

where $\tilde{f} \in C^k(C_{n+r,\tau}, \mathbb{R}^{n+r})$, $\tilde{f}(0) = 0$, and the linearized operator $\tilde{L}$ of $\tilde{f}$ with respect to u_t at the equilibrium point $u = 0$ is given by

$$\tilde{L} = \begin{bmatrix} 0 & 0 \\ f_\alpha(0,0) & L \end{bmatrix}.$$

Similarly, we can define the formal adjoint equation of the linearized equation of (3.40) by

$$\dot{\beta}(t) = -\beta(t)f_\alpha(0,0),$$
$$\dot{y}(t) = -\int_{-\tau}^0 y(t-\theta)d\eta(\theta).$$

Thus, the bilinear form in $C^*_{n+r,\tau} \times C_{n+r,\tau}$ is given by

$$\ll \tilde{\psi}, \tilde{\varphi} \gg = \beta(0)\alpha(0) + \langle \psi, \varphi \rangle \tag{3.42}$$

for $\tilde{\psi} = (\beta, \psi) \in C^*_{n+r,\tau}$ and $\tilde{\varphi} = (\alpha^T, \varphi^T)^T \in C_{n+r,\tau}$. Note that $(\alpha^T, \varphi^T)^T \in C_{n+r,\tau}$ denotes a generic point in $C_{n+r,\tau}$ with $\alpha \in C_{r,\tau}$ and $\varphi \in C_{n,\tau}$.

Similarly, we have to enlarge the phase space $C_{n+r,\tau}$ of (3.41) to $BC_{n+r} = BC_r \times BC_n$ (which can be identified with $C_{n+r,\tau} \times \mathbb{R}^{n+r}$). Thus, the abstract ODE in BC_{n+r} associated with (3.41) can be rewritten in the form

$$\frac{d}{dt}u = \tilde{\mathscr{A}}u + \tilde{X}_0\tilde{F}(u), \qquad u \in BC_{n+r}, \tag{3.43}$$

where $\tilde{\mathscr{A}}\psi = \dot{\psi} + \tilde{X}_0[\tilde{L}\psi - \dot{\psi}(0)]$, $\tilde{F}(u) = \tilde{f}(u) - \tilde{L}u$, $\tilde{X}_0 = \text{diag}(Y_0, X_0)$, and

$$Y_0(\theta) = \begin{cases} \text{Id}_r, & \theta = 0, \\ 0, & \theta \in [-\tau, 0). \end{cases}$$

Namely, for $\tilde{\varphi} = (\alpha^T, \varphi^T)^T \in C^1_{n+r,\tau}$,

$$\tilde{\mathscr{A}}\tilde{\varphi} = \begin{bmatrix} \dot{\alpha} - Y_0\dot{\alpha}(0) \\ \dot{\varphi} + X_0[f_\alpha(0,0)\alpha + L\varphi - \dot{\varphi}(0)] \end{bmatrix}.$$

The spectrum of $\tilde{\mathscr{A}}$ is $\sigma(\tilde{\mathscr{A}}) = \{\lambda \in \mathbb{C}: \tilde{\Delta}(\lambda)v = 0 \text{ for some } v \in \mathbb{C}^{n+r} \setminus \{0\}\}$, where

$$\tilde{\Delta}(\lambda) = \begin{bmatrix} \lambda \text{Id}_r & 0 \\ -f_\alpha(0,0) & \Delta(\lambda) \end{bmatrix}.$$

Hence $0 \in \sigma(\tilde{\mathscr{A}})$, with a multiplicity at least r. Moreover, it is easy to see that $\sigma(\mathscr{A}) \subseteq \sigma(\tilde{\mathscr{A}})$. Let $\tilde{E}_c$, $\tilde{E}_u$, and let $\tilde{E}_s$ be defined as in the previous section. Then, we have the following result.

Lemma 3.1. $\tilde{E}^s = \{(0,v^T)^T : v \in E^s\}$ and $\tilde{E}^u = \{(0,v^T)^T : v \in E^u\}$.

Proof. Clearly, we have $\{(0,v^T)^T : v \in E^s\} \subset \tilde{E}^s =$ and $\{(0,v^T)^T : v \in E^u\} \subset \tilde{E}^u$.

For a fixed $\delta \in \sigma(\tilde{\mathscr{A}})$, consider the equation $(\tilde{\mathscr{A}} - \delta \mathrm{Id})\varphi = 0$. Then, $\varphi(t) = ue^{\delta t}$ with $u = (\alpha^T, x^T)^T$ satisfying

$$\tilde{\Delta}(\delta)u = 0, \tag{3.44}$$

namely, $\delta \alpha = 0$ and $-f_\alpha(0,0)\alpha + \Delta(\delta)x = 0$. Thus, when $\delta \neq 0$, solutions of (3.44) are given by $\{(0,x^T)^T \in \mathbb{R}^{n+r} : \Delta(\delta)x = 0\}$.

When the generalized eigenspace of δ for $\tilde{\mathscr{A}}$ is larger than the eigenspace, we first consider the equation $(\tilde{\mathscr{A}} - \delta \mathrm{Id})\psi = \varphi$, where $\varphi(t) = ue^{\delta t}$ with $u = (0,x^T)^T$ satisfying $\Delta(\delta)x = 0$. Then $\psi(t) = (v + tu)e^{\delta t}$, where $v = (\alpha^T, y^T)^T$ satisfies

$$\tilde{\Delta}_\lambda(\delta)u = -\tilde{\Delta}(\delta)v. \tag{3.45}$$

When $\delta \neq 0$, it follows from (3.45) that $\alpha = 0$ and $\Delta_\lambda(\delta)x = -\Delta(\delta)y$.

This process can be repeated if δ is a nonzero eigenvalue of $\mathscr{A}$ with higher multiplicity to conclude that $\tilde{E}^s \subset \{(0,v^T)^T : v \in E^s\}$ and $\tilde{E}^u \subset \{(0,v^T)^T : v \in E^u\}$. $\square$

Now we consider the structure of $\tilde{E}^c$. Obviously, $\tilde{E}^c = \tilde{E}_0^c \oplus \tilde{E}_1^c$, where $\tilde{E}_0^c$ and $\tilde{E}_1^c$ are generalized eigenspaces associated with the eigenvalue 0 and the purely imaginary eigenvalues different from 0, respectively. The above calculation proves that $\tilde{E}_1^c = \{(0,v^T)^T : v \in E_1^c\}$.

For every $\tilde{\varphi} = (\alpha^T, \varphi^T)^T \in \tilde{E}^c$, we have $\dot{\alpha} = 0$ and

$$\pi_h \dot{\varphi} + \pi_h X_0 [f_\alpha(0,0)\alpha + L\varphi - \dot{\varphi}(0)] = 0. \tag{3.46}$$

It follows from (3.39) and (3.46) that

$$\dot{\varphi}_h + X_0 [f_\alpha(0,0)\alpha + L\varphi - \dot{\varphi}(0)] - \Phi\overline{\Psi}(0)f_\alpha(0,0)\alpha = 0. \tag{3.47}$$

Notice that

$$\dot{\varphi}(0) = \dot{\varphi}_h(0) + \Phi(0)B\langle \Psi, \varphi \rangle$$

and

$$L\varphi = L\varphi_h + \Phi(0)B\langle \Psi, \varphi \rangle.$$

Then it follows from (3.47) that

$$\begin{cases} \dot{\varphi}_h = \Phi\overline{\Psi}(0)f_\alpha(0,0)\alpha, \\ L\varphi_h = \Phi(0)\overline{\Psi}(0)f_\alpha(0,0)\alpha - f_\alpha(0,0)\alpha. \end{cases} \tag{3.48}$$

Note that the restriction of L on $\mathrm{Ker}\,\pi$ is invertible. Then we may solve the above system for a unique solution φ_h, which linearly depends on α. Thus, we write

$\varphi_h = \Omega\alpha$, where Ω is a linear operator such that $\varphi_h = \Omega\alpha$ is the solution to (3.48). Thus, every $u \in \tilde{E}^c$ takes the form

$$u = \begin{bmatrix} \alpha \\ x + \Omega\alpha \end{bmatrix},$$

where $x \in E^c$ and $\alpha \in \mathbb{R}^r$. In particular, every $u \in \tilde{E}_0^c$ takes the form

$$u = \begin{bmatrix} \alpha \\ x + \Omega\alpha \end{bmatrix},$$

where $x \in E_0^c$ and $\alpha \in \mathbb{R}^r$.

We can then consider a basis for $\tilde{E}^{c*}$ in a similar way. Here we present another approach. For $\delta \in \sigma(\tilde{\mathscr{A}})$, consider $(\tilde{\mathscr{A}}^* + \delta\mathrm{Id})\psi = 0$. Then $\psi(t) = (v_1, v_2)e^{\delta t}$ with $v_1 \in \mathbb{R}^{r*}$ and $v_2 \in \mathbb{R}^{n*}$ satisfying $(v_1, v_2)\tilde{\Delta}(-\delta) = 0$, that is,

$$\delta v_1 = -v_2 f_\alpha(0,0), \quad v_2\Delta(-\delta) = 0. \tag{3.49}$$

When $\delta \neq 0$, then (3.49) has solutions $\{(-\delta^{-1}v f_\alpha(0,0), v): v\Delta(-\delta) = 0\}$. If δ is nonsemisimple, then we consider the equation $(\tilde{\mathscr{A}}^* + \delta\mathrm{Id})\psi = \varphi$, where $\varphi(t) = (v_1, v_2)e^{\delta t}$ with (v_1, v_2) satisfying (3.49). Then $\psi(t) = (u_1 - tv_1, u_2 - tv_2)e^{\delta t}$, where $u_1 \in \mathbb{R}^{r*}$ and $u_2 \in \mathbb{R}^{n*}$ satisfy

$$(v_1, v_2)\tilde{\Delta}_\lambda(-\delta) = -(u_1, u_2)\tilde{\Delta}(-\delta). \tag{3.50}$$

When $\delta \neq 0$, it follows from (3.50) that $u_1 = \delta^{-1}(v_1 - u_2 f_\alpha(0,0))$ and $v_2\Delta_\lambda(-\delta)x = -u_2\Delta(-\delta)$. This process can be repeated if δ is a nonzero nonsemisimple multiple eigenvalue of $\mathscr{A}$. Finally, if $\delta = 0$, then (3.49) has solutions $\{(u,0): u \in \mathbb{R}^{r*}\}$. This implies that $(u,0) \in E_0^{c*}$ for all $u \in \mathbb{R}^{r*}$. Obviously, $\tilde{\mathscr{A}}^*(u_1, u_2) = (u_2 f_\alpha(0,0), 0)$ for all $u_1 \in \mathbb{R}^{r*}$ and $u_2 \in \mathbb{R}^{n*}$ satisfying $u_2\Delta(0) = 0$. This process can be repeated if 0 is a nonsemisimple multiple eigenvalue of $\mathscr{A}$.

In summary, there exists some $m \times r$ matrix Ω^* such that the bases of the center space $\tilde{E}^c$ and its adjoint space $\tilde{E}^{c*}$ are given by

$$\tilde{\Phi} = \begin{bmatrix} \mathrm{Id}_r & 0 \\ \Omega & \Phi \end{bmatrix} \qquad \text{and} \qquad \tilde{\Psi} = \begin{bmatrix} \mathrm{Id}_r & 0 \\ \Omega^* & \Psi \end{bmatrix},$$

respectively. Obviously, $\ll \tilde{\Psi}, \tilde{\Phi} \gg = \mathrm{Id}_{m+r}$. Moreover, $\dot{\tilde{\Phi}} = \tilde{\Phi}\tilde{B}$, where

$$\tilde{B} = \begin{bmatrix} 0 & 0 \\ \overline{\Psi}(0)f_\alpha(0,0) & B \end{bmatrix}.$$

Consider the projection $\tilde{\pi}: BC_{n+r} \to \tilde{E}^c$ given by

$$\tilde{\pi}(\tilde{\varphi} + \tilde{X}_0\eta) = \tilde{\Phi}[\ll \tilde{\Psi}, \tilde{\varphi} \gg + \overline{\tilde{\Psi}}(0)\eta] \qquad \text{for all } \tilde{\varphi} \in C_{n+r,\tau} \text{ and } \eta \in \mathbb{R}^{n+r}.$$

For $\tilde{\varphi} \in C_{n+r,\tau}$ and $\eta = (\eta_1^T, \eta_2^T)^T \in \mathbb{R}^{n+r}$ with $\eta_1 \in \mathbb{R}^r$ and $\eta_2 \in \mathbb{R}^n$, there exist $\beta \in \mathbb{R}^r$ and $\varphi \in C_{n,\tau}$ such that $\tilde{\varphi} = (\beta^T, \varphi^T)^T$, and hence

$$\tilde{\pi}(\tilde{\varphi} + \tilde{X}_0 \eta) = \begin{bmatrix} \beta + \eta_1 \\ (\Omega + \Phi\Omega^*)(\beta + \eta_1) + \pi(\varphi + X_0\eta_2) \end{bmatrix}.$$

Due to the decomposition of BC_{n+r}, we can decompose $\tilde{u}$ in (3.43) in the form $u = \tilde{\Phi}\tilde{z} + \tilde{y}$, with $\tilde{z} \in \mathbb{R}^{m+r}$, $\tilde{y} \in \tilde{Q} \overset{\text{def}}{=} \text{Ker}\,\tilde{\pi} \cap C_{n+r,\tau}^1$. Then (3.43) is equivalent to the following system:

$$\begin{aligned} \dot{\tilde{z}} &= \tilde{B}\tilde{z} + \overline{\tilde{\Psi}}(0)\tilde{F}(\tilde{\Phi}\tilde{z} + \tilde{y}), \\ \tfrac{d}{dt}\tilde{y} &= \mathscr{A}_{\tilde{Q}}\tilde{y} + (\tilde{I} - \tilde{\pi})\tilde{X}_0\tilde{F}(\tilde{\Phi}\tilde{z} + \tilde{y}), \end{aligned} \tag{3.51}$$

where $\mathscr{A}_{\tilde{Q}}$ is the operator from $\tilde{Q}$ to $\text{Ker}\,\tilde{\pi}$, i.e.,

$$\mathscr{A}_{\tilde{Q}}\psi = \dot{\psi} + \tilde{X}_0[\tilde{L}\psi - \dot{\psi}(0)] \qquad \text{for } \psi \in \tilde{Q}.$$

Namely, for $\psi = (\alpha^T, \varphi^T)^T$ with $\alpha \in \mathbb{R}^r$ and $\varphi \in Q \overset{\text{def}}{=} \text{Ker}\,\pi \cap C_{n,\tau}^1$,

$$\mathscr{A}_{\tilde{Q}}\psi = \begin{bmatrix} \dot{\alpha} - Y_0\dot{\alpha}(0) \\ \mathscr{A}_Q\varphi + X_0 f_\alpha(0,0)\alpha \end{bmatrix},$$

where $\mathscr{A}_Q : Q \subset \text{Ker}\,\pi \to \text{Ker}\,\pi$ is given by $\mathscr{A}_Q\varphi = \dot{\varphi} + X_0[L\varphi - \dot{\varphi}(0)]$ for $\varphi \in Q$.

We rewrite $\tilde{z} = (\alpha^T, z^T)^T$ with $\alpha \in \mathbb{R}^r$ and $z \in \mathbb{R}^m$. Since $\tilde{\pi}\tilde{y} = 0$, it follows that $\tilde{y} = (0, y)^T$ with $y \in Q$. Then system (3.51) is equivalent to the following system:

$$\begin{aligned} \dot{\alpha} &= 0, \\ \dot{z} &= Bz + \overline{\Psi}(0)F(\alpha, \Phi z + \Omega\alpha + y), \\ \frac{dy}{dt} &= \mathscr{A}_Q y - (I - \pi)X_0 f_\alpha(0,0)\alpha + (I - \pi)X_0 F(\alpha, \Phi z + \Omega\alpha + y), \end{aligned} \tag{3.52}$$

where $F(\alpha, \varphi) = f(\alpha, \varphi) - L\varphi$. Applying the center manifold theorem (Theorem 3.2), there exists $w \in C^k(\mathbb{R}^r \times \mathbb{R}^m, Q)$ with $w(0,0) = 0$, $D_\alpha w(0,0) = 0$, and $D_z w(0,0) = 0$ such that for (α, z) close to $(0,0)$, the manifold

$$\{\Phi z + \Omega\alpha + w(\alpha, z) : \alpha \in \mathbb{R}^r, z \in \mathbb{R}^m\}$$

is locally invariant with respect to system (3.52) and contains all solutions of (3.52) remaining near $(\alpha, z, y) = (0,0,0)$ for all $t \in \mathbb{R}$. In general, $f_\alpha(0,0) \neq 0$, so we incorporate $\Omega\alpha$ and $w(\alpha, z)$, and then have the following theorem.

Theorem 3.4. *There exists* $W \in C^k(\mathbb{R}^r \times \mathbb{R}^m, Q)$ *with* $W(0,0) = 0$ *and* $D_z W(0,0) = 0$ *such that for* (α, z) *close to* $(0,0)$, *the manifold*

$$M_{\text{loc}}^{c,\alpha} = \{\Phi z + W(\alpha, z) : z \in \mathbb{R}^m\}$$

is locally invariant with respect to system (3.35) and contains all solutions of (3.35) remaining near $x = 0$ *for all* $t \in \mathbb{R}$.

Remark 3.2. If $f(\alpha,0) = 0$ for all $\alpha \in \mathbb{R}^r$, then $\Omega\alpha = 0$ and $W \in C^k(\mathbb{R}^r \times \mathbb{R}^m, Q)$ may satisfy $W(\alpha,0) = 0$ for all $\alpha \in \mathbb{R}^r$ and $D_z W(0,0) = 0$.

Note that $W(\alpha,z) = w(\alpha,z) + \varphi_h$ and that $\varphi_h = \Omega\alpha$ is the unique solution to (3.48). Then we have

$$\mathscr{A}_Q W = \dot{W} + X_0[LW - \dot{W}(0)]$$
$$= \mathscr{A}_Q y + \dot{\varphi}_h + X_0[L\varphi_h - \dot{\varphi}_h(0)]$$
$$= \mathscr{A}_Q y - (I - \pi)X_0 f_\alpha(0,0)\alpha.$$

Thus, it follows from the second equation of (3.52) that the reduced equation is

$$\dot{z} = Bz + G(\alpha,z), \tag{3.53}$$

where $G(\alpha,z) = \overline{\Psi}(0)F(\alpha, \Phi z + W(\alpha,z))$, where $W(\alpha,z)$ satisfies

$$\frac{\mathrm{d}}{\mathrm{d}t}W = \mathscr{A}_Q W + H(\alpha,z), \tag{3.54}$$

where $H(\alpha,z) = [X_0 - \Phi\overline{\Psi}(0)]F(\alpha, \Phi z + W(\alpha,z))$. We therefore need to calculate W, which depends on the parameter α.

3.6 Preservation of Symmetry

Throughout this section, we further assume that there exists an invertible linear map T on $\mathbb{R}^n$ commuting with (3.35), that is,

$$f(\alpha, T\phi) = Tf(\alpha,\phi) \tag{3.55}$$

for $(\alpha,\phi) \in \mathbb{R}^r \times C_{n,\tau}$, where $T\phi \in C_{n,\tau}$ is given by $(T\phi)(\theta) = T\phi(\theta)$ for $\theta \in [-\tau,0]$. Condition (3.55) is equivalent to saying that (3.35) is invariant under the transformation $(u,t) \to (Tu,t)$ in the sense that $u(t)$ is a solution of (3.35) if and only if $Tu(t)$ is. Obviously,

$$\mathscr{A}T = T\mathscr{A} \tag{3.56}$$

and

$$\langle \psi, T\varphi \rangle = \langle \psi T, \varphi \rangle \qquad \text{for } \psi \in C_{n,\tau}^* \text{ and } \varphi \in C_{n,\tau}. \tag{3.57}$$

Furthermore, we have the following result.

Lemma 3.2. *The spaces E^c, Q, and E^{c*} are invariant under T, while the matrix B commutes with M, where M is the matrix representation of the restriction of T on E^c.*

Proof. For every $\varphi \in E^c$, there exist $\lambda \in \sigma^c$ and a positive integer j such that $(\mathscr{A} - \lambda \mathrm{Id})^j \varphi = 0$, where Id is the identity mapping on $C_{n,\tau}$. This, together with (3.56), implies that $(\mathscr{A} - \lambda \mathrm{Id})^j T \varphi = 0$. Therefore, E^c is invariant under T. A similar argument shows that E^{c*} is also invariant under T. This implies that there exist invertible $m \times m$ matrices M and N such that

$$T\Phi = \Phi M \qquad \text{and} \qquad \Psi T = N \Psi. \tag{3.58}$$

It follows from (3.57) that

$$\begin{aligned} M &= \langle \Psi, \Phi \rangle M = \langle \Psi, \Phi M \rangle = \langle \Psi, T\Phi \rangle \\ &= \langle \Psi T, \Phi \rangle = \langle N\Psi, \Phi \rangle = N. \end{aligned}$$

Therefore, M is a matrix representation of T on the spaces E^c and E^{c*}, i.e., $M = T_c$. Since $Q = \mathrm{Ker}\, \pi = \{\varphi \in C_{n,\tau} : \langle \Psi, \varphi \rangle = 0\}$, for each $\varphi \in Q$, we have

$$\langle \Psi, T\varphi \rangle = \langle \Psi T, \varphi \rangle = \langle M\Psi, \varphi \rangle = M \langle \Psi, \varphi \rangle = 0,$$

i.e., $T\varphi \in Q$. This means that Q is invariant under T.

It follows from (3.58) that $T\dot{\Phi} = \dot{\Phi} M$. This, combined with $\dot{\Phi} = \Phi B$, gives $T\Phi B = \Phi B M$, i.e., $\Phi M B = \Phi B M$. Therefore, $MB = BM$, that is, B commutes with T_c. $\qquad \square$

Next, we consider the symmetry of $\pi : BC_n \to E^c$ given by (3.38). In view of Lemma 3.2, we have

$$\begin{aligned} \pi(T\varphi + TX_0\xi) &= \Phi[\langle \Psi, T\varphi \rangle + \overline{\Psi}(0)T\xi] \\ &= \Phi[\langle \Psi T, \varphi \rangle + \overline{\Psi}(0)T\xi] \\ &= \Phi[\langle M\Psi, \varphi \rangle + M\overline{\Psi}(0)\xi] \\ &= \Phi M[\langle \Psi, \varphi \rangle + \overline{\Psi}(0)\xi] \\ &= T\Phi[\langle \Psi, \varphi \rangle + \overline{\Psi}(0)\xi] \\ &= T\pi(\varphi + X_0\xi). \end{aligned}$$

Thus, we have proved the following result.

Lemma 3.3. *The projection operator $\pi: BC_n \to E^c$ commutes with T.*

In what follows, we assume that T_c is unitary, i.e., $T_c^* = T_c^{-1}$. Then we have the following result:

Theorem 3.5 (Symmetric center manifold theorem). *The map W in Theorem 3.4 may be chosen such that $W(\alpha, T_c z) = T_Q W(\alpha, z)$, where T_Q and T_c are the restrictions of T on Q and E^c, respectively. As a result, T_c commutes with the reduced system (3.53).*

Chapter 4
Normal Form Theory

4.1 Introduction

Normal forms theory provides one of the most powerful tools in the study of nonlinear dynamical systems, in particular in stability and bifurcation analysis. In the context of finite-dimensional ordinary differential equations (ODEs), this theory can be traced back as far as Euler. However, Poincaré [247] and Birkhoff [33] were the first to bring forth the theory in a more definite form. Since then, many systematic procedures for constructing normal forms have been developed. A method of Lie brackets is given by Chow and Hale [65], Takens [274], and Ushiki [280]; a method using an inner product in the space of homogeneous polynomials is given by Elphick et al. [87] and Ashkenazi and Chow [22]; a method for direct computations is given by Bruno [48], Sanders [259], and Chen and Della Dora [57]; a method using the Carleman linearization is given by Tsiligiannis and Lyberatos [277] and Chen and Della Dora [56]. The nilpotent case is treated by Cushman and Sanders [78, 79] using the representation theory of $sl_2(\mathbb{R})$. Recently, the normal form for a generalized Hopf bifurcation is expressed as a four-dimensional real system by Cushman and Sanders [77] and as a two-dimensional complex system by Elphick et al. [87] and Iooss and Adelmeyer [175].

The basic idea of normal form theory consists of employing successive near-identity nonlinear transformations that lead to a differential equation in a simpler form, qualitatively equivalent to the original system in the vicinity of a fixed equilibrium point, thus, one hopes, greatly simplifying the dynamics analysis. As we develop the method, three important characteristics should become apparent: (i) The method is local in the sense that the coordinates are generated in a neighborhood of a known solution. For our purposes, the known solution will be an equilibrium. (ii) In general, the coordinate transformations will be nonlinear in the dependent variables. However, the important point is that coordinate transformations are found by solving a sequence of problems. (iii) The structure of the norm form is determined entirely by the linear part of the vector field. A key notion in normal form reduction is that of resonance. In particular, the Jacobian matrix of

S. Guo and J. Wu, *Bifurcation Theory of Functional Differential Equations*,
Applied Mathematical Sciences 184, DOI 10.1007/978-1-4614-6992-6_4,
© Springer Science+Business Media New York 2013

the system evaluated at the equilibrium point determines which monomials in the formal expansion of the system are resonant and cannot be removed by any smooth coordinate transformation.

Concerning functional differential equations, the principal difficulty in developing a normal form theory is the fact that the phase space is not finite-dimensional. The first work in the direction of overcoming this difficulty is due to Faria and Magalhães [91, 92], who considered retarded functional differential equation (RFDE) as an abstract ODE in an adequate infinite-dimensional phase space, which was first presented in the work of Chow and Mallet-Paret [68]. This infinite-dimensional ODE was then handled in a similar way as in the finite-dimensional case. Through a recursive process of nonlinear transformations, Faria and Magalhães [91, 92] succeeded in reducing to a simpler infinite-dimensional ODE defined as a normal form of the original RFDE. Faria and Magalhães [91, 92] illustrated that their method provides an efficient algorithm for approximating normal forms for an RFDE directly without computing beforehand a local center manifold near the singularity. This is important because this approach does not lead to the loss of the explicit relationships between the coefficients in the normal form obtained and the coefficients in the original RFDE. We shall see in this chapter how calculating certain Taylor coefficients of the center manifold can be used to carry out this algorithm. In addition, normal form have also been developed by Guo [128] and Weedermann [292] for neutral functional differential equations.

In this chapter, we are concerned with RFDEs having a general singularity, and we assume the existence of a local center manifold by requiring finitely many eigenvalues with zero real part.

4.2 Unperturbed Vector Fields

In this section, we present a basic framework for the normal form theory of the following vector field:

$$\dot{x} = Bx + G(x), \quad x \in \mathbb{R}^m, \tag{4.1}$$

where B is in Jordan canonical form, G is C^k, with k to be made specific as we go along, $G(0) = 0$, and $D_x G(0) = 0$. The next step is to transform the vector field to a simpler form. The resulting *simplified* vector field is a normal form of (4.1).

For the sake of convenience, we introduce the following notation. For each $j \geq 2$, let $\mathscr{H}_j(\mathbb{R}^m)$ denote the linear space of homogeneous polynomials of degree j in m variables, $x = (x_1, x_2, \ldots, x_m)$, with coefficients in $\mathbb{R}^m$, i.e.,

$$\mathscr{H}_j(\mathbb{R}^m) = \left\{ \sum_{|q|=j} c_{(q)} x^q : q \in \mathbb{N}_0^m, c_q \in \mathbb{R}^m \right\},$$

which is equipped with the norm $|\sum_{|q|=j} c_q x^q| = \sum_{|q|=j} |c_q|$. Here and in the sequel, for a given positive integer p, a p-tube $\beta = (\beta_1, \ldots, \beta_p) \in \mathbb{N}_0^p$, and p variables $w = (w_1, \ldots, w_p)$, we define $|\beta| = \sum_{s=1}^p \beta_s$ and $w^\beta = w_1^{\beta_1} \cdots w_p^{\beta_p}$.

Example 4.1. $\mathscr{H}_2(\mathbb{R}^2) = \mathrm{span}\{x_1^2 e_1, x_1 x_2 e_1, x_2^2 e_1, x_1^2 e_2, x_1 x_2 e_2, x_2^2 e_2\}$, where $\{e_1, e_2\}$ is the canonical basis for $\mathbb{R}^2$. Usually, $e_1 = (0,1)^T$ and $e_2 = (1,0)^T$.

4.2.1 The Poincaré–Birkhoff Normal Form Theorem

First, we have the Taylor expansion of $G(z)$, so that (4.1) becomes

$$\dot{z} = Bz + \sum_{j=2}^{k-1} G_j(z) + O(|z|^k), \tag{4.2}$$

where $G_j \in \mathscr{H}_j(\mathbb{R}^m)$, $j = 2, \ldots, k-1$. The basic idea of the normal form theory is to use a near-identical transformation at an equilibrium to form a Lie bracket operator and then repeatedly employ the operator to remove as many higher-order nonlinear terms as possible.

We next introduce the coordinate transformation

$$z = x + h_2(x), \tag{4.3}$$

where $h_2 \in \mathscr{H}_2(\mathbb{R}^m)$ is to be determined later. Substituting (4.3) into (4.2) gives

$$(\mathrm{Id}_m + Dh_2(x))\dot{x} = Bx + Bh_2(x) + \sum_{j=2}^{k-1} G_j(x + h_2(x)) + O(|x|^k). \tag{4.4}$$

Notice that $G_j(x + h_2(x)) = G_j(x) + O(|x|^{j+1})$ for each $2 \le j \le k-1$. Moreover, when x is sufficiently small, $(\mathrm{Id}_m + Dh_2(x))^{-1}$ exists and can be represented in a series expansion as follows:

$$(\mathrm{Id}_m + Dh_2(x))^{-1} = \mathrm{Id}_m - Dh_2(x) + O(|x|^2). \tag{4.5}$$

Thus, (4.4) gives

$$\dot{x} = Bx + Bh_2(x) - Dh_2(x)Bx + G_2(x) + \sum_{j=3}^{k-1} \tilde{G}_j(x) + O(|x|^k), \tag{4.6}$$

where $\tilde{G}_s \in \mathscr{H}_s(\mathbb{R}^m)$, $s = 3, 4, \ldots, k$. Up to this point, $h_2(x)$ has been completely arbitrary. However, now we will choose a specific form for $h_2(x)$ to simplify the $O(|x|^2)$ terms as much as possible. In fact, this would be possible if we chose $h_2(x)$ such that

$$Dh_2(x)Bx - Bh_2(x) = G_2(x), \tag{4.7}$$

which would eliminate $G_2(x)$ from (4.6). Equation (4.7) can be regarded as an equation for the unknown $h_2(x)$. We want to motivate the fact that when viewed in an appropriate way, it is in fact a linear operator acting on a linear vector space. This will be accomplished by defining the appropriate linear vector space as well as the linear operator on the vector space, and also by describing the equation to be solved in this linear vector space (which will turn out to be (4.7)).

Now let us reconsider (4.7). It should be clear that h_2 is in $\mathcal{H}_2(\mathbb{R}^m)$. The reader should easily be able to verify that the map $h_2(x) \mapsto Dh_2(x)Bx - Bh_2(x)$ is a linear map of $\mathcal{H}_2(\mathbb{R}^m)$ to $\mathcal{H}_2(\mathbb{R}^m)$. Indeed, for $h_j \in \mathcal{H}_j$, it similarly follows that

$$h_j(x) \mapsto Dh_j(x)Bx - Bh_j(x)$$

is a linear map of $\mathcal{H}_j(\mathbb{R}^m)$ to $\mathcal{H}_j(\mathbb{R}^m)$. Therefore, we define a linear map $\mathcal{L}_j$: $\mathcal{H}_j(\mathbb{R}^m) \to \mathcal{H}_j(\mathbb{R}^m)$ by

$$(\mathcal{L}_j p)(x) = [B, p](x),$$

where $[\cdot, \cdot]$ denotes the Lie bracket, in fact, $[B, p](x) = D_x p(x)Bx - Bp(x)$. Here $\mathcal{L}_j$ is a homological operator. Now (4.7) takes the form

$$\mathcal{L}_2 h_2 = G_2.$$

From elementary linear algebra, we know that $\mathcal{H}_2(\mathbb{R}^m)$ can be (nonuniquely) represented as

$$\mathcal{H}_2(\mathbb{R}^m) = \mathrm{Ran}\mathcal{L}_2 \oplus (\mathrm{Ran}\mathcal{L}_2)^c,$$

where $(\mathrm{Ran}\mathcal{L}_2)^c$ represents a space complementary to $\mathrm{Ran}\mathcal{L}_2$. If $G_2 \in \mathrm{Ran}\mathcal{L}_2$, then (4.7) can be solved for $h_2 \in \mathcal{H}_2(\mathbb{R}^m)$, and hence all $O(|x|^2)$ terms can be eliminated from (4.6). In any case, we can choose $h_2(x)$ to take away from $G_2(x)$ its component in $\mathrm{Ran}\mathcal{L}_j$ such that only the component in $(\mathrm{Ran}\mathcal{L}_2)^c$, denoted by $g_2(x)$, remains. Thus, (4.6) can be simplified to

$$\dot{x} = Bx + g_2(x) + \sum_{j=3}^{k-1} \tilde{G}_j(x) + O(|x|^k). \tag{4.8}$$

Using a similar argument as above, we can simplify the vector field step by step.

We assume that after computing the vector field up to terms of order $j - 1$, the equations become

$$\dot{x} = Bx + \sum_{s=2}^{j-1} g_s(x) + \sum_{l=j}^{k-1} f_l(x) + O(|x|^k), \tag{4.9}$$

where $g_s \in (\mathrm{Ran}\mathcal{L}_s)^c$ and $f_l \in \mathcal{H}_s(\mathbb{R}^m)$ for $2 \le s \le j - 1$ and $j \le l \le k - 1$. Next, let us simplify the $O(|x|^j)$ terms. Introducing the coordinate change

$$x \mapsto x + h_j(x), \tag{4.10}$$

where $h_j \in \mathcal{H}_j(\mathbb{R}^m)$, and performing the same algebraic manipulations as with the second-order terms, we see that (4.9) becomes

$$\dot{x} = Bx + \sum_{s=2}^{j-1} g_s(x) + f_j(x) - \mathcal{L}_j h_j(x) + \sum_{l=j+1}^{k-1} \tilde{f}_l(x) + O(|x|^k), \tag{4.11}$$

where $\tilde{f}_s \in \mathcal{H}_s(\mathbb{R}^m)$, $j + 1 \le s \le k - 1$. Decompose $f_j \in \mathcal{H}_j(\mathbb{R}^m)$ as

$$f_j(x) = p_j(x) + g_j(x) \quad \text{with} \quad p_j \in \mathrm{Ran}\mathcal{L}_j \quad \text{and} \quad g_j \in (\mathrm{Ran}\mathcal{L}_j)^c.$$

Therefore, we choose $h_j \in \mathscr{H}_j(\mathbb{R}^m)$ such that $\mathscr{L}_j h_j = p_j$, and hence (4.11) becomes

$$\dot{x} = Bx + g_2(x) + \cdots + g_j(x) + \tilde{f}_{j+1}(x) + \cdots + \tilde{f}_{k-1}(x) + O(|x|^k). \qquad (4.12)$$

Thus, by induction, we obtain the following normal form theorem.

Theorem 4.1 (Poincaré–Birkhoff normal form theorem). *By a sequence of analytic coordinate changes, (4.1) can be transformed into*

$$\dot{x} = Bx + \sum_{j=2}^{k} g_j(x) + o(|x|^k), \qquad (4.13)$$

where $g_j(x) \in (\mathrm{Ran}\,\mathscr{L}_j)^c$, $2 \leq j \leq k$, are called the resonant terms. Equation (4.13) is said to be in normal form through order k.

We call the system

$$\dot{x} = Bx + \sum_{j=2}^{k} g_j(x) \qquad (4.14)$$

the (kth-)order truncated Birkhoff normal form of (4.1). The dynamics of the truncated normal form (4.14) are related to, but not identical with, the local dynamics of the system (4.1) around the equilibrium point $x = 0$.

4.2.2 Computation of Normal Forms

The key part of computing the normal form is to find $(\mathrm{Ran}\,\mathscr{L}_j)^c$, which represents a space complementary to $\mathrm{Ran}\,\mathscr{L}_j$ in $\mathscr{H}_j(\mathbb{R}^m)$.

Definition 4.1. Suppose that matrix B has eigenvalues $\lambda_1, \lambda_2, \ldots, \lambda_m$ (including multiple eigenvalues). Eigenvalues $\lambda_1, \lambda_2, \ldots, \lambda_m$ are called *resonant* if there exists a tuple $q = (q_1, \ldots, q_m) \in \mathbb{N}_0^m$ satisfying $|q| \geq 2$ (which is called the *order* of resonance) such that

$$(q, \overline{\lambda}) = \lambda_s \qquad (4.15)$$

for some $1 \leq s \leq m$, where $\overline{\lambda} = (\lambda_1, \ldots, \lambda_m)$ and $(q, \overline{\lambda}) = \sum_{j=1}^{m} q_j \lambda_j$. Eigenvalues λ_1, $\lambda_2, \ldots, \lambda_m$ are called *nonresonant of order j* ($j \geq 2$) if (4.15) does not hold for each s and $q \in \mathbb{N}_0^m$ satisfying $|q| = j$. Eigenvalues $\lambda_1, \lambda_2, \ldots, \lambda_m$ are called *nonresonant* if (4.15) doesn't hold for each s and $q \in \mathbb{N}_0^m$ satisfying $|q| \geq 2$.

It is well known that the spectrum of $\mathscr{L}_j$ is

$$\sigma(\mathscr{L}_j) = \{(q, \overline{\lambda}) - \lambda_s : s = 1, \ldots, m, q \in \mathbb{N}_0^m, |q| = j\}.$$

Therefore, if the eigenvalues of B are nonresonant of order j, then 0 is not an eigenvalue of $\mathscr{L}_j$, and so $\mathscr{L}_j$ is invertible on $\mathscr{H}_j(\mathbb{R}^m)$. This means that for every $G_j \in \mathscr{H}_j(\mathbb{R}^m)$, equation $\mathscr{L}_j h = G_j$ can be solved for $h \in \mathscr{H}_j(\mathbb{R}^m)$. Thus, $\mathrm{Ran}\,\mathscr{L}_j =$

$\mathcal{H}_j(\mathbb{R}^m)$, and the complementary space $(\text{Ran}\mathcal{L}_j)^c$ equals $\{0\}$. Hence, there are no j-order terms in the normal form. In particular, if the eigenvalues of B are nonresonant, then for all $j \geq 2$, $(\text{Ran}\mathcal{L}_j)^c = \{0\}$, and so $g_j = 0$. In this case, the normal form of (4.1) becomes

$$\dot{x} = Bx + O(|x|^k).$$

Therefore, only if eigenvalues of B are resonant do some nonlinear terms of the normal form remain. While computing the normal form, we need to know about the structure of the complementary spaces $(\text{Ran}\mathcal{L}_j)^c$ of $\text{Ran}\mathcal{L}_j$ in $\mathcal{H}_j(\mathbb{R}^m)$, $j = 2, 3, \ldots$. If $(\text{Ran}\mathcal{L}_j)^c$ has a basis $\{\vartheta_1, \ldots, \vartheta_l\}$, then in the normal form (4.13),

$$g_j = a_1 \vartheta_1 + \cdots + a_l \vartheta_l \in (\text{Ran}\mathcal{L}_j)^c,$$

where $l = \dim(\text{Ran}\mathcal{L}_j)^c$, $a_1, \ldots, a_l$, are constants. We should note that the complementary space $(\text{Ran}\mathcal{L}_j)^c$ is not unique; a different choice of $(\text{Ran}\mathcal{L}_j)^c$ leads to a different normal form.

4.2.2.1 The Matrix Method

Note that $\mathcal{H}_j(\mathbb{R}^m)$ is finite-dimensional. Assume that $\{e_i\}$ is a basis of $\mathcal{H}_j(\mathbb{R}^m)$, on which L_j is the matrix representation of the linear operator $\mathcal{L}_j$. It follows from the Fredholm alternative theorem that $\text{Ker}L_j^* = (\text{Ran}L_j)^\perp$, where L_j^* denotes the complex conjugate transpose of L_j and $\text{Ker}(L_j^*)$ is the null space of L_j^*. Thus, we can choose $(\text{Ran}\mathcal{L}_j)^c \cong \text{Ker}L_j^*$ to compute normal forms. In what follows, we illustrate this in system (4.1) with $m = 2$ and B equal to each of the following matrices:

$$B_1 = \begin{bmatrix} 0 & 0 \\ 0 & 0 \end{bmatrix}, \quad B_2 = \begin{bmatrix} 0 & 1 \\ 0 & 0 \end{bmatrix}, \quad B_3 = \begin{bmatrix} 0 & -\omega \\ \omega & 0 \end{bmatrix}, \tag{4.16}$$

where $\omega \in \mathbb{R}$, $\omega > 0$.

Example 4.2. Consider

$$\dot{x} = B_1 x + f_2(x) + O(|x|^3), \tag{4.17}$$

where $x = (x_1, x_2)^T \in \mathbb{R}^2$, B_1 is given in (4.16), and $f_2 \in \mathcal{H}_2(\mathbb{R}^2)$.

In view of Example 4.1, consider a basis $\{e_1, \ldots, e_6\}$ of $\mathcal{H}_2(\mathbb{R}^2)$, where $e_1(x) = x_1^2 e_1$, $e_2(x) = x_1 x_2 e_1$, $e_3(x) = x_2^2 e_1$, $e_4(x) = x_1^2 e_2$, $e_5(x) = x_1 x_2 e_2$, $e_6(x) = x_2^2 e_2$. It is easy to see that $\mathcal{L}_2 = 0$, and hence its representation matrix L_2 is equal to 0. So, $\text{Ker}L_2^* = \mathbb{R}^6$ and $(\text{Ran}\mathcal{L}_2)^c = \mathcal{H}_2(\mathbb{R}^2)$. We regard $\{e_1, \ldots, e_6\}$ as a basis of $(\text{Ran}\mathcal{L}_2)^c$. Then every $g_2 \in (\text{Ran}\mathcal{L}_2)^c$ can be written as

$$g_2(y) = \sum_{i=1}^{6} a_i e_i(y).$$

Thus, the 2-order normal form of (4.17) is

$$\dot{y} = B_1 y + g_2(y) + O(|y|^3), \quad y = (y_1, y_2)^T \in \mathbb{R}^2,$$

or

$$\dot{y}_1 = a_4 y_1^2 + a_5 y_1 y_2 + a_6 y_2^2 + O(|y|^3),$$
$$\dot{y}_2 = a_1 y_1^2 + a_2 y_1 y_2 + a_3 y_2^2 + O(|y|^3),$$

where $a_1, \ldots, a_6$ are constants. Using a similar argument as above, we have

$$(\text{Ran} \mathscr{L}_j)^c = \mathscr{H}_j(\mathbb{R}^2) \quad \text{for all } j \geq 3.$$

Therefore, the k-order normal form is exactly the k-order Taylor expansion.

Example 4.3. Consider

$$\dot{x} = B_2 x + f_2(x) + O(|x|^3), \tag{4.18}$$

where $x = (x_1, x_2)^T \in \mathbb{R}^2$, B_2 is given in (4.16), and $f_2 \in \mathscr{H}_2(\mathbb{R}^2)$.

Similarly to Example 4.2, we still regard $\{\mathbf{e}_1, \ldots, \mathbf{e}_6\}$ as a basis of $\mathscr{H}_2(\mathbb{R}^2)$. For $p = (p_1, p_2)^T \in \mathscr{H}_2(\mathbb{R}^2)$,

$$\mathscr{L}_2 p(x) = [Dp(x)]B_2 x - B_2 p(x) = \begin{bmatrix} x_2 \frac{\partial p_1}{\partial x_1} - p_2 \\ x_2 \frac{\partial p_2}{\partial x_1} \end{bmatrix}.$$

So we have

$$\mathscr{L}_2 \mathbf{e}_1 = 2\mathbf{e}_2 - \mathbf{e}_4, \quad \mathscr{L}_2 \mathbf{e}_2 = \mathbf{e}_3 - \mathbf{e}_5,$$
$$\mathscr{L}_2 \mathbf{e}_3 = -\mathbf{e}_6, \qquad \mathscr{L}_2 \mathbf{e}_4 = 2\mathbf{e}_5,$$
$$\mathscr{L}_2 \mathbf{e}_5 = \mathbf{e}_6, \qquad \mathscr{L}_2 \mathbf{e}_6 = 0.$$

Thus, on the basis $\{\mathbf{e}_1, \ldots, \mathbf{e}_6\}$, the matrix representation of $\mathscr{L}_2$ is

$$L_2 = \begin{bmatrix} 0 & 0 & 0 & 0 & 0 & 0 \\ 2 & 0 & 0 & 0 & 0 & 0 \\ 0 & 1 & 0 & 0 & 0 & 0 \\ -1 & 0 & 0 & 0 & 0 & 0 \\ 0 & -1 & 0 & 2 & 0 & 0 \\ 0 & 0 & -1 & 0 & 1 & 0 \end{bmatrix}.$$

In order to find $\text{Ker} L_2^*$, we need to solve the following linear algebraic equation:

$$L_2^* \xi = 0, \quad \xi \in \mathbb{R}^6. \tag{4.19}$$

In view of $L_2^* = L_2^T$, we obtain a group of fundamental solutions: $\{\mathbf{e}_2 + 2\mathbf{e}_4, \mathbf{e}_1\}$, which is a basis of $\text{Ker}(L_2^*)$ and corresponds to a basis of $(\text{Ran} \mathscr{L}_2)^c$, denoted by $\{\tilde{\mathbf{e}}_1, \tilde{\mathbf{e}}_2\}$. Here,

$$\tilde{\mathbf{e}}_1(x) = \begin{bmatrix} 2x_1^2 \\ x_1 x_2 \end{bmatrix}, \quad \tilde{\mathbf{e}}_2(x) = \begin{bmatrix} 0 \\ x_1^2 \end{bmatrix}.$$

Hence, we can choose a suitable transformation $x = y + p_2(y)$ with $p_2 \in \mathcal{H}_2(\mathbb{R}^2)$ to reduce (4.18) to a 2-order normal form

$$\dot{y} = B_2 y + g_2(y) + O(|y|^3), \quad y = (y_1, y_2) \in \mathbb{R}^2,$$

that is,

$$\dot{y}_1 = y_2 + 2a_1 y_1^2 + O(|y|^3),$$
$$\dot{y}_2 = a_1 y_1 y_2 + a_2 y_1^2 + O(|y|^3),$$

where a_1 and a_2 are constants. Using a similar argument as above, we can obtain a higher-order normal form.

As we know, the choice of $(\mathrm{Ran}\mathcal{L}_2)^c$ is not unique. If $\tilde{\mathbf{e}}_1$ and $\tilde{\mathbf{e}}_2$ are supplemented by some vectors in $\mathrm{Ran}\mathcal{L}_2$ to form a new basis for a new complementary space $(\mathrm{Ran}\mathcal{L}_2)^c$, then we can obtain another 2-order normal form. For example, consider a vector

$$\varepsilon(y) = 2\mathcal{L}_2 \mathbf{e}_1(y) = (-2y_1^2, 4y_1 y_2)^T,$$

in $\mathrm{Ran}\mathcal{L}_2$ and let $\hat{\mathbf{e}}_1(y) = \tilde{\mathbf{e}}_1(y) + \varepsilon(y)$, $\hat{\mathbf{e}}_2(y) = \tilde{\mathbf{e}}_2(y)$. Then, using $\{\hat{\mathbf{e}}_1, \hat{\mathbf{e}}_2\}$ as a basis of the new space complementary to $\mathrm{Ran}\mathcal{L}_2$ yields the following 2-order normal form:

$$\dot{y}_1 = y_2 + O(|y|^3),$$
$$\dot{y}_2 = a_1 y_1 y_2 + a_2 y_1^2 + O(|y|^3).$$

By a similar argument as that given above, we can obtain another normal form of (4.18), such as

$$\dot{y}_1 = y_2 + a_1 y_1^2 + O(|y|^3),$$
$$\dot{y}_2 = a_2 y_1^2 + O(|y|^3).$$

Example 4.4. Consider

$$\dot{x} = B_3 x + f_2(x) + O(|x|^3), \tag{4.20}$$

where $x = (x_1, x_2)^T \in \mathbb{R}^2$, B_3 is given in (4.16), and $f_2 \in \mathcal{H}_2(\mathbb{R}^2)$.

Similarly to Example 4.2, we still regard $\{\mathbf{e}_1, \ldots, \mathbf{e}_6\}$ as a basis of $\mathcal{H}_2(\mathbb{R}^2)$. For $p = (p_1, p_2)^T \in \mathcal{H}_2(\mathbb{R}^2)$,

$$\mathcal{L}_2 p(x) = \omega \begin{bmatrix} -x_2 \frac{\partial p_1}{\partial x_1} + x_1 \frac{\partial p_1}{\partial x_2} + p_2 \\ -x_2 \frac{\partial p_2}{\partial x_1} + x_1 \frac{\partial p_2}{\partial x_2} - p_1 \end{bmatrix}.$$

Thus, on the basis $\{\mathbf{e}_1, \ldots, \mathbf{e}_6\}$, the matrix representation of $\mathcal{L}_2$ is

$$L_2 = \begin{bmatrix} 0 & \omega & 0 & -\omega & 0 & 0 \\ -2\omega & 0 & 2\omega & 0 & -\omega & 0 \\ 0 & -\omega & 0 & 0 & 0 & -\omega \\ \omega & 0 & 0 & 0 & \omega & 0 \\ 0 & \omega & 0 & -2\omega & 0 & 2\omega \\ 0 & 0 & \omega & 0 & -\omega & 0 \end{bmatrix}.$$

It is easy to see that L_2 and therefore also L_2^* are invertible. Hence, $L_2^* \xi = 0$ has only one solution $\xi = 0$. Namely, $(\text{Ran}\mathscr{L}_2)^c \cong \text{Ker}L_2^* = \{0\}$. That is, for every $g_2 \in (\text{Ran}\mathscr{L}_2)^c$, $g_2 = 0$. Therefore, the 2-order normal form of (4.20) is

$$\dot{y} = B_3 y + O(|y|^3), \quad y = (y_1, y_2) \in \mathbb{R}^2,$$

that is,

$$\dot{y}_1 = -\omega y_2 + O(|y|^3), \tag{4.21}$$
$$\dot{y}_2 = \omega y_1 + O(|y|^3),$$

which does not contain the two-order terms. Obviously, the truncated equation of (4.21) is linear and cannot inherit the qualitative properties of (4.20) near the non-hyperbolic equilibria $x = 0$. Thus, we need to compute the 3-order normal form.

The space $\mathscr{H}_3(\mathbb{R}^2)$ has a basis $\{x_1^3 e_1, x_1^2 x_2 e_1, x_1 x_2^2 e_1, x_2^3 e_1, x_1^3 e_2, x_1^2 x_2 e_2, x_1 x_2^2 e_2, x_2^3 e_2\}$, on which the operator $\mathscr{L}_3$ has the following matrix representation

$$L_3 = \begin{bmatrix} 0 & \omega & 0 & 0 & -\omega & 0 & 0 & 0 \\ -3\omega & 0 & 2\omega & 0 & 0 & -\omega & 0 & 0 \\ 0 & -2\omega & 0 & 3\omega & 0 & 0 & -\omega & 0 \\ 0 & 0 & -\omega & 0 & 0 & 0 & 0 & -\omega \\ \omega & 0 & 0 & 0 & 0 & \omega & 0 & 0 \\ 0 & \omega & 0 & 0 & -3\omega & 0 & 2\omega & 0 \\ 0 & 0 & \omega & 0 & 0 & -2\omega & 0 & 3\omega \\ 0 & 0 & 0 & \omega & 0 & 0 & -\omega & 0 \end{bmatrix}.$$

Equation $L_3^* \xi = 0$ has a group of fundamental solutions in $\mathbb{R}^6$, which corresponds to a basis of $(\text{Ran}\mathscr{L}_3)^c$:

$$\begin{bmatrix} x_1(x_1^2 + x_2^2) \\ x_2(x_1^2 + x_2^2) \end{bmatrix}, \quad \begin{bmatrix} -x_2(x_1^2 + x_2^2) \\ x_1(x_1^2 + x_2^2) \end{bmatrix}.$$

Hence, the 3-order normal form of (4.20) is

$$\dot{y}_1 = -\omega y_2 + (ay_1 - by_2)(y_1^2 + y_2^2) + O(|y|^3),$$
$$\dot{y}_2 = \omega y_1 + (ay_2 + by_1)(y_1^2 + y_2^2) + O(|y|^3).$$

In polar coordinates, the above normal form can be rewritten as

$$\dot{r} = ar^3 + O(r^4),$$
$$\dot{\theta} = 1 + br^2 + O(r^3).$$

Finally, we remark that the absence of two-order terms of (4.21) can be concluded from the nonresonance of order 2. Indeed, matrix B_3 has two eigenvalues $\pm i\omega$, which are nonresonant of order 2.

In conclusion, the method of matrix representation is practical in computing low-order resonant terms of a normal form. However, the dimension of $\mathscr{H}_j(\mathbb{R}^m)$ increases rapidly as m and j increase, which brings more and more difficulties in computing normal forms. Moreover, when computing resonant terms of a normal form, we have to resort to a lengthy computation of different linear algebraic equations. See Guckenheimer and Holmes [125], Arrowsmith and Place [21], Wiggins [298] for further readings.

4.2.2.2 The Adjoint Operator Method

We define a suitable inner product in $\mathscr{H}_j(\mathbb{R}^m)$. Let

$$f(x) = \sum_{|\beta|=j} c_\beta x^\beta, \quad x \in \mathbb{R}^m,$$

where $\beta \in \mathbb{N}_0^m$. Define

$$f(\partial) = \sum_{|\beta|=j} c_\beta \frac{\partial^j}{\partial x_1^{\beta_1} \cdots \partial x_m^{\beta_m}}.$$

Define an inner product $\ll \cdot, \cdot \gg: \mathscr{H}_j(\mathbb{R}^m) \times \mathscr{H}_j(\mathbb{R}^m) \to \mathbb{R}$ as

$$\ll p, q \gg = \sum_{i=1}^m p_i(\partial) q_i(x) \Big|_{x=0}$$

for $p = (p_1, \ldots, p_m)^T$, $q = (q_1, \ldots, q_m)^T \in \mathscr{H}_j(\mathbb{R}^m)$. Let $\mathscr{L}_j^*$ denote the adjoint operator of $\mathscr{L}_j$ with respect to the inner product $\ll \cdot, \cdot \gg$. It turns out that $\mathscr{L}_j^*$ is just the homological operator associated with the adjoint of B. In other words, we have the following result.

Lemma 4.1. $(\mathscr{L}_j^* p)(x) = [B^*, p](x)$ for all $p \in \mathscr{H}_j(\mathbb{R}^m)$.

Proof. Notice that for every $p = \sum_{|q|=j} c_q x^q \in \mathscr{H}_j(\mathbb{R}^m)$,

$$
\begin{aligned}
[B, p](x) &= D_x \left[\sum_{|q|=j} c_q x^q \right] Bx - B \sum_{|q|=j} c_q x^q \\
&= \frac{d}{dt} \left[e^{-tB} \sum_{|q|=j} c_q (e^{tB} x)^q \right] \Big|_{t=0}.
\end{aligned}
\tag{4.22}
$$

For all $p_1, p_2 \in \mathscr{H}_j(\mathbb{R}^m)$, we have

$$\ll e^{-tB}(p_1 \circ e^{tB}), p_2 \gg = \ll p_1 \circ e^{tB}, (e^{-tB})^* p_2 \gg$$
$$= \ll p_1, (e^{-tB})^*(p_2 \circ (e^{tB})^*) \gg$$
$$= \ll p_1, e^{-tB^*}(p_2 \circ e^{tB^*}) \gg .$$

Differentiating with respect to t at $t = 0$ gives us

$$\ll [B, p_1], p_2 \gg = \ll p_1, [B^*, p_2] \gg,$$

which yields the conclusion of this lemma.   □

Due to the Fredholm alternative theorem, we have

$$\mathscr{H}_j(\mathbb{R}^m) = \text{Ran}\mathscr{L}_j \oplus \text{Ker}\mathscr{L}_j^*, \quad \text{Ran}\mathscr{L}_j \perp \text{Ker}\mathscr{L}_j^*,$$

and

$$\mathscr{H}_j(\mathbb{R}^m) = \text{Ker}\mathscr{L}_j \oplus \text{Ran}\mathscr{L}_j^*, \quad \text{Ker}\mathscr{L}_j \perp \text{Ran}\mathscr{L}_j^*.$$

Let P_j be the projection from $\mathscr{H}_j(\mathbb{R}^m)$ onto $\text{Ran}\mathscr{L}_j$. Therefore, in order to compute the normal form, we can choose $(\text{Ran}\mathscr{L}_j)^c = \text{Ker}\mathscr{L}_j^*$, which consists of all the solutions to

$$[B^*, p] = 0, \quad p \in \mathscr{H}_j(\mathbb{R}^m). \tag{4.23}$$

For example, we consider the computation of the 2-order normal form of (4.18). Obviously, $B_2^* = B_2^T$. For $x = (x_1, x_2)^T \in \mathbb{R}^2$ and $p = (p_1, p_2)^T \in \mathscr{H}_2(\mathbb{R}^2)$, (4.23) is

$$\begin{bmatrix} \frac{\partial p_1}{\partial x_1} & \frac{\partial p_1}{\partial x_2} \\ \frac{\partial p_2}{\partial x_1} & \frac{\partial p_2}{\partial x_2} \end{bmatrix} \begin{bmatrix} 0 \\ x_1 \end{bmatrix} - \begin{bmatrix} 0 \\ p_1 \end{bmatrix} = 0,$$

that is,

$$x_1 \frac{\partial p_1}{\partial x_2} = 0, \quad x_1 \frac{\partial p_2}{\partial x_2} = p_1, \tag{4.24}$$

all of whose solutions in $\mathscr{H}_2(\mathbb{R}^2)$ are

$$p_1(x) = ax_1^2, \quad p_2(x) = ax_1 x_2 + bx_1^2, \tag{4.25}$$

where a and b are arbitrary constants. Since the family of solutions $p = (p_1, p_2)^T$ given by (4.25) is the space $\text{Ker}\mathscr{L}_2^*$, we can choose $g_2(x) = p(x)$ to obtain the 2-order normal form

$$\dot{x}_1 = x_2 + ax_1^2 + O(|x|^3),$$
$$\dot{x}_2 = ax_1 x_2 + bx_1^2 + O(|x|^3).$$

Here, the choice of $(\text{Ran}\mathscr{L}_2)^c$ is different from those in Example 4.3, and as a consequence, the normal form is different from those given in Example 4.3.

4.2.3 *Internal Symmetry*

We now try to study the characteristics of $\mathrm{Ker}\mathscr{L}_j^*$ and those of g_j. First, recall that every linear operator A on a finite-dimensional vector space has a unique Jordan–Chevalley decomposition into commuting semisimple and nilpotent parts: $A = A_S + A_N$, where $A_S A_N = A_N A_S$. The semisimple part A_S is diagonalizable (over $\mathbb{C}$), and the nilpotent part A_N satisfies the condition that $A_N^k = 0$ for some positive integer k (see Humphreys [173]). Furthermore, the operator A induces the following Lie group:

$$G_A = \overline{\{e^{tA} : t \in \mathbb{R}\}}.$$

Next, we can show that g_j is equivariant with respect to the action of the Lie group $G_{B^*} = G_{B_S^*} \times G_{B_N^*}$. More precisely, we may state the following lemma.

Lemma 4.2. *For $j \geq 2$, each $g_j \in \mathrm{Ker}\mathscr{L}_j^*$ is G_{B^*}-equivariant, that is, $e^{tB^*}g_j(x) = g_j(e^{tB^*}x)$ for all $x \in \mathbb{R}^m$ and $t \in \mathbb{R}$.*

Proof. For $j \geq 2$, since $g_j \in \mathrm{Ker}\mathscr{L}_j^*$, we have $[B^*, g_j](x) = 0$ for $x \in \mathbb{R}^m$. It follows that

$$\frac{\mathrm{d}}{\mathrm{d}t}\left[e^{-tB^*}g_j(e^{tB^*}x)\right] = e^{-tB^*}[B^*, g_j](e^{tB^*}x) = 0$$

for all x and t. Therefore, for fixed x, $e^{-tB^*}g_j(e^{tB^*}x)$ is a constant, which is $g_j(x)$ by taking $t = 0$. Then we have $g_j(e^{tB^*}x) = e^{tB^*}g_j(x)$, as required. □

Remark 4.1. Lemma 4.2 implies that:

(i) If B is semisimple, then we can choose a suitable inner product $< \cdot, \cdot >$ on $\mathbb{R}^m$ such that B is skew-symmetric. That is, $B_S^* = -B$ and $B_N^* = 0$. Then $G_{B^*} = G_{-B}$ is a torus $\mathbb{T}^k$ for some k. Therefore, g_j, $j \geq 2$, is $\mathbb{T}^k$-equivariant, i.e., $e^{tB}g_j(x) = g_j(e^{tB}x)$ for all $x \in \mathbb{R}^m$ and $t \in \mathbb{R}$.

(ii) If B is not semisimple, then we can choose a suitable inner product $< \cdot, \cdot >$ on $\mathbb{R}^m$ such that $B_S^* = -B_S$. It follows that for all $t \in \mathbb{R}$, we have

$$e^{tB_S}g_j(x) = g_j(e^{tB_S}x) \quad \text{and} \quad e^{tB_N^*}g_j(x) = g_j(e^{tB_N^*}x).$$

In summary, we obtain another version of the normal form theorem as follows.

Theorem 4.2 (Elphick et al. [87]). *For every $j \geq 2$, there are polynomials*

$$p, h : \mathbb{R}^m \to \mathbb{R}^m, \quad p(0) = h(0) = 0, \quad Dp(0) = Dh(0) = 0,$$

of degree less than j such that by the change of variable $x \mapsto x + h(x)$, (4.1) becomes

$$\dot{x} = Bx + p(x) + O(|x|^{j+1}), \tag{4.26}$$

where p satisfies $e^{tB^*}p(x) = p(e^{tB^*}x)$ for all $x \in \mathbb{R}^m$ and $t \in \mathbb{R}$, or equivalently, $Dp(x)B^*x = B^*p(x)$ for all $x \in \mathbb{R}^m$.

Proposition 4.2.1 *For system (4.1) with $m = 2$ and B equal to B_2 given in (4.16), the normal form (4.26) is*

$$\dot{x}_1 = x_2 + x_1 \varphi_1(x_1), \qquad (4.27)$$
$$\dot{x}_2 = x_2 \varphi_1(x_1) + \varphi_2(x_1),$$

where φ_1 and φ_2 are polynomials such that $\varphi_1(0) = \varphi_2(0) = D\varphi_2(0) = 0$.

Proof. By Theorem 4.2, p is characterized by $Dp(x)B_2^* x = B_2^* p(x)$, where $x = (x_1, x_2)^T \in \mathbb{R}^2$ and $p(x) = (p_1(x), p_2(x))^T$. Namely, (4.24) holds. We immediately have $p_1(x) = \chi(x_1)$. Since p_1 is a polynomial in x_1, x_2, χ is a polynomial in x_1. So

$$\frac{\partial p_2}{\partial x_2} = \frac{\chi(x_1)}{x_1}.$$

Since p_2 is a polynomial, so is $\partial p_2/\partial x_2$. Hence χ takes the form of $\chi(x_1) = x_1 \varphi_1(x_1)$, where φ_1 is a polynomial. Integration now yields $p_2(x) = x_2 \varphi_1(x_1) + \varphi_2(x_1)$. Since p_2 and $x_2 \varphi_1$ are polynomials, so is φ_2. Therefore, our normal form is (4.27). □

In Proposition 4.2.1, $\mathrm{Ker}\,\mathcal{L}_j^*$ is two-dimensional and is spanned by

$$\begin{bmatrix} x_1^j \\ x_1^{j-1} x_2 \end{bmatrix}, \quad \begin{bmatrix} 0 \\ x_1^j \end{bmatrix}.$$

Changing the projection P_j onto $\mathrm{Ran}\,\mathcal{L}_j$ corresponds to adding a certain vector of $\mathrm{Ran}\,\mathcal{L}_j$ to our vector field $\mathrm{Ker}\,\mathcal{L}_j^*$. Indeed, let P_j and P_j' be two projections from $\mathcal{H}_j(\mathbb{R}^2)$ onto $\mathrm{Ran}\,\mathcal{L}_j$. Then for $R_j \in \mathcal{H}_j(\mathbb{R}^2)$,

$$g_j - g_j' = (I - P_j)R_j - (I - P_j')R_j \in \mathrm{Ran}\,\mathcal{L}_j.$$

Notice that $(-x_1^j, jx_2 x_1^{j-1})^T \in \mathrm{Ran}\,\mathcal{L}_j$ for all $j \geq 2$, since it is orthogonal to $\mathrm{Ker}\,\mathcal{L}_j^*$. Indeed,

$$\ll (-x_1^j, jx_2 x_1^{j-1})^T, (x_1^j, x_2 x_1^{j-1})^T \gg = 0$$

and

$$\ll (-x_1^j, jx_2 x_1^{j-1})^T, (0, x_1^j)^T \gg = 0.$$

Thus, it is possible to choose a projection P_j' such that

$$p_1'(x) = 0, \quad p_2'(x) = x_2 \varphi(x_1) + \phi(x_1),$$

where φ and ϕ are polynomials such that $\varphi(0) = \phi(0) = D\phi(0) = 0$. Hence, the system

$$\dot{x}_1 = x_2,$$
$$\dot{x}_2 = x_2 \varphi(x_1) + \phi(x_1)$$

is also a normal form of system (4.1) with $m = 2$ and $B = B_2$.

Proposition 4.2.2 *For system (4.1) with* $m = 3$ *and*

$$B = \begin{bmatrix} 0 & 1 & 0 \\ 0 & 0 & 1 \\ 0 & 0 & 0 \end{bmatrix},$$

the normal form (4.26) is

$$\begin{aligned}
\dot{x}_1 &= x_2 + x_1 \varphi_1(x_1, x_2^2 - 2x_1 x_3), \\
\dot{x}_2 &= x_1 + x_2 \varphi_1(x_1, x_2^2 - 2x_1 x_3) + x_1 \varphi_2(x_1, x_2^2 - 2x_1 x_3), \\
\dot{x}_3 &= x_3 \varphi_1(x_1, x_2^2 - 2x_1 x_3) + x_2 \varphi_2(x_1, x_2^2 - 2x_1 x_3) + \varphi_3(x_1, x_2^2 - 2x_1 x_3),
\end{aligned} \tag{4.28}$$

where φ_1, φ_2, *and* φ_3 *are polynomials in their arguments.*

Proof. By Theorem 4.2, p is characterized by $Dp(x)B^*x = B^*p(x)$, where $x = (x_1, x_2, x_3)^T \in \mathbb{R}^3$ and $p(x) = (p_1(x), p_2(x), p_3(x))^T$. Thus, we have

$$x_1 \frac{\partial p_1}{\partial x_2} + x_2 \frac{\partial p_1}{\partial x_3} = 0, \quad x_1 \frac{\partial p_2}{\partial x_2} + x_2 \frac{\partial p_2}{\partial x_3} = p_1, \quad x_1 \frac{\partial p_3}{\partial x_2} + x_2 \frac{\partial p_3}{\partial x_3} = p_2. \tag{4.29}$$

Hence the characteristic system is

$$\frac{dx_1}{0} = \frac{dx_2}{x_1} = \frac{dx_3}{x_2} = \frac{dp_1}{0} = \frac{dp_2}{p_1} = \frac{dp_3}{p_2},$$

and the first integrals are

$$x_1, \quad x_2^2 - 2x_1 x_3, \quad p_1, \quad x_1 p_2 - x_2 p_1, \quad x_1 p_3 + x_3 p_1 - x_2 p_2.$$

Introduce new variables

$$u_1 = x_1, \quad u_2 = x_2^2 - 2x_1 x_3, \quad u_3 = x_2,$$

and define $\tilde{p}_j$ for $j = 1, 2, 3$ by

$$\tilde{p}_j(u_1, u_2, u_3) = p_j(x_1, x_2, x_3).$$

Then the partial differential system (4.29) can be written as

$$u_1 \frac{\partial \tilde{p}_1}{\partial u_3} = 0, \quad u_1 \frac{\partial \tilde{p}_2}{\partial u_3} = \tilde{p}_1, \quad u_1 \frac{\partial \tilde{p}_3}{\partial u_3} = \tilde{p}_2.$$

The equation for $\tilde{p}_1$ yields $p_1(x) = \phi(u_1, u_2)$. It is easy to see that ϕ is a polynomial in u_1 and u_2, and so can be rewritten as $\phi(u_1, u_2) = u_1 \varphi_1(u_1, u_2) + \psi_1(u_2)$, where φ_1 and ψ_1 are polynomials and $\psi_1(u_2) = \phi(0, u_2)$. Solving the equation for $\tilde{p}_2$, we obtain

$$p_2(x) = x_2 \varphi_1(u_1, u_2) + \frac{x_2 \psi_1(u_2)}{u_1} + u_1 \varphi_2(u_1, u_2) + \psi_2(u_2),$$

where φ_1 and ψ_2 are polynomials. Multiplying by u_1 and setting $u_1 = 0$, we obtain $x_2 \psi_1(x_2^2) = 0$ for all x_2. This implies that $\psi_1(u_2) = 0$ for all u_2. Thus,

$$p_2(x) = x_2 \varphi_1(u_1, u_2) + x_1 \varphi_2(u_1, u_2) + \psi_2(u_2),$$

where φ_1 and ψ_2 are polynomials. Solving the equation for $\tilde{p}_3$, we now obtain

$$p_3(x) = \left(x_3 + \frac{u_2}{2u_1}\right)\varphi_1(u_1,u_2) + u_3\varphi_2(u_1,u_2) + \frac{u_3\psi_2(u_2)}{u_1} + \phi(u_1,u_2),$$

where ϕ is a polynomial. Multiplying by u_1 and setting $u_1 = 0$, we have $x_2^2\varphi_1(0,x_2^2) + 2x_2\psi_2(x_2^2) = 0$ for all x_2. This implies that $\varphi_1(0,u) = \psi_2(u) = 0$ for all $u \in \mathbb{R}$. Thus, $\varphi_1(u_1,u_2)/u_1$ is a polynomial in u_1 and u_2. By writing

$$\frac{u_2}{2u_1}\varphi_1(u_1,u_2) + \phi(u_1,u_2) = \varphi_3(u_1,u_2),$$

we obtain $p_3(x) = x_3\varphi_1(u_1,u_2) + x_2\varphi_2(u_1,u_2) + \varphi_3(u_1,u_2)$, where φ_3 is a polynomial. Therefore, our normal form is (4.28). $\quad\square$

Similarly, for the system described in Proposition 4.2.2, it is possible to choose an alternative projection P'_j such that our normal form is

$$\begin{aligned}
\dot{x}_1 &= x_2, \\
\dot{x}_2 &= x_1, \\
\dot{x}_3 &= x_3\varphi_1(u_1,u_2) + x_2\varphi_2(u_1,u_2) + \varphi_3(u_1,u_2),
\end{aligned}$$

where u_1 and u_2 are given as above, and φ_1, φ_2, and φ_3 are polynomials in u_1 and u_2.

Proposition 4.2.3 *For system (4.1) with $m = 4$ and*

$$B = \begin{bmatrix} 0 & 1 & 0 & 0 \\ 0 & 0 & 0 & 0 \\ 0 & 0 & 0 & 1 \\ 0 & 0 & 0 & 0 \end{bmatrix},$$

the normal form (4.26) is

$$\begin{aligned}
\dot{x}_1 &= x_2 + x_1\varphi_1(x_1,x_3,v) + x_3\varphi_2(x_1,x_3,v), \\
\dot{x}_2 &= x_2\varphi_1(x_1,x_3,v) + x_4\varphi_2(x_1,x_3,v) + \varphi_3(x_1,x_3), \\
\dot{x}_3 &= x_4 + x_3\varphi_4(x_1,x_3,v) + x_1\varphi_5(x_1,x_3,v), \\
\dot{x}_4 &= x_4\varphi_4(x_1,x_3,v) + x_2\varphi_5(x_1,x_3,v) + \varphi_6(x_1,x_3),
\end{aligned} \qquad (4.30)$$

where $v = x_2x_3 - x_1x_4$ and φ_j ($j = 1,\ldots,6$) are polynomials in their arguments. Furthermore, by choosing an alternative projection P'_j, we can obtain the following normal form:

$$\begin{aligned}
\dot{x}_1 &= x_2, \\
\dot{x}_2 &= x_2h_1(x_1,x_3,v) + x_4h_2(x_1,x_3,v) + w_1(x_1,x_3), \\
\dot{x}_3 &= x_4, \\
\dot{x}_4 &= x_4h_3(x_1,x_3,v) + x_2h_4(x_1,x_3,v) + w_2(x_1,x_3),
\end{aligned}$$

where h_j ($j = 1,\ldots,4$) are polynomials in their arguments starting at degree 1, and w_1 and w_2 are polynomials in x_1 and x_2 starting at degree 2.

Proof. By Theorem 4.2, p is characterized by $Dp(x)B^*x = B^*p(x)$, where $x = (x_1,x_2,x_3,x_4)^T \in \mathbb{R}^4$ and $p(x) = (p_1(x),p_2(x),p_3(x),p_4(x))^T$. Hence, the characteristic system is

$$\frac{dx_1}{0} = \frac{dx_2}{x_1} = \frac{dx_3}{0} = \frac{dx_4}{x_3} = \frac{dp_1}{0} = \frac{dp_2}{p_1} = \frac{dp_3}{0} = \frac{dp_4}{p_3},$$

and the first integrals are

$$x_1, \quad x_3, \quad v = x_2x_3 - x_1x_4, \quad p_1, \quad p_3, \quad x_1p_2 - x_2p_1, \quad x_3p_4 + x_4p_3.$$

The remaining part of the proof is similar to that of Proposition 4.2.2 and hence is omitted.   □

Proposition 4.2.4 *For system (4.1) with $m = 2$ and B equal to B_3 given in (4.16), the normal form (4.26) is*

$$\begin{aligned}
\dot{x}_1 &= -\omega x_2 + x_1 Q_1(x_1^2 + x_2^2) - x_2 Q_2(x_1^2 + x_2^2), \\
\dot{x}_2 &= \omega x_1 + x_1 Q_2(x_1^2 + x_2^2) + x_2 Q_1(x_1^2 + x_2^2),
\end{aligned} \tag{4.31}$$

where Q_1 and Q_2 are polynomials such that $Q_1(0) = Q_2(0) = 0$.

Proof. We complexify, i.e., we identify $\mathbb{R}^2$ with $\{(z,\bar{z}); z \in \mathbb{C}\} \subseteq \mathbb{C}^2$ by the map $(x_1,x_2) \mapsto (x_1 + ix_2, x_1 - ix_2)$. We have $B_3 = \mathrm{diag}(i\omega, -i\omega)$. The operator

$$e^{tB_3^*} = \begin{bmatrix} e^{-i\omega t} & 0 \\ 0 & e^{i\omega t} \end{bmatrix}$$

has to commute with $(p(z,\bar{z}), \overline{p(z,\bar{z})})$:

$$p(ze^{-i\omega t}, \bar{z}e^{i\omega t}) = e^{-i\omega t} p(z,\bar{z})$$

for all z and t, where $p(z,\bar{z}) = p_1(x_1,x_2) + ip_2(x_1,x_2)$. This implies that $p(z,\bar{z})$ is $\mathbb{S}^1$-equivariant, where $\mathbb{S}^1 = \{e^{-i\omega t}: t \in \mathbb{R}\}$. Thus, we see that there is a polynomial Q such that $p(z,\bar{z}) = zQ(|z|^2)$. In real coordinates (x_1,x_2), we have

$$\begin{aligned}
p_1(x_1,x_2) &= x_1 Q_1(x_1^2 + x_2^2) - x_2 Q_2(x_1^2 + x_2^2), \\
p_2(x_1,x_2) &= x_1 Q_2(x_1^2 + x_2^2) + x_2 Q_1(x_1^2 + x_2^2),
\end{aligned}$$

where Q_1 and Q_2 are the real and imaginary parts of Q and hence polynomials. Therefore, our normal form is (4.31).   □

Proposition 4.2.5 *For system (4.1) with $m = 4$ and*

$$B = \begin{bmatrix} 0 & 1 & -\omega & 0 \\ 0 & 0 & 0 & -\omega \\ \omega & 0 & 0 & 1 \\ 0 & \omega & 0 & 0 \end{bmatrix},$$

where $\omega \in \mathbb{R}$, $\omega > 0$, the normal form (4.26), in complex coordinates (z_1,z_2), is

$$\dot{z}_1 = i\omega z_1 + z_2 + z_1 \varphi_1(u_1, u_2), \tag{4.32}$$
$$\dot{z}_2 = i\omega z_2 + z_2 \varphi_1(u_1, u_2) + z_1 \varphi_2(u_1, u_2),$$

where $u_1 = |z_1|^2$, $u_2 = z_1 \bar{z}_2 - \bar{z}_1 z_2$, and φ_j $(j = 1, 2)$ are polynomials in their arguments such that $\varphi_j(0,0) = 0$. Furthermore, by choosing an alternative projection P_j', we can obtain the following normal form:

$$\dot{z}_1 = i\omega z_1 + z_2,$$
$$\dot{z}_2 = i\omega z_2 + z_2 \varphi_1(u_1, u_2) + z_1 \varphi_2(u_1, u_2),$$

where φ_j $(j = 1, 2)$ are polynomials in their arguments such that $\varphi_j(0,0) = 0$.

Proof. We complexify, i.e., we identify $\mathbb{R}^4$ with $\{(z_1, z_2, \bar{z}_1, \bar{z}_2); (z_1, z_2) \in \mathbb{C}^2\} \subseteq \mathbb{C}^4$ by the map $(x_1, x_2, x_3, x_4) \mapsto (z_1, z_2) = (x_2 + ix_4, x_1 + ix_3)$. We have

$$B = \begin{bmatrix} i\omega & 1 & 0 & 0 \\ 0 & i\omega & 0 & 0 \\ 0 & 0 & -i\omega & 1 \\ 0 & 0 & 0 & -i\omega \end{bmatrix}.$$

Let us denote by $\mathscr{D}$ the following operator

$$\mathscr{D} = -i\omega z_1 \frac{\partial}{\partial z_1} + (z_1 - i\omega z_2) \frac{\partial}{\partial z_2} + i\omega \bar{z}_1 \frac{\partial}{\partial \bar{z}_1} + (\bar{z}_1 + i\omega \bar{z}_2) \frac{\partial}{\partial \bar{z}_2}.$$

Then the partial differential equation $Dp(z) B^* z = B^* p(z)$ reads

$$\mathscr{D} p_1 = -i\omega p_1, \quad \mathscr{D} p_2 = -i\omega p_2 + p_1,$$

where $p = (p_1, p_2, \bar{p}_1, \bar{p}_2)$ and $z = (z_1, z_2, \bar{z}_1, \bar{z}_2)$. Similarly to the proofs of the previous propositions, we may solve the above partial differential system for p. On the other hand, it follows from $e^{tB^*} p(z) = p(e^{tB^*} z)$ for all $z \in \mathbb{C}^4$ and $G_{B^*} = \mathbb{S}^1 \times \mathbb{R}$ that p is $\mathbb{S}^1 \times \mathbb{R}$-equivariant under the following $\mathbb{S}^1 \times \mathbb{R}$-action on $\mathbb{C}^2$:

$$\theta \cdot (z_1, z_2) = (e^{-i\omega \theta} z_1, e^{-i\omega \theta} z_2), \quad \sigma \cdot (z_1, z_2) = (z_1, z_2 + \sigma z_1)$$

for all $\theta \in \mathbb{S}^1 = \{e^{-i\omega t} : t \in \mathbb{R}\}$ and $\sigma \in \mathbb{R}$. Now we need to find the $\mathbb{S}^1 \times \mathbb{R}$-invariants and -equivariants.

We derive the $\mathbb{S}^1 \times \mathbb{R}$-invariants by starting with the $\mathbb{S}^1$-invariants. The complex $\mathbb{S}^1$-invariants are generated by

$$z_1 \bar{z}_1, \quad z_2 \bar{z}_2, \quad z_2 \bar{z}_1, \quad z_1 \bar{z}_2$$

with the relation $(z_1 \bar{z}_1)(z_2 \bar{z}_2) = (z_2 \bar{z}_1)(z_1 \bar{z}_2)$. Next, the $\mathbb{R}$-action on $(z_1 \bar{z}_1, z_2 \bar{z}_2, z_2 \bar{z}_1)$ is generated by

$$\sigma \cdot (z_1 \bar{z}_1) = z_1 \bar{z}_1, \quad \sigma \cdot (z_2 \bar{z}_2) = (z_2 + \sigma z_1)(\bar{z}_2 + \sigma \bar{z}_1), \quad \sigma \cdot (z_2 \bar{z}_1) = (z_2 + \sigma z_1) \bar{z}_1.$$

Thus, we obtain the following minimal set of generators of $\mathbb{S}^1$-invariants:

$$u_1 = z_1\bar{z}_1, \quad u_2 = z_1\bar{z}_2 - \bar{z}_1z_2.$$

The commutativity of p with $\mathbb{S}^1$ implies that we can write (p_1, p_2) in the form

$$p_1 = az_1 + bz_2, \quad p_2 = cz_1 + dz_2,$$

where a, b, c, and d are complex-valued $\mathbb{S}^1$-invariant functions of $u = (z_1\bar{z}_1, z_2\bar{z}_2, z_2\bar{z}_1)$. Commutativity with $\mathbb{R}$ additionally requires that

$$a(\sigma \cdot u)z_1 + b(\sigma \cdot u)(z_2 + \sigma z_1) = a(u)z_1 + b(u)z_2, \tag{4.33}$$

$$c(\sigma \cdot u)z_1 + d(\sigma \cdot u)(z_2 + \sigma z_1) = [c(u) + \sigma a(u)]z_1 + [d(u) + \sigma b(u)]z_2. \tag{4.34}$$

Hence, $b(u) = 0$ and $a(u) = d(u)$ for all u. Moreover, a and c are $\mathbb{R}$-equivariant. Therefore, there exist polynomials φ_1 and φ_2 such that

$$p_1 = z_1\varphi_1(u_1, u_2), \quad p_1 = z_1\varphi_2(u_1, u_2) + z_2\varphi_1(u_1, u_2).$$

Thus, the normal form reads (4.32).   □

Proposition 4.2.6 *For system (4.1) with $m = 3$ and*

$$B = \begin{bmatrix} 0 & 0 & 0 \\ 0 & 0 & -\omega \\ 0 & \omega & 0 \end{bmatrix},$$

where $\omega \in \mathbb{R}$, $\omega > 0$, the normal form (4.26), in coordinates $(x,z) \in \mathbb{R} \times \mathbb{C}$, is

$$\dot{x} = \varphi_1(x, |z|^2), \tag{4.35}$$
$$\dot{z} = i\omega z + z\varphi_2(x, |z|^2),$$

where φ_1 is a real polynomial such that $\varphi_1(0,0) = (\partial\varphi_1/\partial x)(0,0) = 0$, and φ_2 is a complex polynomial such that $\varphi_2(0,0) = 0$.

Proof. We complexify, i.e., we identify $\mathbb{R}^3$ with $\{(x,z,\bar{z}) : x \in \mathbb{R}, z \in \mathbb{C}\} \subseteq \mathbb{R} \times \mathbb{C}^2$, by the map $(x_1, x_2, x_3) \mapsto (x_1, x_2 + ix_3, x_2 - ix_3)$. We have $B = \mathrm{diag}(0, i\omega, -i\omega)$. The operator

$$e^{tB^*} = \mathrm{diag}(0, e^{-i\omega t}, e^{i\omega t})$$

has to commute with $(p_1(x,z,\bar{z}), p_2(x,z,\bar{z}), \overline{p_2(x,z,\bar{z})})$:

$$p_1(x, ze^{-i\omega t}, \bar{z}e^{i\omega t}) = p_1(x,z,\bar{z}), \quad p_2(x, ze^{-i\omega t}, \bar{z}e^{i\omega t}) = e^{-i\omega t}p_2(x,z,\bar{z})$$

for all x, z, and t. By a similar argument to that in the proof of Proposition 4.2.4, we have

$$p_1(x,z,\bar{z}) = \varphi_1(x, |z|^2), \quad p_2(x,z,\bar{z}) = z\varphi_2(x, |z|^2),$$

where φ_1 is a real polynomial and φ_2 is a complex polynomial. Hence, the normal form reads (4.35).   □

4.3 Perturbed Vector Fields

Consider the following vector field:

$$\dot{x} = Bx + G(\alpha, x), \quad x \in \mathbb{R}^m, \quad \alpha \in \mathbb{R}^r, \qquad (4.36)$$

where B is in Jordan canonical form, and does not depend on α, and G is sufficiently smooth and satisfies $G(\alpha, 0) = 0$ and $D_x G(0, 0) = 0$. Similarly to the discussion in the previous section, the most straightforward way to put (4.36) into normal form would be to follow the same procedure as for systems without parameters except to allow the coefficients of the transformation to depend on the parameters. In the subsequent subsections, we first seek the normal form for Hopf bifurcation and then present a general approach to the normal form of (4.36).

4.3.1 Normal Form for Hopf Bifurcation

As stated earlier, through center manifold reduction we may obtain the following one-parameter family of vector fields on a two-dimensional center manifold:

$$\dot{z} = \lambda(\alpha)z + \sum_{2 \leq j+s \leq k-1} \frac{1}{j!s!} G_{js}^{\alpha} z^j \bar{z}^s + O(|z|^k), \quad z \in \mathbb{C}, \ \alpha \in \mathbb{R}, \qquad (4.37)$$

where $\lambda(\alpha) = i\omega + G_1^{\alpha}$ is continuous in $\alpha \in \mathbb{R}$.

Lemma 4.3. *Equation (4.37) can be transformed by an invertible parameter-dependent change of complex coordinates*

$$z = w + \frac{1}{2} h_{20}^{\alpha} w^2 + h_{11}^{\alpha} w \bar{w} + \frac{1}{2} h_{02}^{\alpha} \bar{w}^2 \qquad (4.38)$$

for all sufficiently small $|\alpha|$ into an equation without quadratic terms: $\dot{w} = \lambda(\alpha)w + O(|w|^3)$.

Proof. The inverse change of variables is given by the expression

$$w = z - \frac{1}{2} h_{20}^{\alpha} z^2 - h_{11}^{\alpha} z\bar{z} - \frac{1}{2} h_{02}^{\alpha} \bar{z}^2 + O(|z|^2).$$

Therefore,

$$\begin{aligned}
\dot{w} &= \dot{z} - h_{20}^{\alpha} z\dot{z} - h_{11}^{\alpha}(\dot{z}\bar{z} + z\dot{\bar{z}}) - h_{02}^{\alpha} \bar{z}\dot{\bar{z}} + \cdots \\
&= \lambda(\alpha)z + [\frac{1}{2}G_{20}^{\alpha} - \lambda(\alpha)h_{20}^{\alpha}]z^2 + [G_{11}^{\alpha} - \lambda(\alpha)h_{11}^{\alpha} - \bar{\lambda}(\alpha)h_{11}^{\alpha}]z\bar{z} \\
&\quad + [\frac{1}{2}G_{02}^{\alpha} - \bar{\lambda}(\alpha)h_{02}^{\alpha}]\bar{z}^2 + \cdots \\
&= \lambda(\alpha)w + \frac{1}{2}[G_{20}^{\alpha} - \lambda(\alpha)h_{20}^{\alpha}]w^2 + [G_{11}^{\alpha} - \bar{\lambda}(\alpha)h_{11}^{\alpha}]w\bar{w} \\
&\quad + \frac{1}{2}[G_{02}^{\alpha} - (2\bar{\lambda}(\alpha) - \lambda(\alpha))h_{02}^{\alpha}]\bar{w}^2 + O(|w|^3).
\end{aligned}$$

Thus, by setting

$$h_{20}^\alpha = \frac{G_{20}^\alpha}{\lambda(\alpha)}, \quad h_{11}^\alpha = \frac{G_{11}^\alpha}{\overline{\lambda}(\alpha)}, \quad h_{02}^\alpha = \frac{G_{02}^\alpha}{2\overline{\lambda}(\alpha) - \lambda(\alpha)}, \tag{4.39}$$

we will remove all the quadratic terms in (4.37). Obviously, h_{20}^α, h_{11}^α, and h_{02}^α are well defined for sufficiently small $|\alpha|$, because $\lambda(0) = i\omega$ and the above denominators are nonzero. $\square$

In view of Lemma 4.3, by the invertible parameter-dependent transformation (4.38) with the h_{jk}^α given by (4.39), (4.37) becomes

$$\dot{w} = \lambda(\alpha)w + \sum_{j+s=3} \frac{1}{j!s!} \widetilde{G_{js}^\alpha} w^j \overline{w}^s + O(|w|^4). \tag{4.40}$$

In particular, we have

$$\widetilde{G_{21}^\alpha} = \frac{G_{20}^\alpha G_{11}^\alpha (2\lambda(\alpha) + \overline{\lambda}(\alpha))}{|\lambda(\alpha)|^2} + \frac{2|G_{11}^\alpha|^2}{\lambda(\alpha)} + \frac{|G_{02}^\alpha|^2}{2\lambda(\alpha) - \overline{\lambda}(\alpha)} + G_{21}^\alpha.$$

Next, let us try to eliminate the cubic terms as well. Namely, we have the following result.

Lemma 4.4. *Equation (4.40) can be transformed by an invertible parameter-dependent change of complex coordinates*

$$w = z + \sum_{j+s=3} \frac{1}{j!s!} h_{js}^\alpha z^j \overline{z}^s$$

for all sufficiently small $|\alpha|$ into an equation with only one cubic term:

$$\dot{z} = \lambda(\alpha)z + \frac{1}{2} C_1(\alpha) z|z|^2 + O(|z|^4). \tag{4.41}$$

Proof. The inverse change of variables is given by the expression

$$z = w - \sum_{j+s=3} \frac{1}{j!s!} h_{js}^\alpha w^j \overline{w}^s + O(|z|^4).$$

Therefore,

$$\dot{z} = \dot{w} - \frac{1}{2} h_{30}^\alpha w^2 \dot{w} - \frac{1}{2} h_{21}^\alpha (2w\overline{w}\dot{w} + w^2 \dot{\overline{w}}) - \frac{1}{2} h_{12}^\alpha (2w\overline{w}\dot{\overline{w}} + \overline{w}^2 \dot{w}) - \frac{1}{2} h_{03}^\alpha \overline{w}^2 \dot{\overline{w}} + \cdots$$

$$= \lambda(\alpha)w + \frac{1}{6}[G_{30}^\alpha - 3\lambda(\alpha)h_{30}^\alpha]w^3 + \frac{1}{2}[G_{21}^\alpha - 2\lambda(\alpha)h_{21}^\alpha - \overline{\lambda}(\alpha)h_{21}^\alpha]w^2\overline{w}$$

$$+ \frac{1}{2}[G_{12}^\alpha - \lambda(\alpha)h_{12}^\alpha - 2\overline{\lambda}(\alpha)h_{12}^\alpha]w\overline{w}^2 + \frac{1}{6}[G_{03}^\alpha - 3\overline{\lambda}(\alpha)h_{03}^\alpha]\overline{w}^3 + \cdots$$

$$= \lambda(\alpha)z + \frac{1}{6}[G_{30}^{\alpha} - 2\lambda(\alpha)h_{30}^{\alpha}]z^3 + \frac{1}{2}[G_{21}^{\alpha} - (\lambda(\alpha) + \overline{\lambda}(\alpha))h_{21}^{\alpha}]z^2\overline{z}$$
$$+ \frac{1}{2}[G_{12}^{\alpha} - 2\overline{\lambda}(\alpha)h_{12}^{\alpha}]z\overline{z}^2 + \frac{1}{6}[G_{03}^{\alpha} + (\lambda(\alpha) - 3\overline{\lambda}(\alpha))h_{03}^{\alpha}]\overline{z}^3 + O(|z|^4).$$

Thus, by setting

$$h_{30}^{\alpha} = \frac{\widetilde{G_{30}^{\alpha}}}{2\lambda(\alpha)}, \quad h_{12}^{\alpha} = \frac{\widetilde{G_{12}^{\alpha}}}{2\overline{\lambda}(\alpha)}, \quad h_{03}^{\alpha} = \frac{\widetilde{G_{02}^{\alpha}}}{3\overline{\lambda}(\alpha) - \lambda(\alpha)},$$

we can annihilate all cubic terms in the resulting equation except the $z^2\overline{z}$-term, which we have to treat separately. The substitutions are valid, since all the involved denominators are nonzero for all sufficiently small α. One can also try to eliminate the $z^2\overline{z}$ term by formally setting

$$h_{21}^{\alpha} = \frac{\widetilde{G_{21}^{\alpha}}}{\overline{\lambda}(\alpha) + \lambda(\alpha)}.$$

This is possible for small $|\alpha| \neq 0$, but the denominator vanishes at $\alpha = 0$: $\overline{\lambda}(0) + \lambda(0) = 0$. To obtain a transformation that is smoothly dependent on α, set $h_{21}^{\alpha} = 0$, which results in the 3-order normal form (4.41) with $C_1(\alpha) = \widetilde{G_{21}^{\alpha}}$. □

We now combine the two previous lemmas.

Theorem 4.3 (Poincaré normal form for the Hopf bifurcation). *Equation (4.37) can be transformed by an invertible parameter-dependent change of complex coordinates, smoothly depending on the parameter,*

$$z \mapsto z + \sum_{2 \leq j+s \leq 3} \frac{1}{j!s!}h_{js}^{\alpha}z^j\overline{z}^s \tag{4.42}$$

for all sufficiently small $|\alpha|$ into an equation with only the resonant cubic term (4.41) with

$$C_1(\alpha) = \frac{G_{20}^{\alpha}G_{11}^{\alpha}(2\lambda(\alpha) + \overline{\lambda}(\alpha))}{|\lambda(\alpha)|^2} + \frac{2|G_{11}^{\alpha}|^2}{\lambda(\alpha)} + \frac{|G_{02}^{\alpha}|^2}{2\lambda(\alpha) - \overline{\lambda}(\alpha)} + G_{21}^{\alpha}.$$

In particular,

$$C_1(0) = \frac{i}{\omega}\left[G_{20}^0 G_{11}^0 - 2|G_{11}^0|^2 - \frac{1}{3}|G_{02}^0|^2\right] + G_{21}^0. \tag{4.43}$$

4.3.2 Norm Form Theorem

Motivated by the previous subsection, we consider the formal Taylor expansions

$$G(\alpha,x) = \sum_{j\geq 2} \frac{1}{j!} G_j(\alpha,x) \tag{4.44}$$

for $\alpha \in \mathbb{R}^r$ and $x \in \mathbb{R}^m$, where G_j is the jth Fréchet derivative of G with respect to $\alpha \in \mathbb{R}^r$ and $x \in \mathbb{R}^m$. Then (4.36) can be rewritten as

$$\dot{x} = Bx + \sum_{j\geq 2} \frac{1}{j!} G_j(\alpha,x), \tag{4.45}$$

where $x \in \mathbb{R}^m$.

For convenience, we introduce the following notation. For a normed space Y, $\mathscr{H}_j^{m+r}(Y)$ denotes the linear space of homogeneous polynomials of degree j in $m+r$ variables, $x = (x_1,x_2,\ldots,x_m)$ and $\alpha = (\alpha_1,\alpha_2,\ldots,\alpha_r)$, with coefficients in Y, i.e.,

$$\mathscr{H}_j^{m+r}(Y) = \left\{ \sum_{|(q,l)|=j} c_{(q,l)} x^q \alpha^l : (q,l) \in \mathbb{N}_0^{m+r}, c_{(q,l)} \in Y \right\},$$

which is equipped with the norm

$$\left| \sum_{|(q,l)|=j} c_{(q,l)} x^q \alpha^l \right| = \sum_{|(q,l)|=j} |c_{(q,l)}|_Y.$$

Define the operators $\mathscr{L}_j \colon \mathscr{H}_j^{m+r}(\mathbb{R}^m) \to \mathscr{H}_j^{m+r}(\mathbb{R}^m)$, $j \geq 2$, by

$$(\mathscr{L}_j p)(\alpha,x) = [B, p(\alpha,\cdot)](x), \tag{4.46}$$

where $[\cdot,\cdot]$ denotes the Lie bracket.

Next, we try to introduce a suitable inner product in $\mathscr{H}_j^{m+r}(\mathbb{R}^m)$ in order to decompose it. Let

$$f(\alpha,x) = \sum_{|(q,l)|=j} c_{(q,l)} x^q \alpha^l, \quad (q,l) \in \mathbb{N}_0^{m+r}, c_{(q,l)} \in \mathbb{R}.$$

Define

$$f(\partial,\partial) = \sum_{|(q,l)|=j} c_{(q,l)} \frac{\partial^j}{\partial x_1^{q_1} \cdots \partial x_m^{q_m} \partial \alpha_1^{l_1} \cdots \partial \alpha_r^{l_r}},$$

and define an inner product $\ll \cdot, \cdot \gg \colon \mathscr{H}_j^{m+r}(\mathbb{R}^m) \times \mathscr{H}_j^{m+r}(\mathbb{R}^m) \to \mathbb{R}$ as

$$\ll p,q \gg = \sum_{i=1}^{m} p_i(\partial,\partial)q_i(\alpha,x)\Big|_{(x,\alpha)=(0,0)}$$

for all $p = (p_1, p_2, \ldots, p_m)^T$, $q = (q_1, q_2, \ldots, q_m)^T \in \mathscr{H}_j^{m+r}(\mathbb{R}^m)$.

It is easy to see that for all $p, q \in \mathscr{H}_j^{m+r}(\mathbb{R}^m)$ and every invertible map ζ on $\mathbb{R}^m$, we have

$$\ll p \circ \zeta, q \gg = \ll p, q \circ \zeta^* \gg,$$

where ζ^* is the adjoint of ζ. Let P_j be the (unique) orthogonal projection from $\mathscr{H}_j^{m+r}(\mathbb{R}^m)$ onto $\text{Ran}\mathscr{L}_j$. Note that

$$\mathscr{H}_j^{m+r}(\mathbb{R}^m) = \text{Ran}\mathscr{L}_j \oplus \text{Ker}\mathscr{L}_j^*, \quad \text{Ran}\mathscr{L}_j \perp \text{Ker}\mathscr{L}_j^*,$$

and

$$\mathscr{H}_j^{m+r}(\mathbb{R}^m) = \text{Ker}\mathscr{L}_j \oplus \text{Ran}\mathscr{L}_j^*, \quad \text{Ker}\mathscr{L}_j \perp \text{Ran}\mathscr{L}_j^*,$$

where $*$ denotes the adjoint map with respect to the inner product $\ll \cdot, \cdot \gg$. Similarly to Lemma 4.1, $\mathscr{L}_j^*$ is just the homological operator associated with the adjoint of B, i.e.,

$$(\mathscr{L}_j^* p)(\alpha, x) = [B^*, p(\alpha, \cdot)](x) \quad \text{for all } p \in \mathscr{H}_j^{m+r}(\mathbb{R}^m). \tag{4.47}$$

Following the approach in the case of no parameters, the normal forms can be obtained by computing at each step the terms of order $j \geq 2$ in the normal form from the terms of the same order in the original equation and the terms of lower orders already computed for the normal form in previous steps, through a transformation of variables

$$x = \hat{x} + \frac{1}{j!}U_j(\alpha, \hat{x}), \tag{4.48}$$

where $x, \hat{x} \in \mathbb{R}^m$, $\alpha \in \mathbb{R}^r$, and $U_j : \mathbb{R}^{m+r} \to \mathbb{R}^m$ is a homogeneous polynomial of degree j in x and α.

We assume that after computing the normal form up to terms of order $j-1$, the equations become

$$\dot{x} = Bx + \sum_{s=2}^{j-1} \frac{1}{s!}g_s(\alpha, x) + \frac{1}{j!}\overline{G}_j(\alpha, x) + \cdots, \tag{4.49}$$

where $g_i \in \mathscr{H}_i^{m+r}(\mathbb{R}^m)$, $\overline{G}_j \in \mathscr{H}_j^{m+r}(\mathbb{R}^m)$. With the change of variables (4.48) and dropping the hats for simplicity of notation, (4.49) becomes

$$\dot{x} = Bx + \sum_{s=2}^{j} \frac{1}{s!}g_s(\alpha, x) + \cdots, \tag{4.50}$$

where $g_j = \overline{G}_j - \mathscr{L}_j U_j$. System (4.50) is called the normal form for (4.36) near $(u, \alpha) = (0, 0)$.

If $\mathscr{L}_j$ is invertible, then we simply put $g_j = 0$ and $U_j = \mathscr{L}_j^{-1}\overline{G}_j$. In this case, all jth-order terms are removed by the normalization procedure. In general, $\mathscr{L}_j$ need not (and will not) be invertible. Note that

$$\mathscr{L}_j U_j = \overline{G}_j - g_j = P_j\overline{G}_j + (I - P_j)\overline{G}_j - g_j.$$

We will choose $U_j \in \mathscr{H}_j^{m+r}(\mathbb{R}^m)$ such that $U_j \in \mathscr{L}_j^{-1}P_j\overline{G}_j$, which allows us to take away from $\overline{G}_j$ its component in $\mathrm{Ran}\mathscr{L}_j$. Therefore, the above equation can be solved by

$$g_j(\alpha,x) = (I - P_j)\overline{G}_j(\alpha,x). \tag{4.51}$$

In particular, we have

$$g_j(\alpha,x) = \mathrm{Proj}_{\mathrm{Ker}\,\mathscr{L}_j^*}\overline{G}_j(\alpha,x). \tag{4.52}$$

Moreover, we have the following results.

Lemma 4.5. *For* $j \geq 2$, $g_j(\alpha,\cdot)$ *is* G_{B^*}-*equivariant, that is,* $e^{tB^*}g_j(\alpha,x) = g_j(\alpha,e^{tB^*}x)$ *for all* $x \in \mathbb{R}^m$ *and* $t \in \mathbb{R}$.

The proof is similar to that of Lemma 4.2 and hence is omitted. Thus, we summarize the above discussion as follows.

Theorem 4.4. *There are polynomials* $U: \mathbb{R}^m \times \mathbb{R}^r \to \mathbb{R}^m$ *and* $g: \mathbb{R}^m \times \mathbb{R}^r \to \mathbb{R}^m$ *of degree* $\leq k$ *with* $U(0,0) = 0$, $D_x U(0,0) = 0$ *such that by the change of variables* $x \mapsto x + U(\alpha,x)$, *(4.1) becomes*

$$\dot{x} = Bx + g(\alpha,x) + O(|(\alpha,x)|^{k+1}), \tag{4.53}$$

where g *satisfies* $e^{tB^*}g(\alpha,x) = g(\alpha,e^{tB^*}x)$ *for all* $\alpha \in \mathbb{R}^r$, $x \in \mathbb{R}^m$, *and* $t \in \mathbb{R}$, *or equivalently,* $D_x g(\alpha,x)B^*x = B^*g(\alpha,x)$ *for all* $\alpha \in \mathbb{R}^r$ *and* $x \in \mathbb{R}^m$.

4.3.3 Preservation of External Symmetry

Assume that there exists a linear invertible operator $T: \mathbb{R}^m \to \mathbb{R}^m$ that commutes with system (4.36), i.e.,

$$BT = TB, \quad G(\alpha,Tx) = TG(\alpha,x)$$

for all $\alpha \in \mathbb{R}^r$ and $x \in \mathbb{R}^m$. We define the linear map

$$\mathscr{T}_j : \mathscr{H}_j^{m+r}(\mathbb{R}^m) \to \mathscr{H}_j^{m+r}(\mathbb{R}^m)$$

by

$$\mathscr{T}_j p(\alpha,x) = T^{-1}p(\alpha,Tx).$$

We have the following result.

Lemma 4.6. $\mathrm{Ker}\mathscr{L}_j$ and $\mathrm{Ran}\mathscr{L}_j$ are invariant under $\mathscr{T}_j$.

Proof. Since B commutes with T, we have

$$
\begin{aligned}
\mathscr{T}_j\mathscr{L}_j p(\alpha,x) &= T^{-1}D_x p(\alpha,Tx)BTx - T^{-1}Bp(\alpha,Tx) \\
&= T^{-1}D_x p(\alpha,Tx)TBx - BT^{-1}p(\alpha,Tx) \\
&= D_x\left(T^{-1}p(\alpha,Tx)\right)Bx - BT^{-1}p(\alpha,Tx) \\
&= \mathscr{L}_j\mathscr{T}_j p(\alpha,x).
\end{aligned}
$$

This means that $\mathscr{T}_j$ commutes with $\mathscr{L}_j$. Therefore, $\mathrm{Ker}\mathscr{L}_j$ and $\mathrm{Ran}\mathscr{L}_j$ are invariant under $\mathscr{T}_j$. $\qquad\square$

In what follows, we always assume that T is unitary. Then $T^* = T^{-1}$ and hence $T^{-1}B = BT^{-1}$. Namely, $B^*T = TB^*$. In addition, using similar arguments as in the proof of Lemma 4.6, $\mathscr{T}_j$ commutes with $\mathrm{Ran}\mathscr{L}_j^*$. Therefore, we have the following lemmas.

Lemma 4.7. $\mathrm{Ker}\mathscr{L}_j^*$ and $\mathrm{Ran}\mathscr{L}_j^*$ are invariant under $\mathscr{T}_j$.

Lemma 4.8. For $j \geq 2$, $Tg_j(\alpha,x) = g_j(\alpha,Tx)$ and $TU_j(\alpha,x) = U_j(\alpha,Tx)$ for all $x \in \mathbb{R}^m$.

Proof. Since $\mathscr{H}_j^{m+r}(\mathbb{R}^m) = \mathrm{Ran}\mathscr{L}_j \oplus \mathrm{Ker}\mathscr{L}_j^*$, $\mathrm{Ran}\mathscr{L}_j$, and $\mathrm{Ker}\mathscr{L}_j^*$ are invariant under $\mathscr{T}_j$, it follows that $\mathscr{T}_j$ commutes with P_j and $I - P_j$, where I denotes the identity on $\mathscr{H}_j^{m+r}(\mathbb{R}^m)$.

Next, we prove inductively that $\overline{G}_j(\alpha,\cdot)$, U_j, and $g_j(\alpha,\cdot)$ commute with T. Since $\overline{G}_2 = G_2$, we immediately have $\mathscr{T}_2\overline{G}_2 = \overline{G}_2$. Assume that $\mathscr{T}_l\overline{G}_l = \overline{G}_l$ for some $l \geq 2$. It follows from (4.51) that

$$
\mathscr{T}_l g_l = (I - P_l)\overline{G}_l \quad \text{and} \quad \mathscr{L}_l\mathscr{T}_l U_l = P_l\overline{G}_l.
$$

Since $\mathrm{Ran}\mathscr{L}_l^*$ is invariant under $\mathscr{T}_l$ and $\mathscr{T}_l U_l \in \mathrm{Ran}\mathscr{L}_l^*$, by uniqueness we have

$$
\mathscr{T}_l U_l = U_l \quad \text{and} \quad \mathscr{T}_l g_l = g_l.
$$

Since $\overline{G}_{l+1}(\alpha,\cdot)$ is composed of B, G_j ($2 \leq j \leq l+1$), $U_j(\alpha,\cdot)$, and $g_j(\alpha,\cdot)$ ($2 \leq j \leq l$), we have

$$
\mathscr{T}_{l+1}\overline{G}_{l+1} = \overline{G}_{l+1}.
$$

This completes the proof. $\qquad\square$

In view of the Lemmas 4.6, 4.7, 4.8, and Theorem 4.4, we arrive at the following theorem.

Theorem 4.5. Suppose that $T: \mathbb{R}^n \to \mathbb{R}^n$ is unitary, and the vector field (4.36) has the symmetry T. Then the normal form (4.53) also has the symmetry T.

4.4 RFDEs with Symmetry

Consider the following parameterized RFDE:

$$\dot{u}(t) = L(\alpha)u_t + f(\alpha, u_t), \tag{4.54}$$

where $\alpha \in \mathbb{R}^r$, the linear operator $L(\alpha): C_{n,\tau} \to \mathbb{R}^n$ is continuous with respect to $\alpha \in \mathbb{R}^r$, $f \in C^k(\mathbb{R}^r \times C_{n,\tau}, \mathbb{R}^n)$ for a large enough integer k, $f(\alpha, 0) = 0$, and $D_\varphi f(\alpha, 0) = 0$ for all $\alpha \in \mathbb{R}^r$. We further assume that Γ is a given topological group and (4.54) is Γ-equivariant. Namely, there exists a representation ρ of Γ such that

$$f(\alpha, \rho(\gamma)\phi) = \rho(\gamma)f(\alpha, \phi) \quad \text{and} \quad L(\alpha)\rho(\gamma)\phi = \rho(\gamma)L(\alpha)\phi \tag{4.55}$$

for $(\alpha, \gamma, \phi) \in \mathbb{R}^r \times \Gamma \times C_{n,\tau}$, where $\rho(\gamma)\phi \in C_{n,\tau}$ is given by $(\rho(\gamma)\phi)(\theta) = \rho(\gamma)\phi(\theta)$ for $\theta \in [-\tau, 0]$. Recall that a representation ρ of Γ is a group homomorphism $\rho : \Gamma \to GL(n, \mathbb{R})$. Condition (4.55) is equivalent to saying that (4.54) is invariant under the transformation $(u, t) \to (\rho(\gamma)u, t)$ in the sense that $u(t)$ is a solution of (4.54) if and only if $\rho(\gamma)u(t)$ is (see [41, 118, 193, 284] for more details).

Since an RFDE generates a semiflow in an infinite-dimensional Banach space, one naturally first reduces the semiflow to a flow in the finite-dimensional center manifold, and then calculates the normal form of the reduced flow. However, it is not necessary to compute the center manifold before evaluating the normal form for the ODE on the center manifold. Faria and Magalhães [91, 92] developed a method for obtaining normal forms for RFDEs directly, which allows us to obtain the coefficients in the normal form explicitly in terms of the original system. Based on results of Faria and Magalhães [91, 92], our purpose in this section is to obtain explicit normal forms for the equation describing the flow on subcenter manifolds, which not only inherit the symmetry of the original system but also are invariant with respect to some torus actions induced by the imaginary roots of the characteristic equation of the linearization at the steady state. The final outcome of our procedure is the normal forms whose coefficients are explicitly given in terms of the parameters of the original RFDE. As shall be seen, the procedure we develop here for the calculation of normal forms can be easily adapted for high-codimension singularities, and therefore this technique for deriving normal forms may find further applications in addition to those addressed here.

4.4.1 Basic Assumptions

By the Riesz representation theorem, there exists an $n \times n$ matrix-valued function $\eta(\alpha, \cdot): [-\tau, 0] \to \mathbb{R}^{n^2}$ whose elements are of bounded variation in $\theta \in [-\tau, 0]$ for each α such that

$$L(\alpha)\varphi = \int_{-\tau}^0 d\eta(\alpha, \theta)\varphi(\theta), \qquad \varphi \in C_{n,\tau}.$$

For each $\alpha \in \mathbb{R}$, let $\mathscr{A}_\alpha$ be the infinitesimal generator associated with the linear system $\dot{u}(t) = L(\alpha)u_t$. The dynamics of (4.54) near the singularity $\alpha = 0$ at the origin can be completely described through the restriction of the flow to the associated center manifold, which is necessarily finite-dimensional. Therefore, it is important to consider a subcenter manifold relative to a subset Λ of $\sigma^c = \{\lambda \in \sigma(\mathscr{A}_0) : \mathrm{Re}\lambda = 0\}$. Without loss of generality, suppose $\Lambda = \{\lambda_1, \lambda_2, \ldots, \lambda_k\}$ is a nonempty set satisfying the following assumptions:

(H1) $\lambda \in \Lambda$ if and only if $\overline{\lambda} \in \Lambda$.
(H2)   If $0 \in \sigma(\mathscr{A}_0)$, then $0 \in \Lambda$.

Assumption (H1) is generic, because we are interested in real subcenter manifolds relative to Λ. There are many reasons for assumption (H2), one of which will be given in Remark 4.5.

Denote by E_Λ the generalized eigenspace of $\mathscr{A}_0$ associated with Λ. If m is the number of eigenvalues of $\mathscr{A}_0$ in Λ counting multiplicities, then $\dim E_\Lambda = m$. In order to construct coordinates on the center submanifold relative to Λ near the origin, we define a bilinear form

$$\langle \psi, \varphi \rangle = \overline{\psi}(0)\varphi(0) - \int_{-\tau}^{0} \int_{0}^{\theta} \overline{\psi}(\xi - \theta)d\eta(0, \theta)\varphi(\xi)d\xi \qquad (4.56)$$

for $\psi \in C_{n,\tau}^*$ and $\varphi \in C_{n,\tau}$. Here and in the sequel, for the sake of convenience, we shall also allow functions with range in $\mathbb{C}^n$. It follows from the Γ-equivariance of the operator $L(0)$ that

$$\langle \psi, \rho(\gamma)\varphi \rangle = \langle \psi\rho(\gamma), \varphi \rangle \qquad (4.57)$$

for $\psi \in C_{n,\tau}^*$, $\varphi \in C_{n,\tau}$, and $\gamma \in \Gamma$. Let Φ be a basis for E_Λ and let Ψ be the basis for the dual space E_Λ^* in $C_{n,\tau}^*$ such that $\langle \Psi, \Phi \rangle = \mathrm{Id}_m$. We denote by B the $m \times m$ constant matrix such that $\dot{\Phi} = \Phi B$. Note that $\sigma(B) = \Lambda$. In order to analyze the symmetry of the normal form on center manifolds relative to Λ, the following result about the Γ-invariance of these spaces is fundamental.

Lemma 4.9. *For each $\gamma \in \Gamma$, there exists an $m \times m$ matrix M_γ such that*

$$\rho(\gamma)\Phi = \Phi M_\gamma, \quad \Psi\rho(\gamma) = M_\gamma\Psi, \quad \text{and} \quad M_\gamma B = B M_\gamma. \qquad (4.58)$$

In other words, the spaces E_Λ, Q_Λ, and E_Λ^* are Γ-invariant, where Q_Λ is defined as in (2.21).

The proof of Lemma 4.9 is similar to Lemma 3.2 of Sect. 3.6 and hence is omitted.

Lemma 4.10. *Let $\pi : BC_n \to E_\Lambda$ be given by*

$$\pi(\varphi + X_0\xi) = \Phi[\langle \Psi, \varphi \rangle + \overline{\Psi}(0)\xi]. \qquad (4.59)$$

Then $BC_n = E_\Lambda \oplus \mathrm{Ker}\,\pi$ and $Q_\Lambda \subset \mathrm{Ker}\,\pi$. Moreover, the projection operator π: $BC_n \to E_\Lambda$ is Γ-equivariant.

Define $\mathscr{A}_Q : Q \subset \mathrm{Ker}\, \pi \to \mathrm{Ker}\, \pi$ by

$$\mathscr{A}_Q \varphi = \dot{\varphi} + X_0[L(0)\varphi - \varphi(0)]$$

for $\varphi \in Q \overset{\mathrm{def}}{=} \mathrm{Ker}\, \pi \cap C_{n,\tau}^1$. Moreover, define

$$F(\alpha, \varphi) = L(\alpha)\varphi - L(0)\varphi + f(\alpha, \varphi)$$

for all $\alpha \in \mathbb{R}^r$ and $\varphi \in C_{n,\tau}$. As stated in Chap. 3, system (4.54) is equivalent to the system

$$\dot{\alpha} = 0,$$
$$\dot{x} = Bx + \overline{\Psi}(0)F(\alpha, \Phi x + y),$$
$$\frac{dy}{dt} = \mathscr{A}_Q y + (I - \pi)X_0 F(\alpha, \Phi x + y),$$

or simply

$$\dot{x} = Bx + \overline{\Psi}(0)F(\alpha, \Phi x + y), \qquad (4.60)$$
$$\frac{dy}{dt} = \mathscr{A}_Q y + (I - \pi)X_0 F(\alpha, \Phi x + y),$$

where $x \in \mathbb{R}^m$ and $y \in Q$.

Lemma 4.11 (Faria and Magalhães [92]). $\sigma(\mathscr{A}_Q) = \sigma_p(\mathscr{A}_Q) = \sigma(\mathscr{A}) \setminus \Lambda$, where $\sigma_p(\mathscr{A}_Q)$ denotes the point spectrum of $\mathscr{A}_Q$.

4.4.2 Computation of Symmetric Normal Forms

We now describe the computation of normal forms using formal series, though we are interested in situations in which only a few terms of those series are computed. We consider the formal Taylor expansions

$$F(\alpha, u) = \sum_{j \geq 2} \frac{1}{j!} F_j(\alpha, u) \qquad (4.61)$$

for $\alpha \in \mathbb{R}^r$ and $u \in C_{n,\tau}$, where F_j is the jth Fréchet derivative of F with respect to $\alpha \in \mathbb{R}^r$ and $u \in C_{n,\tau}$. It follows from the Γ-equivariance of (4.54) that $F_j(\alpha, \cdot)$, $j \geq 2$, is Γ-equivariant. Then (4.60) can be rewritten as

$$\dot{x} = Bx + \sum_{j \geq 2} \frac{1}{j!} f_j^1(\alpha, x, y), \qquad (4.62)$$
$$\frac{dy}{dt} = \mathscr{A}_Q y + \sum_{j \geq 2} \frac{1}{j!} f_j^2(\alpha, x, y),$$

where $x \in \mathbb{R}^m$, $y \in Q$, and

$$f_j^1(\alpha,x,y) = \overline{\Psi}(0)F_j(\alpha,\Phi x+y),$$
$$f_j^2(\alpha,x,y) = (I-\pi)X_0F_j(\alpha,\Phi x+y), \qquad j \geq 2.$$

In particular, $f_j^1(\alpha,x,0) = \overline{\Psi}(0)F_j(\alpha,\Phi x)$ for $x \in \mathbb{R}^m$ and $j \geq 2$. In addition, in view of Lemma 4.10, we have the following result.

Lemma 4.12. $f_j^1(\alpha,\cdot,0)$ *is* Γ-*equivariant. Moreover,*

$$f_j^1(\alpha,M_\gamma x,0) = M_\gamma f_j^1(\alpha,x,0) \quad and \quad f_j^2(\alpha,M_\gamma x,0) = \rho(\gamma)f_j^2(\alpha,x,0).$$

Proof. For $\gamma \in \Gamma$,

$$\begin{aligned}
f_j^1(\alpha,M_\gamma x,0) &= \overline{\Psi}(0)F_j(\alpha,\Phi M_\gamma x)\\
&= \overline{\Psi}(0)F_j(\alpha,\rho(\gamma)\Phi x)\\
&= \overline{\Psi}(0)\rho(\gamma)F_j(\alpha,\Phi x)\\
&= M_\gamma\overline{\Psi}(0)F_j(\alpha,\Phi x)\\
&= M_\gamma f_j^1(\alpha,x,0)
\end{aligned}$$

and

$$\begin{aligned}
f_j^2(\alpha,M_\gamma x,0) &= (I-\pi)X_0F_j(\alpha,\Phi M_\gamma x)\\
&= (I-\pi)X_0F_j(\alpha,\rho(\gamma)\Phi x)\\
&= (I-\pi)X_0\rho(\gamma)F_j(\alpha,\Phi x)\\
&= \rho(\gamma)f_j^2(\alpha,x,0),
\end{aligned}$$

that is, $f_j^1(\alpha,\cdot,0)$ is Γ-equivariant and $f_j^2(\alpha,M_\gamma x,0) = \rho(\gamma)f_j^2(\alpha,x,0)$. $\qquad\square$

Define the operators $\mathbf{M}_j(p,h) = (\mathbf{M}_j^1,\mathbf{M}_j^2)$, $j \geq 2$, by

$$\begin{aligned}
&\mathbf{M}_j^1 : \mathscr{H}_j^{m+r}(\mathbb{R}^m) \to \mathscr{H}_j^{m+r}(\mathbb{R}^m),\\
&\mathbf{M}_j^2 : \mathscr{H}_j^{m+r}(Q) \subset \mathscr{H}_j^{m+r}(\mathrm{Ker}\,\pi) \to \mathscr{H}_j^{m+r}(\mathrm{Ker}\,\pi), \qquad (4.63)\\
&(\mathbf{M}_j^1 p)(\alpha,x) = [B,p(\alpha,\cdot)](x),\\
&(\mathbf{M}_j^2 h)(\alpha,x) = D_x h(\alpha,x)Bx - \mathscr{A}_Q h(\alpha,x),
\end{aligned}$$

where $[\cdot,\cdot]$ denotes the Lie bracket.

Next, we need to introduce a suitable inner product in $\mathscr{H}_j^{m+r}(\mathbb{R}^m \times \mathrm{Ker}\,\pi)$ for the convenience of decomposition. In fact, we can extend the definition of the inner product $\ll \cdot,\cdot \gg: \mathscr{H}_j^{m+r}(\mathbb{R}^m) \to \mathscr{H}_j^{m+r}(\mathbb{R}^m)$, which was given in the previous sections, to $\mathscr{H}_j^{m+r}(\mathbb{R}^m \times \mathrm{Ker}\,\pi)$ and choose P_j as the (unique) orthogonal projection from $\mathscr{H}_j^{m+r}(\mathbb{R}^m \times \mathrm{Ker}\,\pi)$ onto $\mathrm{Ran}\mathbf{M}_j$. Let $P_j = (P_j^1,P_j^2)$, where P_j^1 and P_j^2 are projections from $\mathscr{H}_j^{m+r}(\mathbb{R}^m \times \mathrm{Ker}\,\pi)$ to $\mathrm{Ran}\mathbf{M}_j^1$ and $\mathrm{Ran}\mathbf{M}_j^2$, respectively. Note that

$$\mathscr{H}_j^{m+r}(\mathbb{R}^m \times \mathrm{Ker}\,\pi) = \mathrm{Ran}\mathbf{M}_j \oplus \mathrm{Ker}\mathbf{M}_j^*, \qquad \mathrm{Ran}\mathbf{M}_j \perp \mathrm{Ker}\mathbf{M}_j^*,$$

and

$$\mathscr{H}_j^{m+r}(\mathbb{R}^m \times \operatorname{Ker}\pi) = \operatorname{Ker}\mathbf{M}_j \oplus \operatorname{Ran}\mathbf{M}_j^*, \quad \operatorname{Ker}\mathbf{M}_j \perp \operatorname{Ran}\mathbf{M}_j^*,$$

where $*$ denotes the adjoint map with respect to the inner product $\ll \cdot, \cdot \gg$. It turns out that $\mathbf{M}_j^{1*}$ is just the homological operator associated with the adjoint B^* of B. In other words, we have

$$(\mathbf{M}_j^{1*}p)(\alpha,x) = [B^*, p(\alpha,\cdot)](x) \qquad \text{for all } p \in \mathscr{H}_j^{m+r}(\mathbb{R}^m). \tag{4.64}$$

The normal forms can be obtained by computing at each step the terms of order $j \geq 2$ in the normal form from the terms of the same order in the original equation and the terms of lower orders already computed for the normal form in previous steps, through a transformation of variables

$$(x,y) = (\hat{x},\hat{y}) + \frac{1}{j!}(U_j^1(\alpha,\hat{x}), U_j^2(\alpha,\hat{x})), \tag{4.65}$$

where x, $\hat{x} \in \mathbb{R}^m$, y, $\hat{y} \in Q$, $\alpha \in \mathbb{R}^r$, $U_j^1 : \mathbb{R}^{m+r} \to \mathbb{R}^m$, and $U_j^2 : \mathbb{R}^{m+r} \to Q$ are homogeneous polynomials of degree j.

We assume that after computing the normal form up to terms of order $j-1$, the equations become

$$\dot{x} = Bx + \sum_{s=2}^{j-1} \frac{1}{s!} g_s^1(\alpha,x,y) + \frac{1}{j!}\overline{f}_j^1(\alpha,x,y) + \cdots, \tag{4.66}$$

$$\frac{dy}{dt} = \mathscr{A}_Q y + \sum_{s=2}^{j-1} \frac{1}{s!} g_s^2(\alpha,x,y) + \frac{1}{j!}\overline{f}_j^2(\alpha,x,y) + \cdots,$$

where $g_s^1 \in \mathscr{H}_i^{m+r}(\mathbb{R}^m)$, $\overline{f}_j^1 \in \mathscr{H}_j^{m+r}(\mathbb{R}^m)$, $g_s^2 \in H_i^{m+r}(\operatorname{Ker}\pi)$, $\overline{f}_j^2 \in H_j^{m+r}(\operatorname{Ker}\pi)$. With the change of variables (4.65) and dropping the hats for simplicity of notation, (4.66) becomes

$$\dot{x} = Bx + \sum_{s=2}^{j} \frac{1}{s!} g_s^1(\alpha,x,y) + \cdots, \tag{4.67}$$

$$\frac{dy}{dt} = \mathscr{A}_Q y + \sum_{s=2}^{j} \frac{1}{s!} g_s^2(\alpha,x,y) + \cdots,$$

where $g_j = \overline{f}_j - \mathbf{M}_j U_j$, $g_j = (g_j^1,g_j^2)$, $\overline{f}_j = (\overline{f}_j^1, \overline{f}_j^2)$. System (4.67) is called the normal form for (4.54) near $(u,\alpha) = (0,0)$ relative to the center space E_Λ.

If $\mathbf{M}_j$ is invertible, then we simply put $g_j = 0$ and $U_j = \mathbf{M}_j^{-1}\overline{f}_j$. In this case, all the jth-order terms are removed by the normalization procedure. In general, $\mathbf{M}_j$ need not (and will not) be invertible. Note that

$$\mathbf{M}_j U_j = \overline{f}_j - g_j = P_j \overline{f}_j + (I - P_j)\overline{f}_j - g_j.$$

We will choose $U_j = (U_j^1, U_j^2) \in V_j^{m+r}(\mathbb{R}^m) \times V_j^{m+r}(Q)$ such that

$$U_j \in \mathbf{M}_j^{-1} P_j \overline{f}_j, \tag{4.68}$$

which allows us to take away from $\overline{f}_j$ its component in $\mathrm{Ran}\mathbf{M}_j$. Therefore, the above equation can be solved by

$$g_j(\alpha, x, y) = (I - P_j)\overline{f}_j(\alpha, x, y). \tag{4.69}$$

In particular, we have

$$g_j^1(\alpha, x, 0) = \mathrm{Proj}_{\mathrm{Ker}\mathbf{M}_j^{1*}} \overline{f}_j^1(\alpha, x, 0). \tag{4.70}$$

We first mention the following important results about the equivariance of g_j^1 and U_j^1, $j \geq 2$. In fact, by a similar argument to that in the proof of Lemma 4.8, we have the following.

Lemma 4.13. *For $j \geq 2$, $g_j^1(\alpha, \cdot, 0)$ and $U_j^1(\alpha, \cdot)$ are Γ-equivariant, i.e.,*

$$M_\gamma g_j^1(\alpha, x, 0) = g_j^1(\alpha, M_\gamma x, 0) \quad and \quad M_\gamma U_j^1(\alpha, x) = U_j^1(\alpha, M_\gamma x)$$

for all $\gamma \in \Gamma$ and $x \in \mathbb{R}^m$.

Lemma 4.14. *For $j \geq 2$, $g_j^1(\alpha, \cdot, 0)$ is G_{B^*}-equivariant, that is,*

$$e^{tB^*} g_j^1(\alpha, x, 0) = g_j^1(\alpha, e^{tB^*} x, 0) \qquad for\ all\ x \in \mathbb{R}^m\ and\ t \in \mathbb{R}.$$

The proof of Lemma 4.14 is almost the same as that of Lemma 4.2.

Lemma 4.15. *For $j \geq 2$, $U_j^2(\alpha, \cdot)$ satisfies $\rho(\gamma)U_j^2(\alpha, x) = U_j^2(\alpha, M_\gamma x)$ for $\gamma \in \Gamma$ and $x \in \mathbb{R}^m$.*

Proof. We define the linear map

$$\mathscr{R}_j : V_j^{m+r}(Q) \to V_j^{m+r}(\mathrm{Ker}\,\pi)$$

by

$$\mathscr{R}_j h(\alpha, x) = \rho(\gamma^{-1})h(\alpha, M_\gamma x).$$

Since B commutes with M_γ for all $\gamma \in \Gamma$, we have

$$\begin{aligned}
\mathscr{R}_j \mathbf{M}_j^2 h(\alpha, x) &= \rho(\gamma^{-1})D_x h(\alpha, M_\gamma x)BM_\gamma x - \rho(\gamma^{-1})\mathscr{A}_Q h(\alpha, M_\gamma x) \\
&= \rho(\gamma^{-1})D_x h(\alpha, M_\gamma x)M_\gamma Bx - \mathscr{A}_Q \rho(\gamma^{-1})h(\alpha, M_\gamma x) \\
&= D_x \left(\rho(\gamma^{-1})h(\alpha, M_\gamma x) \right) Bx - \mathscr{A}_Q \rho(\gamma^{-1})h(\alpha, M_\gamma x) \\
&= \mathbf{M}_j^2 \mathscr{R}_j h(\alpha, x).
\end{aligned}$$

This means that $\mathscr{R}_j$ commutes with $\mathbf{M}_j^2$. Therefore, $\mathrm{Ker}\mathbf{M}_j^2$ and $\mathrm{Ran}\mathbf{M}_j^2$ are invariant under $\mathscr{R}_j$. Then, using similar arguments to those in the proof of Lemma 4.13, we can show inductively that $\overline{f}_j^2(\alpha, \cdot, 0)$, U_j^2, and $g_j^2(\alpha, \cdot, 0)$ are fixed points of $\mathscr{R}_j$, $j \geq 2$. This completes the proof. $\qquad \square$

4.4.3 Nonresonance Conditions

If we want to obtain normal forms in a finite-dimensional local center submanifold tangent to the center subspace E_Λ of the linearized equation at $(u,\alpha) = (0,0)$, then in the normal form, all the terms $g_j^2(\alpha,x,y)$ must vanish at $y = 0$, i.e., $g_j^2(\alpha,x,0) = 0$ for all $j \geq 2$. Therefore, we require that $\mathrm{Ran}\mathbf{M}_j^2$ be the whole space $\mathscr{H}_j^{m+r}(\mathrm{Ker}\,\pi)$, $j \geq 2$. This situation can be characterized in spectral terms by nonresonance conditions appropriate for guaranteeing that $y = 0$ in the normal form (4.67) is a locally invariant manifold. It is well known that

$$\sigma(\mathbf{M}_j^1) = \{(\tilde{q},\tilde{\lambda}) - \lambda_s : s = 1,\dots,m+r, \tilde{q} \in \mathbb{N}_0^{m+r}, |\tilde{q}| = j\},$$

where $\lambda_1, \lambda_2, \dots, \lambda_m$ are the elements of Λ, each of them appearing as many times as its multiplicity as a root of the associated characteristic equation, $\lambda_j = 0$ for $j = m+1, m+2, \dots, m+r$, $\overline{\lambda} = (\lambda_1,\dots,\lambda_m)$, $\tilde{\lambda} = (\lambda_1,\dots,\lambda_{m+r})$, $(q,\overline{\lambda}) = \sum_{j=1}^m q_j\lambda_j$, $(\tilde{q},\tilde{\lambda}) = \sum_{j=1}^{m+r} q_j\lambda_j$. Obviously, $(\tilde{q},\tilde{\lambda}) = (q,\overline{\lambda})$. We expect that the nonresonance conditions appropriate for guaranteeing that $y = 0$ in the normal form (4.67) is a locally invariant manifold can be expressed by relationships between the spectral values of $\mathscr{A}_Q$ and B. So it is essential to pay due attention to the topology of the space on which $\mathscr{A}_Q$ acts and to its domain as an operator in that space. In what follows, we first establish relationships between the spectra $\sigma(\mathscr{A}_0)$ and $\mathscr{A}_Q$. For the sake of convenience, for an operator A, let $\sigma_p(A)$ denote the point spectrum of A. Clearly, $\sigma(\mathscr{A}_Q) = \sigma_p(\mathscr{A}_Q)$.

Lemma 4.16. *The spectra of the operator* $\mathbf{M}_j^2$, $j \geq 2$, *are*

$$\sigma(\mathbf{M}_j^2) = \{(\tilde{q},\tilde{\lambda}) - \mu : \mu \in \sigma(\mathscr{A}_0) \setminus \Lambda, q \in \mathbb{N}_0^{m+r}, |q| = j\}.$$

We refer to Faria and Magalhães [91, 92] for the proof of Lemma 4.16. Therefore, the appropriate nonresonance conditions are the following.

Definition 4.2. System (4.54) satisfies the *nonresonance conditions relative to* Λ if

$$\sum_{k=1}^m q_k\lambda_k \notin \sigma^c \setminus \Lambda \quad \text{for } (q_1,\dots,q_m) \in \mathbb{N}_0^m,$$

where $\lambda_1, \lambda_2, \dots, \lambda_m$ are all eigenvalues in Λ, each repeated as many times as its multiplicity.

We then have the following results.

Theorem 4.6. *For the Γ-equivariant system (4.54), suppose that Λ is a nonempty subset of σ^c satisfying (H1) and (H2). If system (4.54) further satisfies the nonresonance conditions relative to Λ, then there exists a formal change of variables $x = \hat{x} + p(\alpha,\hat{x})$, $y = \hat{y} + h(\alpha,\hat{x})$, where p and h satisfy $p(\alpha,M_\gamma x) = M_\gamma p(\alpha,x)$ and $h(\alpha,M_\gamma x) = \rho(\gamma)h(\alpha,x)$ for all $\gamma \in \Gamma$ and $x \in \mathbb{R}^m$, such that*

(i) *system (4.54) is transformed into a normal form relative to Λ in the form of (4.67), where $g_j^1(\alpha,\cdot,0)$ is $\Gamma \times G_{B^*}$-equivariant and g_j^2 satisfies $g_j^2(\alpha,x,0) = 0$ for all $\alpha \in \mathbb{R}^r$ and $x \in \mathbb{R}^m$, $j \geq 2$;*

(ii) *there exists a local center submanifold for (4.54) at zero satisfying $y = 0$, and the flow on it is given by the m-dimensional $\Gamma \times G_{B_S^*}$-equivariant ODEs*

$$\dot{x}(t) = Bx(t) + \tfrac{1}{2}g_2^1(\alpha,x,0) + \tfrac{1}{3!}g_3^1(\alpha,x,0) + \text{h.o.t.}, \qquad (4.71)$$

which is in normal form (in the usual sense of ODEs), where m is the sum of the multiplicities of the elements in Λ as eigenvalues of $\mathscr{A}_0$.

Remark 4.2. Theorem 4.6 means that if the finite-dimensional system (4.71) has a (periodic) solution $x(t)$ with symmetry $\Sigma \leq \Gamma \times \mathbb{S}^1$, then system (4.60) has a (periodic) solution $(x + p(x,\alpha), h(x,\alpha))$, and hence system (4.54) has a (periodic) solution $u(t)$ with symmetry Σ.

Remark 4.3. In Theorem 4.6, we obtain an alternative characterization in terms of additional equivariance conditions similar to those in Elphick et al. [87], which has advantages as described below. First, we can choose coordinates in E_Λ so that B is in Jordan normal form and B commutes with B_S^T. Then the normal form (4.71) (including the linear terms) is $\Gamma \times G_{B_S^*}$-equivariant. However, the $G_{B_N^*}$-equivariance applies only to the nonlinear terms of (4.71). This $\Gamma \times G_{B_S^*}$-equivariance of the normal form is important understanding the local dynamics, such as generic local branching patterns of equilibria and periodic solutions. Finally, the normal form procedure does not converge, and terms in the tail (beyond all polynomial orders) may affect the qualitative dynamics (see, for example, Guckenheimer and Holmes [125, Sects. 7.4 and 7.5]).

Remark 4.4. If $\Lambda = \sigma^c$, then system (4.54) obviously satisfies the nonresonance conditions relative to Λ, and hence Theorem 4.6 applies to the whole center manifold of system (4.54).

Remark 4.5. Assumption (H2) is a necessary condition in ensuring the nonresonance conditions in Definition 4.2. In fact, if $\pm i\omega \in \Lambda$, $0 \notin \Lambda$ but $0 \in \sigma(\mathscr{A}_0)$, then obviously $i\omega + (-i\omega) = 0 \in \Lambda_0 \setminus \Lambda$, which implies that system (4.54) is resonant.

Chapter 5
Lyapunov–Schmidt Reduction

The main objective of this chapter is to introduce the Lyapunov–Schmidt reduction method and show how this reduction can be performed in a way compatible with symmetries. The Lyapunov–Schmidt reduction results in the so-called bifurcation equations, a finite set of equations equivalent to the original problem. This finite set of equations may inherit the symmetry properties of the original system if the reduction is done properly.

5.1 The Lyapunov–Schmidt Method

Let X, Y, and Λ be real Banach spaces. Let $F\colon X \times \Lambda \to Y$ be a C^k map $(k \geq 1)$ such that $F(0,0) = 0$ (possibly after a change of origin in $X \times \Lambda$). We want to study the solution set of the equation

$$F(x,\alpha) = 0 \tag{5.1}$$

in a neighborhood of $(0,0)$ in $X \times \Lambda$.

Define $\mathscr{L} = D_x F(0,0)$. If the linear operator $\mathscr{L}$ is invertible and $\mathscr{L}^{-1}$ is bounded (i.e., continuous) from Y to X, then the implicit function theorem applies. Therefore in this case, in a neighborhood of the point $(0,0)$, there exists a unique solution branch $x = \phi(\alpha)$ for the equation, and ϕ is a C^k function of α. Note that if $\mathscr{L}$ is bounded and invertible, then $\mathscr{L}^{-1}$ is bounded, thanks to the closed graph theorem; see Kato [186]. The challenging case is that in which $\mathscr{L}$ is not invertible. Since we are considering maps in Banach spaces, not just finite-dimensional vector spaces, we need to be more precise about the way $\mathscr{L}$ is noninvertible.

Denote by $\operatorname{Ker}\mathscr{L}$ and $\operatorname{Ran}\mathscr{L}$ the kernel and range of $\mathscr{L}$, respectively. Assume that $\operatorname{Ker}\mathscr{L}$ has a topological complement X_0 in X, while $\operatorname{Ran}\mathscr{L}$ is closed and has a topological complement Y_0 in Y. Thus, we have the following decompositions:

$$X = \operatorname{Ker}\mathscr{L} \oplus X_0, \quad Y = \operatorname{Ran}\mathscr{L} \oplus Y_0. \tag{5.2}$$

S. Guo and J. Wu, *Bifurcation Theory of Functional Differential Equations*,
Applied Mathematical Sciences 184, DOI 10.1007/978-1-4614-6992-6_5,
© Springer Science+Business Media New York 2013

In particular, if L is a Fredholm operator,[1] then (5.2) holds obviously.

It follows from (5.2) that there exist two continuous projections $P \in L(X, \text{Ker}\mathscr{L})$ and $Q \in L(Y, Y_0)$ such that

$$\text{Ker}\mathscr{L} = \text{Ran}\,P, \quad \text{Ran}\mathscr{L} = \text{Ker}\,Q. \tag{5.3}$$

We can write $x \in \Omega$ in the form $x = u + v$, where $u = Px \in \text{Ker}\mathscr{L}$ and $v = (I - P)x \in X_0$. Then we can rewrite (5.1) as

$$(I - Q)F(u + v, \alpha) = 0, \quad QF(u + v, \alpha) = 0. \tag{5.4}$$

Thus, the first equation of (5.4) can be rewritten as

$$G(u, v, \alpha) \equiv (I - Q)F(u + v, \alpha) = 0.$$

Notice that $G(0,0,0) = (I - Q)F(0,0) = 0$ and $D_v G(0,0,0) = (I - Q)\mathscr{L} = \mathscr{L}$. When $\mathscr{L}$ is restricted in $\text{Ker}\,P$, it is an isomorphism between $\text{Ker}\,P$ and $\text{Ran}\mathscr{L}$, and then so is $D_v G(0,0,0)$. Applying the implicit function theorem, we obtain an open neighborhood Ω of the origin in $\text{Ker}\mathscr{L}$, an open neighborhood Ξ of the origin in Λ, and a C^k map $W : \Omega \times \Xi \to \text{Ker}\,P$ such that $W(0,0) = 0$ and

$$(I - Q)F(u + W(u, \alpha), \alpha) \equiv 0 \tag{5.5}$$

for all $(u, \alpha) \in \Omega \times \Xi$. Substituting $w = W(u, \alpha)$ into the second equation of (5.4), we have

$$\mathscr{B}(u, \alpha) \stackrel{\text{def}}{=} QF(u + W(u, \alpha), \alpha) = 0, \tag{5.6}$$

where $\mathscr{B}$ is a C^k map from $\Omega \times \Xi$ to Y_0. Moreover, it follows that $\mathscr{B}(0,0) = 0$ and $\mathscr{B}_u(0,0) = 0$. The following theorem summarizes the essential result of Lyapunov–Schmidt reduction.

Theorem 5.1. *There exists a neighborhood U of $(0,0) \in \text{Ker}\mathscr{L} \times \Omega$ such that each solution to $\mathscr{B}(u, \alpha) = 0$ in U corresponds one to one some solution to (5.1).*

Equation (5.6) is called the *bifurcation equation* corresponding to (5.1), and $\mathscr{B}$ the *bifurcation map*. In particular, it would be interesting to know for what values of α solutions disappear or are created. These particular values of α are called bifurcation values. Now there exists an extensive mathematical machinery called singularity theory (see Golubitsky et al. [115–118] and Sattinger [260, 261]) that deals with such questions. Singularity theory is concerned with the local properties of smooth functions near a zero of the function. It provides a classification of the various cases based on codimension. The reason this is possible is that the codimension-k submanifolds in the space of all smooth functions having zeros can be described algebraically by imposing conditions on derivatives of the functions. This gives us a way of classifying the various possible bifurcations and of computing the proper unfoldings.

---

[1] L is a Fredholm operator if (i) the kernel $\text{Ker}\mathscr{L}$ is finite-dimensional, (ii) the range $\text{Ran}\mathscr{L}$ is closed, and (iii) $\text{Ran}\mathscr{L}$ has finite codimension in Y. The index of a Fredholm operator $\mathscr{L}$ is defined to be the integer $\text{Ind}\mathscr{L} = \dim \text{Ker}\mathscr{L} - \text{codim}\text{Ran}\mathscr{L}$.

5.2 Derivatives of the Bifurcation Equation

In applications, to solve and study (5.6), it is important to choose a suitable subspace Y_0 and to choose suitable coordinates in $\text{Ker}\,\mathscr{L}$ and Y_0. Throughout this section, we always assume the following:

(i) $\alpha = (\alpha_1, \alpha_2, \ldots, \alpha_m) \in \Lambda = \mathbb{R}^m$.
(ii) There is an inner product $< \cdot, \cdot >$ in Y, and $Y_0 = (\text{Ran}\,\mathscr{L})^{\perp}$, that is,

$$Y_0 = \{y \in Y : < y, z > = 0 \text{ for all } z \in \text{Ran}\,\mathscr{L}\}.$$

(iii) $\dim(\text{Ran}\,\mathscr{L})^{\perp} = \dim \text{Ker}\,\mathscr{L} = n$.

If $\mathscr{L}$ is a Fredholm operator with index 0, then assumption (iii) holds. Suppose that $\{v_1, v_2, \ldots, v_n\}$ and $\{v_1^*, v_2^*, \ldots, v_n^*\}$ are bases for $\text{Ker}\,\mathscr{L}$ and $(\text{Ran}\,\mathscr{L})^{\perp}$, respectively. For $u \in \text{Ker}\,\mathscr{L}$, $u = \sum_{i=1}^{n} x_i v_i$ with scalars x_i, $i = 1, 2, \ldots, n$. Substituting $u = \sum_{i=1}^{n} x_i v_i$ into (5.6) and calculating the inner product with v_j^*, we have

$$
\begin{aligned}
0 &= \left\langle v_j^*, \mathscr{B}(\textstyle\sum_{i=1}^{n} x_i v_i, \alpha) \right\rangle \\
&= \left\langle v_j^*, QF(\textstyle\sum_{i=1}^{n} x_i v_i + W(\sum_{i=1}^{n} x_i v_i, \alpha), \alpha) \right\rangle \\
&= \left\langle v_j^*, F(\textstyle\sum_{i=1}^{n} x_i v_i + W(\sum_{i=1}^{n} x_i v_i, \alpha), \alpha) \right\rangle \\
&\quad - \left\langle v_j^*, (I - Q)F(\textstyle\sum_{i=1}^{n} x_i v_i + W(\sum_{i=1}^{n} x_i v_i, \alpha), \alpha) \right\rangle \\
&= \left\langle v_j^*, F(\textstyle\sum_{i=1}^{n} x_i v_i + W(\sum_{i=1}^{n} x_i v_i, \alpha), \alpha) \right\rangle.
\end{aligned}
$$

Hence, the bifurcation equation can be rewritten as the following system of n equations:

$$g(x, \alpha) = 0, \tag{5.7}$$

where $x = (x_1, \ldots, x_n)^T \in \mathbb{R}^n$, $g(x, \alpha) = (g_1(x, \alpha), \ldots, g_n(x, \alpha))^T$, and for $j = 1, 2, \ldots, n$,

$$g_j(x, \alpha) = \left\langle v_j^*, F(\sum_{i=1}^{n} x_i v_i + W(\sum_{i=1}^{n} x_i v_i, \alpha), \alpha) \right\rangle. \tag{5.8}$$

Obviously, (5.7) is equivalent to (5.6). Hence, we also refer to (5.7) as the bifurcation equation of the system (5.1).

To find zeros of g in a neighborhood of the origin, it is not necessary to figure out a concrete expression for g. In fact, it is enough to know about some low-order terms of g. In what follows, we aim to relate the derivatives of the reduced functions $g_j(x, \alpha)$, $j = 1, 2, \ldots, n$, to the derivatives of the original equation (5.1). If we know the derivatives of the bifurcation function $\mathscr{B}$, then we can find the derivatives of g_j by substitution into (5.8). Calculation of derivatives of $\mathscr{B}$ is a straightforward application of the chain rule. However, the resulting formulas contain derivatives of W, and these must be determined by implicit differentiation of (5.5).

We will use the notation for the k-fold differential of F as follows. For $v_1, v_2, \ldots,$ $v_k \in X$, we define

$$\mathscr{F}^k_{(u,\alpha)}(v_1, v_2, \ldots, v_k) = \frac{\partial^k}{\partial t_1 \partial t_2 \cdots \partial t_k} F\left(u + \sum_{j=1}^k t_j v_j, \alpha\right)\Big|_{t_1=t_2=\cdots=t_k=0}.$$

Obviously, $\mathscr{F}^k_{(u,\alpha)}(v_1, v_2, \ldots, v_k)$ is a symmetric k-linear function of $(v_1, v_2, \ldots, v_k)$. Define $F_{\alpha_j} = \partial F(0,0)/\partial \alpha_j$.

Using similar arguments to those in Golubitsky and Schaeffer [117], we can obtain the following results:

$$\frac{\partial}{\partial x_j} g_i(0,0) = 0,$$

$$\frac{\partial^2}{\partial x_j \partial x_k} g_i(0,0) = \left\langle v_i^*, \mathscr{F}^2_{(0,0)}(v_j, v_k) \right\rangle,$$

$$\frac{\partial^3}{\partial x_j \partial x_k \partial x_l} g_i(0,0) = \left\langle v_i^*, V_{jkl} \right\rangle,$$

$$\frac{\partial}{\partial \alpha_j} g_i(0,0) = \left\langle v_i^*, F_{\alpha_j} \right\rangle,$$

$$\frac{\partial^2}{\partial x_j \partial \alpha_k} g_i(0,0) = \left\langle v_i^*, F_{\alpha_k} \cdot v_j - \mathscr{F}^2_{(0,0)}(v_j, \mathscr{L}^{-1}(I-Q)F_{\alpha_k}) \right\rangle,$$

where

$$V_{jkl} = \mathscr{F}^3_{(0,0)}(v_j, v_k, v_l) + \mathscr{F}^2_{(0,0)}(v_j, W_{lk}) + \mathscr{F}^2_{(0,0)}(v_k, W_{lj}) + \mathscr{F}^2_{(0,0)}(v_l, W_{kj}), \quad (5.9)$$

and $W_{sk} = -\mathscr{L}^{-1}(I-Q)\mathscr{F}^2(v_s, v_k)$, $\mathscr{L}^{-1}: \operatorname{Ran}\mathscr{L} \to X_0$ is the inverse of $\mathscr{L}|_{X_0}$.

5.3 Equivariant Equations

Let Γ be a compact topological group, let $\rho: \Gamma \to L(X)$ and $\tilde{\rho}: \Gamma \to L(Y)$ be the representations of Γ over X and Y, respectively. We say that $F: X \times \Lambda \to Y$ is equivariant with respect to some triple $(\Gamma, \rho, \tilde{\rho})$ if

$$F(\rho(\gamma)x, \alpha) = \tilde{\rho}(\gamma)F(x, \alpha) \qquad (5.10)$$

for all $\gamma \in \Gamma$ and $\alpha \in \Lambda$. Although in many applications, X will be a subspace of Y, we have not made such an assumption here. This forces us to consider two different representations ρ and $\tilde{\rho}$ of Γ. Here, we reconsider the Lyapunov–Schmidt reduction for the equation

$$F(x, \alpha) = 0 \qquad (5.11)$$

as given in Sect. 5.1, under the supplementary condition that F is equivariant with respect to some triple $(\Gamma, \rho, \tilde{\rho})$ in the sense of (5.10). It follows that the linear operator $\mathscr{L}$ satisfies $\mathscr{L}\rho(\gamma) = \tilde{\rho}(\gamma)\mathscr{L}$ for all $\gamma \in \Gamma$. Moreover, for each $\gamma \in \Gamma$, $\operatorname{Ker}\mathscr{L}$ and $\operatorname{Ran}\mathscr{L}$ are invariant under $\rho(\gamma)$ and $\tilde{\rho}(\gamma)$, respectively. In particular, if we define $X^\Gamma = \{x \in X, \rho(\gamma)x = x \text{ for all } \gamma \in \Gamma\}$ and $Y^\Gamma = \{y \in Y, \tilde{\rho}(\gamma)y = y \text{ for all } \gamma \in \Gamma\}$, then $F(x, \alpha) \in Y^\Gamma$ for $x \in X^\Gamma$ and $\alpha \in \Lambda$.

We further assume that the projections $P \in L(X, \text{Ker}\mathscr{L})$ and $Q \in L(Y, Y_0)$ can be chosen to satisfy (5.3) and also

$$\rho(\gamma)P = P\rho(\gamma), \quad \tilde{\rho}(\gamma)Q = Q\tilde{\rho}(\gamma) \tag{5.12}$$

for all $\gamma \in \Gamma$. Therefore, in what follows, we always assume that (5.12) holds. In fact, the case of Hilbert spaces provides a first, simple situation in which projections can be chosen to be equivariant. Namely, we have the following result.

Lemma 5.1 (Chossat and Lauterbach [62]). *Suppose the spaces* X *and* Y *are Hilbert spaces, and the group* Γ *is compact. Then an inner product can be found in* X *and in* Y *such that* Γ *acts isometrically in each space. In this case, the orthogonal projections* $P \in L(X, \text{Ker}\mathscr{L})$ *and* $Q \in L(Y, Y_0)$ *are* Γ*-equivariant.*

Due to Vanderbauwhede [284], the continuously differentiable map $W: \Omega \times \Xi \to \text{Ker}P$ given by (5.5) is also Γ-equivariant, that is,

$$W(\rho(\gamma)u, \alpha) = \rho(\gamma)W(u, \alpha) \tag{5.13}$$

for $\gamma \in \Gamma$ and $(u, \alpha) \in \Omega \times \Xi$.

Finally, it is easy to see that the bifurcation map $\mathscr{B}$ given in (5.6) is Γ-equivariant. Namely,

$$\mathscr{B}(\rho(\gamma)u, \alpha) = \tilde{\rho}(\gamma)\mathscr{B}(u, \alpha) \tag{5.14}$$

for all $(u, \alpha) \in \Omega \times \Xi$. Therefore, we have the following result.

Theorem 5.2. *There exists a* Γ*-invariant neighborhood* U *of* $(0,0) \in \text{Ker}\mathscr{L} \times \Lambda$ *such that each zero of the* Γ*-equivariant map* $\mathscr{B}(u, \alpha)$ *in* U *corresponds one-to-one to some solution to (5.11).*

5.4 The Steady-State Equivariant Branching Lemma

Suppose $F: X \times \Lambda \to Y$ is a C^k map $(k > 1)$ in Banach spaces $(X \subset Y)$. One can define the spectrum for the linear operator $\mathscr{L} = D_x F(0,0)$ in the Banach space X: this is the set of complex numbers λ such that $\mathscr{L} - \lambda I$ is not invertible. An eigenvalue is an element of the spectrum such that $\text{Ker}(\mathscr{L} - \lambda I) \neq \{0\}$. Finite multiplicity of an eigenvalue λ means that $(\mathscr{L} - \lambda I)^k = 0$ for some integer k. That the eigenvalues are isolated means that there exists a closed curve $\mathscr{C}$ that separates λ from the rest of the spectrum.

Throughout this section, we always assume that 0 is an isolated eigenvalue of $\mathscr{L}$ with finite multiplicity. Due to Chossat and Lauterbach [62], we first have the following.

Lemma 5.2. *If 0 is an isolated eigenvalue of* $\mathscr{L}$ *with finite multiplicity, then* $\mathscr{L}$ *is a Fredholm operator with index 0. If* $\mathscr{L}$ *is* Γ*-equivariant, then the projections P and Q can be chosen to be* Γ*-equivariant.*

Let us now apply the equivariant Lyapunov–Schmidt reduction to (5.1). Let n be the dimension of $\text{Ker}\mathscr{L}$. As stated in Sect. 5.2, the bifurcation equation has the general form

$$\mathscr{B}(u, \alpha) = 0,$$

where $\mathscr{B}$ is a C^k map from $\text{Ker}\mathscr{L} \times \Lambda$ to $\text{Ker}\mathscr{L}$. Since $\text{Ker}\mathscr{L}$ is a real space of dimension n, we may as well regard $\mathscr{B}$ as a map $\mathbb{R}^n \times \Lambda \to \mathbb{R}^n$ that is equivariant for the induced action of Γ on $\text{Ker}\mathscr{L}$.

Suppose now that the action of Γ on $\mathbb{R}^n$ possesses an isotropy subgroup Σ with a one-dimensional fixed-point space $\text{Fix}(\Sigma)$. If we look for solutions in $\text{Fix}(\Sigma)$, we consider the restriction mapping $g : \text{Fix}(\Sigma) \times \mathbb{R}^m \to \mathbb{R}^n$ of $\mathscr{B}: \mathbb{R}^n \times \mathbb{R}^m \to \mathbb{R}^n$. In view of the Γ-equivariance of F and W, it is easy to see that $g: \text{Fix}(\Sigma) \times \mathbb{R}^m \to \text{Fix}(\Sigma)$ is also Γ-equivariant and $\text{Ran}\, g \subset \text{Fix}(\Sigma)$. Namely, g maps $\text{Fix}(\Sigma) \times \mathbb{R}^m$ to $\text{Fix}(\Sigma)$. So g is a scalar function. Now we can state the following equivariant branching lemma.

Theorem 5.3. *Suppose* $F: \text{X} \times \Lambda \to \text{Y}$ *is a* C^k *map* $(k > 1)$ *in Banach spaces* $(\text{X} \subset \text{Y})$. *Suppose that the compact group* Γ *acts linearly in* Y *(and in* X *by restriction) and that* F *is* Γ-*equivariant. Suppose, finally, that* F *satisfies the following bifurcation conditions: (i)* $F(0,0) = 0$, *(ii)* $\mathscr{L} = D_x F(0,0)$ *has 0 as an isolated eigenvalue with finite multiplicity. Then for each isotropy subgroup* Σ *of* Γ *such that* $\dim\text{Fix}(\Sigma) = 1$ *in* $\text{Ker}\mathscr{L}$, *either one of the following situations occurs (where* $g(x, \alpha) = 0$ *denotes the bifurcation equation in* $\text{Fix}(\Sigma)$):

(i) $\Sigma = \Gamma$. *If* $D_{\alpha}g(0,0) \neq 0$, *there exists one branch of the solution* $x(\alpha)$. *If in addition,* $D_{xx}g(0,0) \neq 0$, *then* $x^2 = O(|\alpha|)$ *(saddle-node bifurcation).*

(ii) $\Sigma \leq \Gamma$, *and the normalizer* $N(\Sigma)$ *of* Σ *in* Γ, *i.e., the group* $\{\gamma \in \Gamma: \gamma\Sigma\gamma^{-1} = \Sigma\}$, *acts trivially in* $\text{Fix}(\Sigma)$. *Then* $g(x, \alpha) = xh(x, \alpha)$, *and if* $D_{x\alpha}g(0,0) \neq 0$, *there exists a branch of solutions* $x(\alpha)$. *If in addition,* $D_{xx}(0,0) \neq 0$, *then* $x = O(|\alpha|)$ *(transcritical bifurcation).*

(iii) $\Sigma < \Gamma$, *and the normalizer* $N(\Sigma)$ *of* Σ *in* Γ *acts as* $-\text{I}$ *in* $\text{Fix}(\Sigma)$. *Then* $g(x, \alpha) = xh(x^2, \alpha)$, *and if* $D_{x\alpha}g(0,0) \neq 0$, *there exist two branches of solutions* $\pm x(\alpha)$ *satisfying* $g(x, \alpha) = 0$. *If in addition,* $D_{xxx}(0,0) \neq 0$, *then* $x^2 = O(|\alpha|)$ *(pitchfork bifurcation).*

The proof is a direct application of the implicit function theorem to the bifurcation equation in $\text{Fix}(\Sigma)$, and can be found in the book by Chossat and Lauterbach [62]. Theorem 5.3, known as the *equivariant branching lemma* for steady-state bifurcation, was first stated by Cicogna [75] and Vanderbauwhede [283]. In the case of variational problems, a nice geometric characterization for the existence of extrema with a certain isotropy was stated by Michel [223]. This condition is closely related to the equivariant branching lemma.

Remark 5.1. If $\dim\text{Fix}(\Sigma) = 1$, then Σ is a maximal isotropy subgroup, meaning that there is no proper isotropy subgroup containing Σ. However the converse is not true: there exist maximal isotropy subgroups for which $\dim\text{Fix}(\Sigma) > 1$. We may encounter this situation in examples involving spherical symmetry; see Chossat and Lauterbach [62].

Remark 5.2. When $\Sigma \leq \Gamma$, the bifurcating solutions in dim $\text{Fix}(\Sigma)$ have lower symmetry than the basic solution $(x = 0)$. This effect is called *spontaneous symmetry breaking*.

5.5 Generalized Hopf Bifurcation of RFDE

We consider the following parameterized RFDE:

$$\dot{u}(t) = L(\alpha)u_t + f(\alpha, u_t), \tag{5.15}$$

where the linear operator $L(\alpha) : C_{n,\tau} \to \mathbb{R}^n$ is continuous with respect to $\alpha \in \mathbb{R}$, $f \in C^l(\mathbb{R} \times C_{n,\tau}, \mathbb{R}^n)$ for a large enough integer l such that $f(\alpha, 0) = 0$, and $D_\varphi f(\alpha, 0) = 0$ for all $\alpha \in \mathbb{R}$. As usual, there exists an $n \times n$ matrix-valued function $\eta(\alpha, \cdot) : [-\tau, 0] \to \mathbb{R}^{n^2}$ whose elements are of bounded variation such that

$$L(\alpha)\varphi = \int_{-\tau}^0 d\eta(\alpha, \theta)\varphi(\theta), \qquad \varphi \in C_{n,\tau}.$$

Denote by $\mathscr{A}_\alpha$ the infinitesimal generator associated with the linear system $\dot{u} = L(\alpha)u_t$. In this section, we introduce the work [141] to investigate the nonsemisimple resonant case. Namely, throughout this section, we always assume that

(NS) $\mathscr{A}_0$ has a pair of purely imaginary eigenvalues $\pm i\omega$ and there exists some $k \geq 1$ such that $\dim_{\mathbb{C}} \text{Ker}((\mathscr{A}_0 - i\omega\text{Id})^j) = \min(j, k)$ for all $j \in \mathbb{N}$. Moreover, all other eigenvalues of $\mathscr{A}_0$ are not integer multiples of $i\omega$.

Assumption (NS) implies that eigenvalues $\pm i\omega$ are of geometric multiplicity one and algebraic multiplicity $k \in \mathbb{N}$. In particular, if $k = 1$, then eigenvalues $\pm i\omega$ are simple, and then the classical Hopf bifurcation theory applies. Here, our main concern is the case $k > 1$. Thus, the generalized eigenspace is $\text{Ker}((\mathscr{A}_0 - i\omega\text{Id})^k)$, which is k-dimensional. Therefore, we have the following direct sum decomposition:

$$C_{n,\tau} = \text{Ker}((\mathscr{A}_0 - i\omega\text{Id})^k) \oplus \text{Ran}((\mathscr{A}_0 - i\omega\text{Id})^k).$$

Moreover, $\text{Ker}((\mathscr{A}_0 - i\omega\text{Id})^k)$ is k-dimensional and satisfies $\mathscr{A}_0\text{Ker}((\mathscr{A}_0 - i\omega\text{Id})^k) \subseteq \text{Ker}((\mathscr{A}_0 - i\omega\text{Id})^k)$. Let $\{\varphi_1, \ldots, \varphi_k\}$ be a basis for $\text{Ker}((\mathscr{A}_0 - i\omega\text{Id})^k)$ such that

$$(\mathscr{A}_0 - i\omega\text{Id})\varphi_1 = 0, \quad (\mathscr{A}_0 - i\omega\text{Id})\varphi_j = \varphi_{j-1} \tag{5.16}$$

for all $j = 2, 3, \ldots, k$. In fact, we have the following result.

Lemma 5.3. $\varphi_j(t) = \sum_{s=0}^{j-1} \frac{1}{s!} t^s u_{j-s} e^{i\omega t}$, $j = 1, 2, \ldots, k$, where $u_j \in \mathbb{C}^n$ $(j = 1, 2, \ldots, k)$ satisfy

$$\sum_{s=0}^{j-1} \frac{1}{s!} \Delta_s(0, i\omega) u_{j-s} = 0, \tag{5.17}$$

where $\Delta_s(\alpha, \lambda)$ denotes the sth partial derivative of $\Delta(\alpha, \lambda)$ with respect to λ, and $\Delta_0(\alpha, \lambda) = \Delta(\alpha, \lambda)$.

Proof. In view of $(\mathscr{A}_0 - i\omega\mathrm{Id})\varphi_1 = 0$, we can choose $\varphi_1(t) = e^{i\omega t}u_1$, where $u_1 \in \mathbb{C}^n$ satisfies $\Delta(0,i\omega)u_1 = 0$. Assume that the statement of Lemma 5.3 holds for $j = 2,3,\ldots,m$, where the integer m is less than or equal to k. In view of $(\mathscr{A}_0 - i\omega\mathrm{Id})\varphi_{m+1} = \varphi_m$, φ_{m+1} satisfies the differential equation

$$\dot{\varphi}(t) - i\omega\varphi(t) = \varphi_m(t) \tag{5.18}$$

with the boundary condition

$$L(0)\varphi - i\omega\varphi(0) = \varphi_m(0). \tag{5.19}$$

It follows from (5.18) that $\varphi(t) = \varphi_{m+1}(t)$, where $u_{m+1} \in \mathbb{C}^n$. Substituting it into (5.19) yields

$$
\begin{aligned}
0 &= \sum_{s=0}^{m}\tfrac{1}{s!}\left[\int_{-\tau}^{0}(t+\theta)^s d\eta(0,\theta)u_{m+1-s}e^{i\omega\theta} - i\omega t^s u_{m+1-s}\right] - \sum_{s=0}^{m-1}\tfrac{1}{s!}t^s u_{m-s}\\
&= -\Delta(0,i\omega)u_{m+1} + \sum_{s=0}^{m-1}\left[\int_{-\tau}^{0}\tfrac{(t+\theta)^{s+1}}{(s+1)!}d\eta(0,\theta)u_{m-s}e^{i\omega\theta} - i\omega\tfrac{t^{s+1}}{(s+1)!}u_{m-s} - \tfrac{t^s}{s!}u_{m-s}\right]\\
&= -\Delta(0,i\omega)u_{m+1} - \sum_{s=0}^{m-1}\sum_{l=0}^{s+1}\tfrac{t^l}{(s+1-l)!l!}\Delta_{s+1-l}(0,i\omega)u_{m-s}\\
&= -\sum_{s=0}^{m}\tfrac{1}{s!}\Delta_s(0,i\omega)u_{m+1-s} - \sum_{s=0}^{m-1}\sum_{l=0}^{s}\tfrac{t^{l+1}}{(s-l)!l!}\Delta_{s-l}(0,i\omega)u_{m-s}\\
&= -\sum_{s=0}^{m}\tfrac{1}{s!}\Delta_s(0,i\omega)u_{m+1-s} - \sum_{l=0}^{m-1}\left[\tfrac{t^{l+1}}{l!}\sum_{s=l}^{m-1}\tfrac{1}{(s-l)!}\Delta_{s-l}(0,i\omega)u_{m-s}\right]\\
&= -\sum_{s=0}^{m}\tfrac{1}{s!}\Delta_s(0,i\omega)u_{m+1-s} - \sum_{l=0}^{m-1}\left[\tfrac{t^{l+1}}{l!}\sum_{s=0}^{m-l-1}\tfrac{1}{s!}\Delta_s(0,i\omega)u_{m-l-s}\right]\\
&= -\sum_{s=0}^{m}\tfrac{1}{s!}\Delta_s(0,i\omega)u_{m+1-s}.
\end{aligned}
$$

Therefore, this lemma has been proved by induction. $\square$

In order to coincide with the inner product introduced later, we here define the following bilinear form (which is a little bit different from what we defined previously):

$$(\psi,\varphi) = \overline{\psi}^T(0)\varphi(0) - \int_{-\tau}^{0}\int_{0}^{\theta}\overline{\psi}^T(\xi-\theta)d\eta(0,\theta)\varphi(\xi)d\xi$$

for $\psi \in C_{n,\tau}^{*} \overset{\text{def}}{=} C([0,\tau],\mathbb{C}^n)$ and $\varphi \in C_{n,\tau}$. Here and in the sequel, for the sake of convenience, we shall also allow functions with range in $\mathbb{C}^n$. The adjoint operator $\mathscr{A}_0^{*}$ of $\mathscr{A}_0$ is defined by

$$(\mathscr{A}^{*}\psi)(\xi) = \begin{cases} -d\psi(\xi)/d\xi, & \text{if } \xi \in (0,\tau],\\ \int_{-\tau}^{0}d\eta^T(0,\theta)\psi(-\theta), & \text{if } \xi = 0. \end{cases}$$

Similarly, $\mathrm{Ker}((\mathscr{A}_0^{*} + i\omega\mathrm{Id})^k)$ has a basis $\{\psi_1,\ldots,\psi_k\}$ such that

$$(\mathscr{A}_0^{*} + i\omega\mathrm{Id})\psi_k = 0, \quad (\mathscr{A}_0^{*} + i\omega\mathrm{Id})\psi_j = \psi_{j+1} \tag{5.20}$$

for all $j = 1,2,\ldots,k-1$. Similarly to the proof of Lemma 5.3, we have the following.

Lemma 5.4. $\psi_j(t) = \sum_{s=0}^{k-j} \frac{1}{s!}(-t)^s v_{j+s}e^{i\omega t}$, $j = 1,2,\ldots,k$, where $v_j \in \mathbb{C}^n$ ($j = 1,2,\ldots,k$) satisfy

$$\sum_{s=0}^{k-j} \frac{1}{s!} v_{j+s}^T \Delta_s(0,-i\omega) = 0. \tag{5.21}$$

Since $\varphi_k \notin \mathrm{Ran}(\mathscr{A}_0 - i\omega \mathrm{Id})$ and $\psi_k \in \mathrm{Ran}(\mathscr{A}_0^* + i\omega \mathrm{Id})$, we have $(\psi_k, \varphi_k) = v \neq 0$. Moreover, it is easy to see that $(\psi_j, \varphi_s) = 0$ for all $j \neq s$, and $(\psi_j, \varphi_j) = v$ for all $j = 1,2,\ldots,k$. In fact,

$$v = \sum_{s=0}^{k-1} \bar{v}_{1+s}^T \left[\frac{(-t)^s}{s!}\mathrm{Id}_n + \int_{-\tau}^0 \frac{(-t)^{s+1} - (-t+\theta)^{s+1}}{(s+1)!} d\eta(0,\theta)e^{i\omega\theta} \right] u_1$$

$$= \sum_{s=0}^{k-1} \bar{v}_{1+s}^T \left[\frac{(-t)^s}{s!}\mathrm{Id}_n - \sum_{m=0}^s \frac{(-t)^m}{m!(s+1-m)!} \int_{-\tau}^0 \theta^{s+1-m} d\eta(0,\theta)e^{i\omega\theta} \right] u_1$$

$$= \sum_{s=0}^{k-1} \sum_{m=0}^s \frac{(-t)^m}{m!(s+1-m)!} \bar{v}_{1+s}^T \Delta_{s+1-m}(0,i\omega)u_1$$

$$= \sum_{m=0}^{k-1} \sum_{s=m}^{k-1} \frac{(-t)^m}{m!(s+1-m)!} \bar{v}_{1+s}^T \Delta_{s+1-m}(0,i\omega)u_1$$

$$= \sum_{m=0}^{k-1} \left[\frac{(-t)^m}{m!} \sum_{s=m}^{k-1} \frac{1}{(s+1-m)!} \bar{v}_{1+s}^T \Delta_{s+1-m}(0,i\omega)u_1 \right]$$

$$= \sum_{m=0}^{k-1} \left[\frac{(-t)^m}{m!} \sum_{s=1}^{k-m} \frac{1}{s!} \bar{v}_{m+s}^T \Delta_s(0,i\omega)u_1 \right]$$

$$= \sum_{s=1}^k \frac{1}{s!} \bar{v}_s^T \Delta_s(0,i\omega)u_1 + \sum_{m=1}^{k-1} \left[\frac{(-t)^m}{m!} \sum_{s=1}^{k-m} \frac{1}{s!} \bar{v}_{m+s}^T \Delta_s(0,i\omega)u_1 \right]$$

$$= \sum_{s=1}^k \frac{1}{s!} \bar{v}_s^T \Delta_s(0,i\omega)u_1 - \sum_{m=1}^{k-1} \left[\frac{(-t)^m}{m!} \bar{v}_m^T \Delta(0,i\omega)u_1 \right]$$

$$= \sum_{s=1}^k \frac{1}{s!} \bar{v}_s^T \Delta_s(0,i\omega)u_1.$$

In what follows, we develop the Lyapunov–Schmidt procedure for (5.15) to find a periodic solution with period near the constant $\frac{2\pi}{\omega}$. We start with the normalization of the period. Let $\beta \in (-1,1)$, $x(t) = u((1+\beta)t)$. Then (5.15) can be rewritten as

$$(1+\beta)\dot{u}(t) = L(\alpha)u_{t,\beta} + f(\alpha, u_{t,\beta}),$$

where $u_{t,\beta}(\theta) = u(t + (1+\beta)\theta)$, $\theta \in [-\tau, 0]$. To fix a functional setting for the above equation, consider the Banach subspace $\mathscr{C}_\omega$ (respectively, $\mathscr{C}_\omega^1$) of $C(\mathbb{R}, \mathbb{R}^n)$, $\frac{2\pi}{\omega}$-periodic continuous (respectively, differentiable) functions equipped with their usual sup-norms. It is easy to see that $\mathscr{C}_\omega$ is an isometric Banach representation of the group $\mathbb{S}^1$ with the action given by

$$\theta \cdot u(t) = u(t + \theta) \quad \text{for} \quad \theta \in \mathbb{S}^1.$$

Here and in what follows, we do not distinguish $\theta \in \mathbb{S}^1$ and its realization χ such that $\theta = \exp\{i\chi\}$. Define a scalar product on the complexification of $\mathscr{C}_\omega$ by $\langle \cdot, \cdot \rangle$: $\mathscr{C}_\omega \times \mathscr{C}_\omega \to \mathbb{R}$ defined by

$$\langle v, u \rangle = \frac{\omega}{2\pi} \int_0^{2\pi/\omega} \bar{v}^T(t)u(t)dt \tag{5.22}$$

for $u, v \in \mathscr{C}_\omega$. Define $F : \mathscr{C}_\omega^1 \times \mathbb{R}^2 \to \mathscr{C}_\omega$ by

$$F(u, \alpha, \beta) = -(1 + \beta)\dot{u}(t) + L(\alpha)u_{t,\beta} + f(\alpha, u_{t,\beta}). \qquad (5.23)$$

By varying the newly introduced small variable β, one keeps track not only of so-lutions of (5.15) with period ω but also of solutions with nearby period. In fact, solutions to $F(u, \alpha, \beta) = 0$ correspond to $\frac{2\pi}{\omega(1+\beta)}$-periodic solutions of (5.15). It is easy to see that F is $\mathbb{S}^1$-equivariant:

$$\theta \cdot F(u, \alpha, \beta) = F(\theta \cdot u, \alpha, \beta),$$

for all $\theta \in \mathbb{S}^1$. The operator $\mathscr{L}u = -\dot{u} + L(0)u_t$ is the linearization of F at the origin. Obviously, the elements of $\mathrm{Ker}\mathscr{L}$ correspond to solutions of the linear system $\dot{u} = L(0)u_t$ satisfying $u(t) = u(t + \frac{2\pi}{\omega})$. It is easy to see that the adjoint operator of $\mathscr{L}$ is given by

$$\mathscr{L}^*u = \dot{u} + \int_{-\tau}^0 d\eta^T(0, \theta)u(t - \theta),$$

that is, $\langle v, \mathscr{L}u \rangle = \langle \mathscr{L}^*v, u \rangle$ for all $u, v \in \mathscr{C}_\omega^1$. It follows from condition (NS) that

$$\mathrm{Ker}\mathscr{L} = \{\mathrm{Re}(z\varphi_1); z \in \mathbb{C}\}, \quad \mathrm{Ker}\mathscr{L}^* = \{\mathrm{Re}(z\psi_k); z \in \mathbb{C}\}.$$

Let P and Q be projections defined by

$$P\phi = 2\mathrm{Re}\{\langle v_1 e^{i\omega t}, \phi \rangle \varphi_1\}, \quad Q\phi = 2\mathrm{Re}\{\langle v_k e^{i\omega t}, \phi \rangle \varphi_k\}$$

for $\phi \in \mathscr{C}_\omega$. Obviously, P and Q are $\mathbb{S}^1$-equivariant and $\mathrm{Ker}\mathscr{L} = \mathrm{Ran}P$ and $\mathrm{Ran}\mathscr{L} = \mathrm{Ker}Q$.

The equation $F(u, \alpha, \beta) = 0$ is equivalent to the following system:

$$\begin{aligned} (I - Q)F(v + w, \alpha, \beta) &= 0, \\ QF(v + w, \alpha, \beta) &= 0. \end{aligned} \qquad (5.24)$$

Here we have written $u \in \mathscr{C}_\omega$ in the form $u = v + w$, with $v = Pu \in \mathrm{Ker}\mathscr{L}$ and $w = (I - P)u \in \mathscr{C}_\omega \cap \mathrm{Ker}P$. Near the critical point $(u, \alpha, \beta) = (0, 0, 0)$, the implicit function theorem implies that the first equation of (5.24) can be solved for $w = W(v, \alpha, \beta)$, where $W : \mathrm{Ker}\mathscr{L} \times \mathbb{R}^2 \to \mathscr{C}_\omega \cap \mathrm{Ker}P$ is a continuously differentiable $\mathbb{S}^1$-equivariant map satisfying $W(0, 0, 0) = 0$. Substituting $w = W(v, \alpha, \beta)$ into the second equation of (5.24), we have

$$\vartheta(v, \alpha, \beta) \stackrel{\mathrm{def}}{=} QF(v + W(v, \alpha, \beta), \alpha, \beta) = 0. \qquad (5.25)$$

Thus, we can reduce our bifurcation problem to the problem of finding zeros of the map $\vartheta : \mathrm{Ker}\mathscr{L} \times \mathbb{R}^2 \to \mathrm{Ran}Q$. We refer to ϑ as the bifurcation map of the system (5.15). It follows from the $\mathbb{S}^1$-equivariance of F and W that the bifurcation map ϑ is also $\mathbb{S}^1$-equivariant. Moreover,

$$\vartheta(0, 0, 0) = 0, \quad \vartheta_v(0, 0, 0) = 0.$$

Therefore, we obtain that *there exists a* $\mathbb{S}^1$-*invariant neighborhood* U *of* $(0,0,0) \in$ Ker$\mathscr{L} \times \mathbb{R} \times \mathbb{R}$ *such that each solution to* $F(u,\alpha,\beta) = 0$ *in* U *corresponds one-to-one to some zero of the* $\mathbb{S}^1$-*equivariant map* ϑ *defined in* (5.25). In other words, small-amplitude periodic solutions of (5.15), of period near $\frac{2\pi}{\omega}$, correspond to solutions to system (5.25).

For $v = z\varphi_1 + \overline{z}\overline{\varphi}_1 \in \text{Ker}\mathscr{L}$, with $z = \langle v_1 e^{i\omega t}, v \rangle$, using this in (5.25) and then calculating the inner product with ψ_k, we have

$$g(z,\alpha,\beta) = 0, \tag{5.26}$$

where $g \colon \mathbb{C} \times \mathbb{R}^2 \to \mathbb{C}$ is explicitly given by

$$g(z,\alpha,\beta) = \langle \psi_k, F(\sigma(z,\alpha,\beta),\alpha,\beta) \rangle, \tag{5.27}$$

and $\sigma(z,\alpha,\beta) = z\varphi_1 + \overline{z}\overline{\varphi}_1 + W(z\varphi_1 + \overline{z}\overline{\varphi}_1, \alpha, \beta)$. Obviously, $g(\cdot, \alpha, \beta)$ is also $\mathbb{S}^1$-equivariant and so can be written as

$$g(z,\alpha,\beta) = zh(|z|^2, \alpha, \beta),$$

where the smooth function $h \colon \mathbb{R}^3 \to \mathbb{C}$ is $\mathbb{Z}_2$-equivariant. It follows that for the non-trivial solutions, the bifurcation problem reduces to the equation $h(r^2, \alpha, \beta) = 0$, where $r = |z|$.

First, we notice that

$$\langle v_1 e^{i\omega t}, \sigma(z,\alpha,\beta) \rangle \equiv z \tag{5.28}$$

and

$$(I - Q)F(\sigma(z,\alpha,\beta),\alpha,\beta) \equiv 0 \tag{5.29}$$

for all $(z,\alpha,\beta) \in \mathbb{C} \times \mathbb{R}^2$. Differentiation of (5.28) and (5.29) at $z = 0$ gives

$$\langle v_1 e^{i\omega t}, \sigma_z(0,\alpha,\beta) \rangle = 1$$

and

$$(I - Q)F_u(0,\alpha,\beta) \cdot \sigma_z(0,\alpha,\beta) = 0. \tag{5.30}$$

In particular, $\sigma_z(0,0,0) = \varphi_1$.

It follows from (5.27) that $zh(|z|^2, \alpha, \beta) = \langle \psi_k, F(\sigma(z,\alpha,\beta),\alpha,\beta) \rangle$. Differentiation at $z = 0$ gives

$$h(0,\alpha,\beta) = \langle \psi_k, F_u(0,\alpha,\beta) \cdot \sigma_z(0,\alpha,\beta) \rangle. \tag{5.31}$$

Thus, we obtain the following results.

Lemma 5.5. $h_\alpha(0,0,0) = -\overline{v}_k^T \Delta_\alpha(0,i\omega)u_1$ *and* $h(0,0,\beta) = -(i\beta\omega)^k v + O(|\beta|^{k+1})$.

Proof. In view of (5.31), we have $h(0,0,0) = 0$. Differentiation of (5.31) at $\alpha = 0$ gives

$$\begin{aligned} h_\alpha(0,0,0) &= \langle \psi_k, F_{u\alpha}(0,0,0) \cdot \sigma_z(0,0,0) \rangle \\ &= \overline{v}_k^T \int_{-\tau}^0 d\eta_\alpha(0,\theta)\varphi_1(\theta) \\ &= -\overline{v}_k^T \Delta_\alpha(0,i\omega)u_1. \end{aligned}$$

Letting $\alpha = 0$ in (5.30) and taking the inner product with $v_j e^{i\omega t}$, we have

$$0 = \langle v_j e^{i\omega t}, F_u(0,0,\beta) \cdot \sigma_z(0,0,\beta) \rangle$$

$$= \langle (1+\beta)i\omega v_j e^{i\omega t} + \int_{-\tau}^0 d\eta^T(0,\theta)v_j e^{i\omega(t-\theta-\beta\theta)}, \sigma_z(0,0,\beta) \rangle$$

$$= \langle (1+\beta)i\omega v_j e^{i\omega t} + \sum_{s=0}^{\infty} \frac{(-i\omega\beta)^s}{s!} \int_{-\tau}^0 \theta^s d\eta^T(0,\theta)v_j e^{i\omega(t-\theta)}, \sigma_z(0,0,\beta) \rangle$$

$$= -\langle \sum_{s=0}^{k-1} \frac{(-i\beta\omega)^s}{s!} \Delta_s^T(0,-i\omega)v_j e^{i\omega t}, \sigma_z(0,0,\beta) \rangle + O(|\beta|^k).$$

Hence,

$$\langle \Delta^T(0,-i\omega)v_j e^{i\omega t}, \sigma_z(0,0,\beta) \rangle = -\langle \sum_{s=1}^{k-1} \frac{(-i\beta\omega)^s}{s!} \Delta_s^T(0,-i\omega)v_j e^{i\omega t}, \sigma_z(0,0,\beta) \rangle + O(|\beta|^k).$$

$$(5.32)$$

Similarly, we can use (5.31) to get

$$h(0,0,\beta) = -\langle \sum_{s=1}^{k} \frac{(-i\beta\omega)^s}{s!} \Delta_s^T(0,-i\omega)v_k e^{i\omega t}, \sigma_z(0,0,\beta) \rangle + O(|\beta|^{k+1}).$$

In view of (5.32), we have

$$\langle \sum_{l=1}^{k-1} (-i\beta\omega)^l \Delta^T(0,-i\omega)v_{k-l} e^{i\omega t}, \sigma_z(0,0,\beta) \rangle$$

$$= -\langle \sum_{l=1}^{k-1} \sum_{s=1}^{k-1} \frac{(-i\beta\omega)^{l+s}}{s!} \Delta_s^T(0,-i\omega)v_{k-l} e^{i\omega t}, \sigma_z(0,0,\beta) \rangle$$

$$= -\langle \sum_{s=1}^{k-1} \sum_{m=s+1}^{k-1+s} \frac{(-i\beta\omega)^m}{s!} \Delta_s^T(0,-i\omega)v_{k-m+s} e^{i\omega t}, \sigma_z(0,0,\beta) \rangle$$

$$= -\langle \sum_{s=1}^{k-1} \sum_{m=s+1}^{k} \frac{(-i\beta\omega)^m}{s!} \Delta_s^T(0,-i\omega)v_{k-m+s} e^{i\omega t}, \sigma_z(0,0,\beta) \rangle + O(|\beta|^{k+1})$$

$$= -\langle \sum_{m=2}^{k} \sum_{s=1}^{m-1} \frac{(-i\beta\omega)^m}{s!} \Delta_s^T(0,-i\omega)v_{k-m+s} e^{i\omega t}, \sigma_z(0,0,\beta) \rangle + O(|\beta|^{k+1})$$

$$= -\langle \sum_{l=2}^{k} \sum_{s=1}^{l-1} \frac{(-i\beta\omega)^l}{s!} \Delta_s^T(0,-i\omega)v_{k-l+s} e^{i\omega t}, \sigma_z(0,0,\beta) \rangle + O(|\beta|^{k+1}).$$

Thus, it follows from (5.21) and the expression of v that

$$h(0,0,\beta) = \langle \sum_{l=1}^{k-1} \sum_{s=0}^{l-1} \frac{(-i\beta\omega)^l}{s!} \Delta_s^T(0,-i\omega)v_{k-l+s} e^{i\omega t}, \sigma_z(0,0,\beta) \rangle + O(|\beta|^{k+1})$$

$$= \langle \sum_{l=2}^{k-1} \sum_{s=1}^{l-1} \frac{(-i\beta\omega)^l}{s!} \Delta_s^T(0,-i\omega)v_{k-l+s} e^{i\omega t}, \sigma_z(0,0,\beta) \rangle$$

$$\quad + \langle \sum_{l=1}^{k-1} (-i\beta\omega)^l \Delta^T(0,-i\omega)v_{k-l} e^{i\omega t}, \sigma_z(0,0,\beta) \rangle + O(|\beta|^{k+1})$$

$$= \langle \sum_{l=2}^{k-1} \sum_{s=1}^{l-1} \frac{(-i\beta\omega)^l}{s!} \Delta_s^T(0,-i\omega)v_{k-l+s} e^{i\omega t}, \sigma_z(0,0,\beta) \rangle$$

$$\quad - \langle \sum_{l=2}^{k} \sum_{s=1}^{l-1} \frac{(-i\beta\omega)^l}{s!} \Delta_s^T(0,-i\omega)v_{k-l+s} e^{i\omega t}, \sigma_z(0,0,\beta) \rangle + O(|\beta|^{k+1})$$

$$= -(i\beta\omega)^k \langle \sum_{s=1}^{k-1} \frac{1}{s!} \Delta_s^T(0,-i\omega)v_s e^{i\omega t}, \sigma_z(0,0,\beta) \rangle + O(|\beta|^{k+1})$$

$$= -(i\beta\omega)^k v + O(|\beta|^{k+1}).$$

$\square$

In order to consider the first-order partial derivative $h_1(0,0,0)$ of $h(u,\alpha,\beta)$ with respect to u at $(u,\alpha,\beta) = (0,0,0)$, we need to write the function $f(\alpha,\varphi)$ of (5.15) in the form of a Taylor expansion in φ at $\alpha = 0$:

$$f(0,\varphi) = \frac{1}{2}\mathscr{B}(\varphi,\varphi) + \frac{1}{6}\mathscr{E}(\varphi,\varphi,\varphi) + o(\|\varphi\|^3) \tag{5.33}$$

for $\varphi \in C_{n,\tau}$, where $\mathscr{B}(\cdot,\cdot)$ and $\mathscr{E}(\cdot,\cdot,\cdot)$ are second- and third-order derivatives of $f(0,\cdot)$, and so are symmetric 2- and 3-linear functions, respectively. Obviously,

$$h_1(0,0,0) = \frac{\partial^3}{\partial z^2 \partial \bar{z}} g(0,0,0).$$

Using a similar argument to that in [117], we have

$$h_1(0,0,0) = \langle \psi_k, \mathscr{E}(\varphi_1,\varphi_1,\overline{\varphi}_1)\rangle + 2\langle \psi_k, \mathscr{B}(\varphi_1,W_{11})\rangle + \langle \psi_k, \mathscr{B}(\overline{\varphi}_1,W_{20})\rangle, \tag{5.34}$$

where W_{11} and W_{20} are the coefficients of $z\bar{z}$ and $\frac{z^2}{2}$ in the Taylor expansion of $\sigma(z,0,0)$, respectively. In view of (5.29), we have $W_{20} = -\mathscr{L}^{-1}(I-Q)\mathscr{B}(\varphi_1,\varphi_1)$ and $W_{11} = -\mathscr{L}^{-1}(I-Q)\mathscr{B}(\varphi_1,\overline{\varphi}_1)$. Note that $\mathscr{B}(\varphi_1,\varphi_1), \mathscr{B}(\varphi_1,\overline{\varphi}_1) \in \text{Ran}\mathscr{L}$. Then the projections $(I-Q)$ on $\mathscr{B}(\varphi_1,\varphi_1)$ and $\mathscr{B}(\varphi_1,\overline{\varphi}_1)$ act as the identity. Therefore,

$$\mathscr{L}W_{20} + \mathscr{B}(\varphi_1,\varphi_1) = 0, \quad \mathscr{L}W_{11} + \mathscr{B}(\varphi_1,\overline{\varphi}_1) = 0.$$

In addition, it follows from $\text{Ran}W \subseteq \text{Ker}P$ that

$$\langle v_1 e^{i\omega t}, W_{20}\rangle = 0, \quad \langle v_1 e^{i\omega t}, W_{11}\rangle = 0.$$

Therefore, we have

$$W_{20} = \Delta^{-1}(0,2i\omega)\mathscr{B}(\varphi_1,\varphi_1), \quad W_{11} = \Delta^{-1}(0,0)\mathscr{B}(\varphi_1,\overline{\varphi}_1). \tag{5.35}$$

In view of (5.34) and Lemma 5.5, the reduced equation $h(r^2,\alpha,\beta) = 0$ takes the form

$$-(i\beta\omega)^k v - \bar{v}_k^T \Delta_\alpha(0,i\omega)u_1\alpha + r^2 h_1(0,0,0) + \text{h.o.t.} = 0.$$

Since $v \neq 0$, we can rewrite the above equation as

$$\beta^k - A\alpha + Br^2 + \text{h.o.t.} = 0, \tag{5.36}$$

where

$$A = \frac{\bar{v}_k^T \Delta_\alpha(0,i\omega)u_1}{-(i\omega)^k v}, \quad B = \frac{h_1(0,0,0)}{-(i\omega)^k v}. \tag{5.37}$$

Separating the real and imaginary parts of equation (5.36) gives us

$$\beta^k - \alpha\text{Re}\{A\} + r^2\text{Re}\{B\} + \text{h.o.t.} = 0 \tag{5.38}$$

and

$$- \alpha \text{Im}\{A\} + r^2 \text{Im}\{B\} + \text{h.o.t.} = 0. \tag{5.39}$$

If $\text{Im}\{A\} \neq 0$, i.e.,

$$\text{Im}\{i^k \overline{v} v_k^T \Delta_\alpha(0, i\omega) u_1\} \neq 0, \tag{5.40}$$

then by the implicit function theorem, we may solve (5.39) for α to get

$$\alpha = \alpha(r^2, \beta) := \frac{\text{Im}\{B\}}{\text{Im}\{A\}} r^2 + O(r^4, \beta). \tag{5.41}$$

Substituting this into (5.38) gives an equation of the form

$$\beta^k = \mu_0 r^2 + O(r^4, \beta), \tag{5.42}$$

where

$$\mu_0 = \frac{\text{Im}(B\overline{A})}{\text{Im}(A)}. \tag{5.43}$$

Let $\beta = \varepsilon \sqrt[k]{r^2}$. Then (5.42) becomes $\varepsilon^k r^2 = \mu_0 r^2 + O(r^4, \varepsilon)$, which yields

$$\varepsilon^k = \mu_0 + O(r^2, \varepsilon). \tag{5.44}$$

When k is odd and $\mu_0 \neq 0$, then (5.44) has for $r = 0$ the unique solution $\varepsilon = \sqrt[k]{\mu_0}$. Thus, the implicit function theorem implies that this solution can be continued for r near 0, giving a unique solution branch $\varepsilon = \varepsilon(r)$ of (5.44), satisfying $\varepsilon(0) = \sqrt[k]{\mu_0}$. If we define $\beta(r) = \sqrt[k]{r^2} \varepsilon(r)$ and $\alpha(r) = \alpha(r^2, \beta(r))$, where the function $\alpha(r^2, \beta)$ is given in (5.41), then $h(r^2, \alpha, \beta) = 0$ has unique solution branches of the form $(r, \alpha(r), \beta(r))$, which pass through $(0, 0, 0)$ for $r = 0$ and exist for sufficiently small $r > 0$.

When k is even and $\mu_0 < 0$, then (5.44) has no solutions for $r = 0$ and hence also no solutions for $|r|$ sufficiently small. This implies that no bifurcation occurs. On the other hand, if k is even and $\mu_0 > 0$m then (5.44) has for $r = 0$ two solutions $\varepsilon = \pm \sqrt[k]{\mu_0}$, each of which can be continued for small r, giving rise to two solution branches $\varepsilon = \varepsilon_\pm(r)$, where the functions $\varepsilon_\pm : [0, \infty) \to \mathbb{R}$ are smooth in a neighborhood of the origin and satisfy $\varepsilon_\pm(0) = \pm \sqrt[k]{\mu_0}$. If we define $\beta_\pm(r) = \sqrt{r^2} \varepsilon_\pm(r)$ and $\alpha_\pm(r) = \alpha(r^2, \beta_\pm(r))$, where the function $\alpha(r^2, \beta)$ is given in (5.41), then $h(r^2, \alpha, \beta) = 0$ has two solution branches of the form $(r, \alpha_\pm(r), \beta_\pm(r))$ that pass through $(0, 0, 0)$ for $r = 0$ and exist for sufficiently small $r > 0$. Therefore, we have the following results.

Theorem 5.4. *In addition to (NS), assume that inequality (5.40) holds and that A, B, and μ_0 are defined by (5.37) and (5.43), respectively.*

(i) *If k is odd and $\mu_0 \neq 0$, then for (u, α, β) near $(0, 0, 0)$, (5.15) has exactly one branch of nontrivial $\frac{2\pi}{(1+\beta)\omega}$-periodic solutions, which exists for $\alpha > 0$ (respectively, $\alpha < 0$) when $\text{Im}\{A\}\text{Im}\{B\} > 0$ (respectively, $\text{Im}\{A\}\text{Im}\{B\} < 0$).*

(ii) If k is even and $\mu_0 < 0$, then (5.15) has no nontrivial $\frac{2\pi}{(1+\beta)\omega}$-periodic solutions with (u, α, β) near $(0,0,0)$.

(iii) If k is even and $\mu_0 > 0$, then for (u, α, β) near $(0,0,0)$, (5.15) has exactly two branches of nontrivial $\frac{2\pi}{(1+\beta)\omega}$-periodic solutions, which exist for $\alpha > 0$ (respectively, $\alpha < 0$) when $\text{Im}\{A\}\text{Im}\{B\} > 0$ (respectively, $\text{Im}\{A\}\text{Im}\{B\} < 0$).

If assumption (NS) holds for $k = 1$, then the infinitesimal generator $\mathscr{A}_\alpha$ has a pair of simple complex conjugate eigenvalues $\lambda(\alpha)$ and $\overline{\lambda}(\alpha)$ satisfying $\lambda(0) = i\omega$. Moreover, there exists a C^1-continuous function $u(\alpha)$ such that $u(0) = u_1$ and $\Delta(\alpha, \lambda(\alpha))u(\alpha) \equiv 0$ for all sufficiently small α. We differentiate it with respect to α and obtain

$$\left[\Delta_\alpha(\alpha, \lambda(\alpha)) + \lambda'(\alpha)\Delta_1(\alpha, \lambda(\alpha))\right] u(\alpha) + \Delta(\alpha, \lambda(\alpha))u'(\alpha) = 0.$$

In particular, we have

$$\left[\Delta_\alpha(0, i\omega) + \lambda'(0)\Delta_1(0, i\omega)\right] u_1 + \Delta(0, i\omega)u'(0) = 0.$$

This, together with the fact that $\overline{v}_1^T \Delta(0, i\beta_0) = 0$ and $\overline{v}_1^T \Delta_1(0, i\omega)u_1 = (\psi_1, \varphi_1) = v$, implies that $\overline{v}_1^T \Delta_\alpha(0, i\omega)u_1 + \lambda'(0)v = 0$. Thus, the quantities A and B in (5.37) can be figured out:

$$A = \frac{\lambda'(0)}{i\omega}, \quad B = \frac{h_1(0,0,0)}{-i\omega v}.$$

Hence, (5.40) is equivalent to $\text{Re}\{\lambda'(0)\} \neq 0$ and

$$\text{sgn}\{\text{Im}\{A\}\text{Im}\{B\}\} = -\text{sgn}\{\text{Re}\{\lambda'(0)\}\text{Re}\{\overline{v}h_1(0,0,0)\}\}.$$

Moreover,

$$\text{sgn}\{\mu_0\} = -\text{sgn}\{\text{Re}\{\lambda'(0)\}\text{Im}\{\lambda'(0)v\overline{h}_1(0,0,0)\}\}.$$

Therefore, Theorem 5.4 reduces to the following standard Hopf bifurcation theorem.

Corollary 5.1. If assumption (NS) holds for $k = 1$ and $\text{Re}\{\lambda'(0)\} \neq 0$, then there exists a unique branch of periodic solutions, parameterized by α, bifurcating from the trivial solution $x = 0$ of (5.15). Moreover,

(i) $\text{Re}\{\lambda'(0)\}\text{Re}\{\overline{v}h_1(0,0,0)\}$ determines the direction of the bifurcation: the bifurcation is supercritical (respectively, subcritical), i.e., the bifurcating periodic solutions exist for $\alpha > 0$ (respectively, < 0), if $\text{Re}\{\lambda'(0)\}\text{Re}\{\overline{v}h_1(0,0,0)\} < 0$ (respectively, > 0);

(ii) $\text{Re}\{\lambda'(0)\}\text{Im}\{\lambda'(0)v\overline{h}_1(0,0,0)\}$ determines the period of the bifurcating periodic solutions along the branch: the period is greater than (respectively, smaller than) $\frac{2\pi}{\omega}$ if it is positive (respectively, negative).

5.6 Equivariant Hopf Bifurcation of NFDEs

In this section, we introduce the work [136] on equivariant Hopf bifurcation for the following parameterized system of NFDEs:

$$\frac{d}{dt}h(\alpha, x_t) = f(\alpha, x_t), \tag{5.45}$$

where $h, f : \mathbb{R} \times C_{n,\tau} \to \mathbb{R}^n$ are two continuously differentiable mappings satisfying $f(\alpha, 0) = 0$ for all $\alpha \in \mathbb{R}$. We say that (5.45) is equivariant with respect to a group Γ if there exists a representation ρ of Γ such that

$$h(\alpha, \rho(\gamma)\phi) = \rho(\gamma)h(\alpha, \phi), \quad f(\alpha, \rho(\gamma)\phi) = \rho(\gamma)f(\alpha, \phi) \tag{5.46}$$

for $(\alpha, \gamma, \phi) \in \mathbb{R} \times \Gamma \times C([-\tau, 0]; \mathbb{R}^n)$, where $\rho(\gamma)\phi \in C([-\tau, 0]; \mathbb{R}^n)$ is given by $(\rho(\gamma)\phi)(s) = \rho(\gamma)\phi(s)$ for $s \in [-\tau, 0]$. Recall that a representation ρ of a group Γ is a group homomorphism $\rho : \Gamma \to GL(n, \mathbb{R})$. Condition (5.46) implies that system (5.45) is invariant under the transformation $(x, t) \to (\rho(\gamma)x, t)$. Namely, $x(t)$ is a solution of (5.45) if and only if $\rho(\gamma)x(t)$ is a solution. Throughout this section, we always assume that Γ is a compact Lie group and system (5.45) is Γ-equivariant.

Linearizing (5.45) at the equilibrium point $x = 0$ yields

$$\frac{d}{dt}D(\alpha)x_t = L(\alpha)x_t. \tag{5.47}$$

Without loss of generality, we assume that there exist two $n \times n$ matrix-valued functions $\mu, \eta : [-\tau, 0] \to \mathbb{R}^{n^2}$ whose components each have bounded variation in $\theta \in [-\tau, 0]$ for each α and such that for $\varphi \in C_{n,\tau}$,

$$D(\alpha)\varphi = \varphi(0) - \int_{-\tau}^{0} d\mu(\alpha, \theta)\varphi(\theta), \quad L(\alpha)\varphi = \int_{-\tau}^{0} d\eta(\alpha, \theta)\varphi(\theta).$$

Moreover, we assume that $D(\alpha)$ is atomic at zero, that is, $\mathrm{Var}_{[s,0]}\mu(\alpha, \theta) \to 0$ as $s \to 0$ (see Hale and Verduyn Lunel [154] for more details). Denote by $\mathscr{A}_\alpha$ the infinitesimal generator associated with the linear system (5.47). The spectrum of $\mathscr{A}_\alpha$, denoted by $\sigma(\mathscr{A}_\alpha)$, is the point spectrum. Moreover, λ is an eigenvalue of $\mathscr{A}_\alpha$, i.e., $\lambda \in \sigma(\mathscr{A}_\alpha)$, if and only if λ satisfies $\det \Delta(\alpha, \lambda) = 0$, where the characteristic matrix $\Delta(\alpha, \lambda)$ is given by

$$\Delta(\alpha, \lambda) = \lambda D(\alpha)(e^{\lambda(\cdot)}\mathrm{Id}) - L(\alpha)(e^{\lambda(\cdot)}\mathrm{Id}).$$

It is well known that $\phi \in C_{n,\tau}$ is an eigenvector of $\mathscr{A}_\alpha$ associated with the eigenvalue λ if and only if $\phi(\theta) = e^{\lambda\theta}b$ for $\theta \in [-\tau, 0]$ and some vector $b \in \mathbb{R}^n$ such that $\Delta(\alpha, \lambda)b = 0$. Let $E_{\alpha,\lambda}$ be the eigenspace of $\mathscr{A}_\alpha$ associated with the eigenvalues λ and $\bar{\lambda}$. Assume that $\mathscr{A}_0$ has a pair of purely imaginary eigenvalues $\pm i\omega$. The symmetry group Γ often causes purely imaginary eigenvalues to be multiple. So, we always assume the following:

(NHB1) $\mathscr{A}_0$ has a pair of purely imaginary eigenvalues $\pm i\omega$, each of multiplicity m, and no other eigenvalue of $\mathscr{A}_0$ is an integer multiple of $i\omega$.

In studying the bifurcation problem, we wish to consider how the eigenvalues of $\mathscr{A}_\alpha$ cross the imaginary axis at $\alpha = 0$ and to describe the structure of the associated eigenspace $E_{\alpha,\lambda}$. We consider the following nontrivial restrictions on the corresponding imaginary eigenspace of $\mathscr{A}_0$:

(NHB2)   The imaginary eigenspace $E_{0,i\omega}$ of $\mathscr{A}_0$ is Γ-simple.

Thus, we make use of the implicit function theorem and Lemma 1.5 on Page 265 of Golubitsky et al. [118] and obtain the following results about the multiplicity of this eigenvalue and its associated eigenvectors of $\mathscr{A}_\alpha$.

Theorem 5.5. *Under conditions (NHB1)–(NHB2), for sufficiently small α, the infinitesimal generator $\mathscr{A}_\alpha$ has one pair of complex conjugate eigenvalues $\sigma(\alpha) \pm i\rho(\alpha)$, each of multiplicity m. Moreover, σ and ρ are smooth functions of α and satisfy $\sigma(0) = 0$ and $\rho(0) = \omega$.*

In view of (NHB1), the purely imaginary eigenvalues of $\mathscr{A}_0$ have high multiplicity, so the standard Hopf bifurcation theorem cannot be applied directly. So, we first develop the equivariant Lyapunov–Schmidt reduction for (5.45) to consider the existence of periodic solutions. Let $\mathscr{C}_\omega$ (respectively, $\mathscr{C}_\omega^1$) be the Banach spaces of continuous (respectively, differentiable) n-dimensional $\frac{2\pi}{\omega}$-periodic functions equipped with their usual sup-norms. It is easy to see that $\mathscr{C}_\omega$ is a Banach representation of the group $\Gamma \times \mathbb{S}^1$ with the action given by

$$(\gamma, \theta)u(t) = \rho(\gamma)u(t + \theta), \quad \text{for} \quad (\gamma, \theta) \in \Gamma \times \mathbb{S}^1.$$

In view of the complexity in analyzing NFDEs, we introduce two kinds of bilinear forms. One is the inner product $\langle \cdot, \cdot \rangle : \mathscr{C}_\omega \times \mathscr{C}_\omega \to \mathbb{R}$ defined by (5.22). The other is $(\cdot, \cdot) : C_{n,\tau} \times C_{n,\tau} \to \mathbb{R}$ defined by

$$
\begin{aligned}
(\psi, \varphi) = \overline{\psi}^T(0)\varphi(0) - \int_{-\tau}^0 \left[\frac{d}{ds} \int_0^s \overline{\psi}^T(\xi - s)d\mu(0,\theta)\varphi(\xi)d\xi \right]_{s=\theta} \\
- \int_{-\tau}^0 \int_0^\theta \overline{\psi}^T(\xi - \theta)d\eta(0,\theta)\varphi(\xi)d\xi
\end{aligned}
\tag{5.48}
$$

for $\psi \in C_{n,\tau}$ and $\varphi \in C_{n,\tau}$. Let $\beta \in (-1,1)$, $x(t) = u((1 + \beta)t)$. Then (5.45) can be rewritten as

$$(1 + \beta)\frac{d}{dt}h(\alpha, u_{t,\beta}) = f(\alpha, u_{t,\beta}),$$

where $u_{t,\beta}(\theta) = u(t + (1 + \beta)\theta)$ for $\theta \in [-\tau, 0]$. Define $F : \mathscr{C}_\omega^1 \times \mathbb{R}^2 \to \mathscr{C}_\omega$ by

$$F(u, \alpha, \beta) = -(1 + \beta)\frac{d}{dt}h(\alpha, u_{t,\beta}) + f(\alpha, u_{t,\beta}), \tag{5.49}$$

so solutions to $F(u, \alpha, \beta) = 0$ correspond to $\frac{2\pi}{(1+\beta)\omega}$-periodic solutions of (5.45). It follows that the Γ-equivariance of L and f that F is $\Gamma \times \mathbb{S}^1$-equivariant:

$$(\gamma, \theta)F(u, \alpha, \beta) = F((\gamma, \theta)u, \alpha, \beta),$$

for all $(\gamma, \theta) \in \Gamma \times \mathbb{S}^1$. The linearized operator $\mathscr{L}$ of F with respect to u at $(u, \alpha, \beta) = (0, 0, 0)$ is given by

$$\mathscr{L}u = -\frac{d}{dt}D(0)u_t + L(0)u_t.$$

With respect to the inner product $\langle \cdot, \cdot \rangle : \mathscr{C}_\omega \times \mathscr{C}_\omega \to \mathbb{R}$, the adjoint operator of $\mathscr{L}$ is

$$\mathscr{L}^*u = \frac{d}{dt}\left[u(t) - \int_{-\tau}^0 d\mu^T(0, \theta)u(t - \theta)\right] + \int_{-\tau}^0 d\eta^T(0, \theta)u(t - \theta).$$

It follows from (NHB1) that $\mathrm{Ker}\mathscr{L} \cong E_{0,i\omega}$ and $\mathrm{Ker}\mathscr{L}^* \cong E_{0,i\omega}^*$, both of which are $2m$-dimensional. Furthermore, we have the following result.

Lemma 5.6. *Spaces* $\mathrm{Ker}\mathscr{L}$, $\mathrm{Ran}\mathscr{L}$, *and* $\mathscr{W} = (\mathrm{Ker}\mathscr{L}^*)^\perp \cap \mathscr{C}_\omega^1$ *are* $\Gamma \times \mathbb{S}^1$-*invariant subspaces of* $\mathscr{C}_\omega$. *Moreover,* $\mathscr{C}_\omega = \mathrm{Ker}\mathscr{L} \oplus \mathrm{Ran}\mathscr{L}$ *and* $\mathscr{C}_\omega^1 = \mathrm{Ker}\mathscr{L} \oplus \mathscr{W}$.

Let P and $I - P$ denote the projection operators defined by

$$P : \mathscr{C}_\omega \to \mathrm{Ran}\mathscr{L}, \qquad I - P : \mathscr{C}_\omega \to \mathrm{Ker}\mathscr{L}.$$

Obviously, P and $I - P$ are $\Gamma \times \mathbb{S}^1$-equivariant. Thus, $F(u, \alpha, \beta) = 0$ is equivalent to the following system:

$$PF(u, \alpha, \beta) = 0, \qquad (I - P)F(u, \alpha, \beta) = 0. \tag{5.50}$$

According to the above direct sum decomposition, for each $u \in \mathscr{C}_\omega^1$, there is a unique decomposition such that $u = v + w$, where $v \in \mathrm{Ker}\mathscr{L}$ and $w \in \mathscr{W}$. Applying the implicit function theorem, we obtain a continuously differentiable $\Gamma \times \mathbb{S}^1$-equivariant map $W : \mathrm{Ker}\mathscr{L} \times \mathbb{R}^2 \to \mathscr{W}$ such that $W(0, 0, 0) = 0$ and

$$PF(v + W(v, \alpha, \beta), \alpha, \beta) \equiv 0. \tag{5.51}$$

Substituting $w = W(v, \alpha, \beta)$ into the second equation of (5.50), we have

$$\vartheta(v, \alpha, \beta) \stackrel{\text{def}}{=} (I - P)F(v + W(v, \alpha, \beta), \alpha, \beta) = 0. \tag{5.52}$$

Thus, we reduce our Hopf bifurcation problem to the problem of finding zeros of the map $\vartheta : \mathrm{Ker}\mathscr{L} \times \mathbb{R}^2 \to \mathrm{Ker}\mathscr{L}$. We refer to ϑ as the bifurcation map of the system (5.45). It follows from the $\Gamma \times \mathbb{S}^1$-equivariance of F and W that the bifurcation map ϑ is also $\Gamma \times \mathbb{S}^1$-equivariant. Moreover, $\vartheta(0, 0, 0) = 0$ and $\vartheta_v(0, 0, 0) = 0$.

Finding periodic solutions to (5.45) rests on prescribing in advance the symmetry of the solution we seek. This can often be used to select a subspace on which the eigenvalues are simple. In addition, we should take temporal phase-shift symmetries in terms of the circle group $\mathbb{S}^1$ into account as well as spatial symmetries. Here, we place emphasis on two-dimensional fixed-point subspaces and assume the following:

(NHB3) $\dim \text{Fix}(\Sigma, E_{0,i\beta}) = 2$ for some subgroup Σ of $\Gamma \times \mathbb{S}^1$.
(NHB4) $\sigma'(0) \neq 0$.

Assumption (NHB4) is the transversality condition analogous to those of the standard Hopf bifurcation theorem. Now we can present our main results about equivariant Hopf bifurcation.

Theorem 5.6. *Under conditions* (NHB1)–(NHB4), *in every neighborhood of* $(x = 0, \alpha = 0)$, *system* (5.45) *has a bifurcation of periodic solutions whose spatiotemporal symmetry can be completely characterized by* Σ.

Proof. We consider the restriction mapping $\tilde{\vartheta}$: $\text{Fix}(\Sigma, \text{Ker}\mathscr{L}) \times \mathbb{R}^2 \to \text{Ker}\mathscr{L}$ of ϑ: $\text{Ker}\mathscr{L} \times \mathbb{R}^2 \to \text{Ker}\mathscr{L}$ on $\text{Fix}(\Sigma, \text{Ker}\mathscr{L}) \times \mathbb{R}^2$, i.e.,

$$\tilde{\vartheta}(v, \alpha, \beta) = (I - P)F(v + W(v, \alpha, \beta), \alpha, \beta)$$

for $v \in \text{Fix}(\Sigma, \text{Ker}\mathscr{L})$, $\alpha \in \mathbb{R}$, and $\beta \in \mathbb{R}$. Clearly, $\tilde{\vartheta}$ is also $\Gamma \times \mathbb{S}^1$-equivariant and satisfies

$$\tilde{\vartheta}(0,0,0) = 0, \quad \tilde{\vartheta}_v(0,0,0) = 0. \tag{5.53}$$

Moreover, it is easy to see that $\text{Ran}\tilde{\vartheta} \subseteq \text{Fix}(\Sigma, \text{Ker}\mathscr{L})$. Namely, $\tilde{\vartheta}$ maps $\text{Fix}(\Sigma, \text{Ker}\mathscr{L}) \times \mathbb{R}^2$ to $\text{Fix}(\Sigma, \text{Ker}\mathscr{L})$. Therefore, we only need to consider the existence of nontrivial zeros of $\tilde{\vartheta}$.

Without loss of generality, assume that $\text{Fix}(\Sigma, \text{Ker}\mathscr{L}) = \text{span}\{q, \bar{q}\}$, where $q(\theta) = Ae^{i\omega\theta}$ and $A \in \mathbb{C}^n$ satisfies $\Delta(0, i\beta)A = 0$. Thus, there exists $p \in \text{Fix}(\Sigma, \text{Ker}\mathscr{L}^*)$ such that $(p, q) = 1$, where $p(\theta) = Be^{i\omega\theta}$ and $B \in \mathbb{C}^n$ satisfies $\bar{B}^T \Delta_{\lambda}(0, i\omega)A = 1$ and $\bar{B}^T \Delta(0, i\beta) = 0$. Obviously, $\text{Fix}(\Sigma, \text{Ker}\mathscr{L})^* = \text{Fix}(\Sigma, \text{Ker}\mathscr{L}^*) = \text{span}\{p, \bar{p}\}$. As stated in Theorem 5.5, for sufficiently small α, the infinitesimal generator $\mathscr{A}_\alpha$ has one pair of complex conjugate eigenvalues $\lambda(\alpha)$ and $\bar{\lambda}(\alpha)$, each of multiplicity m, satisfying $\lambda(0) = i\omega$. By a similar argument to that in the proof of Corollary 5.1, we have

$$\bar{B}^T \Delta_\alpha(0, i\omega)A + \lambda'(0) = 0. \tag{5.54}$$

For each $\phi \in \text{Fix}(\Sigma, \text{Ker}\mathscr{L})$, $\phi = zq + \bar{z}\bar{q}$, where $z = \langle p, \phi \rangle$. Let

$$g(z, \alpha, \beta) \overset{\text{def}}{=} \langle p, \tilde{\vartheta}(zq + \bar{z}\bar{q}, \alpha, \beta) \rangle.$$

Thus, we only need to consider the existence of nontrivial solutions to $g(z, \alpha, \beta) = 0$. It follows from (5.53) that

$$g_z(0,0,0) = 0, \quad g_{\bar{z}}(0,0,0) = 0. \tag{5.55}$$

It is easy to see that $g(z, \alpha, \beta))$ is $\mathbb{S}^1$-equivariant. Thus, we can find two functions $\mathfrak{R}, \mathfrak{I} : \mathbb{R}^3 \to \mathbb{R}$ such that

$$g(z, \alpha, \beta) = \mathfrak{R}(|z|^2, \alpha, \beta)z + \mathfrak{I}(|z|^2, \alpha, \beta)iz. \tag{5.56}$$

It follows from $g_z(0,0,0) = 0$ that $\mathfrak{R}(0,0,0) = 0$ and $\mathfrak{I}(0,0,0) = 0$. Let $z = re^{i\theta}$. Then solving g is equivalent to solving either $r = 0$ or $\mathfrak{R}(r^2, \alpha, \beta) = 0$ and

$\Im(r^2, \alpha, \beta) = 0$. In view of the implicitly defined function $W(v, \alpha, \beta)$, which vanishes through first order in $v = zq + \overline{z}\overline{q}$, we have

$$F(v + W(v, \alpha, \beta), \alpha, \beta) = -(1 + \beta)\tfrac{d}{dt}D(\alpha)v_{t,\beta} + L(\alpha)v_{t,\beta} + O(|z|^2).$$

$$F_\alpha(v + W(v, 0, 0), 0, 0) = \Omega v_t + O(|z|^2)$$
$$F_\beta(v + W(v, 0, 0), 0, 0) = \Xi v_t + O(|z|^2),$$

where we have $\Omega = -\tfrac{d}{dt}D'(0) + L'(0)$, $\Xi v_t = \tfrac{\partial}{\partial\beta}[-(1+\beta)\tfrac{d}{dt}D(\alpha)v_{t,\beta} + L(\alpha)v_{t,\beta}]$ $\beta{=}0$, $D'(0) = \tfrac{d}{d\alpha}D(\alpha)|_{\alpha=0}$, and $L'(0) = \tfrac{d}{d\alpha}L(\alpha)|_{\alpha=0}$. Notice that

$$\langle p, \Omega q \rangle = \langle Be^{i\omega t}, i\omega \int_{-\tau}^{0} d\mu_\alpha(0, \theta)Ae^{i\omega(t+\theta)} + \int_{-\tau}^{0} d\eta_\alpha(0, \theta)Ae^{i\omega(t+\theta)} \rangle$$

$$= \overline{B}^T \int_{-\tau}^{0}[i\omega d\mu_\alpha(0, \theta) + d\eta_\alpha(0, \theta)]Ae^{i\omega\theta}$$

$$= -\overline{B}^T \Delta_\alpha(0, i\omega)A,$$

$$\langle p, \Xi q \rangle = -\langle Be^{i\omega t}, i\omega e^{i\omega t} - i\omega \int_{-\tau}^{0} d\mu(0, \theta)A[1 + i\omega\theta]e^{i\omega(t+\theta)} \rangle$$

$$+i\omega \overline{B}^T \int_{-\tau}^{0} d\eta(0, \theta)A\theta e^{i\omega\theta}$$

$$= -i\omega \overline{B}^T A + i\omega \overline{B}^T \int_{-\tau}^{0} d\mu(0, \theta)A[1 + i\omega\theta]e^{i\omega\theta}$$

$$+i\omega \overline{B}^T \int_{-\tau}^{0} d\eta(0, \theta)A\theta e^{i\omega\theta}$$

$$= -i\omega.$$

It follows from (5.54) that $\langle p, \Omega q \rangle = \lambda'(0)$. Similarly, we have $\langle p, \Omega\overline{q}\rangle = \langle p, \Xi\overline{q}\rangle = 0$. Therefore,

$$g_\alpha(z, 0, 0) = \langle p, F_\alpha(v, 0, 0) \rangle = z\lambda'(0) + O(|z|^2),$$
$$g_\beta(z, 0, 0) = \langle p, F_\beta(v, 0, 0) \rangle = -i\omega z + O(|z|^2).$$

Then

$$\Re_\alpha(0, 0, 0) = \mathrm{Re}\{\lambda'(0)\}, \quad \Im_\alpha(0, 0, 0) = \mathrm{Im}\{\lambda'(0)\},$$
$$\Re_\beta(0, 0, 0) = 0, \quad \Im_\beta(0, 0, 0) = -\omega.$$

So the Jacobi determinant of the functions $\Re$ and $\Im$ with respect to α and β is

$$\det \begin{bmatrix} \Re_\alpha(0,0,0) & \Re_\beta(0,0,0) \\ \Im_\alpha(0,0,0) & \Im_\beta(0,0,0) \end{bmatrix} = -\omega\,\mathrm{Re}\{\lambda'(0)\}.$$

Thus, under condition (NHB4), the above Jacobi determinant is nonzero. The implicit function theorem implies that there exists a unique function $\alpha = \alpha(r^2)$ and $\beta = \beta(r^2)$ satisfying $\alpha(0) = 0$ and $\beta(0) = 0$ such that

$$\Re(r^2, \alpha(r^2), \beta(r^2)) \equiv 0, \quad \Im(r^2, \alpha(r^2), \beta(r^2)) \equiv 0 \tag{5.57}$$

for all sufficiently small r. Therefore, $g(z, \alpha(|z|^2), \beta(|z|^2)) \equiv 0$ for z sufficiently near 0. Therefore, system (5.45) has a bifurcation of periodic solutions whose spatiotemporal symmetry can be completely characterized by Σ. This completes the proof of Theorem 5.6. □

Remark 5.3. Theorem 5.6 implies that a Hopf bifurcation for (5.45) occurs at $\alpha = 0$. Namely, in every neighborhood of $(x = 0, \alpha = 0)$, there is a branch of Σ-symmetric periodic solutions $x(t, \alpha)$ with $x(t, \alpha) \to 0$ as $\alpha \to 0$. The period T_α of $x(t, \alpha)$ satisfies $T_\alpha \to \frac{2\pi}{\omega}$ as $\alpha \to 0$. Moreover, Γ-equivariance implies that there are $(\Gamma \times \mathbb{S}^1)/\Sigma$ different periodic solutions, which have isotropy subgroups conjugate to Σ in $\Gamma \times \mathbb{S}^1$.

In what follows, we consider the bifurcation direction. Assuming sufficient smoothness of h and f, we write

$$h(0, \varphi) = D(0)\varphi + \frac{1}{2}\mathscr{H}^2(\varphi, \varphi) + \frac{1}{6}\mathscr{H}^3(\varphi, \varphi, \varphi) + o(\|\varphi\|^3)$$

$$f(0, \varphi) = L(0)\varphi + \frac{1}{2}\mathscr{F}^2(\varphi, \varphi) + \frac{1}{6}\mathscr{F}^3(\varphi, \varphi, \varphi) + o(\|\varphi\|^3).$$

In view of (5.51), we have $PF(zq + \overline{z}\overline{q} + W(zq + \overline{z}\overline{q}, \alpha, \beta), \alpha, \beta) \equiv 0$. Write $W(zq + \overline{z}\overline{q}, 0, 0)$ and $g(z, 0, 0)$ as

$$W(zq + \overline{z}\overline{q}, 0, 0) = \sum_{s+l \geq 2} \frac{1}{s!l!} W_{sl} z^s \overline{z}^l \quad g(z, 0, 0) = \sum_{s+l \geq 2} \frac{1}{s!l!} g_{sl} z^s \overline{z}^l.$$

It follows from (5.56) that $g_{21} = \mathfrak{R}_1(0,0,0) + i\mathfrak{I}_1(0,0,0)$, where $\mathfrak{R}_1(u, \alpha, \beta) = \mathfrak{R}_u(u, \alpha, \beta)$ and $\mathfrak{I}_1(u, \alpha, \beta) = \mathfrak{I}_u(u, \alpha, \beta)$. Therefore, $\mathfrak{R}_1(0,0,0) = \text{Re}\{g_{21}\}$ and $\mathfrak{I}_1(0,0,0) = \text{Im}\{g_{21}\}$. From (5.57), we can calculate the derivatives of $\alpha(r^2)$ and $\beta(r^2)$ and evaluate at $r = 0$:

$$\alpha'(0) = -\frac{\text{Re}\{g_{21}\}}{\text{Re}\{\lambda'(0)\}}, \quad \beta'(0) = -\frac{\text{Im}\{\lambda'(0)\overline{g_{21}}\}}{\text{Re}\{\lambda'(0)\}}.$$

The bifurcation direction is determined by $\text{sign}\,\alpha'(0)$, and the monotonicity of the period of the bifurcating closed invariant curve depends on $\text{sign}\,\beta'(0)$. Using a similar argument to that in [117], we have

$$\begin{aligned} g_{21} = &\langle p, \mathscr{F}^3(q, q, \overline{q}) - \tfrac{d}{dt}\mathscr{H}^3(q, q, \overline{q})\rangle \\ &+ 2\langle p, \mathscr{F}^2(q, W_{11}) - \tfrac{d}{dt}\mathscr{H}^2(q, W_{11})\rangle \\ &+ \langle p, \mathscr{F}^2(\overline{q}, W_{20}) - \tfrac{d}{dt}\mathscr{H}^2(\overline{q}, W_{20})\rangle. \end{aligned}$$

We still need to compute W_{11} and W_{20}. In fact, it follows that

$$W_{20} = -\mathscr{L}^{-1}P\left\{-\tfrac{d}{dt}\mathscr{H}^2(q, q) + \mathscr{F}^2(q, q)\right\},$$

$$W_{11} = -\mathscr{L}^{-1}P\left\{-\tfrac{d}{dt}\mathscr{H}^2(q, \overline{q}) + \mathscr{F}^2(q, \overline{q})\right\}.$$

In order to evaluate the function W_{20}, we must solve the following differential equations:

$$\frac{\mathrm{d}}{\mathrm{d}t}D(0)W_{20} - L(0)W_{20} = P\left\{-\frac{\mathrm{d}}{\mathrm{d}t}\mathcal{H}^2(q,q) + \mathcal{F}^2(q,q)\right\}. \tag{5.58}$$

Note that

$$\mathcal{H}^2(q,q) = \mathcal{H}^2(Ae^{i\omega(\cdot)},Ae^{i\omega(\cdot)})e^{2i\omega t}$$

and

$$\mathcal{F}^2(q,q) = \mathcal{F}^2(Ae^{i\omega(\cdot)},Ae^{i\omega(\cdot)})e^{2i\omega t}.$$

So, $g_{20} = \langle p, -\frac{\mathrm{d}}{\mathrm{d}t}\mathcal{H}^2(q,q) + \mathcal{F}^2(q,q)\rangle = 0$. Namely, $-\frac{\mathrm{d}}{\mathrm{d}t}\mathcal{H}^2(q,q) + \mathcal{F}^2(q,q) \in$ Ran$\mathcal{L}$. Hence, the projection P on $-\frac{\mathrm{d}}{\mathrm{d}t}\mathcal{H}^2(q,q) + \mathcal{F}^2(q,q)$ acts as the identity, and (5.58) is an inhomogeneous difference equation with constant coefficients. Thus, there is a particular solution of (5.58) of the form $W_{20}^*(t) = D_2 e^{2i\omega t}$. Substituting W_{20}^* into (5.58) and comparing the coefficients, we obtain

$$D_2 = \Delta^{-1}(0,2i\omega)\left\{\mathcal{F}^2(Ae^{i\omega(\cdot)},Ae^{i\omega(\cdot)}) - 2i\omega\mathcal{H}^2(Ae^{i\omega(\cdot)},Ae^{i\omega(\cdot)})\right\}. \tag{5.59}$$

In addition, W_{20}^* is orthogonal to p, so it belongs to Ran$\mathcal{L}$. Thus $W_{20}(0,0,0)$ is equal to W_{20}^* with D_2 determined by (5.59). Similarly, we have

$$g_{02} = g_{11} = 0, \quad W_{02} = \overline{D}_2 e^{-2i\omega t}, \quad W_{11} = D_0,$$

where $D_0 = \Delta^{-1}(0,0)\mathcal{F}^2(Ae^{i\omega(\cdot)},\overline{A}e^{-i\omega(\cdot)})$. Therefore,

$$\begin{aligned}
g_{21} = &\ \overline{B}^T\mathcal{F}^3(Ae^{i\omega(\cdot)},Ae^{i\omega(\cdot)},\overline{A}e^{-i\omega(\cdot)}) - i\omega\overline{B}^T\mathcal{H}^3(Ae^{i\omega(\cdot)},Ae^{i\omega(\cdot)},\overline{A}e^{-i\omega(\cdot)}) \\
&+ 2\overline{B}^T\mathcal{F}^2(Ae^{i\omega(\cdot)},D_0) - 2i\omega\overline{B}^T\mathcal{H}^2(Ae^{i\omega(\cdot)},D_0) \\
&+ \overline{B}^T\mathcal{F}^2(\overline{A}e^{-i\omega(\cdot)},D_2 e^{2i\omega(\cdot)}) - i\omega\overline{B}^T\mathcal{H}^2(\overline{A}e^{-i\omega(\cdot)},D_2 e^{2i\omega(\cdot)}).
\end{aligned}$$

We summarize the above discussion as follows.

Theorem 5.7. *In addition to conditions* (NHB1)–(NHB4), *assume that $L(\alpha)$ and $f(\alpha,\cdot)$ are sufficiently smooth. Then there exists a branch of Σ-symmetric periodic solutions, parameterized by α, bifurcating from the trivial solution $x = 0$ of* (5.45). *Moreover,*

(i) $\mathrm{Re}\{\lambda'(0)\}\mathrm{Re}\{g_{21}\}$ *determines the direction of the bifurcation: the bifurcation is supercritical (respectively, subcritical), i.e., the bifurcating periodic solutions exist for $\alpha > 0$ (respectively, < 0), if $\mathrm{Re}\{\lambda'(0)\}\mathrm{Re}\{g_{21}\} < 0$ (respectively, > 0);*

(ii) $\mathrm{Re}\{\lambda'(0)\}\mathrm{Im}\{\lambda'(0)\overline{g_{21}}\}$ *determines the period of the bifurcating periodic solutions along the branch: the period is greater than (respectively, smaller than) $\frac{2\pi}{\omega}$ if it is positive (respectively, negative).*

5.7 Application to a Delayed van der Pol Oscillator

The van der Pol oscillator is an oscillator with nonlinear damping governed by the second-order differential equation

$$\ddot{x} - \varepsilon(1 - x^2)\dot{x} + x = 0, \tag{5.60}$$

where x is the dynamical variable and $\varepsilon > 0$ a parameter. This model was proposed by Balthasar van der Pol [281], and the dynamics of the van del Pol oscillator with delayed feedback

$$\ddot{x} - \varepsilon(1 - x^2)\dot{x} + x = f(x(t - \tau)) \tag{5.61}$$

has recently been studied. For convenience, throughout this section, we always assume that the function $f\colon \mathbb{R} \to \mathbb{R}$ is a C^3-smooth odd function satisfying

$$f(0) = f''(0) = 0, \ f'(0) = \gamma, \text{ and } f'''(0) = \delta. \tag{5.62}$$

The characteristic equation of the linearization of (5.61) about the equilibrium point $x = 0$ is

$$\lambda^2 - \varepsilon\lambda + 1 = \gamma e^{-\lambda\tau}. \tag{5.63}$$

Thus, the Hopf bifurcation surface is given by

$$\mathcal{H} = \{(\varepsilon, \tau, \gamma) : 1 - \omega^2 = \gamma\cos\tau\omega, \ \varepsilon\omega = \gamma\sin\tau\omega, \ \omega \in \mathbb{R} \setminus \{0\}\}.$$

Now we seek the nonsemisimple $1 : 1$ resonant Hopf bifurcation points of (5.61) on the surface $\mathcal{H}$. Differentiating both sides of (5.63) with respect to λ, we have

$$2\lambda - \varepsilon = -\gamma\tau e^{-\lambda\tau}. \tag{5.64}$$

Substituting $\lambda = i\omega$ ($\omega > 0$) into (5.63) and (5.64), we obtain the following system:

$$\begin{aligned}
1 - \omega^2 - i\varepsilon\omega &= \gamma\cos\tau\omega - i\gamma\sin\tau\omega, \\
2\omega i - \varepsilon &= i\tau\gamma\sin\tau\omega - \tau\gamma\cos\tau\omega.
\end{aligned} \tag{5.65}$$

This system can be written by equating real and imaginary parts to yield the system of equations

$$\begin{aligned}
1 - \omega^2 &= \gamma\cos\tau\omega, \\
\varepsilon\omega &= \gamma\sin\tau\omega, \\
\varepsilon &= \tau\gamma\cos\tau\omega, \\
2\omega &= \tau\gamma\sin\tau\omega.
\end{aligned} \tag{5.66}$$

From this system, we have

$$\begin{aligned}
\varepsilon\tau &= 2, \\
\tau(1 - \omega^2) &= \varepsilon, \\
\tau\omega &= \xi,
\end{aligned} \tag{5.67}$$

where $x = \xi$ is a solution to the equation $x = \tan x$. Obviously, it follows from (5.67) that $\tau = \sqrt{2 + \xi^2}$, $\varepsilon = 2/\sqrt{2 + \xi^2}$, and $\omega = \xi/\sqrt{2 + \xi^2}$. Let $\{\xi_n\}_{n=1}^{\infty}$ be the monotonic increasing sequence of positive solutions of $x = \tan x$, and

$$\varepsilon_n = \frac{2}{\sqrt{2 + \xi_n^2}}, \quad \tau_n = \sqrt{2 + \xi_n^2}, \quad \gamma_n = \frac{2}{(2 + \xi_n^2)\cos \xi_n}, \quad \omega_n = \frac{\xi_n}{\sqrt{2 + \xi_n^2}} \quad (5.68)$$

for all $n \in \mathbb{N}$. Then we have the following result.

Lemma 5.7. *At and only at* $(\varepsilon, \tau, \gamma) = (\varepsilon_n, \tau_n, \gamma_n)$ *for some* $n \in \mathbb{N}$ *does* (5.63) *have a pair of double purely imaginary solutions, which are* $\pm i\omega_n$.

We rewrite the van der Pol equation (5.61) as the system

$$\dot{x} = y, \quad \dot{y} = -x + f(x(t - \tau)) - \varepsilon(x^2 - 1)y. \quad (5.69)$$

Then the linearization equation and characteristic matrix at the trivial equilibrium of (5.69) are

$$\dot{x} = y, \quad \dot{y} = -x + \gamma x(t - \tau) + \varepsilon y \quad (5.70)$$

and

$$\Delta(\varepsilon, \tau, \gamma, \lambda) = \begin{bmatrix} \lambda & -1 \\ 1 - \gamma e^{-\lambda \tau} & \lambda - \varepsilon \end{bmatrix}.$$

Consider a fixed $(\varepsilon_0, \tau_0, \gamma_0) \in \mathscr{H} \setminus \{(\varepsilon_n, \tau_n, \gamma_n)\}_{n=1}^{\infty}$, which corresponds to $\omega_0 > 0$ satisfying

$$\begin{cases} 1 - \omega_0^2 = \gamma_0 \cos \tau_0 \omega_0, \\ \varepsilon_0 \omega_0 = \gamma_0 \sin \tau_0 \omega_0, \\ 2i\omega_0 - \varepsilon_0 + \tau_0 \gamma_0 e^{-i\omega_0 \tau_0} \neq 0. \end{cases} \quad (5.71)$$

This means that (5.63) has a pair of simple purely imaginary solutions $\pm i\omega_0$. Thus, $\det \Delta(\varepsilon_0, \tau_0, \gamma_0, \pm i\omega_0) = 0$. It is easy to see that $u_1 = (1, i\omega_0)^T$ and $v_1 = (-i\omega_0 - \varepsilon_0, 1)^T$ satisfy $\Delta(\varepsilon_0, \tau_0, \gamma_0, i\omega_0)u_1 = 0$ and $\bar{v}_1^T \Delta(\varepsilon_0, \tau_0, \gamma_0, i\omega_0) = 0$. Moreover, $v = \bar{v}_1^T \Delta_1(0, i\omega_0)u_1 = 2i\omega_0 - \varepsilon_0 + \tau_0 \gamma_0 e^{-i\omega_0 \tau_0} \neq 0$. In addition, we may write (5.69) in the form of (5.15) and (5.33) with

$$\mathscr{B}(\varphi, \psi) = \begin{bmatrix} 0 \\ 0 \end{bmatrix}, \quad \mathscr{E}(\varphi, \psi, \phi) = \begin{bmatrix} 0 \\ \mathscr{C}_2(\varphi, \psi, \phi) \end{bmatrix}$$

for $\varphi = (\varphi_1, \varphi_2)^T$, $\psi = (\psi_1, \psi_2)^T$, $\phi = (\phi_1, \phi_2)^T \in C([-\tau_0, 0], \mathbb{R}^2)$, where

$$\mathscr{C}_2(\varphi, \psi, \phi) = \delta \varphi_1(-\tau_0)\psi_1(-\tau_0)\phi_1(-\tau_0) - 2\varepsilon_0 \varphi_1(0)\psi_1(0)\phi_2(0)$$
$$- 2\varepsilon_0 \varphi_1(0)\phi_1(0)\psi_2(0) - 2\varepsilon_0 \phi_1(0)\psi_1(0)\varphi_2(0).$$

Thus, $h_1(0, 0, 0) = \delta e^{-i\omega_0 \tau_0} - 2i\omega_0 \varepsilon_0$. Regarding ε (respectively, τ or γ) as a bifurcation parameter and fixing $(\tau, \gamma) = (\tau_0, \gamma_0)$ (respectively, $(\varepsilon, \gamma) = (\varepsilon_0, \gamma_0)$, or $(\varepsilon, \tau) = (\varepsilon_0, \tau_0)$), we have $\lambda'(\varepsilon_0) = i\omega_0/v$ (respectively, $\lambda'(\tau_0) = -i\omega_0 \gamma_0 e^{-i\omega_0 \tau_0}/v$, or $\lambda'(\gamma_0) = e^{-i\omega_0 \tau_0}/v$). It follows that

$$\text{sgn}\{\text{Re}[\lambda'(\varepsilon_0)]\} = \text{sgn}\{2 - \tau_0\varepsilon_0\},$$
$$\text{sgn}\{\text{Re}[\lambda'(\tau_0)]\} = \text{sgn}\{\varepsilon_0^2 + 2\omega_0^2 - 2\},$$
$$\text{sgn}\{\text{Re}[\lambda'(\gamma_0)]\} = \text{sgn}\{\gamma_0[\tau_0\gamma_0^2 - \varepsilon_0(1 + \omega_0^2)]\}.$$

In addition, we have

$$\text{sgn}\{\text{Re}[\overline{v}h_1(0,0,0)]\} = \text{sgn}\{\frac{\delta}{\gamma_0}[\tau_0\gamma_0^2 - \varepsilon_0(1 + \omega_0^2)] + 2\varepsilon_0\omega_0^2(\tau_0\varepsilon_0 - 2)\}$$

and

$$\text{sgn}\{\text{Im}[\lambda'(\varepsilon_0)v\overline{h}_1(0,0,0)]\} = \text{sgn}\{\delta\gamma_0(1 - \omega_0^2)\},$$
$$\text{sgn}\{\text{Im}[\lambda'(\tau_0)v\overline{h}_1(0,0,0)]\} = -\text{sgn}\{\gamma_0\delta + 2\varepsilon_0^2\omega_0^2\},$$
$$\text{sgn}\{\text{Im}[\lambda'(\gamma_0)v\overline{h}_1(0,0,0)]\} = \text{sgn}\{\gamma_0(1 - \omega_0^2)\}.$$

Therefore, in view of Corollary 5.1, we have the following results.

Theorem 5.8. *For fixed* $(\varepsilon_0, \tau_0, \gamma_0) \in \mathscr{H} \setminus \{(\varepsilon_n, \tau_n, \gamma_n)\}_{n=1}^{\infty}$ *with* $\omega_0 > 0$ *satisfying* (5.71), *we have the following:*

(1) *If* $(\tau, \gamma) = (\tau_0, \gamma_0)$, *then there exists a unique branch of periodic solutions, parameterized by* ε, *bifurcating from the trivial solution* $x = 0$ *of* (5.69), *which exists for* $\varepsilon > \varepsilon_0$ *(respectively,* $< \varepsilon_0$*) when*

$$\frac{\delta}{\gamma_0}[\tau_0\gamma_0^2 - \varepsilon_0(1 + \omega_0^2)](2 - \tau_0\varepsilon_0) - 2\varepsilon_0\omega_0^2(2 - \tau_0\varepsilon_0)^2 < 0$$

(respectively, > 0*), and whose period is greater than (respectively, smaller than)* $\frac{2\pi}{\omega_0}$ *when* $\delta\gamma_0(1 - \omega_0^2)(2 - \tau_0\varepsilon_0) > 0$ *(respectively,* < 0*).*

(2) *If* $(\varepsilon, \gamma) = (\varepsilon_0, \gamma_0)$, *then there exists a unique branch of periodic solutions, parameterized by* τ, *bifurcating from the trivial solution* $x = 0$ *of* (5.69), *which exists for* $\tau > \tau_0$ *(respectively,* $< \tau_0$*) when*

$$\left\{\frac{\delta}{\gamma_0}[\tau_0\gamma_0^2 - \varepsilon_0(1 + \omega_0^2)] + 2\varepsilon_0\omega_0^2(\tau_0\varepsilon_0 - 2)\right\}(\varepsilon_0^2 + 2\omega_0^2 - 2) < 0$$

(respectively, > 0*), and whose period is greater than (respectively, smaller than)* $\frac{2\pi}{\omega_0}$ *when* $(\varepsilon_0^2 + 2\omega_0^2 - 2)(\gamma_0\delta + 2\varepsilon_0^2\omega_0^2) < 0$ *(respectively,* > 0*).*

(3) *If* $(\varepsilon, \tau) = (\varepsilon_0, \tau_0)$, *then there exists a unique branch of periodic solutions, parameterized by* γ, *bifurcating from the trivial solution* $x = 0$ *of* (5.69), *which exists for* $\gamma > \gamma_0$ *(respectively,* $< \gamma_0$*) when*

$$\delta[\tau_0\gamma_0^2 - \varepsilon_0(1 + \omega_0^2)]^2 + 2\varepsilon_0\omega_0^2\gamma_0(\tau_0\varepsilon_0 - 2)[\tau_0\gamma_0^2 - \varepsilon_0(1 + \omega_0^2)] < 0$$

(respectively, > 0*), and whose period is greater than (respectively, smaller than)* $\frac{2\pi}{\omega_0}$ *when* $(1 - \omega_0^2)[\tau_0\gamma_0^2 - \varepsilon_0(1 + \omega_0^2)] > 0$ *(respectively,* < 0*).*

In view of Lemma 5.7, $\det\Delta(\varepsilon_n, \tau_n, \gamma_n, \pm i\omega_n) = 0$. Moreover, it is easy to see that $u_1 = (1, i\omega_n)^T$, $v_1 = (1,0)^T$, and $v_2 = (-i\omega_n - \varepsilon_n, 1)^T$ satisfy

$$\Delta(\varepsilon_n, \tau_n, \gamma_n, i\omega_n)u_1 = 0,$$
$$\bar{v}_2^T \Delta(\varepsilon_n, \tau_n, \gamma_n, i\omega_n) = 0,$$
$$\bar{v}_1^T \Delta(\varepsilon_n, \tau_n, \gamma_n, i\omega_n) = -\bar{v}_2^T \Delta_\lambda(\varepsilon_n, \tau_n, \gamma_n, i\omega_n).$$

Thus, it follows that $v = i\xi_n$. Moreover,

$$\bar{v}_2^T \Delta_\varepsilon(\varepsilon_n, \tau_n, \gamma_n, i\omega_n)u_1 = -i\omega_n,$$
$$\bar{v}_2^T \Delta_\tau(\varepsilon_n, \tau_n, \gamma_n, i\omega_n)u_1 = i\omega_n\gamma_n e^{-i\xi_n},$$
$$\bar{v}_2^T \Delta_\gamma(\varepsilon_n, \tau_n, \gamma_n, i\omega_n)u_1 = -e^{-i\xi_n}.$$

Similarly, we have $h_1(0,0,0) = \delta e^{-i\xi_n} - 2i\omega_n\varepsilon_n$. It follows from (5.37) that $B = -(2\omega_n\varepsilon_n + i\delta e^{-i\xi_n})/(\xi_n\omega_n^2)$ and $(-1)^n\delta\text{Im}(B) > 0$.

We denote by A_ε (respectively, A_τ and A_γ) the corresponding quantity A in (5.37) when ε (respectively, τ and γ) is regarded as a bifurcation parameter. In fact, we have

$$A_\varepsilon = \frac{-1}{\xi_n\omega_n}, \quad A_\tau = \frac{\varepsilon_n(1-i\xi_n)}{\xi_n^2}, \quad A_\gamma = \frac{(\xi_n+i)\cos\xi_n}{\xi_n\omega_n^2}.$$

It follows that $\text{Im}(A_\varepsilon) = 0$, $\text{Im}(A_\tau) < 0$, but $(-1)^n\text{Im}(A_\gamma) < 0$. Therefore, Theorem 5.4 is applicable only to the case in which either τ or γ is a bifurcation parameter.

In what follows, we calculate μ_0 given by (5.43) by regarding either τ or γ as a parameter. In the case in which τ is a parameter, we have

$$\text{sgn}\{\mu_0\} = -\text{sgn}\left\{\text{Im}(h_1(0,0,0)v_2^T\Delta_\tau(\varepsilon_n, \tau_n, \gamma_n, -i\omega_n)\bar{u}_1)\right\}$$
$$= \text{sgn}\{\text{Im}[i\gamma_n e^{i\xi_n}(\delta e^{-i\xi_n} - 2i\omega_n\varepsilon_n)]\}$$
$$= \text{sgn}\{\delta\gamma_n + 2\varepsilon_n^2\omega_n^2\}.$$

In the case in which γ is regarded as a parameter, we have

$$\text{sgn}\{\mu_0\} = (-1)^{n-1}\text{sgn}\left\{\text{Im}(h_1(0,0,0)v_2^T\Delta_\gamma(\varepsilon_n, \tau_n, \gamma_n, -i\omega_n)\bar{u}_1)\right\}$$
$$= (-1)^n\text{sgn}\{\text{Im}[e^{i\xi_n}(\delta e^{-i\xi_n} - 2i\omega_n\varepsilon_n)]\}$$
$$= 1.$$

Applying Theorem 5.4, we have the following results.

Theorem 5.9. *For a fixed $n \in \mathbb{N}$, at $(\varepsilon, \tau, \gamma) = (\varepsilon_n, \tau_n, \gamma_n)$, (5.61) undergoes a Hopf bifurcation with nonsemisimple 1 : 1 resonance. More precisely:*

(i) *For γ near γ_n, (5.61) with $(\varepsilon, \tau) = (\varepsilon_n, \tau_n)$ has exactly two nontrivial periodic solutions bifurcated from the equilibrium point $x = 0$, which exist for $\gamma > \gamma_n$ (respectively, $\gamma < \gamma_n$) if $\delta < 0$ (respectively, $\delta > 0$).*

(ii) *If $\delta \gamma_n + 2\varepsilon_n^2 \omega_n^2 < 0$, then for τ near τ_n, (5.61) with $(\varepsilon, \gamma) = (\varepsilon_n, \gamma_n)$ has no nontrivial periodic solutions bifurcated from the equilibrium point $x = 0$.*

(iii) *If $\delta \gamma_n + 2\varepsilon_n^2 \omega_n^2 > 0$, then for τ near τ_n, (5.61) with $(\varepsilon, \gamma) = (\varepsilon_n, \gamma_n)$ has exactly two nontrivial periodic solutions bifurcated from the equilibrium point $x = 0$, which exist for $\tau > \tau_n$ (respectively, $\tau < \tau_n$) if $(-1)^n \delta < 0$ (respectively, $(-1)^n \delta > 0$).*

5.8 Applications to a Ring Network

To illustrate the general results presented in the previous sections of this chapter, we now consider a ring network consisting of n identical elements with time-delayed nearest-neighbor coupling:

$$[u_j(t) - cu_j(t-1)]' = g(u_{j+1}(t-1)) + g(u_{j-1}(t-1)) - 3g(u_j(t-1)), \quad (5.72)$$

where $i \pmod n$, $g \in C^3(\mathbb{R}; \mathbb{R})$ with $g(0) = g''(0) = 0$ and $g'(0) = b > 0$, and $c \in [0, 1)$ is the bifurcation parameter. Define the action of the dihedral group $\mathbb{D}_n$ on $\mathbb{R}^n$ by

$$(\rho \cdot u)_j = u_{j+1} \qquad \text{and} \qquad (\kappa \cdot u)_j = u_{2-j} \qquad (5.73)$$

for all $j \pmod n$ and $u \in \mathbb{R}^N$. It is easy to see that system (5.72) is $\mathbb{D}_n$-equivariant (see [128, 133]). Let $\mathscr{A}(c)$ be the infinitesimal generator of the linear operator generated by the linearization of (5.72) about the trivial solution $u = 0$. It can be shown that $\lambda \in \mathbb{C}$ is an eigenvalue of $\mathscr{A}(c)$ if and only if $\prod_{j=0}^{n-1} p_j(\lambda, c) = 0$, where $p_j(\lambda, c) = \lambda + (\vartheta_j - c\lambda)e^{-\lambda}$ and $\vartheta_j = b + 4b \sin^2(2j\pi/n) > 0$.

For a given j, $p_j(\cdot, c)$ has a pair of purely imaginary zeros $\pm\beta_0$ if $c = \cos\beta_0$ and $\vartheta_j = \beta_0 \sin\beta_0$. This results in a family of bifurcation values $c_{j,k}$ in the interval $[0, 1)$, where $c_{j,k} = \cos\beta_{j,k}$ for $k \in \mathbb{N}$, and $\{\beta_{j,k}\}_{k=1}^{\infty}$ is a strictly increasing sequence of positive numbers satisfying $\vartheta_j = \beta_{j,k} \sin\beta_{j,k}$ for all $k \in \mathbb{N}$ and $\lim_{k \to \infty} \beta_{j,k} = \infty$. Moreover, if $\lambda(c)$ is a smooth curve of zeros of $p_j(\cdot, c)$ with $\lambda(c_{j,k}) = i\beta_{j,k}$, it is easy to see that

$$\lambda'(c_{j,k}) = D(c_{j,k}\beta_{j,k}^2 - i\vartheta_j\beta_{j,k}), \qquad (5.74)$$

where $D = |(\cos\beta_{j,k} + i\sin\beta_{j,k})(1 + \beta_{j,k}) - c_{j,k}|^{-2}$. Therefore, for fixed j and k, $\mathscr{A}(c_{j,k})$ has a pair of purely imaginary eigenvalues $\pm i\beta_{j,k}$ with the associated eigenspace E_0 spanned by the eigenvectors $e^{i\beta_{j,k}(\cdot)}v_j$, $e^{i\beta_{j,k}(\cdot)}\overline{v_j}$, $e^{-i\beta_{j,k}(\cdot)}v_j$, and $e^{-i\beta_{j,k}(\cdot)}\overline{v_j}$, where $v_j = (1, e^{2i\pi j/n}, \ldots, e^{2i(n-1)j\pi/n})$. Thus, assumptions (NHB1), (NHB2), and (NHB4) hold. If $j \neq 0$ and $j \neq n/2$, it furthermore follows [133] that

$$\text{Fix}(\Sigma_\kappa^+) = \text{span}\{w_1 \cos(\beta_{j,k}t), w_1 \sin(\beta_{j,k}t)\},$$
$$\text{Fix}(\Sigma_\kappa^-) = \text{span}\{w_2 \cos(\beta_{j,k}t), w_2 \sin(\beta_{j,k}t)\},$$
$$\text{Fix}(\Sigma_\rho^+) = \text{span}\{\text{Re}(\overline{v_j}e^{i\beta_{j,k}t}), \text{Im}(\overline{v_j}e^{i\beta_{j,k}t})\}, \tag{5.75}$$
$$\text{Fix}(\Sigma_\rho^-) = \text{span}\{\text{Re}(v_j e^{i\beta_{j,k}t}), \text{Im}(v_j e^{i\beta_{j,k}t})\},$$

where $w_1 = \text{Re}(v_j)$, $w_2 = \text{Im}(v_j)$, and $\Sigma_\kappa^\pm = (\kappa, \pm 1)$ and $\Sigma_\rho^\pm = (\rho, e^{\pm 2ij\pi/(n\beta_{j,k})})$ are subgroups of $\mathbb{D}_n \times \mathbb{S}^1$. Thus, all conditions of the equivariant Hopf bifurcation theorem (Theorem 5.6) are satisfied. Therefore, we apply Theorem 5.6 to system (5.72) and obtain the following results.

Theorem 5.10. *(i) Near $c = c_{0,k}$ for each $k \in \mathbb{N}$, there exists a branch of synchronous periodic solutions of period ω near $(2\pi/\beta_{0,k})$ bifurcated from the zero solution of the system. (ii) Near $c = c_{j,k}$, for each $j \in \{1, 2, \ldots, [(n-1)/2]\}$ and $k \in \mathbb{N}$, there exist $2(n+1)$ branches of asynchronous periodic solutions of period ω near $(2\pi/\beta_{j,k})$ bifurcated from the zero solution of the system, and these are two phase-locked waves, n mirror-reflecting waves, and n standing waves.*

In what follows, we start with the two phase-locked oscillations mentioned above, which are characterized by $\Sigma_\rho^\pm$. In view of (5.75), for the vectors A and B defined in Sect. 5.6, we choose $A = nB = \overline{v}_j$ or $A = nB = v_j$. We have

$$\text{Re}\{\overline{(p,q)}g_{21}\} = \vartheta_j^2 g'''(0) \quad \text{and} \quad \text{Im}\{\lambda'(0)(p,q)\overline{g_{21}}\} = -g'''(0)\vartheta_j\beta_{j,k}. \tag{5.76}$$

Similarly, we choose $A = \frac{n}{2}B = w_1$ for the mirror-reflecting waves characterized by Σ_κ^+, and $A = \frac{n}{2}B = w_2$ for the mirror-reflecting waves characterized by Σ_κ^-. By a direct computation, we have

$$\text{Re}\{\overline{(p,q)}g_{21}\} = \frac{m}{4}\vartheta_j^2 g'''(0) \quad \text{and} \quad \text{Im}\{\lambda'(0)(p,q)\overline{g_{21}}\} = -\frac{m}{4}g'''(0)\vartheta_j\beta_{j,k}, \tag{5.77}$$

where $m = 4$ if $4j = 0 \pmod{n}$, and $m = 3$ otherwise.

Finally, for the synchronous periodic solution mentioned in Theorem 5.10, we can show that (5.76) holds for $j = 0$. Thus, applying Theorem 5.7, we have the following results.

Theorem 5.11. *Near $c = c_{j,k}$, for each $j \in \{0, 1, \ldots, n-1\}$ and $k \in \mathbb{N}$, system (5.72) undergoes a Hopf bifurcation, whereby both the bifurcation direction and the period of bifurcating periodic solutions are determined by the sign of $g'''(0)$. More precisely, if $g'''(0) < 0$ (or > 0), then (i) the Hopf bifurcation is supercritical (respectively, subcritical), and all the bifurcating periodic solutions exist for $c > c_{j,k}$ (respectively, $< c_{j,k}$); (ii) the period of each branch of bifurcating periodic solutions is greater (respectively, less) than $(2\pi/\beta_{j,k})$.*

5.9 Coupled Systems of NFDEs and Lossless Transmission Lines

Coupled systems of neutral functional differential equations also arise very naturally from distributed transmission lines, specially lossless transmission lines [37–40]. The main idea used to obtain such neutral difference–differential equations is the reduction of the classical telegrapher's partial differential equation, which describes the voltage and current changes in a transmission line by introducing the d'Alembert solution of the wave equation and using the boundary condition at terminals. This idea goes back at least as far as Abolinia and Mishkis [1, 2], who demonstrated the existence and uniqueness of solutions to a mixed problem for hyperbolic systems by converting them to integral–functional equations, with integration along characteristics. Self-sustained periodic solutions of small amplitude were established by Brayton [37, 38] for a single lossless transmission, which has been widely investigated in the literature.

Often used in industrial applications are, however, multiconductor lines. As an electric circuit, a self-contained single transmission line is assumed to be removed far enough from other lines so that it is not affected by any electrical changes occurring in the latter. As soon as a second transmission line is placed close to the first one, the fields of the first line induce a voltage and a current on the second. Capacitive coupling is then produced by the electric field, and inductive coupling results from the magnetic field. The classical applications of telephone (or telegraph) line and high-voltage power transmission line are examples of coupling. The coupling phenomenon is also used in practice to realize directional couplers and interdigital filters. Moreover, in modern high-speed integrated circuit (IC) technology, coupling among a group of physically close transmission lines is very common, and interconnects in high-density ICs are usually treated as transmission lines. We refer to the work [305] for some relevant references for coupled electric circuits and transmission lines. Here, we present how a coupled system of neutral equations can be derived.

Following [305], we consider a ring array of mutually coupled lossless transmission lines. For simplicity, we assume that the transmission lines are resistively coupled and the capacitive and inductive couplings among the system are neglected. We also assume that each linked transmission line is identical and terminates at each end by a lumped linear or nonlinear circuit element. By employing a telegrapher's equation at each line together with a coupling term in the initial–boundary condition, we derive a symmetric difference–differential system of neutral type that is equivalent to the original partial differential equations governing the coupled lines.

Let N be a positive integer. We consider a ring of N mutually coupled lossless transmission line (LLTL) networks that are interconnected by a common resistor R. We assume that all coupled LLTL networks are identical, each of which is a uniformly distributed lossless transmission line with series inductance L_s and parallel capacitance C_s per unit length of the line. To derive the network equations, let us take an x-axis in the direction of the line, with two ends of the normalized line at $x = 0$ and $x = 1$. Let $i_k(x,t)$ denote the current flowing in the kth line at time t and distance x down the line and let $v_k(x,t)$ denote the voltage across the line at t and x.

Then we obtain the following partial differential equations (telegrapher's equations)

$$L_s \frac{\partial i_k}{\partial t} = -\frac{\partial v_k}{\partial x},$$

$$C_s \frac{\partial v_k}{\partial t} = -\frac{\partial i_k}{\partial x}, \qquad k = 1, \cdots N.$$

We now couple the network resistively so that the middle lines have coupling terms from the preceding and succeeding lines. At two ends $x = 0$ and $x = 1$, the line gives rise to the boundary conditions

$$0 = E - v_k(0,t) - R_0 i_k(0,t),$$

$$-C\frac{\mathrm{d}}{\mathrm{d}t} v_k(1,t) = -i_k(1,t) + f(v_k(1,t)) - (I_{k-1} - I_k),$$

$$v_k(1,t) - v_{k+1}(1,t) = RI_k(t),$$

where E is the constant DC bias voltage, $f(v_k(1,t))$ is the current $(V - I$ characteristic) through the nonlinear resistor, and I_k is the network current coupling term. Under equilibrium conditions, we have $i_k(0,t) = i_k(1,t)$ and $v_k(0,t) = v_k(1,t)$. Thus, we have the following equilibrium equations:

$$E - v_k - R_0 i_k = 0,$$

$$i_k = f(v_k) - \frac{1}{R}(v_{k+1} - 2v_k + v_{k-1}),$$

which are assumed to have a unique homogeneous solution (v^*, i^*), homogeneous for $1 \leq k \leq N$. By changing variables, the equilibrium can be shifted from (v^*, i^*) to $(0,0)$, and we obtain

$$L_s \frac{\partial i_k}{\partial t} = -\frac{\partial v_k}{\partial x},$$

$$C_s \frac{\partial v_k}{\partial t} = -\frac{\partial i_k}{\partial x}, k = 1, \cdots N,$$

subject to the boundary condition

$$0 = v_k(0,t) + R_0 i_k(0,t),$$

$$-C\frac{\mathrm{d}}{\mathrm{d}t} v_k(1,t) = -i_k(1,t) + \tilde{g}(v_k(1,t)) - \frac{1}{R}(v_{k+1} - 2v_k + v_{k-1})(1,t),$$

with $\tilde{g}(v) = f(v + v^*) - f(v^*)$.

The above problem has the unique solution (d'Alembert solution) $i_k(x,t)$ and $v_k(x,t)$, which are of the form

$$v_k(x,t) = \frac{1}{2}[\phi_k(x - \sigma t) + \psi_k(x + \sigma t)],$$

$$i_k(x,t) = \frac{1}{2Z}[\phi_k(x - \sigma t) - \psi_k(x + \sigma t)],$$

with

$$\sigma = \sqrt{1/L_s C_s}, \quad Z = \sqrt{L_s/C_s}$$

respectively the propagation velocity of waves and the characteristic impedance of the line, and both ϕ_k and ψ_k are C^1-smooth.

Let $\phi_{k_1} = \phi_k(1 - \sigma t), \phi_{k_0} = \phi_k(-\sigma t), \psi_{k_1} = \psi_k(1 + \sigma t), \phi_{k_0} = \phi_k(\sigma t)$, and let $V_k(t) = v_k(1,t)$. Then we have

$$\phi_{k_1}(t) = V_k(t) + Zi_k(1,t), \quad \phi_{k_0}(t) = v_k(0,t) + Zi_k(0,t),$$
$$\psi_{k_1}(t) = V_k(t) - Zi_k(1,t), \quad \psi_{k_0}(t) = v_k(0,t) - Zi_k(0,t).$$

Note that $\phi_{k_1}(t) = \phi_{k_0}(t - 1/\sigma)$ and $\psi_{k_1}(t) = \psi_{k_0}(t + 1/\sigma)$. We thereby obtain

$$V_k(t) + Zi_k(1,t) = -q\psi_{k_1}(t - r),$$
$$V_k(t) - Zi_k(1,t) = \psi_{k_1}(t),$$

with

$$r = \frac{2}{\sigma}, \quad q = \frac{Z - R_0}{Z + R_0}.$$

The second boundary condition gives

$$i_k(1,t) = CV_k'(t) + \tilde{g}(V_k(t)) - \frac{1}{R}(V_{k+1}(t) - 2V_k(t) + V_{k-1}(t)).$$

Substituting this into the equation for $V_k(t) + Zi_k(1,t)$ and eliminating $\psi_{k-1}(t - r)$ leads to

$$V_k(t) + Z[CV_k' + \tilde{g}(V_k) - \frac{1}{R}(V_{k+1}(t) - 2V_k(t) + V_{k-1}(t))]$$
$$= -qV_k(t - r) + qZ[CV_k'(t - r) + \tilde{g}(V_k(t - r))]$$
$$- \frac{qZ}{R}[V_{k+1}(t - r) - 2V_k(t - r) + V_{k-1}(t - r)].$$

Finally, we arrive at the following neutral functional differential equations:

$$\frac{d}{dt}[V_k(t) - qV_k(t - r)] = -\frac{1}{ZC}V_k(t) - \frac{q}{ZC}V_k(t - r)$$
$$- g(V_k) + qg(V_k(t - r))$$
$$+ \frac{1}{RC}[V_{k+1}(t) - qV_{k+1}(t - r) - 2(V_k(t)$$
$$- qV_k(t - r)) + V_{k-1}(t) - qV_{k-1}(t - r)]$$

with $g(V_k) = (1/C)\tilde{g}(V_k)$.

Under suitable conditions, Wu and Xia [305, 306] proved that there is a sequence of critical values $q_1 < q_2 < \cdots$ in $(0,1)$ at which the neutral system has a Hopf bifurcation of periodic solutions bifurcating from the trivial equilibrium, and these periodic solutions take the form of synchronized or phase-locked, mirror-reflecting, and standing waves as discussed in the previous section. It should be possible to conduct the stability and bifurcation direction analysis as well, and this is left to the interested reader. An important problem of interest to applications is whether coupling can generate stable asynchronous periodic solutions. This issue has also been addressed by Hale in [148–150]. Finally, notice that the amplitude of the bifurcated periodic solution obtained using the theory developed so far must be small. To obtain periodic solutions of potentially large amplitudes, the bifurcation parameter should be away from the bifurcation values, and we are led to a discussion of global Hopf bifurcation theory, which will be discussed in the next chapter.

5.10 Wave Trains in the FPU Lattice

The FPU lattice was introduced in [97] as a model for a nonlinear string formed by identical point masses that interact with their nearest neighbors. It consists of an infinite set of ordinary differential equations for the particle positions q_j:

$$\ddot{q}_j = W'(q_{j+1} - q_j) - W'(q_j - q_{j-1}) , \ j \in \mathbb{Z} . \tag{5.78}$$

Note that (5.78) are Hamiltonian with respect to the formal Hamiltonian function

$$H = \sum_{j \in \mathbb{Z}} \frac{1}{2} \dot{q}_j^2 + W(q_{j+1} - q_j) .$$

Usually, we assume that the interaction potential has a Taylor expansion of the form

$$W(z) = \frac{1}{2} z^2 + \frac{\alpha}{3!} z^3 + \frac{\beta}{4!} z^4 + \cdots .$$

We note that the FPU lattice is $\mathbb{Z}$-equivariant with respect to the group of simultaneous particle shifts: $\{q_j(t)\}_{j \in \mathbb{Z}}$ is a solution of the equations of motion if and only if $\{\tilde{q}_j(t)\}_{j \in \mathbb{Z}}$ defined by $\tilde{q}_j(t) = q_{j+1}(t)$ is a solution of the equations of motion. We now say that a solution to (5.78) is a wave train if it is a time-periodic solution that is relatively periodic with respect to the maximal particle-shift symmetry. In other words, it satisfies

- $\exists\, T > 0$, such that $q_j(t) = q_j(t+T)$.
- $\exists\, \tau > 0$, such that $q_{j+1}(t) = q_j(t+\tau)$.

Such solutions have the form

$$q_j(t) = u(\omega t - kj), \tag{5.79}$$

where $\omega = 1/T > 0$, $k = \omega\tau$, and u is a one-periodic function. One sees that the ansatz (5.79) produces solutions of (5.78) precisely when u satisfies the advance–delay differential equations

$$\omega^2 \frac{\mathrm{d}^2 u(s)}{\mathrm{d}s^2} = W'(u(s-k) - u(s)) - W'(u(s) - u(s+k)). \tag{5.80}$$

It is easy to see that wave trains exist in the *linear* FPU lattice, i.e., the lattice for which $\alpha = \beta = \ldots = 0$. Indeed, for every $\varepsilon > 0$ and $\phi_0 \in \mathbb{R}/\mathbb{Z}$, the functions

$$q_j(t) = \varepsilon \cos(2\pi\omega t - 2\pi k j + \phi_0) \tag{5.81}$$

are solutions of the linear FPU equations of motion, exactly when ω and k are related by the dispersion relation

$$\omega = \pm\omega(k) \stackrel{\text{def}}{=} \pm\frac{1}{\pi}\sin(k\pi).$$

The above wave trains are *monochromatic*, and it follows from a Fourier transformation that every motion of the linear lattice are a superposition of such monochromatic wave trains. Some of these superpositions are actually wave trains themselves, for instance if k is "resonant" in the sense that

$$\frac{\sin(qk\pi)}{\sin(pk\pi)} = \frac{q}{p}$$

for some integers $p, q \in \mathbb{Z}$. Writing $\omega \stackrel{\text{def}}{=} \frac{\sin(qk\pi)}{q} = \frac{\sin(pk\pi)}{p}$, we have that

$$q_j(t) = \varepsilon_1 \cos(2\pi p\omega t - 2\pi pkj + \phi_0) + \varepsilon_2 \cos(2\pi q\omega t - 2\pi qkj + \phi_1)$$

are wave train solutions of the linear lattice with temporal period $T = qp/\omega$ and relative spatial period $\tau = k/\omega$. We call these *bichromatic* wave trains.

An elementary question is whether the monochromatic and bichromatic wave trains of the linear FPU lattice continue to exist in the nonlinear lattice. Guo–Lamb –Rink [137] have addressed this elementary question by means of a Lyapunov–Schmidt reduction. This is a way of reducing the advance–delay differential equations (5.80) to a finite-dimensional bifurcation equation. The work [137] also shows how the particle-shift $\mathbb{Z}$-equivariance, the time reversal symmetry, and the Hamiltonian structure manifest themselves in the reduced bifurcation equation, following ideas set out in [116]. Other existence results for wave trains in the FPU lattice can be found in Iooss [174], where, among others, wave trains of small amplitude and long wave length are found by means of a center manifold reduction. Nonperturbative existence results for wave trains also exist; cf. Filip et al. [104] for the case that the potential energy function W is convex. In the latter paper, both a variational proof and a degree-theoretic argument are given.

Chapter 6
Degree Theory

6.1 Introduction

Many applications, including some bifurcation problems of functional differential equations, lead to the problem of finding all zeros of a mapping $f\colon U \subseteq X \to X$, where X is some (real) Banach space. In this type of nonlinear problem, we are interested in the solutions of

$$f(x) = 0, \quad x \in U. \tag{6.1}$$

In most cases, it turns out that it is too much to ask to determine the zeros analytically and explicitly. Hence one looks for a more qualitative study of the zeros, such as the number, location, and multiplicity.

To illustrate this and to motivate the topological degree, we consider the case $f \in \mathscr{H}(\mathbb{C})$, where $\mathscr{H}(\mathbb{C})$ denotes the set of holomorphic functions on a domain $U \subset \mathbb{C}$. Recall that the winding number of a path $\gamma\colon [0,1] \to \mathbb{C}$ around a point $z_0 \in \mathbb{C}$ is defined by

$$n(\gamma, z_0) = \frac{1}{2\pi i} \int_\gamma \frac{dz}{z - z_0} \in \mathbb{Z}. \tag{6.2}$$

It gives the number of times that z_0 is encircled, taking orientation into account (that is, encirclings in opposite directions are counted with opposite signs).

In particular, if we pick $f \in \mathscr{H}(\mathbb{C})$, we compute (assuming $0 \notin f(\gamma)$)

$$n(f(\gamma), 0) = \frac{1}{2\pi i} \int_\gamma \frac{f'(z)}{f(z)} dz = \sum_k n(\gamma, z_k) \alpha_k, \tag{6.3}$$

where z_k denotes zeros of f, and α_k their respective multiplicity. Moreover, if γ is a Jordan curve encircling a simply connected domain $U \subset \mathbb{C}$, then $n(\gamma, z_k) = 0$ if $z_k \notin U$ and $n(\gamma, z_k) = 1$ if $z_k \in U$. Hence $n(f(\gamma), 0)$ counts the number of zeros inside U.

Let us also recall how we compute complex integrals along complicated paths using homotopy invariance (see [23, 240, 241]). In this approach, we look for a

S. Guo and J. Wu, *Bifurcation Theory of Functional Differential Equations*,
Applied Mathematical Sciences 184, DOI 10.1007/978-1-4614-6992-6_6,
© Springer Science+Business Media New York 2013

simpler path along which the integral can be computed that is homotopic to the original one. In particular, if $f: \gamma \to \mathbb{C} \setminus \{0\}$ and $g: \gamma \to \mathbb{C} \setminus \{0\}$ are homotopic, we have $n(f(\gamma),0) = n(g(\gamma),0)$ (which is known as Rouché's theorem). More explicitly, we need to find a mapping g for which $n(g(\gamma),0)$ can be computed and a homotopy $H: [0,1] \times \gamma \to C \setminus \{0\}$ such that $H(0,z) = f(z)$ and $H(1,z) = g(z)$ for $z \in \gamma$. For example, to see how many zeros of $f(z) = \frac{1}{2}z^6 + z - \frac{1}{3}$ lie inside the unit circle, we consider $g(z) = z$. Then $H(t,z) = (1-t)f(z) + tg(z)$ is the required homotopy, since $|f(z) - g(z)| < |g(z)|$, $|z| = 1$, implying $H(t,z) \neq 0$ on $[0,1] \times \gamma$. Hence $f(z)$ has one zero inside the unit circle.

To summarize, given a (sufficiently smooth) domain U with enclosing Jordan curve ∂U, we have defined a degree $\deg(f,U,z_0) = n(f(\partial U),z_0) = n(f(\partial U) - z_0,0) \in \mathbb{Z}$ that counts the number of solutions of $f(z) = z_0$ inside U. The invariance of this degree with respect to certain deformations of f allow us to explicitly compute $\deg(f,U,z_0)$ even in nontrivial cases. Degree theory has been developed for various classes of mappings, not all of which are mentioned in the chapter. For relevant results on topological degree, see, for example, [24, 25, 177–182, 191–195]. Moreover, similar ideas also appears in the definitions of Fuller index. See, for example, Chow and Mallet-Paret [69].

6.2 The Brouwer Degree

In 1912, Brouwer [47] introduced the so-called Brouwer degree in $\mathbb{R}^n$. See Brouwder [46], Alexander et al. [8–10], Chow et al. [71], Krasnosel'skii [191], Sieberg [265] for historical developments. In this section, we introduce Brouwer degree theory. Throughout this section, U will be a bounded open subset of $\mathbb{R}^n$. For $f \in C^1(U,\mathbb{R}^n)$, the Jacobi matrix of f at $x \in U$ is $f'(x) = (\frac{\partial f_j(x)}{\partial x_i})_{1 \leq i,j \leq n}$, and the Jacobi determinant of f at $x \in U$ is

$$J_f(x) = \det f'(x).$$

The set of regular values is

$$\mathrm{RV}(f) = \{y \in \mathbb{R}^n : J_f(x) \neq 0 \text{ for all } x \in f^{-1}(y)\}.$$

Its complement $\mathrm{CV}(f) = \mathbb{R}^n \setminus \mathrm{RV}(f)$ is called the set of critical values. Set $C^r(\bar{U},\mathbb{R}^n) = \{f \in C^r(U,\mathbb{R}^n) : d^j f \in C(\bar{U},\mathbb{R}^n) \text{ for all } 0 \leq j \leq r\}$ and

$$D_y^r(\bar{U},\mathbb{R}^n) = \{f \in C^r(\bar{U},\mathbb{R}^n) : y \notin f(\partial U)\},$$
$$D_y^0(\bar{U},\mathbb{R}^n) = \{f \in C(\bar{U},\mathbb{R}^n) : y \notin f(\partial U)\}$$

for $y \in \mathbb{R}^n$.

Lemma 6.1 (Sard's lemma). *Let $U \subset \mathbb{R}^n$ be open and $f \in C^1(U,\mathbb{R}^n)$. Then μ_n $(f(\mathrm{CV}(f))) = 0$, where μ_n is the n-dimensional Lebesgue measure.*

A function deg that assigns each $f \in D_y^0(U, \mathbb{R}^n)$, $y \in \mathbb{R}^n$, a real number $\deg(f, U, y)$ will be called a degree if it satisfies the following conditions:

(BD1) *(translation invariance)* $\deg(f, U, y) = \deg(f - y, U, 0)$.

(BD2) *(normalization)* $\deg(\mathbb{I}, U, y) = 1$ if $y \in U$, where $\mathbb{I}$ denotes the identity operator when the space involved is clear.

(BD3) *(additivity)*   If U_1 and U_2 are open, disjoint subsets of U such that $y \notin f(U \setminus (U_1 \cup U_2))$, then $\deg(f, U, y) = \deg(f, U_1, y) + \deg(f, U_2, y)$.

(BD4) *(homotopy invariance)*   If $H : [0,1] \times \bar{U} \to \mathbb{R}^n$ is continuous, so that $y \notin H(t, \partial U)$ for evert $t \in [0,1]$, and $f = H(0, \cdot), g = H(1, \cdot)$, then $\deg(f, U, y) = \deg(g, U, y)$.

To compute the degree of a nonsingular matrix, we need the following lemma.

Lemma 6.2. *Two nonsingular matrices $M_1, M_2 \in GL(n)$ are homotopic in $GL(n)$ if and only if* $\operatorname{sgn} \det M_1 = \operatorname{sgn} \det M_2$.

Using this lemma, we can prove the following theorem.

Theorem 6.1. *Suppose $f \in D_y^1(\bar{U}, \mathbb{R}^n)$ and $y \notin CV(f)$. Then a degree satisfying* (BD1)–(BD4) *satisfies*

$$\deg(f, U, y) = \sum_{x \in f^{-1}(y)} \operatorname{sgn} J_f(x), \tag{6.4}$$

where the sum is finite.

In fact, the determinant formula (6.4) can be extended to all $f \in D_y^0(\bar{U}, \mathbb{R}^n)$, that is, $\deg(f, U, y)$ as defined in (6.4) is locally constant with respect to both y and f. In particular, we have the following result.

Theorem 6.2. *There is a unique degree* deg *satisfying* (BD1)–(BD4). *Moreover, for each given $f \in D_y^0(\bar{U}, \mathbb{R}^n)$, we have*

$$\deg(f, U, y) = \sum_{x \in \tilde{f}^{-1}(y)} \operatorname{sgn} J_{\tilde{f}}(x), \tag{6.5}$$

where $\tilde{f} \in D_y^2(\bar{U}, \mathbb{R}^n)$ is sufficiently close to f (with respect to the sup-norm topology in $C^r(\bar{U}, \mathbb{R}^n)$), and $y \in RV(\tilde{f})$, and the above calculation is independent of the choice of $\tilde{f}$.

To extend the formula (6.4) to all $f \in D_y^0(\bar{U}, \mathbb{R}^n)$, we first note that $\varepsilon := \min\{|f(x) - y| : x \in \partial U\} > 0$, and then apply the Weierstrass theorem to obtain $\tilde{g} \in C^2(\bar{U}, R^n)$, so that $\max\{|f(x) - \tilde{g}(x)| : x \in \bar{U}\} < \varepsilon/2$. We then use Sard's theorem to find a regular value $y_0 \in \mathbb{R}^n$ of $\tilde{g}$ such that $|y - y_0| < \varepsilon/2$. We then define $g : \bar{U} \to \mathbb{R}^n$ as $g(x) = \tilde{g}(x) - y_0$. Then $g \in C^2(\bar{U}; \mathbb{R}^n), \max\{|g(x) - f(x)|\} < \varepsilon$, and 0 is a regular value of g. We can define

$$\deg(f, U, y) = \sum_{x \in g^{-1}(0)} \operatorname{sgn} J_g(x),$$

and we need to check that this definition is independent of the choice of such g.

6.3 The Leray–Schauder Degree

In 1934, Leray and Schauder [207] generalized Brouwer degree theory to an infinite Banach space and established the so-called Leray–Schauder degree. It turns out that the Leray–Schauder degree is a powerful tool in proving various existence results for nonlinear differential equations (see, for example, [89, 90]). The objective of this section is to extend the Brouwer degree to general Banach spaces. We first extend the Brouwer degree to general finite-dimensional spaces.

Let X be a (real) Banach space of dimension n, and let n be an isomorphism between X and $\mathbb{R}^n$. Then for $f \in D_y(\bar{U}, X)$, $U \subset X$ open, $y \in X$, we can define

$$\deg(f, U, y) = \deg(\phi \circ f \circ \phi^{-1}, \phi(U), \phi(y)), \tag{6.6}$$

provided this definition is independent of the basis chosen. To see this, let ψ be a second isomorphism. Then $A = \psi \circ \phi^{-1} \in \mathrm{GL}(n)$. Abbreviate $f^* = \phi \circ f \circ \phi^{-1}$, $y^* = \phi(y)$, and pick $\tilde{f}^* \in D_y^1(\phi(\bar{U}), \mathbb{R}^n)$ in the same component of $D_y(\phi(\bar{U}), \mathbb{R}^n)$ as f^* such that $y^* \in \mathrm{RV}(f^*)$. Then $A \circ \tilde{f}^* \circ A^{-1} \in D_y^1(\psi(U), R^n)$ is the same component of $D_y(\psi(\bar{U}), \mathbb{R}^n)$ as $A \circ f^* \circ A^{-1} = \psi \circ f \circ \psi$ (since A is also a homeomorphism) and

$$J_{A \circ \tilde{f}^* \circ A^{-1}}(Ay^*) = \det(A) J_{\tilde{f}^*}(y^*) \det(A^{-1}) = J_{\tilde{f}^*}(y^*) \tag{6.7}$$

by the chain rule. Thus we have $\deg(\psi \circ f \circ \psi^{-1}, \psi(U), \psi(y)) = \deg(\phi \circ f \circ \phi^{-1}, \phi(U), \phi(y))$, and our definition is independent of the basis chosen. In addition, it inherits all properties from the mapping degree in $\mathbb{R}^n$. Note also that the reduction property holds if $\mathbb{R}^m$ is replaced by an arbitrary subspace X_1, since we can always choose $\phi \colon X \to \mathbb{R}^n$ such that $\phi(X_1) = \mathbb{R}^m$.

Our next aim is to tackle the infinite-dimensional case. The general idea is to approximate F by finite-dimensional operators (in the same spirit as we approximated continuous f by smooth functions). To do this, we need to know which operators can be approximated by finite-dimensional operators. Hence we have to recall some basic facts first.

Let X and Y be Banach spaces and $U \subset X$. An operator $F \colon U \subset X \to Y$ is called finite-dimensional if its range is finite-dimensional. In addition, it is called *compact* (completely continuous) if it is continuous and maps bounded sets into relatively compact ones. The set of all compact operators is denoted by $\mathscr{C}(U, Y)$, and the set of all compact finite-dimensional operators is denoted by $\mathscr{F}(U, Y)$. Both sets are normed linear spaces, and we have $\mathscr{F}(U, Y) \subseteq \mathscr{C}(U, Y) \subseteq C(U, Y)$. If U is compact, then $\mathscr{C}(U, Y) = C(U, Y)$ (since the continuous image of a compact set is compact), and if $\dim(Y) < \infty$, then $\mathscr{F}(U, Y) = \mathscr{C}(U, Y)$. In particular, if $U \subset \mathbb{R}^n$ is bounded, then $\mathscr{F}(U, Y) = \mathscr{C}(U, \mathbb{R}^n) = C(U, \mathbb{R}^n)$.

For $U \subset X$, we set $\mathscr{D}_y(\bar{U}, X) = \{F \in \mathscr{C}(\bar{U}, X) \colon y \notin (\mathbb{I} + F)(\partial U)\}$ and $\mathscr{F}_y(\bar{U}, X) = \{F \in \mathscr{F}(\bar{U}, X) \colon y \notin (\mathbb{I} + F)(\partial U)\}$. Note that for $F \in \mathscr{D}_y(\bar{U}, X)$, we have $\rho = \mathrm{dist}(y, (\mathbb{I} + F)(\partial U)) > 0$, since $\mathbb{I} + F$ maps closed sets to closed sets.

Pick $F_1 \in \mathscr{F}(\bar{U},X)$ such that $|F - F_1| < \rho$. Hence, $F_1 \in \mathscr{F}_y(\bar{U},X)$. Next, let X_1 be a finite-dimensional subspace of X such that $F_1(U) \subset X_1$, $y \in X_1$, and set $U_1 = U \cap X_1$. Then we have $F_1 \in \mathscr{F}_y(\bar{U}_1,X_1)$, and we may define

$$\deg(\mathbb{I}+F,U,y) = \deg(\mathbb{I}+F_1,U_1,y). \tag{6.8}$$

It is easy to verify that this definition is independent of F_1 and X_1.

Theorem 6.3. *Let U be a bounded open subset of a (real) Banach space X and let $F \in \mathscr{F}_y(\bar{U},X)$, $y \in X$. Then the following hold.*

(i) $\deg(\mathbb{I}+F,U,y) = \deg(\mathbb{I}+F-y,U,0)$.
(ii) $\deg(\mathbb{I},U,y) = 1$ if $y \in U$.
(iii) If $U_{1,2}$ are open, disjoint subsets of U such that $y \notin f(\bar{U} \setminus (U_1 \cup U_2))$, then $\deg(\mathbb{I}+F,U,y) = \deg(\mathbb{I}+F,U_1,y) + \deg(\mathbb{I}+F,U_2,y)$.
(iv) If $H: [0,1] \times \bar{U} \to X$ and $y: [0,1] \to X$ are both continuous such that $H(t) \in D_{y(t)}(U,\mathbb{R}^n)$, $t \in [0,1]$, then $\deg(\mathbb{I}+H(0),U,y(0)) = \deg(\mathbb{I}+H(1),U,y(1))$.

6.4 Global Bifurcation Theorem

As in Sect. 5.1, we study the nonlinear parameter-dependent problem

$$F(u,\alpha) = 0, \tag{6.9}$$

where $F: E \times \mathbb{R} \to X$ is a C^1-map such that $F(0,\alpha) = 0$ for all $\alpha \in \mathbb{R}$, and $E \subseteq X$ is an open neighborhood of 0 (possibly $E = X$). Note that (6.9) has the trivial solution for all values of α. We shall now consider the question of bifurcation from this trivial branch of solutions and demonstrate the existence of global branches of nontrivial solutions bifurcating from the trivial branch. If $X = \mathbb{R}^n$, then we use the Brouwer degree; if X is an infinite-dimensional (real) Banach space, then we assume that $F(x,\alpha) = x + f(x,\alpha)$ and that $f: E \times \mathbb{R} \to X$ is completely continuous. Thus for $F(\cdot,\alpha)$, the Leray–Schauder degree is applicable. The application of degree theory to bifurcation theory goes back to Krasnosel'skii [191]. Global bifurcation theorem of the following type were first proved by Rabinowitz [251]. Several generalizations have been given by Ize et al. [177–182], Krawcewicz et al. [192–195], and Nussbaum et al. [232–236].

Theorem 6.4. *Let there exist a, $b \in \mathbb{R}$ with $a < b$ such that $u = 0$ is an isolated solution of (6.9) for $\alpha = a$ and $\alpha = b$, where a and b are not bifurcation points. Furthermore, assume that*

$$\deg(F(\cdot,a),B_r(0),0) \neq \deg(F(\cdot,b),B_r(0),0), \tag{6.10}$$

where $B_r(0) = \{u \in E: \|u\| < r\}$ is an isolating neighborhood of the trivial solution. Let

$$\mathscr{S} = \overline{\{(u,\alpha): (u,\alpha) \text{ solves } (6.9) \text{ with } u \neq 0\}} \cup \{0\} \times [a,b],$$

and let $\mathscr{C}$ be the maximal connected subset of $\mathscr{S}$ that contains $\{0\} \times [a,b]$. Then either

(i) $\mathscr{C}$ is unbounded in $E \times \mathbb{R}$,

or else

(ii) $\mathscr{C} \cap \{0\} \times (\mathbb{R} \setminus [a,b]) \neq \emptyset$.

Proof. Define a class $\mathscr{U}$ of subsets of $E \times \mathbb{R}$ as follows:

$$\mathscr{U} = \{\Omega \subset E \times \mathbb{R} : \Omega = \Omega_0 \cup \Omega_\infty\},$$

where $\Omega_0 = B_r(0) \times [a,b]$, and Ω_∞ is a bounded open subset of $(E \setminus \{0\}) \times \mathbb{R}$. We shall first show that (6.9) has a nontrivial solution $(u,\alpha) \in \partial \Omega$ for every such $\Omega \in \mathscr{U}$. To accomplish this, let us consider the following sets:

$$\begin{cases} K = F^{-1}(0) \cap \bar{\Omega}, \\ A = \{0\} \times [a,b], \\ B = F^{-1}(0) \cap \{\partial \Omega \setminus [B_r(0) \times \{a\} \cup B_r(0) \times \{b\}]\}. \end{cases} \quad (6.11)$$

We observe that K may be regarded as a compact metric space, and A and B are compact subsets of K. We hence may apply Whyburn's lemma to deduce that either there exists a continuum in K connecting A to B, or else there is a separation K_A, K_B of K, with $A \subset K_A$, $B \subset K_B$. If the latter holds, we may find open sets U, V in $E \times \mathbb{R}$ such that $K_A \subset U$, $K_B \subset V$, with $U \cap V = \emptyset$. We let $\Omega^* = \Omega \cap (U \cup V)$ and observe that $\Omega^* \in \mathscr{U}$. It follows by construction that there are no nontrivial solutions of (6.9) that belong to $\partial \Omega^*$; this, however, is impossible, since it would imply, by the generalized homotopy and the excision principle of the Leray–Schauder degree, that $\deg(F(\cdot,a), B_r(0), 0) = \deg(F(\cdot,b), B_r(0), 0)$, contradicting (6.10). We hence have that, for each $\Omega \in \mathscr{U}$, there is a continuum C of solutions of (6.9) that intersects $\partial \Omega$ in a nontrivial solution.

We assume now that neither of the alternatives of the theorem holds, that is, we assume that $\mathscr{C}$ is bounded and $\mathscr{C} \cap \{0\} \times (\mathbb{R} \setminus [a,b]) = \emptyset$. In this case, we may, using the boundedness of $\mathscr{C}$, construct a set $\Omega \in \mathscr{U}$ containing no nontrivial solutions in its boundary, thus arriving once more at a contradiction. $\qquad \square$

6.5 $\mathbb{S}^1$-Equivariant Degree

Let $\mathbb{E}$ be a real isometric Banach representation of the group $G = \mathbb{S}^1$. The isotypical direct sum decomposition is denoted by

$$\mathbb{E} = \mathbb{E}_0 \oplus \mathbb{E}_1 \oplus \cdots \oplus \mathbb{E}_k \oplus \cdots, \quad (6.12)$$

where $\mathbb{E}_0 = \mathbb{E}^G \overset{\text{def}}{=} \{x \in \mathbb{E}; gx = x \text{ for all } g \in G\}$ is the subspace of G-fixed points, and for $k \geq 1$, $x \in \mathbb{E}_k \setminus \{0\}$ implies that G_x, the isotropy group of x, is $\mathbb{Z}_k \overset{\text{def}}{=} \{g \in G; g^k = 1\}$. Throughout this section, we assume the following:

(SD1) For each integer $k = 0, 1, \ldots$, the subspace $\mathbb{E}_k$ is of finite dimension.

All subspaces $\mathbb{E}_k$, $k \geq 1$, admit a natural structure of complex vector spaces such that an $\mathbb{R}$-linear operator $A : \mathbb{E}_k \to \mathbb{E}_k$ is G-equivariant if and only if it is $\mathbb{C}$-linear with respect to this complex structure. Therefore, by choosing a $\mathbb{C}$ basis in $\mathbb{E}_k, k \geq 1$, we can define an isomorphism between the group of all G-equivariant automorphisms of $\mathbb{E}_k$, denoted by $GL_G(\mathbb{E}_k)$, and the general linear group $GL(m_k, \mathbb{C})$, where $m_k = \dim_{\mathbb{C}} \mathbb{E}_k$.

Let $\mathbb{F}$ be another Banach isometric representation of G, and $L \colon \mathbb{E} \to \mathbb{F}$ a given equivariant linear bounded Fredholm operator of index zero. We say that an equivariant compact operator $K \colon \mathbb{E} \to \mathbb{F}$ is an *equivariant compact resolvent* of L if $L + K \colon \mathbb{E} \to \mathbb{F}$ is an isomorphism. We shall denote by $CR^G(L)$ the set of all equivariant compact resolvents of L, and assume that

(SD2) $CR^G(L) \neq \emptyset$.

In what follows, a point of the Banach space $\mathbb{E} \times \mathbb{R}^2$ is denoted by (x, λ) with $x \in \mathbb{E}$ and $\lambda \in \mathbb{R}^2$, and the action of G on $\mathbb{E} \times \mathbb{R}^2$ is defined by $g(x, \lambda) = (gx, \lambda)$ for every $g \in G$.

We consider a G-equivariant continuous map $f \colon \mathbb{E} \times \mathbb{R}^2 \to \mathbb{F}$ such that

$$f(u, \lambda) = Lu - Q(u, \lambda), \quad (u, \lambda) \in \mathbb{E} \times \mathbb{R}^2, \tag{6.13}$$

where $Q \colon \mathbb{E} \times \mathbb{R}^2 \to \mathbb{F}$ is a completely continuous map and the following assumption is satisfied:

(SD3)   There exists a two-dimensional submanifold $M \subset \mathbb{E}_0 \times \mathbb{R}^2$ such that (i) $M \subset f^{-1}(0)$; (ii) if $(u_0, \lambda_0) \in M$, then there exist an open neighborhood U_{λ_0} of λ_0 in $\mathbb{R}^2$, an open neighborhood U_{u_0} of u_0 in $\mathbb{E}_0$, and a C^1-map $\eta \colon U_{\lambda_0} \to \mathbb{E}_0$ such that $M \cap (U_{u_0} \times U_{\lambda_0}) = \{(\eta(\lambda), \lambda); \lambda \in U_{\lambda_0}\}$.

In relation to the bifurcation problem of (6.13), we consider the structure of the set of solutions to the following equation:

$$f(u, \lambda) = 0, \ (u, \lambda) \in \mathbb{E} \times \mathbb{R}^2. \tag{6.14}$$

All points $(u, \lambda) \in M$ are called *trivial solutions* of (6.13) or (6.14), and all other solutions in $f^{-1}(0) \setminus M$ are called *nontrivial solutions*. A point $(u_0, \lambda_0) \in M$ is called a *bifurcation point* if in every neighborhood of $(u_0, \lambda_0) \in M$ there is a nontrivial solution for (6.14).

Equation (6.14) can be transformed into the equivariant fixed-point problem

$$u = (L + K)^{-1} \circ [K + Q(\cdot, \lambda)](u), \quad (u, \lambda) \in \mathbb{E} \times \mathbb{R}^2. \tag{6.15}$$

Let $\mathscr{F}(u, \lambda) = u - (L + K)^{-1} \circ [Q(\cdot, \lambda) + K](u), \ (u, \lambda) \in \mathbb{E} \times \mathbb{R}^2$. Then (6.14) is equivalent to the equation

$$\mathscr{F}(u, \lambda) = 0, (u, \lambda) \in \mathbb{E} \times \mathbb{R}^2. \tag{6.16}$$

The idea of finding nontrivial solutions to (6.16) in an open G-invariant neighborhood $\mathscr{U} \subseteq \mathbb{E} \times \mathbb{R}^2$ of $(u_0, \lambda_0) \in M$ is based on an *auxiliary function* ψ to (6.16), which is introduced to distinguish nontrivial solutions from trivial solutions. Here $\mathscr{U}$ is said to be G-invariant if $(gx, \lambda) \in \mathscr{U}$ for all $g \in G$, $(x, \lambda) \in \mathscr{U}$. An auxiliary function to (6.16) on the set $\mathscr{U}$ is an equivariant function (i.e.,

$\psi(gx) = g\psi(x)$ for all $g \in G$ and $x \in \overline{\mathcal{U}}$, where $\overline{\mathcal{U}}$ denotes the closure of $\mathcal{U}$; here and in what follows G acts on $\mathbb{R}^2$ trivially) $\psi : \overline{\mathcal{U}} \subset \mathbb{E} \times \mathbb{R}^2 \to \mathbb{R}$ satisfying $\psi(u, \lambda) < 0$ for all $(u, \lambda) \in \overline{\mathcal{U}} \cap M$. Then every solution to the system

$$\begin{cases} \mathscr{F}(u, \lambda) = 0, \\ \psi(u, \lambda) = 0, \end{cases} (u, \lambda) \in \overline{\mathcal{U}}, \tag{6.17}$$

is a nontrivial solution to (6.13). This leads to the equivariant map $\mathscr{F}_\psi : \overline{\mathcal{U}} \to \mathbb{E} \times \mathbb{R}$ defined by

$$\mathscr{F}_\psi(u, \lambda) = (\mathscr{F}(u, \lambda), \psi(u, \lambda)), (u, \lambda) \in \overline{\mathcal{U}}, \tag{6.18}$$

and the problem of finding a nontrivial solution to (6.13) in $\mathcal{U}$ can be reduced to the problem of finding a solution to the equation $\mathscr{F}_\psi(u, \lambda) = 0$ in $\mathcal{U}$, which may be solved by the so-called $\mathbb{S}^1$-*equivariant degree* as a topological invariant associated with the problem (6.17).

To describe the definition and basic properties of $\mathbb{S}^1$-degree, we assume that V is an isometric Hilbert representation of $G = \mathbb{S}^1$. If U is an open bounded invariant subset of $V \oplus \mathbb{R}$ (where $\mathbb{S}^1$ acts trivially on $\mathbb{R}$) and $F : (\overline{U}, \partial U) \to (V, V \setminus \{0\})$ is an equivariant compact vector field on $\overline{U}$, then there is defined the $\mathbb{S}^1$-equivariant degree of F with respect to U, which is a sequence of integers

$$\mathbb{S}^1\text{-deg}(F, U) := \{\deg_k(F, U)\}_{k=1}^\infty \in \bigoplus_{k=1}^\infty \mathbb{Z}$$

such that $\deg_k(F, U) \neq 0$ for only a finite number of indices k. The basic properties of $\mathbb{S}^1$-deg are as follows (see [24, 25, 112, 180, 194] for details):

(i) *Existence:* If $\mathbb{S}^1\text{-deg}(F, U) := \{\deg_k(F, U)\}_{k=1}^\infty \neq 0$, i.e., there exists $k \in \mathbb{N}$ such that $\deg_k(F, U) \neq 0$, then $F^{-1}(0) \cap U^H \neq \emptyset$, where $H = \mathbb{Z}_k$ and

$$U^H := \{v \in U : gv = v \text{ for any } g \in H\}.$$

(ii) *Additivity:* If U_1 and U_2 are two open invariant subsets of U such that $U_1 \cap U_2 = \emptyset$ and $F^{-1}(0) \cap U \subseteq U_1 \cup U_2$, then $\mathbb{S}^1\text{-deg}(F, U) = \mathbb{S}^1\text{-deg}(F, U_1) + \mathbb{S}^1\text{-deg}(F, U_2)$.

(iii) *Homotopy invariance:* If $\mathscr{H} : (\overline{U}, \partial U) \times [0, 1] \to (V, V \setminus \{0\})$ is an $\mathbb{S}^1$-equivariant homotopy of compact vector fields, then $\mathbb{S}^1\text{-deg}(\mathscr{H}_0, U) = \mathbb{S}^1\text{-deg}(\mathscr{H}_1, U)$, where $\mathscr{H}_t(\theta) = \mathscr{H}(t, \theta)$ for $t \in [0, 1]$ and $\theta \in \overline{U}$.

(iv) *Contraction:* Suppose that W is another isometric Hilbert representation of $\mathbb{S}^1$ and let Ω be an open, bounded, invariant subset of W such that $0 \in \Omega$. Define $\Phi : \overline{\Omega} \times \overline{U} \to W \oplus V$ by $\Phi(y, x, t) = (y, F(x, t))$. Then $\mathbb{S}^1\text{-deg}(\Phi, U) = \mathbb{S}^1\text{-deg}(F, U)$.

Now we return to the problem (6.17). If the mapping $\mathscr{F}_\psi : \overline{\mathcal{U}} \to \mathbb{E} \times \mathbb{R}$ has no solution on $\partial \mathcal{U}$ and $\mathscr{F} : \mathcal{U} \to \mathbb{E}$ is a condensing field (i.e., $\pi - \mathscr{F}$ is a condensing map, where $\pi : \mathcal{U} \to \mathbb{E}$ is the natural projection on $\mathbb{E}$), then the $\mathbb{S}^1$-equivariant degree $\mathbb{S}^1\text{-deg}(\mathscr{F}_\psi, \mathcal{U})$ is well defined, and its nontriviality implies the existence of solutions of $\mathscr{F}_\psi(u, \lambda) = 0$ in $\mathscr{U}$. Global continuation of the branch of nontrivial solutions (solutions in $f^{-1}(0) \setminus M$) bifurcating from (u_0, λ_0) can be characterized by the above $\mathbb{S}^1$-degree at all bifurcation points along the closure of the branch if such a branch is bounded in $\mathbb{E} \times \mathbb{R}^2$ (the so-called Fuller space).

To describe precisely this $\mathbb{S}^1$-degree-based bifurcation theory, we need some additional information about: (i) the construction of the open neighborhood $\mathscr{U}$, (ii) the auxiliary function ψ, (iii) the computation of $\mathbb{S}^1$-deg$(\mathscr{F}_\psi, \mathscr{U})$.

If $\mathscr{F}(u, \lambda)$ is differentiable with respect to u, we are able to define *singular points* of system (6.16) through its linearization at the trivial solutions of (6.14). This is unfortunately not so for state-dependent DDEs. Therefore, we shall distinguish two cases.

6.5.1 Differentiability Case

Throughout this subsection, we further assume that at all points $(u_0, \lambda_0) \in M$, the derivative $D_u f(u_0, \lambda_0): \mathbb{E} \to \mathbb{F}$ of f with respect to u exists and is continuous on M. We say that $(u_0, \lambda_0) \in M$ is $\mathbb{E}$-*singular* if $D_u f(u_0, \lambda_0): \mathbb{E} \to \mathbb{F}$ is not an isomorphism. An $\mathbb{E}$-singular point (u_0, λ_0) is *isolated* if there are no other $\mathbb{E}$-singular points in some neighborhood of (u_0, λ_0). It follows from the implicit function theorem that if (u_0, λ_0) is a bifurcation point, then (u_0, λ_0) is an $\mathbb{E}$-singular point.

We start with the construction of the open neighborhood $\mathscr{U}$. We consider the open neighborhood of $(u_0, \lambda_0) \in M$ defined by

$$B_M(u_0, \lambda_0; r, \rho) \stackrel{\text{def}}{=} \{(u, \lambda) \in \mathbb{E} \times \mathbb{R}^2 : |\lambda - \lambda_0| < \rho, \|u - \eta(\lambda)\| < r\}, \quad (6.19)$$

where $r > 0$ is chosen such that

(i) $\mathscr{F}(u, \lambda) \neq 0$ for all $(u, \lambda) \in \overline{B_M(u_0, \lambda_0; r, \rho)}$ such that $|\lambda - \lambda_0| = \rho, \|u - \eta(\lambda)\| \neq 0$;

(ii) (u_0, λ_0) is the only $\mathbb{E}$-singular point of $\mathscr{F}$ in $\overline{B_M(u_0, \lambda_0; r, \rho)}$.

We call $B_M(u_0, \lambda_0; r, \rho)$ a *special neighborhood* of $\mathscr{F}$ determined by r and ρ. The existence of a special neighborhood $\mathscr{U} \stackrel{\text{def}}{=} B_M(u_0, \lambda_0; r, \rho)$ follows from the assumption that the $\mathbb{E}$-singular point (u_0, λ_0) of $\mathscr{F}$ is isolated. Note that the equivariant version of Dugundji's extension theorem (see [193, p. 197]) implies that there exists a continuous $\mathbb{S}^1$-equivariant function $\theta: \overline{\mathscr{U}} \to \mathbb{R}$ such that

(i) $\theta(\eta(\lambda), \lambda) = -|\lambda - \lambda_0|$ for all $(\eta(\lambda), \lambda) \in \overline{\mathscr{U}} \cap M$;

(ii) $\theta(u, \lambda) = r$ if $\|u - \eta(\lambda)\| = r$.

Such a function θ is called a *completing function* (or Ize's function). Clearly, if θ is a completing function, then $\psi_\delta(u, \lambda) \stackrel{\text{def}}{=} \theta(u, \lambda) - \delta$ is negative on the subset of trivial solutions $\mathscr{U} \cap M$, provided that $\delta > 0$. So ψ_δ is an auxiliary function to (6.16). For $\delta > 0$ small enough, we can define $\mathscr{F}_{\psi_\delta}: \overline{\mathscr{U}} \to \mathbb{E} \times \mathbb{R}$ by

$$\mathscr{F}_{\psi_\delta}(u, \lambda) \stackrel{\text{def}}{=} (\mathscr{F}(u, \lambda), \psi_\delta(u, \lambda)),$$

and define the $\mathbb{S}^1$-equivariant degree $\mathbb{S}^1\text{-}\deg(\mathscr{F}_{\psi_\delta}, \mathscr{U})$. By the homotopy invariance of the $\mathbb{S}^1$-degree, $\mathbb{S}^1\text{-}\deg(\mathscr{F}_{\psi_\delta}, \mathscr{U}) = \mathbb{S}^1\text{-}\deg(\mathscr{F}_\theta, \mathscr{U})$. Therefore, the nontriviality of $\mathbb{S}^1\text{-}\deg(\mathscr{F}_\theta, \mathscr{U})$ implies the existence of a nontrivial solution of (6.14) in $\mathscr{U}$.

We now turn to the computation of $\mathbb{S}^1\text{-}\deg(\mathscr{F}_\theta, \mathscr{U})$. We identify $\mathbb{R}^2$ with $\mathbb{C}$, and for sufficiently small $\rho > 0$, we define $\alpha \colon D \to M, D \overset{\text{def}}{=} \{z \in \mathbb{C} : |z| \leq 1\}$, by

$$\alpha(z) = (\eta(\lambda_0 + \rho z), \lambda_0 + \rho z) \in \mathbb{E}_0 \times \mathbb{R}^2.$$

Since we have assumed that $(x_0, \lambda_0) = (\eta(\lambda_0), \lambda_0) \in M$ is an isolated $\mathbb{E}$-singular point, it is clear that we can choose sufficiently small $\rho > 0$ such that $\alpha(D)$ contains only one $\mathbb{E}$-singular point, namely (x_0, λ_0). Consequently, the formula $\Psi(z) \overset{\text{def}}{=} D_u\mathscr{F}(\alpha(z)), z \in \mathbb{S}^1 \subseteq D$, defines a continuous map $\Psi \colon \mathbb{S}^1 \to GL_G(\mathbb{E})$, which has the decomposition (see [88] for details) $\Psi = \Psi_0 \oplus \Psi_1 \oplus \cdots \oplus \Psi_k \oplus \cdots$, where $\Psi_k = \Psi|_{\mathbb{E}_k} \colon \mathbb{S}^1 \to GL_G(\mathbb{E}_k)$ for $k = 1, 2, \cdots$ and $\Psi_0 \colon \mathbb{S}^1 \to GL(\mathbb{E}_0)$, with $GL(\mathbb{E}_0)$ the set of linear automorphisms of $\mathbb{E}_0$. We now define

$$\begin{cases} \varepsilon_0(u_0, \lambda_0) = \operatorname{sgn} \det \Psi_0(z), \\ \mu_k(u_0, \lambda_0) = \deg_B(\det_{\mathbb{C}}[\Psi_k]), k = 1, 2, \cdots, \\ \mu(u_0, \lambda_0) = \{\mu_k(u_0, \lambda_0)\} \in \oplus_{k=1}^{\infty} \mathbb{Z}. \end{cases} \quad (6.20)$$

It is clear that ε_0 does not depend on the choice of $z \in \mathbb{S}^1$.

We need one more notion, the crossing number, to calculate $\deg_B(\det_{\mathbb{C}}[\Psi_k])$:

Lemma 6.3 ([88]). *Suppose $\alpha_0, \beta_0, \delta, \varepsilon$ are given numbers with $\alpha_0, \delta, \varepsilon > 0$. Let $\Omega \overset{\text{def}}{=} (0, \alpha_0) \times (\beta_0 - \varepsilon, \beta_0 + \varepsilon) \subseteq \mathbb{R}^2$. Assume that $H \colon [\sigma_0 - \delta, \sigma_0 + \delta] \times \bar{\Omega} \to \mathbb{R}^2$ is a continuous function satisfying:*

(i) $H(\sigma, \alpha, \beta) \neq 0$ for all $\sigma \in [\sigma_0 - \delta, \sigma_0 + \delta]$ and $(\alpha, \beta) \in \partial\Omega \setminus \{(0, \beta); \beta \in (\beta_0 - \varepsilon, \beta_0 + \varepsilon)\}$;
(ii) if $(\alpha, \beta) \in \Omega$ and $H_{\sigma_0 \pm \delta}(\alpha, \beta) = 0$, then $\alpha \neq 0$.

Let $\Omega_1 \overset{\text{def}}{=} (\sigma_0 - \delta, \sigma_0 + \delta) \times (\beta_0 - \varepsilon, \beta_0 + \varepsilon)$ and define the function $\Psi_H \colon \bar{\Omega}_1 \to \mathbb{R}^2$ by $\Psi_H(\sigma, \beta) = H(\sigma, 0, \beta)$, for $\sigma \in [\sigma_0 - \delta, \sigma_0 + \delta]$, and $\beta \in [\beta_0 - \varepsilon, \beta_0 + \varepsilon]$. Then $\Psi_H(\sigma, \beta) \neq 0$ for $(\sigma, \beta) \in \partial\Omega_1$ and $\deg_B(\Psi_H, \Omega_1) = \gamma$, where γ is the crossing number given by

$$\gamma \overset{\text{def}}{=} \deg_B(H_{\sigma_0 - \delta}, \Omega) - \deg_B(H_{\sigma_0 + \delta}, \Omega).$$

Lemma 6.4. *Let $\mathscr{U} = B_M(u_0, \lambda_0; r', \rho) \subseteq \mathbb{E} \times \mathbb{R}^2$ be a special neighborhood of $\mathscr{F}$, and θ a completing function. Then the $\mathbb{S}^1$ degree $\mathbb{S}^1\text{-}\deg(\mathscr{F}_\theta, \mathscr{U})$ is well defined, and*

$$\mathbb{S}^1\text{-}\deg(\mathscr{F}_\theta, \mathscr{U}) = \varepsilon_0 \cdot \mu(u_0, \lambda_0).$$

That is,

$$\mathbb{S}^1\text{-}\deg_k(\mathscr{F}_\theta, \mathscr{U}) = \varepsilon_0 \cdot \mu_k(u_0, \lambda_0), k = 1, 2, \cdots,$$

where $\mu(u_0, \lambda_0)$ is defined by (6.20).

By Lemma 6.4, we have the following local bifurcation theorem of Krasnosel'skii type [191].

Theorem 6.5. *Suppose that* $f \colon \mathbb{E} \oplus \mathbb{R}^2 \to \mathbb{F}$ *is a G-equivariant continuous map that is continuously differentiable with respect to* x *at points* $(x, \lambda) \in M$ *and satisfies* (SD1)–(SD3). *If* $(u_0, \lambda_0) \in M$ *is an isolated* $\mathbb{E}$-*singular point such that* $\varepsilon_0 \mu_k(u_0, \lambda_0) \neq 0$ *for some* $k \geq 1$, *then* (u_0, λ_0) *is a bifurcation point of* (6.13). *More precisely, there exists a sequence* (u_n, λ_n) *of nontrivial solutions to* (6.13) *such that the isotropy group of* u_n *contains* $\mathbb{Z}_k$ *and* $(u_n, \lambda_n) \to (u_0, \lambda_0)$ *as* $n \to \infty$.

We remark that the above results hold when $\mathbb{R}^2$ is replaced by an open subset of $\mathbb{R}^2$. Geba and Marzantowicz [111] established the following global bifurcation theorem of Rabinowitz type [251] by applying the $\mathbb{S}^1$-degree theory due to Dylawerski, Geba, Jodel, and Marzantowicz [85].

Theorem 6.6. *Suppose that* $f \colon \mathbb{E} \oplus \mathbb{R}^2 \to \mathbb{F}$ *is as in Theorem 6.5 and suppose further that* M *is complete and every* $\mathbb{E}$-*singular point in* M *is isolated. Let* $\mathscr{S}(f)$ *denote the closure of the set of all nontrivial solutions of* (6.13). *Then for each bounded component* C *of* $\mathscr{S}(f)$, *the set* $C \cap M$ *is a finite set, i.e.,*

$$C \cap M = \{(u_1, \lambda_1), (u_2, \lambda_2), \cdots, (u_q, \lambda_q)\},$$

and

$$\sum_{i=1}^{q} \mathbb{S}^1\text{-}\deg(\mathscr{F}_{\theta_i}, \mathscr{U}_i) = \sum_{i=1}^{q} \varepsilon_i \cdot \mu(u_i, \lambda_i) = 0,$$

where $\underline{\mathscr{U}_i}$ *is a special neighborhood of* (u_i, λ_i), θ_i *is a completing function defined on* $\overline{\mathscr{U}}_i$, *and* ε_i *and* $\mu(u_i, \lambda_i)$ *are defined by* (6.20).

Proof. If C is a bounded component of $\mathscr{S}(f)$, then every point of $C \cap M$ is a bifurcation point that is also a $\mathbb{E}$-singular point of f. Since every $\mathbb{E}$-singular point of f is isolated and M is complete, $C \cap M$ is a bounded and closed subset of $\mathbb{E}_0 \times \mathbb{R}^2 \supset M$. By (SD1), $\mathbb{E}_0 \times \mathbb{R}^2$ is finite-dimensional, and hence $C \cap M$ is compact. Therefore, $C \cap M$ is a finite set.

Choose $r, \rho > 0$ sufficiently small that for each $i = 1, 2, \cdots, q$, $U_i = B_M(u_i, \lambda_i; r, \rho)$ is a special neighborhood of (u_i, λ_i) for f and $U_i \cap U_j = \emptyset$ if $i \neq j$. Let $U = U_1 \cup U_2 \cup \cdots \cup U_q$ and find a bounded open set $\Omega_1 \subset \mathbb{E} \times \mathbb{R}^2$ such that $C \setminus U \subseteq \Omega_1$ and $\Omega_1 \cap M = \emptyset$. Put $\Omega_2 = U \cup \Omega_1$. Then $C \subseteq \Omega_2$. We can (e.g., see [193, p. 174]) find an open invariant subset $\Omega \subseteq \mathbb{E} \times \mathbb{R}^2$ such that $C \subseteq \Omega \subseteq \Omega_2$ and $\partial\Omega \cap \mathscr{S}(f) = \emptyset$.

Note that Ω is an open, bounded, invariant subset. We now choose $r_0 \in (0, r)$ and $\rho_0 \in (0, \rho)$ such that for every $i = 1, 2, \cdots, q$, we have

(i) $B_M(u_i, \lambda_i; r_0, \rho_0) \subseteq \Omega$;

(ii) $\mathscr{U}_i \stackrel{\text{def}}{=} B_M(u_i, \lambda_i; r_0, \rho_0)$ is a special neighborhood of (u_i, λ_i) for f.

Set $\mathscr{U} = \mathscr{U}_1 \cup \mathscr{U}_2 \cup \cdots \cup \mathscr{U}_q$ and

$$\partial \mathscr{U}_{r_0} \stackrel{\text{def}}{=} \{(u,\lambda) \in \bar{\Omega} : \|u - \eta(\lambda)\| = r_0, (\eta(\lambda), \lambda) \in \overline{\mathscr{U}} \cap M\}.$$

We note that $r_0 > 0$, and define an invariant function by

$$\theta(u,\lambda) = \begin{cases} |\lambda - \lambda_i| \frac{\|u - \eta(\lambda)\| - r_0}{r_0} + \|u - \eta(\lambda)\|, & \text{if } (u,\lambda) \in \bar{\mathscr{U}}_i, \\ r_0, & \text{if } (u,\lambda) \in C \setminus \mathscr{U}. \end{cases} \tag{6.21}$$

Now, $\mathscr{U}_i$ is a special neighborhood, and hence we have $(C \setminus \mathscr{U}) \cap \bar{\mathscr{U}}_i = C \cap \partial \mathscr{U}_i \subseteq \partial \mathscr{U}_{r_0}$, where we have $\theta(u,\lambda) = r_0$. Then by (6.21), $\theta(u,\lambda)$ is continuous on $(C \setminus \mathscr{U}) \cap \bar{\mathscr{U}}_i$. Also, by the construction of $\mathscr{U}_i$, we have $\bar{\mathscr{U}}_i \cap \bar{\mathscr{U}}_j = \emptyset$ if $i \neq j$. Therefore, $\theta : C \cup \bar{\mathscr{U}} \to \mathbb{R}$ is continuous.

By the equivariant version of Dugundji's extension theorem (see [193, p. 197]), we can extend $\theta : C \cup \bar{\mathscr{U}} \to \mathbb{R}$ to a continuous invariant function $\theta : \bar{\Omega} \to \mathbb{R}$ such that

(iii) $\theta(u,\lambda) = -|\lambda - \lambda_i|$ if $(u,\lambda) \in \overline{\mathscr{U}}_i \bigcap M$;
(iv) $\theta(u,\lambda) = r_0$ if $(u,\lambda) \in (C \setminus \mathscr{U}) \cup \partial \mathscr{U}_{r_0}$.

Let $\mathscr{F}_\theta(u,\lambda) = (\mathscr{F}(u,\lambda), \theta(u,\lambda))$. Then $\mathscr{F}_\theta^{-1}(0) = \mathscr{F}^{-1}(0) \bigcap \theta^{-1}(0)$. By (iii), we know that $\mathscr{F}_\theta^{-1}(0) \subseteq C$. Since $C \cap \partial \Omega = \emptyset$, $\mathscr{F}_\theta^{-1}(0) \cap \partial \Omega = \emptyset$. Therefore, $\mathbb{S}^1\text{-deg}(\mathscr{F}_\theta, \Omega)$ is well defined.

We now construct a homotopy $H : \bar{\Omega} \times [0,1] \to \mathbb{E} \times \mathbb{R}$ as follows:

$$H(u,\lambda,\alpha) = (\mathscr{F}(u,\lambda), (1-\alpha)\theta(u,\lambda) - \alpha\rho_0), (u,\lambda,\alpha) \in \bar{\Omega} \times [0,1].$$

Note that trivial solutions $(u,\lambda) \in \bar{\Omega}$ outside $\mathscr{S}(f)$ are contained in $\bar{\mathscr{U}}_i \cap M$ for some $i = 1, 2, \cdots, q$, and by (iii), we have

$$(1-\alpha)\theta(u,\lambda) - \alpha\rho_0 = -(1-\alpha)|\lambda - \lambda_i| - \alpha\rho_0 < 0.$$

Then by the fact that $\partial \Omega \cap \mathscr{S}(f) = \emptyset$, we have $H(u,\lambda,\alpha) \neq 0$ for all $(u,\lambda,\alpha) \in \partial \Omega \times [0,1]$. Note that θ is invariant and $\mathscr{F}$ is equivariant. So H is an $\mathbb{S}^1$-homotopy. Since $H(u,\lambda,0) = \mathscr{F}_\theta(u,\lambda)$ and $H(u,\lambda,1) = (\mathscr{F}(u,\lambda), -\rho_0) \neq 0$ for all $(u,\lambda) \in \bar{\Omega} \times [0,1]$, by the existence and homotopy invariance of the $\mathbb{S}^1$-degree, we have $\mathbb{S}^1\text{-deg}(\mathscr{F}_\theta, \Omega) = 0$. But (i)–(iv) imply that $\mathscr{F}_\theta^{-1}(0) \subseteq C \cap \mathscr{U}$. Then it follows from the excision property of the $\mathbb{S}^1$-degree that

$$\mathbb{S}^1\text{-deg}(\mathscr{F}_\theta, \mathscr{U}) = \mathbb{S}^1\text{-deg}(\mathscr{F}_\theta, \Omega) = 0.$$

On the other hand, by the additivity property of the $\mathbb{S}^1$-degree, we have

$$\sum_{i=1}^q \mathbb{S}^1\text{-deg}(\mathscr{F}_\theta, \mathscr{U}_i) = \mathbb{S}^1\text{-deg}(\mathscr{F}_\theta, \mathscr{U}) = 0.$$

Let $\theta_i(u, \lambda) = \theta(u, \lambda)|_{\overline{\mathscr{U}}_i}$. Note that $\mathscr{U} \subseteq \overline{\Omega}$ implies that $((C \setminus \mathscr{U}) \cup \partial \mathscr{U}_{r_0}) \cap \overline{\mathscr{U}}_i = \partial \mathscr{U}_i \cap \partial \mathscr{U}_{r_0}$. Then $\theta_i(u, \lambda)$ is a completing function on $\overline{\mathscr{U}}_i$, and we have

$$\sum_{i=1}^{q} \mathbb{S}^1\text{-}\deg(\mathscr{F}_{\theta_i}, \mathscr{U}_i) = \mathbb{S}^1\text{-}\deg(\mathscr{F}_{\theta}, \mathscr{U}) = 0.$$

Therefore, it follows from Lemma 6.4 that

$$\sum_{i=1}^{q} \varepsilon_i \cdot \mu(u_i, \lambda_i) = \sum_{i=1}^{q} \mathbb{S}^1\text{-}\deg(\mathscr{F}_{\theta_i}, \mathscr{U}_i) = 0.$$

The proof is complete. $\square$

6.5.2 Nondifferentiability Case

If $f(u, \lambda)$ is not differentiable with respect to u, then we need to justify that the formal linearization can be utilized to detect the local Hopf bifurcation and to describe the global continuation of periodic solutions for such a system with state-dependent delay. Our approach to this justification of formal linearization is through a simple homotopy argument. Namely, we will consider the equation

$$\tilde{\mathscr{F}}(u, \lambda) = 0, \quad (u, \lambda) \in \tilde{\mathscr{U}}, \tag{6.22}$$

for an $\mathbb{S}^1$-equivariant C^1-map $\tilde{\mathscr{F}}: \overline{\mathscr{U}} \to \mathbb{E}$ that is $\mathbb{S}^1$-homotopic to $\mathscr{F}$ in a sense to be detailed below. For the functional-analytic setting of the Hopf bifurcation of state-dependent DDEs, such a C^1-map is attained by extending a linear operator obtained through the formal linearization from a C^1-space to a C-space, an idea previously used by Eichmann [86] and Mallet-Paret–Nussbaum–Paraskevopoulos [215]. To be more precise, we assume that such a C^1-map is given by

$$\tilde{\mathscr{F}}(u, \lambda) = u - (L + K)^{-1} \circ [\tilde{Q}(\cdot, \lambda) + K](u), \quad (u, \lambda) \in \overline{\mathscr{U}}, \tag{6.23}$$

where $\tilde{Q}: \overline{\mathscr{U}} \to \mathbb{E}$ is an $\mathbb{S}^1$-equivariant C^1-map and

(SD4) $M \subseteq \tilde{\mathscr{F}}^{-1}(0)$, and for every $\lambda \in \mathbb{R}^2$, $(L + K)^{-1} \circ (\tilde{Q}(\cdot, \lambda) + K): \mathbb{E} \to \mathbb{E}$ is a condensing map.

By the implicit function theorem, if $(u_0, \lambda_0) \in M$ is a bifurcation point of system (6.23), then the derivative $D_u \tilde{\mathscr{F}}(u_0, \lambda_0)$, which is G-equivariant, is not an automorphism of $\mathbb{E}$. Therefore, all bifurcation points of (6.23) are contained in the set

$$\Lambda \overset{\text{def}}{=} \{(u, \lambda) \in M : D_u \tilde{\mathscr{F}}(u, \lambda) \notin GL_G(\mathbb{E})\}.$$

Let (u_0, λ_0) be an isolated $\mathbb{E}$-singular point of $\tilde{\mathscr{F}}$. To tie the $\mathbb{S}^1$-equivariant degree of $\mathscr{F}$ to that of $\tilde{\mathscr{F}}$, we assume that:

(SD5) We can choose the constants $r > 0$ and $\rho > 0$ such that $B_M(u_0, \lambda_0; r, \rho)$ is a special neighborhood of $\tilde{\mathscr{F}}$ and there exists $0 < r' \le r$ such that $\mathscr{F}(u, \lambda) \neq 0$ for all $(u, \lambda) \in B_M(u_0, \lambda_0; r', \rho)$ with $|\lambda - \lambda_0| = \rho$ and $\|u - \eta(\lambda)\| \neq 0$.

If ψ is an auxiliary function to (6.16), then by the construction of the $\mathbb{S}^1$-degree and the assumptions (SD2), (SD4), and (SD5), there exists a special neighborhood $\mathscr{U} \stackrel{\text{def}}{=} B_M(u_0, \lambda_0; r', \rho)$ of $\tilde{\mathscr{F}}$ such that the continuous G-equivariant maps $\mathscr{F}_\psi$ and $\tilde{\mathscr{F}}_\psi$ are nonzero on the boundary of $\mathscr{U}$, and therefore both $\mathbb{S}^1$-deg$(\tilde{\mathscr{F}}_\psi, \mathscr{U})$ and $\mathbb{S}^1$-deg$(\mathscr{F}_\psi, \mathscr{U})$ are well defined.

For a completing function θ defined on $\mathscr{U}$, if $\tilde{\mathscr{F}}_\theta = (\tilde{\mathscr{F}}, \theta)$ is homotopic to $\mathscr{F}_\theta$ on $\mathscr{U}$, then the homotopy invariance of the $\mathbb{S}^1$-degree ensures that $\mathbb{S}^1$-deg$(\mathscr{F}_\theta, \mathscr{U}) = \mathbb{S}^1$-deg$(\tilde{\mathscr{F}}_\theta, \mathscr{U})$. On the other hand, we can follow the approach presented in the previous subsection to calculate $\mathbb{S}^1$-deg$(\tilde{\mathscr{F}}_\theta, \mathscr{U})$.

Finally, in order to exclude bifurcation of solutions of (6.22) in $\mathbb{E}_0 \times \mathbb{R}^2$, we assume that

(SD6) $D_u \tilde{\mathscr{F}}(u_0, \lambda_0)|_{\mathbb{E}_0} : \mathbb{E}_0 \to \mathbb{E}_0$ is an isomorphism.

Theorem 6.7 ([170]). *Assume that* (SD1)–(SD6) *hold and let* $\mathscr{U} = B_M(u_0, \lambda_0; r', \rho) \subseteq \mathbb{E} \times \mathbb{R}^2$ *be a special neighborhood for* $\tilde{\mathscr{F}}$ *and* θ *a completing function. If* $\tilde{\mathscr{F}}_\theta$ *is homotopic to* $\mathscr{F}_\theta$ *on* $\overline{\mathscr{U}}$ *and there exists* $k \ge 1$ *such that* $\mathbb{S}^1$-deg$_k(\tilde{\mathscr{F}}_\theta, \mathscr{U}) \neq 0$, *then* (u_0, λ_0) *is a bifurcation point for* (6.13). *That is, there exists a sequence of nontrivial solutions* (u_n, λ_n) *of* (6.13) *such that the isotropy group of* u_n *contains* $\mathbb{Z}_k$ *and* $(u_n, \lambda_n) \to (u_0, \lambda_0)$ *as* $n \to \infty$.

For global bifurcation, we assume further that both $\mathscr{F}$ and $\tilde{\mathscr{F}}$ are defined on $\mathbb{E} \times \mathbb{R}^2$, and that:

(SD7) Every bifurcation point of (6.13) is a $\mathbb{E}$-singular point of $\tilde{\mathscr{F}}$.
(SD8) $\tilde{\mathscr{F}}_\theta$ is homotopic to $\mathscr{F}_\theta$ on some special neighborhood $\mathscr{U}$ of each isolated $\mathbb{E}$-singular point of $\tilde{\mathscr{F}}$, where θ is a completing function defined on $\mathscr{U}$.

Now we can state the following global bifurcation theorem of Rabinowitz type.

Theorem 6.8 ([170]). *Assume that* (SD1)–(SD8) *hold and* (SD5)–(SD6) *hold for every* $\mathbb{E}$-singular point (u_0, λ_0) *of* $\tilde{\mathscr{F}}$. *Assume further that every* $\mathbb{E}$-singular point of $\tilde{\mathscr{F}}$ *in* M *is isolated and* M *is complete. Let* $\mathscr{S}$ *denote the closure of the set of all nontrivial solutions of* (6.13). *Then for each bounded component* C *of* $\mathscr{S}$, *the set* $C \cap M$ *is a finite set, i.e.,* $C \cap M = \{(u_1, \lambda_1), (u_2, \lambda_2), \cdots, (u_q, \lambda_q)\}$, *and*

$$\sum_{i=1}^{q} \mathbb{S}^1\text{-deg}(\tilde{\mathscr{F}}_{\theta_i}, \mathscr{U}_i) = 0,$$

where $\mathscr{U}_i$ *is a special neighborhood of* (u_i, λ_i), *and* θ_i *is a completing function defined on* $\overline{\mathscr{U}}_i$.

6.6 Global Hopf Bifurcation Theory of DDEs

In this section, we employ the $\mathbb{S}^1$-equivariant degree to establish global Hopf bifurcations for general functional differential equations of mixed type with two parameters. We state our theory in a very general setting to allow for mixed type to ensure that the general theory can be used to address the issue of global bifurcations of bifurcated periodic solutions with additional features, such as spatial–temporal symmetry, for systems of DDEs with special symmetries.

Let X denote the Banach space of bounded continuous mappings $x\colon \mathbb{R} \to \mathbb{R}^n$ equipped with the supremum norm. For reasons mentioned above, we will consider functional differential equations with both delayed and advanced arguments. Therefore, for $x \in X$ and $t \in \mathbb{R}$, we will use x^t to denote an element in X defined by $x^t(s) = x(t+s)$ for $s \in \mathbb{R}$.

Consider the functional differential equation

$$\dot{x}(t) = F(x^t, \alpha, p) \tag{6.24}$$

parameterized by two real numbers $(\alpha, p) \in \mathbb{R} \times \mathbb{R}_+$, where $\mathbb{R}_+ = (0, \infty)$, and $F\colon X \times \mathbb{R} \times \mathbb{R}_+ \to \mathbb{R}^n$ is completely continuous. Identifying the subspace of X consisting of all constant mappings with $\mathbb{R}^n$, we obtain a mapping $\hat{F} = F|_{\mathbb{R}^n \times \mathbb{R} \times \mathbb{R}_+}\colon \mathbb{R}^n \times \mathbb{R} \times \mathbb{R}_+ \to \mathbb{R}^n$. We require the following assumption:

(GHB1) $\hat{F}$ is twice continuously differentiable.

Denote by $\hat{x}_0 \in X$ the constant mapping with the value $x_0 \in \mathbb{R}^n$. We call $(\hat{x}_0, \alpha_0, p_0)$ a *stationary solution* of (6.24) if $\hat{F}(x_0, \alpha_0, p_0) = 0$. We assume that:

(GHB2)   At each stationary solution $(\hat{x}_0, \alpha_0, p_0)$, the derivative of $\hat{F}(x, \alpha, p)$ with respect to the first variable x, evaluated at $(\hat{x}_0, \alpha_0, p_0)$, is an isomorphism of $\mathbb{R}^n$.

Under (GHB1)–(GHB2), for each stationary solution $(\hat{x}_0, \alpha_0, p_0)$, there exist $\varepsilon_0 > 0$ and a continuously differentiable mapping $y\colon B_{\varepsilon_0}(\alpha_0, p_0) \to \mathbb{R}^n$ such that $\hat{F}(y(\alpha, p), \alpha, p) = 0$ for $(\alpha, p) \in B_{\varepsilon_0}(\alpha_0, p_0) = (\alpha_0 - \varepsilon_0, \alpha_0 + \varepsilon_0) \times (p_0 - \varepsilon_0, p_0 + \varepsilon_0)$. We need the following smoothness condition:

(GHB3) $F(\varphi, \alpha, p)$ is differentiable with respect to φ, and the $n \times n$ complex matrix function $\Delta_{(\hat{y}(\alpha,p),\alpha,p)}(\lambda)$ is continuous in $(\alpha, p, \lambda) \in B_{\varepsilon_0}(\alpha_0, p_0) \times \mathbb{C}$. Here, for each stationary solution $(\hat{x}_0, \alpha_0, p_0)$, we have $\Delta_{(\hat{x}_0,\alpha_0,p_0)}(\lambda) = \lambda \operatorname{Id} - DF(\hat{x}_0, \alpha_0, p_0)(e^{\lambda \cdot} \operatorname{Id})$, where $DF(\hat{x}_0, \alpha_0, p_0)$ is the complexification of the derivative of $F(\varphi, \alpha, p)$ with respect to φ, evaluated at $(\hat{x}_0, \alpha_0, p_0)$.

For easy reference, we will again call $\Delta_{(\hat{x}_0,\alpha_0,p_0)}(\lambda)$ the *characteristic matrix* and the zeros of $\det \Delta_{(\hat{x}_0,\alpha_0,p_0)}(\lambda) = 0$ the *characteristic values* of the stationary solution $(\hat{x}_0, \alpha_0, p_0)$. So (GHB2) is equivalent to assuming that 0 is not a characteristic value of any stationary solution of (6.24).

Definition 6.1. A stationary solution $(\hat{x}_0, \alpha_0, p_0)$ is called a *center* if it has purely imaginary characteristic values of the form $im\frac{2\pi}{p_0}$ for some positive integer m. A center $(\hat{x}_0, \alpha_0, p_0)$ is said to be *isolated* if (i) it is the only center in some neighborhood of $(\hat{x}_0, \alpha_0, p_0)$; (ii) it has only finitely many purely imaginary characteristic values of the form $im\frac{2\pi}{p_0}$, m an integer.

Assume now that $(\hat{x}_0, \alpha_0, p_0)$ is an isolated center. Let $J(\hat{x}_0, \alpha_0, p_0)$ denote the set of all positive integers m such that $im\frac{2\pi}{p_0}$ is a characteristic value of $(\hat{x}_0, \alpha_0, p_0)$. We assume that there exists $m \in J(\hat{x}_0, \alpha_0, p_0)$ such that:

(GHB4)   There exist $\varepsilon \in (0, \varepsilon_0)$ and $\delta \in (0, \varepsilon_0)$ such that on $[\alpha_0 - \delta, \alpha_0 + \delta] \times \partial\Omega_{\varepsilon, p_0}$, $\det\Delta_{(\hat{y}(\alpha, p), \alpha, p)}(u + im\frac{2\pi}{p}) = 0$ if and only if $\alpha = \alpha_0$, $u = 0$, $p = p_0$, where $\Omega_{\varepsilon, p_0} = \{(u, p) : 0 < u < \varepsilon, p_0 - \varepsilon < p < p_0 + \varepsilon\}$.

Let

$$H^{\pm}(\hat{x}_0, \alpha_0, p_0)(u, p) = \det\Delta_{(\hat{y}(\alpha_0 \pm \delta, p), \alpha_0 \pm \delta, p)}\left(u + im\frac{2\pi}{p}\right).$$

Then (GHB4) implies that $H_m^{\pm}(\hat{x}_0, \alpha_0, p_0) \neq 0$ on $\partial\Omega_{\varepsilon, p_0}$. Consequently, the integer

$$\gamma_m(\hat{x}_0, \alpha_0, p_0) = \deg_B(H_m^-(\hat{x}_0, \alpha_0, p_0), \Omega_{\varepsilon, p_0}) - \deg_B(H_m^+(\hat{x}_0, \alpha_0, p_0), \Omega_{\varepsilon, p_0})$$

is well defined.

Definition 6.2. $\gamma_m(\hat{x}_0, \alpha_0, p_0)$ is called the *m*th *crossing number* of $(\hat{x}_0, \alpha_0, p_0)$.

We will show that $\gamma_m(\hat{x}_0, \alpha_0, p_0) \neq 0$ implies the existence of a local bifurcation of periodic solutions with periods near p_0/m. More precisely, we have the following:

Theorem 6.9. *Assume that* (GHB1)–(GHB3) *are satisfied, and that there exist an isolated center* $(\hat{x}_0, \alpha_0, p_0)$ *and an integer* $m \in J(\hat{x}_0, \alpha_0, p_0)$ *such that* (GHB4) *holds and* $\gamma_m(\hat{x}_0, \alpha_0, p_0) \neq 0$. *Then there exists a sequence* $(\alpha_k, p_k) \in \mathbb{R} \times \mathbb{R}_+$ *such that*

(i) $\lim_{k\to\infty}(\alpha_k, p_k) = (\alpha_0, p_0)$;
(ii) *at each* $(\alpha, p) = (\alpha_k, p_k)$, (6.24) *has a nonconstant periodic solution* $x_k(t)$ *with period* p_k/m;
(iii) $\lim_{k\to\infty} x_k(t) = \hat{x}_0$, *uniformly for* $t \in \mathbb{R}$.

To describe the global continuation of the local bifurcation obtained in Theorem 6.9, we need to assume that:

(GHB5)   All centers of (6.24) are isolated and (GHB4) holds for each center $(\hat{x}_0, \alpha_0, p_0)$ and each $m \in J(\hat{x}_0, \alpha_0, p_0)$.
(GHB6)   For each bounded set $W \subseteq X \times \mathbb{R} \times \mathbb{R}_+$, there exists a constant $l > 0$ such that $|F(\varphi, \alpha, p) - F(\psi, \alpha, p)| \leq l \sup_{s \in R} |\varphi(s) - \psi(s)|$ for $(\varphi, \alpha, p), (\psi, \alpha, p) \in W$.

Theorem 6.10. *Let*

$$\Sigma(F) = Cl\{(x, \alpha, p); x \text{ is a } p\text{-periodic solution of (6.24)}\} \subset X \times \mathbb{R} \times \mathbb{R},$$
$$N(F) = \{(\hat{x}, \alpha, p); F(\hat{x}, \alpha, p) = 0\}.$$

Assume that $(\hat{x}_0, \alpha_0, p_0)$ is an isolated center satisfying the conditions in Theorem 6.9. Denote by $C(\hat{x}_0, \alpha_0, p_0)$ the connected component of $(\hat{x}_0, \alpha_0, p_0)$ in $\Sigma(F)$. Then either

(i) $C(\hat{x}_0, \alpha_0, p_0)$ is unbounded, or
(ii) $C(\hat{x}_0, \alpha_0, p_0)$ is bounded, $C(\hat{x}_0, \alpha_0, p_0) \cap N(F)$ is finite, and

$$\sum_{(\hat{x}, \alpha, p) \in C(\hat{x}_0, \alpha_0, p_0) \cap N(F)} \gamma_m(\hat{x}, \alpha, p) = 0 \qquad (6.25)$$

for all $m = 1, 2, \ldots$, where $\gamma_m(\hat{x}, \alpha, p)$ is the mth crossing number of $(\hat{x}, \alpha, p)$ if $m \in J(\hat{x}, \alpha, p)$, and zero otherwise.

Proof of Theorems 6.9 and 6.10: Put $\mathbb{S}^1 = \mathbb{R}/2\pi\mathbb{Z}$, $\mathbb{E} = L^1(\mathbb{S}^1; \mathbb{R}^n)$, $\mathbb{F} = L^2(\mathbb{S}^1; \mathbb{R}^n)$. Define $L \colon \mathbb{E} \to \mathbb{F}$ and $Q \colon \mathbb{E} \times \mathbb{R} \times \mathbb{R}_+ \to \mathbb{F}$ by

$$Lz = \dot{z}(t), \quad Q(z, \alpha, p)(t) = \frac{p}{2\pi} F(z_{t,p}, \alpha, p),$$

where

$$z_{t,p}(\theta) = z\left(t + \frac{2\pi}{p}\theta\right), \quad \theta \in \mathbb{R}.$$

Clearly, $x(t)$ is a p-periodic solution of (6.24) if and only if $z(t) = x(\frac{p}{2\pi}t)$ is a solution in $\mathbb{E}$ of the operator equation $Lz = Q(z, \alpha, p)$.

The representations $\mathbb{E}$ and $\mathbb{F}$ are isometric Hilbert representations of the group $\mathbb{S}^1$, where $\mathbb{S}^1$ acts by shifting the argument. With respect to these $\mathbb{S}^1$-actions, L is an equivariant bounded linear Fredholm operator of index zero with an equivariant compact resolvent K, and Q is an $\mathbb{S}^1$-equivariant compact mapping. Moreover, at $(\hat{y}(\alpha, p), \alpha, p)$ with $(\alpha, p) \in \mathscr{D} \overset{\text{def}}{=} (\alpha_0 - \delta, \alpha_0 + \delta) \times (p_0 - \varepsilon, p_0 + \varepsilon)$, the derivative of Q with respect to the first variable is given by

$$D_z Q(\hat{y}(\alpha, p), \alpha, p)z(t) = \frac{p}{2\pi} DF(\hat{y}(\alpha, p), \alpha, p)z_{t,p}.$$

Identifying $\partial\mathscr{D}$ with $\mathbb{S}^1$, since $(\hat{x}_0, \alpha_0, p_0)$ is an isolated center, we can easily show that the mapping $\mathrm{Id} - (L + K)^{-1}[K + D_z F(\hat{y}(\alpha, p), \alpha, p)]$ is an isomorphism of $\mathbb{E}$ and that the mapping $\Psi \colon \mathbb{S}^1 \to GL(\mathbb{E})$ defined by

$$(\alpha, p) \in \partial\mathscr{D} \cong \mathbb{S}^1 \to \mathrm{Id} - (L + K)^{-1}[K + D_z F(\hat{y}(\alpha, p), \alpha, p)] \in GL(\mathbb{E})$$

is continuous.

The representation $\mathbb{E}$ has the well-known isotypical decomposition $\mathbb{E} = \bigoplus_{k=0}^{\infty} \mathbb{E}_k$, where $\mathbb{E}_0 \cong \mathbb{R}^n$ and for each $k \geq 1$, $\mathbb{E}_k$ is spanned by $\cos(kt)\varepsilon_j$ and $\sin(kt)\varepsilon_j$, $1 \leq j \leq n$, where $\{\varepsilon_1, \ldots, \varepsilon_n\}$ is the standard basis of $\mathbb{R}^n$. So we have $\Psi(\alpha, p)\mathbb{E}_k \subseteq \mathbb{E}_k$. Let $\Psi_k(\alpha, p) = \Psi(\alpha, p)|_{\mathbb{E}_k}$. It is not difficult to show that

$$\Psi_k(\alpha, p) = \frac{p}{i2k\pi} \Delta_{(\hat{y}(\alpha, p), \alpha, p)}\left(ik\frac{2\pi}{p}\right).$$

Let

$$\varepsilon = \operatorname{sign} \det \Psi_0(\alpha, p), \qquad (\alpha, p) \in \partial \mathscr{D},$$
$$n_k(\hat{x}_0, \alpha_0, p_0) = \varepsilon \deg_B(\det \Psi_k(\cdot), \mathscr{D}), \quad k = 1, 2, \dots.$$

Then one can show, as in Erbe, Geba, Krawcewicz, and Wu [112], that $\gamma_k(\hat{x}_0, \alpha_0, p_0) = n_k(\hat{x}_0, \alpha_0, p_0)$, and therefore Theorems 6.9 and 6.10 are simply immediate consequences of Theorems 6.5 and 6.6 with $M = \{(\hat{x}_0, \alpha_0, p_0) \in \mathbb{R}^n \times \mathbb{R} \times \mathbb{R}_+; F(\hat{x}_0, \alpha_0, p_0) = 0\}$. This completes the proof. $\qquad \square$

For ease of applications, we describe below the local and global Hopf bifurcation theory for parameterized DDEs. Let $X = C_{n,\tau}$ and consider the following functional differential equation:

$$\dot{x}(t) = F(x_t, \alpha) \tag{6.26}$$

with parameter $\alpha \in \mathbb{R}$, $F: X \times \mathbb{R} \to \mathbb{R}^n$ is completely continuous.

Identifying the subspace of X consisting of all constant mappings with $\mathbb{R}^n$, we obtain a mapping $\hat{F} = F|_{\mathbb{R}^n \times \mathbb{R}}: \mathbb{R}^n \times \mathbb{R} \to \mathbb{R}^n$. We now describe conditions (GHB1)–(GHB6) in relatively simple form:

(SGHB1) $\hat{F}$ is twice continuously differentiable.

Denote by $\hat{x}_0 \in X$ the constant mapping with the value $x_0 \in \mathbb{R}^n$. We call $(\hat{x}_0, \alpha_0)$ a *stationary solution* of (6.26) if $\hat{F}(x_0, \alpha_0) = 0$. We assume that:

(SGHB2)   At each stationary solution $(\hat{x}_0, \alpha_0)$, the derivative of $\hat{F}(x, \alpha)$ with respect to the first variable x, evaluated at $(\hat{x}_0, \alpha_0)$, is an isomorphism of $\mathbb{R}^n$.

Under (SGHB1)–(SGHB2), for each stationary solution $(\hat{x}_0, \alpha_0)$, there exist $\varepsilon_0 > 0$ and a continuously differentiable mapping $y: B_{\varepsilon_0}(\alpha_0) \to \mathbb{R}^n$ such that $\hat{F}(y(\alpha), \alpha) = 0$ for $\alpha \in B_{\varepsilon_0}(\alpha_0) = (\alpha_0 - \varepsilon_0, \alpha_0 + \varepsilon_0)$.

We need the following smoothness condition:

(SGHB3) $F(\varphi, \alpha)$ is differentiable with respect to φ, and the $n \times n$ complex matrix function $\Delta_{(\hat{y}(\alpha), \alpha)}(\lambda)$ is continuous in $(\alpha, \lambda) \in B_{\varepsilon_0}(\alpha_0) \times \mathbb{C}$. Here, for each stationary solution $(\hat{x}_0, \alpha_0)$, we have $\Delta_{(\hat{x}_0, \alpha_0)}(\lambda) = \lambda \operatorname{Id} - DF(\hat{x}_0, \alpha_0)(e^{\lambda \cdot} \operatorname{Id})$, where $DF(\hat{x}_0, \alpha_0)$ is the complexification of the derivative of $F(\varphi, \alpha)$ with respect to φ, evaluated at $(\hat{x}_0, \alpha_0)$.

For easy reference, we will again call $\Delta_{(\hat{x}_0, \alpha_0)}(\lambda)$ the *characteristic matrix* and the zeros of $\det \Delta_{(\hat{x}_0, \alpha_0)}(\lambda) = 0$ the *characteristic values* of the stationary solution $(\hat{x}_0, \alpha_0)$. So (SGHB2) is equivalent to assuming that 0 is not a characteristic value of any stationary solution of (6.26).

The concepts of isolated centers and crossing numbers are now simplified as follows:

Definition 6.3. A stationary solution $(\hat{x}_0, \alpha_0)$ is called a *center* if it has purely imaginary characteristic values $\pm i\beta_0$ for some positive $\beta_0 > 0$. A center $(\hat{x}_0, \alpha_0)$ is said to be *isolated* if it is the only center in some neighborhood of $(\hat{x}_0, \alpha_0)$.

Assume that $(\hat{x}_0, \alpha_0)$ is an isolated center. We assume that:

(SGHB4)   There exist $\varepsilon \in (0, \varepsilon_0)$ and $\delta \in (0, \varepsilon_0)$ such that on $[\alpha_0 - \delta, \alpha_0 + \delta] \times \partial \Omega_{\varepsilon, p_0}$, $\det \Delta_{(\hat{y}(\alpha), \alpha)}(u + i\beta) = 0$ if and only if $\alpha = \alpha_0$, $u = 0$, $\beta = \beta_0$, where $\Omega_{\varepsilon, \beta_0} = \{(u, p) : 0 < u < \varepsilon, \beta_0 - \varepsilon < \beta < \beta_0 + \varepsilon\}$.

Let

$$H^{\pm}(\hat{x}_0, \alpha_0)(u, \beta) = \det \Delta_{(\hat{y}(\alpha_0 \pm \delta), \alpha_0 \pm \delta)}(u + i\beta).$$

Then (SGHB4) implies that $H^{\pm}(\hat{x}_0, \alpha_0, \beta_0) \neq 0$ on $\partial \Omega_{\varepsilon, \beta_0}$. Consequently, the integer

$$\gamma(\hat{x}_0, \alpha_0, \beta_0) = \deg_B(H^-(\hat{x}_0, \alpha_0, \beta_0), \Omega_{\varepsilon, \beta_0}) - \deg_B(H^+(\hat{x}_0, \alpha_0, \beta_0), \Omega_{\varepsilon, \beta_0})$$

is well defined; it is called the first *crossing number* of $(\hat{x}_0, \alpha_0, \beta_0)$.

The local Hopf bifurcation theory below shows that $\gamma(\hat{x}_0, \alpha_0, \beta_0) \neq 0$ implies the existence of a local bifurcation of periodic solutions with periods near $2\pi/\beta_0$. More precisely, we have the following theorem:

Theorem 6.11. *Assume that (SGHB1)–(SGHB3) are satisfied, and that there exists an isolated center $(\hat{x}_0, \alpha_0)$ such that (SGHB4) holds and $\gamma(\hat{x}_0, \alpha_0, \beta_0) \neq 0$. Then there exists a sequence $(\alpha_k, \beta_k) \in \mathbb{R} \times \mathbb{R}_+$ such that*

(i) $\lim_{k \to \infty}(\alpha_k, \beta_k) = (\alpha_0, \beta_0)$;
(ii) *at each $\alpha = \alpha_k$, (6.26) has a nonconstant periodic solution $x_k(t)$ with a period $\frac{2\pi}{\beta_k}$;*
(iii) $\lim_{k \to \infty} x_k(t) = \hat{x}_0$, *uniformly for $t \in \mathbb{R}$.*

The global Hopf bifurcation theorem can now be stated as follows:

(SGHB5)   All centers of (6.26) are isolated and (SGHB4) holds for each center $(\hat{x}_0, \alpha_0)$ with the corresponding β_0.
(SGHB6)   For each bounded set $W \subseteq X \times \mathbb{R}$, there exists a constant $l > 0$ such that $|F(\varphi, \alpha) - F(\psi, \alpha)| \leq l \sup_{s \in \mathbb{R}} |\varphi(s) - \psi(s)|$ for $(\varphi, \alpha), (\psi, \alpha) \in W$.

Theorem 6.12. *Set*

$$\Sigma(F) = Cl\{(x, \alpha, \beta); \ x \text{ is a } 2\pi/\beta\text{-periodic solution of (6.26)}\} \subset X \times \mathbb{R} \times \mathbb{R},$$
$$N(F) = \{(\hat{x}, \alpha, \beta); F(\hat{x}, \alpha) = 0, \det \Delta_{(\hat{y}(\alpha), \alpha)}(i\beta) = 0\}.$$

Assume that $(\hat{x}_0, \alpha_0, \beta_0$ is an isolated center satisfying the conditions in Theorem 6.11. Denote by $C(\hat{x}_0, \alpha_0, \beta_0)$ the connected component of $(\hat{x}_0, \alpha_0, \beta_0)$ in $\Sigma(F)$. Then either

(i) $C(\hat{x}_0, \alpha_0, \beta_0)$ *is unbounded, or*
(ii) $C(\hat{x}_0, \alpha_0, \beta_0)$ *is bounded, $C(\hat{x}_0, \alpha_0, \beta_0) \cap N(F)$ is finite, and*

$$\sum_{(\hat{x}, \alpha, \beta) \in C(\hat{x}_0, \alpha_0, \beta_0) \cap N(F)} \gamma(\hat{x}, \alpha, \beta) = 0, \qquad (6.27)$$

where $\gamma(\hat{x}, \alpha, \beta)$ is the crossing number of $(\hat{x}, \alpha, \beta)$.

6.7 Application to a Delayed Nicholson Blowflies Equation

6.7.1 The Nicholson Blowflies Equation

Gurney et al. [142] proposed the following simple-looking delay differential equation to explain the oscillatory behavior of the observed sheep blowfly *Lucilia cuprina* population in the experimental data collected by the Australian entomologist Nicholson [231]:

$$N'(t) = f(N(t - \tau)) - \gamma N(t)$$

with $f(N) = pNe^{-\alpha N}$, where $N(t)$ denotes the population of sexually mature adults at time t, p is the maximum possible per capita egg production rate, $1/\alpha$ is the population size at which the whole population reproduces at its maximum rate. In the model, τ is the generation time, or the time from egg to sexually mature adult, and γ is the per capita mortality rate of adults. This model is now commonly called Nicholson's blowflies equation. It was used by Oster and Ipatkchi [239] for the development of an insect population, and its modifications have been intensively studied in the literature of theoretical biology and delay differential equations. Notably, it has been shown that a unique positive equilibrium of the model is globally asymptotically stable (with respect to nonnegative and nontrivial initial conditions) for every $\tau \geq 0$, provided that $1 < p/\gamma < e^2$ (see, for example, [198]). In the case $p/\gamma > e^2$, the positive equilibrium loses its local stability, and Hopf bifurcations occur at an unbounded sequence of critical values. In the next subsection, we introduce the work of Wei and Li [294] that uses the global Hopf bifurcation theorem coupled with Bendixson's criterion for higher dimensional ordinary differential equations to establish the existence of periodic solutions when the delay τ is not necessarily near the local Hopf bifurcation values.

6.7.2 The Global Hopf Bifurcation Theorem of Wei–Li

In this subsection, we consider the equation

$$N'(t) = -\gamma N(t) + pN(t - \tau)e^{-aN(t-\tau)}, t \geq 0. \qquad (6.28)$$

We introduce the theorem of Wei–Li [294] that shows that under the assumption $p > \gamma e^2$, as the delay τ increases, the positive equilibrium N^* loses its stability, a sequence of Hopf bifurcations occurs at N^*, and these periodic solutions persist for τ far away from these Hopf bifurcation values. Wei and Li established this theorem using a global Hopf bifurcation result (Theorem 6.12). A key step in establishing the global extension of the local Hopf branch at $\tau = \tau_0$ is to show that (6.28) has no nonconstant periodic solutions of period 4. This is accomplished by applying a higher-dimensional Bendixson criterion for ordinary differential equations due to Li and Muldowney [210].

The positive equilibrium $N^* = \frac{1}{a}\log\frac{p}{\gamma}$ of (6.28) exists if and only if $a > 0$ and $p > \gamma$. These relations are assumed throughout this section. Set $N(t) = N^* + \frac{1}{a}y(t)$. Then $x(t)$ satisfies

$$y'(t) = -\gamma y(t) - a\gamma N^*\left[1 - e^{-y(t-\tau)}\right] + \gamma y(t-\tau)e^{-y(t-\tau)}. \tag{6.29}$$

The linearization of (6.29) at $y = 0$ is

$$Y'(t) = -\gamma Y(t) - \gamma[aN^* - 1]Y(t-\tau),$$

whose characteristic equation is

$$\lambda = \gamma - \gamma[aN^* - 1]e^{-\lambda\tau}. \tag{6.30}$$

For $\tau = 0$, the only root of (6.30) is $\lambda = -aN^* < 0$, since $p > \gamma$. For $\omega \neq 0$, $i\omega$ is a root of (6.30) if and only if

$$i\omega = -\gamma - \gamma[aN^* - 1](\cos\omega\tau - i\sin\omega\tau).$$

Separating the real and imaginary parts, we obtain

$$\gamma(aN^* - 1)\cos\omega\tau = -\gamma,$$
$$\gamma(aN^* - 1)\sin\omega\tau = \omega,$$

which leads to

$$\gamma^2(aN^* - 1)^2 = \gamma^2 + \omega^2,$$

namely,

$$\omega = \pm\gamma\sqrt{aN^*(aN^* - 2)}.$$

This is possible if and only if $aN^* > 2$, or equivalently, if $p > \gamma e^2$.
    For $p > \gamma e^2$, let

$$\tau_k = \frac{1}{\gamma\sqrt{aN^*(aN^* - 2)}}\left[\sin^{-1}\left(\frac{\sqrt{aN^*(aN^* - 2)}}{aN^* - 1}\right) + 2k\pi\right],$$

$k = 0, 1, 2, \cdots$. Set

$$\omega_0 = \gamma\sqrt{aN^*(aN^* - 2)}. \tag{6.31}$$

Let $\lambda_k = \alpha_k(\tau) + i\omega_k(\tau)$ denote a root of (6.30) near $\tau = \tau_k$ such that $\alpha_k(\tau_k) = 0$, $\omega_k(\tau_k) = \omega_0$. Obviously, $\alpha_k'(\tau_k) > 0$. Therefore, we have obtained that when $\gamma < p \leq \gamma e^2$, all roots of the characteristic equation (6.30) have negative real parts; when $p > \gamma e^2$, (6.30) has a pair of simple imaginary roots $\pm i\omega_0$ at $\tau = \tau_k$, $k = 0, 1, 2, \cdots$. Furthermore, if $\tau \in [0, \tau_0)$, then all roots of (6.30) have negative real part; if $\tau = \tau_0$, then all roots of (6.30) except $\pm i\omega_0$ have negative real part; and if $\tau \in (\tau_k, \tau_{k+1})$ for $k = 0, 1, 2, \cdots$, then (6.30) has $2(k+1)$ roots with positive real part. In particular, we have shown that under the condition $p > \gamma e^2$, $N = N^*$ is asymptotically stable for $\tau \in [0, \tau_0)$ and unstable for $\tau > \tau_0$. Furthermore, (6.28) undergoes a Hopf bifurcation at N^* when $\tau = \tau_k$, for $k = 0, 1, 2, \cdots$.

Let $x(t) = y(\tau t)$. Then (6.29) becomes

$$x'(t) = -\gamma\tau \left[x(t) + aN^*(1 - e^{-x(t-1)}) - x(t-1)e^{-x(t-1)} \right]. \tag{6.32}$$

Lemma 6.5. *All periodic solutions to (6.32) are uniformly bounded.*

Proof. Let $x(t)$ be a nonconstant periodic solution to (6.32), and let $x(t_1) = M$, $x(t_2) = m$ be its maximum and minimum, respectively. Then, $x'(t_1) = x'(t_2) = 0$, and by (6.32),

$$M = x(t_1 - 1)e^{-x(t_1-1)} - aN^*[1 - e^{-x(t_1-1)}], \tag{6.33}$$

$$m = x(t_2 - 1)e^{-x(t_2-1)} - aN^*[1 - e^{-x(t_2-1)}]. \tag{6.34}$$

We claim that $x(t_1 - 1) < 0$ and $x(t_2 - 1) > 0$. In fact, if $x(t_1 - 1) = 0$, then (6.33) implies $M = 0$, and thus $m < 0$ and $x(t_2 - 1) \le 0$. Using (6.34), we know that $x(t_2 - 1) < 0$, and thus

$$m > x(t_2 - 1)e^{-x(t_2-1)},$$

which contradicts the fact that m is the minimum. If $x(t_1 - 1) > 0$, then by (6.33), we arrive at

$$M \le M - aN^*(1 - e^{-x(t_1-1)}) < M,$$

a contradiction. Therefore, $x(t_1 - 1) < 0$. A similar argument shows that $x(t_2 - 1) > 0$. Therefore, we have $m < 0$ and $M > 0$. Again by (6.33) and (6.34), we have

$$m > aN^*[e^{-M} - 1] > -aN^*. \tag{6.35}$$

Also by (6.33), we have

$$
\begin{aligned}
M &= -aN^* + (x(t_1 - 1) + aN^*)e^{-x(t_1-1)} \\
&= -aN^* + e^{aN^*}(x(t_1 - 1) + aN^*)e^{-(x(t_1-1)+aN^*)} \\
&\le -aN^* + e^{aN^*}e^{-1} = -aN^* + e^{aN^*-1}.
\end{aligned}
\tag{6.36}
$$

Here we have used the fact that $x(t_1 - 1) + aN^* > m + aN^* > 0$ and that $xe^{-x} < e^{-1}$ for $x \ge 0$. Relations (6.35) and (6.36) imply uniform boundedness of the periodic solutions. $\qquad\square$

Lemma 6.6. *Assume that $\gamma e^2 < p < \sqrt{2}\,\gamma e^2$. Then (6.32) has no periodic solutions of period 4.*

Proof. Let $x(t)$ be a periodic solution to (6.32) of period 4. Set $u_j(t) = x(t - j + 1)$, $j = 1, 2, 3, 4$. Then $u(t) = (u_1(t), u_2(t), u_3(t), u_4(t))$ is a periodic solution of the following system of ordinary differential equations:

$$u_1'(t) = -\gamma\tau[u_1(t) + aN^*(1 - e^{-u_2(t)}) - u_2(t)e^{-u_2(t)}],$$
$$u_2'(t) = -\gamma\tau[u_2(t) + aN^*(1 - e^{-u_3(t)}) - u_3(t)e^{-u_3(t)}],$$
$$u_3'(t) = -\gamma\tau[u_3(t) + aN^*(1 - e^{-u_4(t)}) - u_4(t)e^{-u_4(t)}], \tag{6.37}$$
$$u_4'(t) = -\gamma\tau[u_4(t) + aN^*(1 - e^{-u_1(t)}) - u_1(t)e^{-u_1(t)}],$$

whose orbit belongs to the region

$$G = \{u \in \mathbb{R}^4 : \quad \bar{m} < |u_k| < \bar{M}, \quad k = 1,2,3,4\}, \tag{6.38}$$

where $\bar{m}$ and $\bar{M}$ are a pair of uniform bounds for periodic solutions of (6.32) obtained in Lemma 6.5. To rule out 4-periodic solutions of (6.32), it suffices to prove the nonexistence of nonconstant periodic solutions of (6.37) in the region G. To do the latter, we use a general Bendixson's criterion in higher dimensions developed in Li and Muldowney [210]. More specifically, we will apply Corollary 3.5 in [210]. The Jacobian matrix $J = J(u)$ of (6.37), for $u \in \mathbb{R}^4$, is

$$J(u) = -\gamma\tau \begin{pmatrix} 1 & f(u_2) & 0 & 0 \\ 0 & 1 & f(u_3) & 0 \\ 0 & 0 & 1 & f(u_4) \\ f(u_1) & 0 & 0 & 1 \end{pmatrix},$$

where

$$f(v) = (aN^* + v - 1)e^{-v}. \tag{6.39}$$

The second additive compound matrix $J^{[2]}(u)$ of $J(u)$ is (see [103] and [226])

$$J^{[2]}(u) = -\gamma\tau \begin{pmatrix} 2 & f(u_3) & 0 & 0 & 0 & 0 \\ 0 & 2 & f(u_4) & f(u_2) & 0 & 0 \\ 0 & 0 & 2 & 0 & f(u_2) & 0 \\ 0 & 0 & 0 & 2 & f(u_4) & 0 \\ -f(u_1) & 0 & 0 & 0 & 2 & f(u_3) \\ 0 & -f(u_1) & 0 & 0 & 0 & 2 \end{pmatrix}.$$

Choose a vector norm in $\mathbb{R}^6$ as

$$|(x_1,x_2,x_3,x_4,x_5,x_6)| = \max\{\sqrt{2}|x_1|, |x_2|, \sqrt{2}|x_3|, \sqrt{2}|x_4|, |x_5|, \sqrt{2}|x_6|\}.$$

Then with respect to this norm, the Lozinskiĭ measure $\mu(J^{[2]}(u))$ of the matrix $J^{[2]}(u)$ is, see [73],

$$\mu(J^{[2]}(u)) =$$
$$\max\{\sqrt{2}\gamma\tau(-\sqrt{2}+|f(u_3)|), \sqrt{2}\gamma\tau(-\sqrt{2}+|f(u_4)|/2+|f(u_2)|/2),$$
$$\sqrt{2}\gamma\tau(-\sqrt{2}+|f(u_2)|), \sqrt{2}\gamma\tau(-\sqrt{2}+|f(u_4)|), \tag{6.40}$$
$$\sqrt{2}\gamma\tau(-\sqrt{2}+|f(u_1)|/2+|f(u_3)|/2), \sqrt{2}\gamma\tau(-\sqrt{2}+|f(u_1)|)\}.$$

By Corollary 3.5 in [210], system (6.37) has no periodic orbits in G if $\mu(J^{[2]}(u)) < 0$
for all $u \in G$. From (6.40), we see that $\mu(J^{[2]}(u)) < 0$ if and only if

$$|f(u_j)| < \sqrt{2}, \quad j = 1, 2, 3, 4, \tag{6.41}$$

for $u \in G$.

To establish (6.41), we first use the assumption $e^{aN^*} = p/\gamma < \sqrt{2}e^2$ to improve
the lower bound m given in Lemma 6.5. In (6.34), we now have

$$x(t_2 - 1) + aN^* < M + aN^* < e^{aN^* - 1} < \sqrt{2}e.$$

Using the fact that the function xe^{-x} is monotonically decreasing for $x > 1$ and that
$x(t_2 - 1) + aN^* > 1$, we have

$$\begin{aligned} m &= -aN^* + e^{aN^*}(x(t_2 - 1) + aN^*)e^{-(x(t_2 - 1) + aN^*)} \\ &> -aN^* + e^{aN^*}\sqrt{2}e\,e^{-\sqrt{2}e} > -aN^* + 2e^2\sqrt{2}e\,e^{-\sqrt{2}e} \\ &= -aN^* + 2\sqrt{2}e^{3 - \sqrt{2}e}. \end{aligned}$$

Therefore, $u \in G$ satisfies

$$|u_i| > -aN^* + 2\sqrt{2}e^{3 - \sqrt{2}e}.$$

For $\delta = 2\sqrt{2}e^{3 - \sqrt{2}e} > 1$, we can verify

$$|f(-aN^* + \delta)| = e^{aN^* - \delta}|\delta - 1| = e^{aN^* - 2}e^{2 - \delta}(\delta - 1) < e^{aN^* - 2}.$$

From the graph of $f(v)$, we know that $f(v)$ has a global maximum $e^{aN^* - 2} = e^{-2}p/\gamma$.
Therefore, for $u \in G$,

$$|f(u_k)| \leq \max\{e^{aN^* - 2}, |f(-aN^* + \delta)|\} \leq e^{aN^* - 2} = \frac{p}{\gamma}e^{-2} < \sqrt{2},$$

and (6.41) is satisfied, completing the proof.   □

Lemma 6.7. *Assume that $\gamma e^2 < p$. Then (6.32) has no periodic solutions of period
1 or 2.*

Proof. First note that every nonconstant 1-periodic solution $u(t)$ of (6.32) is also a
nonconstant periodic solution of the ordinary differential equation

$$u'(t) = -\gamma\tau(1 - e^{-u(t)})(u(t) + aN^*). \tag{6.42}$$

A simple phase-line analysis shows that (6.42) has no nonconstant periodic solutions.

As in the proof of Lemma 6.6, if $u(t)$ is a periodic solution of (6.32) of period
2, then $u_1(t) = u(t)$ and $u_2(t) = u(t - 1)$ are periodic solutions of the system of
ordinary differential equations

$$u_1'(t) = -\gamma\tau[u_1(t) + aN^*(1 - e^{-u_2(t)}) - u_2(t)e^{-u_2(t)}]$$
$$u_2'(t) = -\gamma\tau[u_2(t) + aN^*(1 - e^{-u_1(t)}) - u_1(t)e^{-u_1(t)}].$$

$$(6.43)$$

Let $(P(u_1, u_2), Q(u_1, u_2))$ denote the vector field of (6.43). Then

$$\frac{\partial P}{\partial u_1} + \frac{\partial Q}{\partial u_2} = -2\gamma\tau < 0$$

for all (u_1, u_2). Thus the classical Bendixson's negative criterion implies that (6.43) has no nonconstant periodic solutions. $\qquad\square$

Theorem 6.13. *Suppose that* $\gamma e^2 < p < \sqrt{2}\gamma e^2$ *holds. Then for each* $\tau > \tau_k$, $k = 0, 1, 2, \cdots$, *(6.32) has at least* $k + 1$ *periodic solutions.*

Proof. First note that

$$F(x_t, \tau) \stackrel{\text{def}}{=} -\gamma\tau[x(t) + aN^*(1 - e^{x(t-1)}) - y(t-1)e^{-x(t-1)}]$$

satisfies hypotheses (SGHB1), (SGHB2), and (SGHB3) of Sect. 6.6, with

$$(\hat{x}_0, \alpha_0) = (0, \tau_k),$$
$$\Delta_{(0,\tau_k)}(z) = z + \tau\gamma + \tau\gamma[aN^* - 1]e^{-z}.$$

It can also be verified that $(0, \tau_k)$ are isolated centers with the corresponding imaginary characteristic values $\pm i\tau_k\omega_0$. We have shown that there exist $\varepsilon > 0$, $\delta > 0$, and a smooth curve $z : (\tau_k - \delta, \tau_k + \delta) \to \mathbb{C}$ such that $\Delta(z(\tau)) = 0$, $|z(\tau) - i\tau_k\omega_0| < \varepsilon$ for all $\tau \in [\tau_k - \delta, \tau_k + \delta]$, and

$$z(\tau_k) = i\tau_k\omega_0, \quad \frac{d\mathrm{Re}z(\tau)}{d\tau}\bigg|_{\tau=\tau_k} > 0.$$

Set $\beta_k = \tau_k\omega_0$ and let

$$\Omega_\varepsilon = \{(0, \beta) : 0 < u < \varepsilon, |\beta - \beta_k| < \varepsilon\}.$$

Clearly, if $|\tau - \tau_k| \le \delta$ and $(u, p) \in \partial\Omega_\varepsilon$ such that $\Delta_{0,\tau)}(u + i\beta) = 0$, then $\tau = \tau_k$, $u = 0$, and $\beta = \beta_k$. This satisfies assumption (SGHB4) in Sect. 6.6. Moreover, if we put

$$H^\pm(0, \tau_k)(u, \beta) = \Delta_{(0,\tau_k\pm\delta)}(u + i\beta),$$

then we have the cross number

$$\gamma(0, \tau_k) = \deg_B(H^-(0, \tau_k, \tau_k\omega_0), \Omega_\varepsilon)$$
$$- \deg_B(H^+(0, \tau_k, \tau_k\omega_0), \Omega_\varepsilon) = -1.$$

By Theorem 6.12, we conclude that the connected component $C(0, \tau_k, \tau_k\omega_0)$ through $(0, \tau_k, \tau_k\omega_0)$ in $\Sigma(F)$ is nonempty. Meanwhile, we have

$$\sum_{(\hat{x},\tau,\beta)\in C(0,\tau_k,\tau_k\omega_0)\cap N(F)} \gamma(\hat{y},\tau,T) < 0,$$

and hence $C(0,\tau_k,\tau_k\omega_0)$ is unbounded.

Lemma 6.5 implies that the projection of $C(0,\tau_k,\tau_k\omega_0)$ onto the x-space is bounded. It can be verified using a phase-line analysis that when $\tau = 0$, (6.32) has no nonconstant periodic solutions. Therefore, the projection of $C(0,\tau_k,\tau_k\omega_0)$ onto the τ-space is bounded below. From the definitions of τ_k and ω_0, we obtain

$$\tau_k\omega_0 = \sin^{-1}\left(\frac{\sqrt{aN^*(aN^*-2)}}{aN^*-1}\right) + 2k\pi \tag{6.44}$$

for $k \geq 0$. Also, we know that $\sin\omega_0\tau_k > 0$ and $\cos\omega_0\tau_k < 0$, for $k \geq 0$. Hence

$$\frac{\pi}{2} < \omega_0\tau_0 < \pi, \quad \text{and} \quad 2\pi < \omega_0\tau_k < (2k+1)\pi, \; k \geq 1.$$

Therefore

$$2 < \frac{2\pi}{\tau_0\omega_0} < 4, \quad \text{and} \quad \frac{1}{k+1} < \frac{2\pi}{\omega_0\tau_k} < 1, \; k \geq 1. \tag{6.45}$$

Applying Lemmas 6.6 and 6.7, we know that $2 < 2\pi/\beta < 4$ if $(x,\tau,\beta) \in C(0,\tau_0,\tau_0\omega_0)$, and that $1/(k+1) < 2\pi/\beta < 1$ if $(x,\tau,\beta) \in C(0,\tau_k,\tau_k\omega_0)$ for $k \geq 1$. This shows that in order for $C(0,\tau_k,\tau_k\omega_0)$ to be unbounded, its projection onto the τ-space must be unbounded. Consequently, the projection of $C(0,\tau_k,\tau_k\omega_0)$ onto the τ-space includes $[\tau_k,\infty)$. This shows that for each $\tau > \tau_k$, (6.32) has $k+1$ nonconstant periodic solutions, completing the proof of the theorem. □

Remark 6.1. (i) From the proof of Theorem 6.13, we know that the first global Hopf branch contains periodic solutions of period between 2 and 4. These are the slowly oscillating periodic solutions. See [13, 60, 197, 291] for more details about the existence of slowly oscillating periodic solutions in delay differential equations. The τ_k branches, for $k \geq 1$, since the periods are less than 1, contain fast-oscillating periodic solutions.

(ii) For $k \geq 1$,

$$\frac{1}{k+1} < \frac{2\pi}{\tau_k\omega_0} < 1$$

automatically holds. The bounds on the period $2\pi/\beta$ for $(x,\tau,\beta)\in C(0,\tau_k,\tau_k\omega_0)$ hold without resort to Lemma 6.6. Thus, the global extension of the τ_k-branch for $k \geq 1$ can be proved without the restriction $p < \sqrt{2}\gamma e^2$.

6.7.3 Nicholson's Blowflies Equation Revisited: Onset and Termination of Nonlinear Oscillations

In [264], the authors reexamined the Nicholson's blowflies model with natural death rate explicitly incorporated into the delay feedback, obtaining the following delay differential equation with a delay-dependent coefficient

$$N'(t) = e^{-\delta\tau}f(N(t-\tau)) - \gamma N(t), \tag{6.46}$$

where $\delta > 0$ is the death rate of the immature population, and $f(N) = pNe^{-\alpha N}$. One can derive this, as was done in [82, 222], from a structured population model for $u(t,a)$ (the population density at age a and time t) as follows:

$$\frac{\partial}{\partial t}u(t,a) + \frac{\partial}{\partial a}u(t,a) = -\mu(a)u(t,a),$$

with the stage-specific mortality rate

$$\mu(a) = \begin{cases} \gamma, & t > \tau, \\ \delta, & t < \tau. \end{cases}$$

A simple application of the integration along characteristic lines leads to the model equation for the mature population $N(t) = \int_{\tau}^{\infty} u(t,a)da$ with the Ricker's-type birth function f.

The additional term $e^{-\delta\tau}$ is the probability of the immature population surviving τ time units before becoming mature. This addition, as shown in [264], leads to rather different dynamics for model (6.46): as the delay τ increases, the positive equilibrium loses its stability and undergoes local Hopf bifurcations at a *finite even number* of critical values, and as τ passes a critical threshold, the positive equilibrium regains its stability. In other words, as τ keeps increasing and passes another threshold value, the positive equilibrium disappears, and the species becomes extinct (the zero solution is globally asymptotically stable). Shu, Wang, and Wu [264] also observed the coexistence of multiple stable periodic solutions.

As we did in the last subsection, Shu, Wang, and Wu [264] considered the delay a bifurcation parameter and examined the onset and termination of Hopf bifurcations of periodic solutions from a positive equilibrium. They proved that the model has only a finite number of Hopf bifurcation values and that these branches of Hopf bifurcations are paired, so that the existence of periodic solutions with specific oscillation frequencies occurs only in bounded delay intervals. The bifurcation analysis then guided some numerical simulations to identify ranges of parameters for coexisting multiple attractive periodic solutions.

6.8 Rotating Waves and Circulant Matrices

We have noticed that a key step in applying the global Hopf bifurcation theory is to exclude the existence of nonconstant periodic solutions with a certain prescribed period, normally the integer multiplier of the delay if the delay is constant. A general approach outlined in [237] is as follows: If one assumes that $y(t)$ is a periodic solution of a prototype equation $x'(t) = f(x(t), x(t - \tau))$ for some scalar function f, of period $m\tau$ for a certain integer m, and defines $u_j(t) = y(t - (j-1)\tau)$ for $1 \leq j \leq m$, one then discovers that $u(t) = (u_1(t), \dots, u_m(t))$ satisfies a cyclic system of ordinary differential equations $u'(t) = g(u(t))$, and we shall show that solutions of such a cyclic system satisfy $\lim_{t\to\infty}|u(t)| = 0$ or ∞, and the key step in proving the latter statement will be the construction of appropriate Lyapunov functions for the cyclic system. This will normally require the estimation of the spectral radius of a so-called circulant matrix. If we linearize this cyclic system at the trivial solution, we are led to a linear system with a real circulant matrix. Here and in what follows, an $n \times n$ matrix is called circulant if its (i,j)-element is given by a_{j-i+1} for n real numbers $a_1, \cdots a_n$. This matrix will be written as $A = \mathrm{circ}(a_1, a_2, \cdots a_n)$. For such a matrix, it was shown in [237] that

$$\inf\{\langle Ay, y\rangle : y \in \mathbb{R}^n, \sum_{j=1}^{n} y_j^2 = 1\} = \min\{\mathrm{Re}\big(\sum_{j=1}^{n} a_j z^{j-1}\big) : z \in \mathbb{C}; z^n = 1\}.$$

In this section, we will demonstrate the use of the approach outlined by Nussbaum based on the above-mentioned spectral property of circulant matrices.

We consider the following partial NFDE:

$$\frac{\partial}{\partial t}[u(t,x) - qu(t-\tau,x)] = d\frac{\partial^2}{\partial x^2}[u(t,x) - qu(t,x)]$$
$$- au(t,x) - aqu(t-\tau,x) - g[u(t,x) - qu(t-\tau,x)],$$

(6.47)

where $x \in \mathbb{S}^1$, a, d, τ are positive constants, $g : \mathbb{R} \to \mathbb{R}$ is continuously differentiable with $g(0) = 0$, $q \in (0,1)$ is the bifurcation parameter. This partial NFDE can be obtained from the coupled lossless transmission line NFDE introduced in Sect. 5.9 by letting the number of coupled oscillators go to infinity.

We are interested in the Hopf bifurcation of rotating waves from the trivial solution. Rotating wave solutions are solutions that satisfy

$$u(t,x) = u(t + \frac{p}{2\pi}x, 0), \quad u(t+p,x) = u(t,x), \quad (t,x) \in \mathbb{R} \times \mathbb{S}^1, \quad (6.48)$$

where $p > 0$ is a constant.

Let $y(t) = u(t\frac{p}{2\pi}, 0)$. Then using the spatiotemporal relation (6.48) of the rotating waves, we can show that u is a rotating wave if and only if y is a 2π-periodic solution of an NFDE with two parameters (q,p). This two-parameter NFDE is very much similar to (6.24), and a global Hopf bifurcation has been established (see [193] for details). Here we describe how Wu and Xia [306] applied this theory to establish the existence of rotating waves, and how this is related to circulant matrices.

Let $g'(0) = -\gamma$ and assume that $0 < \gamma < a$ in what follows. The characteristic equation of (6.47) at the trivial solution takes the form

$$(\lambda + dk^2 + a - \gamma)e^{\lambda\tau} - q(\lambda + dk^2 - a - \gamma) = 0, \quad k \geq 1. \quad (6.49)$$

Letting $\lambda = i\beta$ in (6.49), we get

$$\begin{cases} -(dk^2 + a - \gamma)\cos\beta\tau + \beta\sin\beta\tau = q(a + \gamma - dk^2), \\ \beta\cos\beta\tau + (dk^2 + a - \gamma)\sin\beta\tau = q\beta, \end{cases}$$

or equivalently,

$$
\begin{cases}
\tan(\beta\tau) = \dfrac{2\alpha\beta}{\beta^2 - (a+dk^2-\gamma)(a-dk^2+\gamma)}, \\[3mm]
q^2 = \dfrac{\beta^2 + (a-\gamma+dk^2)^2}{\beta^2 + (a+\gamma-dk^2)^2}.
\end{cases}
\tag{6.50}
$$

It is easy to show that for a real number $\beta > 0$, the second equation of (6.50) has a solution $q \in (0,1)$ only if

$$
dk^2 < \gamma.
\tag{6.51}
$$

Therefore, there are only finitely many $k \geq 1$ such that (6.50) has a pair of purely imaginary solutions.

For each fixed $k \geq 1$ such that $dk^2 < \gamma$, we can easily show graphically that there exists a sequence of positive numbers $\beta_{k,1} < \beta_{k,2} < \cdots$ such that the first equation of (6.50) is satisfied by $\beta_{k,j}, j = 1, 2, \ldots$. Substituting this $\beta_{k,j}$ into the second equation of (6.50) gives

$$
q_{k,j} = \sqrt{\frac{\beta_{k,j}^2 + (a-\gamma+dk^2)^2}{\beta_{k,j}^2 + (a+\gamma-dk^2)^2}}.
\tag{6.52}
$$

Therefore, we can conclude that the set $\{(q,p) \in (0,1) \times (0,\infty); (6.49) \text{ has a solution } i(2\pi/p)m \text{ for some } m \geq 1\}$ is discrete.

Let $\lambda = \lambda(q)$ be a smooth curve of zeros of (6.49) such that $\lambda(q_{k,j}) = i\beta_{k,j}$. Differentiating (6.49) with respect to q, we get

$$
\lambda'(q)e^\lambda + \tau(\lambda + dk^2 + a - \gamma)e^{\lambda\tau}\lambda'(q) = \lambda + dk^2 - \gamma - a + q\lambda'(q).
$$

That is,

$$
\lambda'(q) = \frac{\lambda + dk^2 - \gamma - a}{\tau(\lambda + dk^2 + a - \gamma)e^{\lambda\tau} + e^\lambda - q}.
$$

This leads to

$$
\begin{aligned}
&\operatorname{sgn}\operatorname{Re}\lambda'(q)|_{q=q_{k,j}} \\
&= \operatorname{sgn}\operatorname{Re}\frac{1}{\lambda'(q)}\Big|_{q=q_{k,j}} \\
&= \operatorname{sgn}\left\{\tau + \frac{2a\beta_{k,j}^2}{[(dk^2 + a - \gamma)^2 + \beta_{k,j}^2][(dk^2 - \gamma - a)^2 + \beta_{k,j}^2]}\right\} = 1 > 0.
\end{aligned}
$$

From the definition of the crossing number in Sect. 6.6, we can see that this will be crucial in ruling out bounded connected components of rotating waves of (6.47).

For the sake of later application, let us look at the location of $\beta_0 = \beta_{1,1}$. We assume that

$$
0 < d < \gamma.
\tag{6.53}
$$

Then β_0 is the first positive solution of

$$\tan(\beta\tau) = \frac{2a\beta}{\beta^2 - (a-\gamma+d)(a+\gamma-d)}, \tag{6.54}$$

and hence $i\beta_0$ is a solution of (6.49) with $k = 1$ and

$$q_0 = q_{1,1} = \sqrt{\frac{\beta_0^2 + (a-\gamma+d)^2}{\beta_0^2 + (a+\gamma-d)^2}}. \tag{6.55}$$

Lemma 6.8. *If*

$$\frac{\pi}{2\tau} < \sqrt{(a+\gamma-d)(a-\gamma+d)}, \tag{6.56}$$

then $\pi/2\tau < \beta_0 < \sqrt{(a+\gamma-d)(a-\gamma+d)}$, *and hence*

$$\frac{2\pi}{\sqrt{(a+\gamma-d)(a-\gamma+d)}} < \frac{2\pi}{\beta_0} < 4\tau. \tag{6.57}$$

In particular, if

$$\frac{\pi}{2\tau} < \sqrt{(a+\gamma-d)(a-\gamma+d)} < \frac{\pi}{\tau}, \tag{6.58}$$

then

$$2\tau < \frac{2\pi}{\beta_0} < 4\tau. \tag{6.59}$$

In order to apply the global bifurcation theorem to establish the global existence of rotating waves, we need to obtain a priori bounds for rotating waves. Assume that $u(t,x)$ is a rotating wave of (6.47) satisfying (6.48). Let $[u(t_0,x_0) - qu(t_0 - \tau,x_0)]^2$ be the maximum value of $[u(t,x) - qu(t-\tau,x)]^2$ over $\mathbb{R} \times \mathbb{S}^1$. Then

$$0 = \frac{\partial}{\partial t}[u(t_0,x_0) - qu(t_0-\tau,x_0)]^2$$

$$= 2[u(t_0,x_0) - qu(t_0-\tau,x_0)]\frac{\partial}{\partial t}[u(t_0,x_0) - qu(t_0-\tau,x_0)],$$

$$0 = \frac{\partial}{\partial x}[u(t_0,x_0) - qu(t_0-\tau,x_0)]^2$$

$$= 2[u(t_0,x_0) - qu(t_0-\tau,x_0)]\frac{\partial}{\partial x}[u(t_0,x_0) - qu(t_0-\tau,x_0)],$$

$$0 \le \frac{\partial^2}{\partial x^2}[u(t_0,x_0) - qu(t_0-\tau,x_0)]^2$$

$$= 2\{\frac{\partial}{\partial x}[u(t_0,x_0) - qu(t_0-\tau,x_0)]\}^2$$

$$+ 2[u(t_0,x_0) - qu(t_0-\tau,x_0)]\frac{\partial^2}{\partial x^2}[u(t_0,x_0) - qu(t_0-\tau,x_0)].$$

Without loss of generality, we may assume that $u(t_0,x_0) - qu(t_0 - \tau,x_0) \neq 0$. Therefore, from (6.47) it follows that

$$[u(t_0,x_0) - qu(t_0 - \tau,x_0)]\{-au(t_0,x_0) - aqu(t_0 - \tau,x_0)$$
$$- g[u(t_0,x_0) - qu(t_0 - \tau,x_0)]\} \geq 0.$$

That is,

$$-2aqu(t_0 - \tau,x_0)[u(t_0,x_0) - qu(t_0 - \tau,x_0)]$$
$$\geq \{a[u(t_0,x_0) - qu(t_0 - \tau,x_0)] \tag{6.60}$$
$$+ g[u(t_0,x_0) - qu(t_0 - \tau,x_0)]\}[u(t_0,x_0) - qu(t_0 - \tau,x_0)].$$

Note that

$$|u(t,x) - qu(t - \tau,x)| \leq |u(t_0,x_0) - qu(t_0 - \tau,x_0)|, \quad t \in \mathbb{R}, \quad x \in \mathbb{S}^1$$

implies

$$|u(t,x)| \leq \frac{1}{1-q}|u(t_0,x_0) - qu(t_0 - \tau,x_0)|, \quad t \in \mathbb{R}, \quad x \in \mathbb{S}^1. \tag{6.61}$$

Therefore, by (6.60), we obtain

$$a + \frac{g[u(t_0,x_0) - qu(t_0 - \tau,x_0)]}{u(t_0,x_0) - qu(t_0 - \tau,x_0)} \leq \frac{2aq}{1-q}. \tag{6.62}$$

If we assume that

$$\lim_{z \to \infty} \frac{g(z)}{z} = \infty, \tag{6.63}$$

then (6.62) implies the existence of $Q = Q(2aq/(1-q))$, so that

$$|u(t_0,x_0) - qu(t_0 - \tau,x_0)| \leq Q,$$

and hence from (6.61), it follows that

$$|u(t,x)| \leq \frac{1}{1-q}Q(\frac{2aq}{1-q}), \quad t \in \mathbb{R}, \quad x \in \mathbb{S}^1. \tag{6.64}$$

Summarizing the above discussion, we get the following.

Lemma 6.9. *If (6.63) is satisfied, then there exists a nondecreasing function* $Q :$ $(0,\infty) \to (0,\infty)$ *such that every rotating wave* $u(t,x)$ *of (6.47) satisfies* $|u(t,x)| \leq$ $(1/(1-q))Q(2aq/(1-q))$ *for* $t \in \mathbb{R}$ *and* $x \in \mathbb{S}^1$. *In particular, for fixed* $q^* \in (0,1)$, *the set of rotating waves of (6.47) corresponding to* $q \in [0,q^*)$ *is uniformly bounded in the sup-norm.*

Now we try to exclude nontrivial 4τ-periodic rotating waves. Assume that $u(t,x)$ is a nontrivial rotating wave of (6.47) satisfying (6.48) with $p = 4\tau$. Then

$$
\begin{aligned}
u(t,0) &= u(t+4\tau,0), \\
u(t,x) &= u(t - \tfrac{4\tau}{2\pi}x,0) = u(t - \tfrac{2}{\pi}x,0), \quad t \in \mathbb{R}, \quad x \in \mathbb{S}^1.
\end{aligned}
$$

So, $v(t) = u(t,0)$ satisfies

$$
\begin{aligned}
&\frac{d}{dt}[v(t) - qv(t-\tau)] \\
&= \left(\frac{2\tau}{\pi}\right)^2 d\frac{d^2}{dt^2}[v(t) - qv(t-\tau)] \\
&\quad - a[v(t) - qv(t-\tau)] - 2aqv(t-\tau) - g[v(t) - qv(t-\tau)],
\end{aligned}
\tag{6.65}
$$

$t \in \mathbb{R}$. Let

$$
\begin{cases}
x_1(t) = v(t) - qv(t-\tau), \\
x_2(t) = v(t-\tau) - qv(t-2\tau), \\
x_3(t) = v(t-2\tau) - qv(t-3\tau), \\
x_4(t) = v(t-3\tau) - qv(t).
\end{cases}
\tag{6.66}
$$

Then

$$
\begin{pmatrix}
v(t-\tau) \\
v(t-2\tau) \\
v(t-3\tau) \\
v(t)
\end{pmatrix}
= \frac{1}{1-q^4} B
\begin{pmatrix}
x_1(t) \\
x_2(t) \\
x_3(t) \\
x_4(t)
\end{pmatrix},
\tag{6.67}
$$

where we have the following circulant matrix:

$$
B =
\begin{pmatrix}
q^3 & 1 & q & q^2 \\
q^2 & q^3 & 1 & q \\
q & q^2 & q^3 & 1 \\
1 & q & q^2 & q^3
\end{pmatrix}.
$$

Substituting (6.66) and (6.67) into (6.65), we get

$$
\dot{x}_i = \left(\frac{2\tau}{\pi}\right)^2 d\ddot{x}_i - ax_i - \frac{2aq}{1-q^4}(Bx)_i - g(x_i), \quad 1 \le i \le 4.
$$

Its similarity to the Liénard equation suggests a transformation that leads to an equivalent system,

$$
\begin{cases}
\dot{x}_i = y_i + \left(\frac{\pi}{2\tau}\right)^2 \frac{1}{d}x_i, \\
\dot{y}_i = \left(\frac{\pi}{2\tau}\right)^2 \frac{1}{d}\left[ax_i + \frac{2aq}{1-q^4}(Bx)_i + g(x_i)\right], \quad 1 \le i \le 4,
\end{cases}
\tag{6.68}
$$

and a related Lyapunov function,

$$
V = \sum_{i=1}^{4} \left[\frac{1}{2}y_i^2 - \left(\frac{\pi}{2\tau}\right)^2 \frac{1}{d} \int_0^{x_i} g(s)\,ds - ax_iy_i - \frac{2aq}{1-q^4}y_i(Bx)_i\right].
$$

The derivative of V along solutions of (6.68) is given by

$$
\begin{aligned}
\dot{V} = &-a\sum_{i=1}^{4} y_i^2 - \frac{2aq}{1-q^4}\sum_{i=1}^{4} y_i(By)_i \\
&-\left[(\frac{\pi}{2\tau})^2\frac{1}{d}\right]^2\sum_{i=1}^{4} x_i g(x_i) - \frac{a}{d}\left(\frac{\pi}{2\tau}\right)^2\sum_{i=1}^{4} x_i g(x_i) \\
&-\frac{a^2}{d}\sum_{i=1}^{4}\left(\frac{\pi}{2\tau}\right)^2 x_i^2 - \frac{4a^2 q}{1-q^4}\left(\frac{\pi}{2\tau}\right)^2\frac{1}{d}\sum_{i=1}^{4} x_i(Bx)_i \\
&-\left(\frac{2aq}{1-q^4}\right)^2\left(\frac{\pi}{2\tau}\right)^2\frac{1}{d}\sum_{i=1}^{4}(Bx)_i(Bx)_i - \frac{2aq}{1-q^4}\left(\frac{\pi}{2\tau}\right)^2\frac{1}{d}\sum_{i=1}^{4} g(x_i)(Bx)_i.
\end{aligned}
$$

We need the following lemma.

Lemma 6.10. $\sum_{i=1}^{4} z_i(Bz)_i \geq -(1-q)(1+q^2)\sum_{i=1}^{4} z_i^2$, $z_i \in \mathbb{R}$, $1 \leq i \leq 4$.

Proof. Using the aforementioned spectral property of circulant matrices, we have $\sum_{i=1}^{4} z_i(Bz)_i \geq \Gamma\sum_{i=1}^{4} z_i^2, z_i \in \mathbb{R}, \leq i \leq 4$, where

$$
\begin{aligned}
\Gamma &= \min\{\operatorname{Re}\sum_{j=1}^{4} a_j z^{j-1} : z^4 = 1, a_1 = q^3, a_2 = 1, a_3 = q, a_4 = q^2\} \\
&= \min\{\operatorname{Re}(q^3 + e^{i(2\pi/4)j} + qe^{i(4\pi/4)j} + q^2 e^{i(6\pi/4)j} : j = 0,1,2,3\} \\
&= \min\{(1+q)(1+q^2), -q(1-q^2), -(1-q)(1+q^2)\} \\
&= -(1-q)(1+q^2).
\end{aligned}
$$

$\square$

We also need to compute the eigenvalues of $B^T B$. While this can be done directly, Wu and Xia [306] have presented an approach that can be extended to general circular matrices.

Lemma 6.11. The minimal eigenvalue of $B^T B$ is $\lambda_{\min}(B^T B) = (1-q^4)^2/(1+q)^2$, and the maximal eigenvalue of $B^T B$ is $\lambda_{\max}(B^T B) = (1-q^4)^2/(1-q)^2$.

Proof. Let

$$
v_j = (1, e^{i(\pi/2)j}, e^{i(2\pi/2)j}, e^{i(3\pi/2)j}), j = 0,1,2,3.
$$

It can be shown that v_j is an eigenvector of B corresponding to the eigenvalue

$$
\alpha_j = e^{i(\pi/2)j}(1 + qe^{i(\pi/2)j} + q^2 e^{i(2\pi/2)j} + q^3 e^{i(3\pi/2)j}) = e^{i(\pi/2)j}\frac{1-q^4}{1-qe^{i(\pi/2)j}}
$$

and an eigenvector of B^T corresponding to the eigenvector

$$
\beta_j = e^{-i(\pi/2)j}(1 + qe^{-i(\pi/2)j} + q^2 e^{-i(2\pi/2)j} + q^3 e^{-i(3\pi/2)j}) = e^{-i(\pi/2)j}\frac{1-q^4}{1-qe^{-i(\pi/2)j}}.
$$

Assume that $x \in \mathbb{C}^4$ is an eigenvector of $B^T B$ corresponding to an eigenvalue $\lambda \in \mathbb{C}$. Then $x = a_0 v_0 + a_1 v_1 + a_2 v_2 + a_3 v_3$, and $B^T B x = \lambda x$ is equivalent to

$$\sum_{j=0}^{3} \alpha_j \beta_j a_j v_j = \lambda \sum_{j=0}^{3} a_j v_j,$$

from which it follows that $\lambda = \alpha_j \beta_j$ for some $j = 0, 1, 2, 3$. Therefore, all eigenvalues of $B^T B$ are given by

$$\frac{(1-q^4)^2}{(1 - q e^{\mathrm{i}(\pi/2)j})(1 - q e^{-\mathrm{i}(\pi/2)j})}, \, j = 0, 1, 2, 3,$$

from which the conclusion follows. $\square$

We also note the following result; see [306]

Lemma 6.12. *Assume that*

$$-K \le \frac{g(x)}{x}, \quad g(-x) = -g(x) \quad for \quad x \ne 0, \tag{6.69}$$

$$\frac{g(x)}{x} \quad is \ nondecreasing \ in \quad x \in (0, \infty). \tag{6.70}$$

Let $x_i(t)$, $i = 1, \ldots, 4$, be given by (6.66). Then

$$\left| \frac{g(x_i(t))}{x_i(t)} \right| \le \max \left\{ K, \frac{a(3q-1)}{1-q} \right\}. \tag{6.71}$$

We now return to the estimation of $\dot{V}$. Using Lemma 6.11, we get

$$\sum_{i=1}^{4} (Bx)_i (Bx)_i \ge \lambda_{\min}(B^T B) \sum_{i=1}^{4} x_i^2 = \frac{(1-q^4)^2}{(1+q)^2} \sum_{i=1}^{4} x_i^2.$$

By Lemma 6.12, we have

$$\left| \sum_{i=1}^{4} g(x_i)(Bx)_i \right| \le \sqrt{\sum_{i=1}^{4} g^2(x_i)} \sqrt{\sum_{i=1}^{4} x_i^2 \lambda_{\max}(B^T B)}$$

$$\le \frac{(1-q^4)^2}{(1+q)^2} \max \left\{ K, \frac{a(3q-1)}{1-q} \right\} \sum_{i=1}^{4} x_i^2.$$

Therefore, using Lemma 6.10, we get

$$\dot{V} \le -a \sum_{i=1}^{4} \left[1 - \frac{2q(1-q)(1+q^2)}{1-q^4} \right] y_i$$

$$- \left(\frac{\pi}{2\tau} \right)^2 \frac{1}{d} \left\{ \left(\frac{1}{d} \left(\frac{\pi}{2\tau} \right)^2 + a \right) \sum_{i=1}^{4} x_i g(x_i) \right.$$

$$- \sum_{i=1}^{4} \frac{1}{d} \left(\frac{\pi}{2\tau} \right)^2 \left\{ a^2 \left[1 - \frac{4q}{1-q^4}(1-q)(1+q^2) \right. \right.$$

$$+ \frac{4q^2}{(1-q^4)^2} \frac{(1-q^4)^2}{(1+q)^2} \Bigg] x_i^2$$

$$- \frac{2aq}{1-q^4} \frac{(1-q^4)^2}{(1-q)^2} \max\left\{ K, \frac{a(3q-1)}{1-q} \right\} \sum_{i=1}^{4} x_i^2 \Bigg\}$$

$$\leq -a \sum_{i=1}^{4} \left(\frac{1-q}{1+q} \right) y_i^2$$

$$- \left(\frac{\pi}{2\tau} \right)^2 \frac{1}{d} \sum_{i=1}^{4} \Bigg[a^2 \left(\frac{1-q}{1+q} \right)^2 - \left(\frac{1}{d} \left(\frac{\pi}{2\tau} \right)^2 + a \right) K$$

$$- \frac{2aq(1+q)(1+q^2)}{1-q} \max\left\{ K, \frac{a(3q-1)}{1-q} \right\} \Bigg] x_i^2.$$

Consequently, if we assume that

$$0 \leq q < 1 - \delta \quad \text{for some} \quad \delta \in (0,1), \tag{6.72}$$

$$\frac{1}{4} a^2 \delta^2 > \left[\frac{1}{d} \left(\frac{\pi}{2\tau} \right)^2 + a \right] K + \frac{8a(1-\delta) \max\left\{ K, \frac{4a}{\delta} \right\}}{\delta}, \tag{6.73}$$

then $\dot{V}$ is a strictly negative function of $(x_1, \ldots, x_4, y_1, \ldots, y_4)$ unless $x_i = y_i = 0$ for $1 \leq i \leq 4$. Therefore, under assumptions (6.69), (6.70), (6.72), and (6.73), system (6.68) has no nontrivial periodic solution. This implies that system (6.47) has no nontrivial rotating wave of period 4τ. That is, we have proved the following.

Lemma 6.13. *Under assumptions (6.69), (6.70), (6.72), and (6.73), the partial neutral functional differential equation (6.47) has no nontrivial 4π-periodic rotating wave for $q \in [0, 1 - \delta)$.*

We can then use global bifurcation theory to obtain the following result, for which we refer to [306] for more details of the proof.

Theorem 6.14. *Assume that*

(i) $g'(0) = -\gamma, d < \gamma < a, \pi/2 < \sqrt{(a+\gamma-d)(a-\gamma+d)}$;
(ii) $\inf_{y \neq 0} g(y)/y > -a, \lim_{y \to \infty} g(y)/y = \infty$;
(iii) $g(-y) = -g(y)$ *for* $y \in \mathbb{R}$ *and* $g(y)/y$ *is nondecreasing in* $y \in (0, \infty)$;
(iv) *there exist constants* $\delta \in (0,1)$ *and* $K \geq 0$ *such that*

$$-K \leq g(x)/x \text{ for } x \neq 0,$$

$$\frac{1}{4} a^2 \delta^2 > \left[\frac{1}{d} \left(\frac{\pi}{2} \right)^2 + a \right] K + 8a \left(\frac{1-\delta}{\delta} \right) \max\left\{ K, \frac{4a}{\delta} \right\},$$

and

$$q_0 \overset{\text{def}}{=} \sqrt{ \frac{\beta_0^2 + (a+\gamma-d)^2}{\beta_0^2 + (a-\gamma+d)^2} } < 1 - \delta,$$

where β_0 is the first solution in $((\pi/2\tau), \sqrt{(a+\gamma-d)(a-\gamma+d)})$ of the equation

$$\tan(\beta\tau) = \frac{2a\beta}{\beta^2 - (a+\gamma-d)(a-\gamma+d)}.$$

Then for each $q \in (q_0, 1-\delta)$, system (6.47) has a rotating wave with a period less than 4. If, in addition, we assume

(iv)

$$\frac{\pi}{2} < \sqrt{(a+\gamma-d)(a-\gamma+d)} < \frac{\pi}{\tau},$$

then for each $q \in (q_0, 1-\delta)$, system (6.47) has a slowly oscillating rotating wave, that is, a rotating wave with a period in $(2\tau, 4\tau)$.

6.9 State-Dependent DDEs

State-dependent DDEs arises from a number of applications such as electrodynamics, automatic and remote control, machine cutting, neutral networks, population biology, mathematical epidemiology, and economics. They describe the evolution of systems in which the rate of change depends on the history of the rate, and the delay depends on the system's status in a complicated manner, such as by an explicit or implicit algebraic equation or a differential or integral equation.

Early results on the existence of periodic solutions for state-dependent DDEs include work by Smith [269] that considered bifurcations of periodic solutions from a stationary state for a system of integral equations with state-dependent delay, and work on the existence of periodic solutions by Nussbaum, Mallet-Paret, and Paraskevopoulos [215]. These studies address the aspect of global continuation of Hopf bifurcations of periodic solutions, especially the existence of periodic solutions in which the bifurcation parameter is away from the critical value where a local Hopf bifurcation is born. The work of Nussbaum et al. [215] focuses on important prototype classes of state-dependent delay differential equations with negative feedback and provides some detailed information on slowly oscillating periodic solutions. Here we introduce the work [170, 171] to provide a general tool and framework for studying the Hopf bifurcation problem, and in particular, the global continuation of local bifurcation of periodic solutions of the following parameterized state-dependent DDEs from an equivariant-degree point of view:

$$\begin{pmatrix} \dot{x}(t) \\ \dot{\tau}(t) \end{pmatrix} = \begin{pmatrix} f(x(t), x(t-\tau(t)), \sigma) \\ g(x(t), \tau(t), \sigma) \end{pmatrix}, \tag{6.74}$$

where $x \in \mathbb{R}^N$, $\tau \in \mathbb{R}$, $t \in \mathbb{R}$ and $\sigma \in \mathbb{R}$, $f : \mathbb{R}^N \times \mathbb{R}^N \times \mathbb{R} \to \mathbb{R}^N$, and $g : \mathbb{R}^N \times \mathbb{R} \times \mathbb{R} \to \mathbb{R}$ are given maps. A stationary state of (6.74) with parameter σ is a vector $(x, \tau) \in \mathbb{R}^N \times \mathbb{R}$ such that $f(x, x, \sigma) = 0$ and $g(x, \tau, \sigma) = 0$.

The major problem in developing such a global Hopf bifurcation theory for the system of state-dependent DDEs (6.74) is that in the spaces of continuous periodic functions $C_T(\mathbb{R};\mathbb{R}^N) = \{x \in C(\mathbb{R};\mathbb{R}^N) : x(t+T) = x(t) \text{ for all } t \in \mathbb{R}\}$ and $C_T(\mathbb{R};\mathbb{R}) = \{\tau \in C(\mathbb{R};\mathbb{R}) : \tau(t+T) = \tau(t) \text{ for all } t \in \mathbb{R}\}$ with a fixed period $T > 0$, the composition operator

$$\begin{aligned} \chi : C_T(\mathbb{R};\mathbb{R}^N) \times C_T(\mathbb{R};\mathbb{R}) &\to C_T(\mathbb{R};\mathbb{R}^N), \\ \chi(x,\tau)(t) &= x(t-\tau(t)), t \in \mathbb{R}, \end{aligned} \tag{6.75}$$

is generally not a C^1 (continuously differentiable) map with respect to τ in the supremum norm. This causes difficulty in formulating linearization at a stationary state, and such a linearization is usually necessary in the functional-analytic setting for Hopf bifurcation problems in which a topological index such as an $\mathbb{S}^1$-equivariant degree can be calculated and applied to investigate the birth and continuation of periodic solutions bifurcating from a stationary state.

In [72], a system of auxiliary equations obtained through a formal linearization technique was used in the study of local stability of state-dependent DDEs in the space of continuously differentiable functions. This formal linearization technique is only heuristic and can be described in the following way: the state-dependent delay $\tau(t)$ in $x(t-\tau(t))$ is first fixed at a given stationary state, and then the resulting nonlinear system with frozen constant delay is linearized. Other applications of systems of auxiliary equations obtained through a formal linearization process can be found in [26, 45, 156] and [157]. None of these results is sufficient for us to develop a global Hopf bifurcation theory based on the $\mathbb{S}^1$-equivariant degree for state-dependent DDEs (6.74). However, the above-mentioned results strongly indicate that the system of auxiliary equations obtained through the heuristic technique of formal linearization can be used to detect the local Hopf bifurcation and to describe its global continuation for state-dependent DDEs.

In this section, we use the homotopy invariance property of the $\mathbb{S}^1$-equivariant degree to relate the Hopf bifurcation problem of (6.74) to the change of stability of stationary states of the corresponding system of auxiliary equations obtained through formal linearization. As such, much of the effort has been dedicated to justifying that the detection of Hopf bifurcation can be achieved through the formal linearization technique: the state-dependent delay $\tau(t)$ in $x(t-\tau(t))$ is first fixed at a given stationary state; then the resulting nonlinear system with frozen constant delay is linearized. This linearization technique is used in the functional-analytic setting that converts the Hopf bifurcation problem of system (6.74) to solving an operator equation (6.13) involving $\mathbb{S}^1$-equivariant maps with two parameters, in the space of periodic functions with a fixed period. Implicitly used is the C^1-smoothness of the operator defined in Lemma 6.17 in the space $\mathbb{E}$ (the space of periodic functions with fixed period 2π). The formal linearization leads to this operator naturally in the space of continuously differentiable periodic functions with period 2π, and the fact that this operator can be extended to a bounded operator in the space $\mathbb{E}$ is essential in our homotopy argument. This technique of extending the linearized operator of a state-dependent delay differential equation from the space C^1 to the space C has previously been used in other contexts; see, for example, Mallet-Paret–Nussbaum–Paraskevopoulos [215], Krisztin [196], Walther [290], and the survey paper by Hartung–Krisztin–Walther–Wu [158].

6.9.1 Local Hopf Bifurcation

We turn to the Hopf bifurcation of (6.74), with its solution denoted by $u(t) = (x(t), \tau(t))$. Denote by $C(\mathbb{R}; \mathbb{R}^N)$ the normed space of continuous functions from $\mathbb{R}$ to $\mathbb{R}^N$ equipped with the usual supremum norm $\|x\| = \sup_{t \in \mathbb{R}} |x(t)|$ for $x \in C(\mathbb{R}; \mathbb{R}^N)$, where $|\cdot|$ denotes the Euclidean norm. We also denote by $C^1(\mathbb{R}; \mathbb{R}^N)$ the normed space of continuously differentiable bounded functions from $\mathbb{R}$ to $\mathbb{R}^N$ equipped with the usual C^1 norm

$$\|x\|_{C^1} = \max\{\sup_{t \in \mathbb{R}} |x(t)|, \sup_{t \in \mathbb{R}} |\dot{x}(t)|\}$$

for $x \in C(\mathbb{R}; \mathbb{R}^N)$. For a stationary state (u_0, τ_0) of (6.74) with the parameter σ_0, we say that (u_0, σ_0) is a *Hopf bifurcation point* of system (6.74) if there exist a sequence $\{(u_k, \sigma_k, T_k)\}_{k=1}^{+\infty} \subseteq C(\mathbb{R}; \mathbb{R}^{N+1}) \times \mathbb{R}^2$ and $T_0 > 0$ such that

$$\lim_{k \to +\infty} \|(u_k, \sigma_k, T_k) - (u_0, \sigma_0, T_0)\|_{C(\mathbb{R}; \mathbb{R}^{N+1}) \times \mathbb{R}^2} = 0,$$

and (u_k, σ_k) is a nonconstant T_k-periodic solution of system (6.74).

We assume that:

(SHB1)   The map $f: \mathbb{R}^N \times \mathbb{R}^N \times \mathbb{R} \ni (\theta_1, \theta_2, \sigma) \to f(\theta_1, \theta_2, \sigma) \in \mathbb{R}^N$ and the map $g: \mathbb{R}^N \times \mathbb{R} \times \mathbb{R} \ni (\gamma_1, \gamma_2, \sigma) \to g(\gamma_1, \gamma_2, \sigma) \in \mathbb{R}$ are C^2 (twice continuously differentiable).

(SHB2)   There exists $L_0 > 0$ such that $g(\gamma_1, \gamma_2, \sigma) < \frac{L_0}{L_0+1}$ for $\gamma_1 \in \mathbb{R}^N$, $\gamma_2 \in \mathbb{R}$, $\sigma \in \mathbb{R}$.

In what follows, we write $\partial_i f = \frac{\partial}{\partial \theta_i} f$ for $i = 1, 2$, and similarly we define $\partial_i g$ for $i = 1, 2$.

We shall study the Hopf bifurcation of (6.74) through its formal linearization. We assume that for a fixed $\sigma_0 \in \mathbb{R}$, $(x_{\sigma_0}, \tau_{\sigma_0})$ (or abusing notation, $(x_{\sigma_0}, \tau_{\sigma_0}, \sigma_0)$) is a stationary state of (6.74). That is,

$$f(x_{\sigma_0}, x_{\sigma_0}, \sigma_0) = 0, \quad g(x_{\sigma_0}, \tau_{\sigma_0}, \sigma_0) = 0.$$

We also assume that

(SHB3) $(\frac{\partial}{\partial \theta_1} + \frac{\partial}{\partial \theta_2}) f(\theta_1, \theta_2, \sigma)|_{\sigma=\sigma_0, \theta_1=\theta_2=x_{\sigma_0}}$ is nonsingular and

$$\frac{\partial}{\partial \gamma_2} g(\gamma_1, \gamma_2, \sigma)|_{\sigma=\sigma_0, \gamma_1=x_{\sigma_0}, \gamma_2=\tau_{\sigma_0}} \neq 0.$$

This assumption implies that there exist $\varepsilon_0 > 0$ and a C^1-smooth curve $(\sigma_0 - \varepsilon_0, \sigma_0 + \varepsilon_0) \ni \sigma \mapsto (x_\sigma, \tau_\sigma) \in \mathbb{R}^{N+1}$ such that (x_σ, τ_σ) is the unique stationary state of (6.74) in a small neighborhood of $(x_{\sigma_0}, \tau_{\sigma_0})$ for σ close to σ_0.

We now consider, for $\sigma \in (\sigma_0 - \varepsilon_0, \sigma_0 + \varepsilon_0)$, the following formal linearization of system (6.74) at the stationary point $\eta(\sigma) = (x_\sigma, z_\sigma)$:

$$
\begin{pmatrix} \dot{x}(t) \\ \dot{\tau}(t) \end{pmatrix} = \begin{bmatrix} \partial_1 f(\sigma) & 0 \\ \partial_1 g(\sigma) & \partial_2 g(\sigma) \end{bmatrix} \begin{pmatrix} x(t) - x_\sigma \\ \tau(t) - \tau_\sigma \end{pmatrix}
$$
$$
+ \begin{bmatrix} \partial_2 f(\sigma) & 0 \\ 0 & 0 \end{bmatrix} \begin{pmatrix} x(t - \tau_\sigma) - x_\sigma \\ \tau(t - \tau_\sigma) - \tau_\sigma \end{pmatrix}, \tag{6.76}
$$

where

$$
\partial_1 f(\sigma) \overset{\text{def}}{=} \partial_1 f(x_\sigma, \tau_\sigma, \sigma), \quad \partial_2 f(\sigma) \overset{\text{def}}{=} \partial_2 f(x_\sigma, \tau_\sigma, \sigma),
$$
$$
\partial_1 g(\sigma) \overset{\text{def}}{=} \partial_1 g(x_\sigma, \tau_\sigma, \sigma), \quad \partial_2 g(\sigma) \overset{\text{def}}{=} \partial_2 g(x_\sigma, \tau_\sigma, \sigma).
$$

Then we obtain the following characteristic equation of the linear system corresponding to the inhomogeneous linear system (6.76):

$$
\det \Delta_{(x_\sigma, \tau_\sigma, \sigma)}(\omega) = 0, \tag{6.77}
$$

where $\Delta_{(x_\sigma, \tau_\sigma, \sigma)}(\omega)$ is an $(N+1) \times (N+1)$ complex matrix defined by

$$
\Delta_{(x_\sigma, \tau_\sigma, \sigma)}(\omega) = \omega I - \begin{bmatrix} \partial_1 f(\sigma) & 0 \\ \partial_1 g(\sigma) & \partial_2 g(\sigma) \end{bmatrix} - \begin{bmatrix} \partial_2 f(\sigma) & 0 \\ 0 & 0 \end{bmatrix} e^{-\omega \tau_\sigma}. \tag{6.78}
$$

A solution ω_0 to the characteristic equation (6.77) is called a *characteristic value* of the stationary state $(x_{\sigma_0}, \tau_{\sigma_0}, \sigma_0)$. We call $(x_{\sigma_0}, \tau_{\sigma_0}, \sigma_0)$ a nonsingular stationary state if and only if zero is not a characteristic value of $(x_{\sigma_0}, \tau_{\sigma_0}, \sigma_0)$. Here $(x_{\sigma_0}, \tau_{\sigma_0}, \sigma_0)$ is a *center* if the set of nonzero purely imaginary characteristic values of $(x_{\sigma_0}, \tau_{\sigma_0}, \sigma_0)$ is nonempty and discrete. We call $(x_{\sigma_0}, \tau_{\sigma_0}, \sigma_0)$ an *isolated center* if it is the only center in some neighborhood of $(x_{\sigma_0}, \tau_{\sigma_0}, \sigma_0)$ in $\mathbb{R}^{N+1} \times \mathbb{R}$.

If $(x_{\sigma_0}, \tau_{\sigma_0}, \sigma_0)$ is an isolated center of (6.76), then there exist $\beta_0 > 0$ and $\delta \in (0, \varepsilon_0)$ such that

$$
\det \Delta_{(x_{\sigma_0}, \tau_{\sigma_0}, \sigma_0)}(i\beta_0) = 0, \quad \det \Delta_{(x_\sigma, \tau_\sigma, \sigma)}(i\beta) \neq 0, \tag{6.79}
$$

for every $\sigma \in (\sigma_0 - \delta, \sigma_0) \cup (\sigma_0, \sigma_0 + \delta)$ and $\beta \in (0, +\infty)$. Hence, we can choose constants $\alpha_0 = \alpha_0(\sigma_0, \beta_0) > 0$ and $\varepsilon = \varepsilon(\sigma_0, \beta_0) > 0$ such that the closure of the set $\Omega \overset{\text{def}}{=} (0, \alpha_0) \times (\beta_0 - \varepsilon, \beta_0 + \varepsilon) \subset \mathbb{R}^2 \cong \mathbb{C}$ contains no other zero of $\det \Delta_{(x_{\sigma_0}, \tau_{\sigma_0}, \sigma_0)}(\cdot)$. The quantity $p_0 = 2\pi/\beta_0$ is called the *virtual period* associated with the center $(x_{\sigma_0}, \tau_{\sigma_0}, \sigma_0)$. We note that $\det \Delta_{(x_\sigma, \tau_\sigma, \sigma)}(\omega)$ is analytic in ω and is continuous in σ. If $\delta > 0$ is small enough, then there is no zero of $\det \Delta_{(x_{\sigma_0 \pm \delta}, \tau_{\sigma_0 \pm \delta}, \sigma_0 \pm \delta)}(\omega)$ in $\partial \Omega$. So we can define the number

$$
\gamma_\pm(x_{\sigma_0}, \tau_{\sigma_0}, \sigma_0, \beta_0) = \deg_B(\det \Delta_{(x_{\sigma_0 \pm \delta}, \tau_{\sigma_0 \pm \delta}, \sigma_0 \pm \delta)}(\cdot), \Omega),
$$

and the crossing number of $(x_{\sigma_0}, \tau_{\sigma_0}, \sigma_0, \beta_0)$ as

$$\gamma(x_{\sigma_0}, \tau_{\sigma_0}, \sigma_0, \beta_0) = \gamma_-(x_{\sigma_0}, \tau_{\sigma_0}, \sigma_0, \beta_0) - \gamma_+(x_{\sigma_0}, \tau_{\sigma_0}, \sigma_0, \beta_0). \qquad (6.80)$$

This crossing number counts the number of characteristic values (with multiplicities) escaping from the region Ω as α increases and crosses α_0. Define the function $H : [\sigma_0 - \delta, \sigma_0 + \delta] \times \overline{\Omega} \to \mathbb{R}^2 \simeq \mathbb{C}$ by

$$H(\alpha, u, \beta) \overset{\text{def}}{=} \det \Delta_{(x_\sigma, \tau_\sigma, \sigma)}(u + i\beta),$$

and

$$\deg(\Psi_H, \mathscr{D}(\sigma_0, \beta_0)) = \gamma(x_{\sigma_0}, \tau_{\sigma_0}, \sigma_0, \beta_0), \qquad (6.81)$$

where $\Psi_H : \mathscr{D}(\sigma_0, \beta_0) \to \mathbb{R}^2 \simeq \mathbb{C}$ is defined by $\Psi_H(\sigma, \beta) = \det \Delta_{(x_\sigma, \tau_\sigma, \sigma)}(i\beta)$ and $\mathscr{D}(\sigma_0, \beta_0) = (\sigma_0 - \delta, \sigma_0 + \delta) \times (\beta_0 - c, \beta_0 + c)$.

Let $\mathbb{E} \overset{\text{def}}{=} C_{2\pi}(\mathbb{R}; \mathbb{R}^n)$ be the normed space of continuous 2π-periodic functions from $\mathbb{R}$ to $\mathbb{R}^n$ equipped with the usual supremum norm. Then $\mathbb{S}^1$ acts on $\mathbb{E}$ by argument shift. Namely, for $\xi = e^{i\nu} \in \mathbb{S}^1$, $u \in \mathbb{E}$, $(\xi u)(t) \overset{\text{def}}{=} u(t + \nu)$. For the isotypical direct sum decomposition (6.12) of $\mathbb{E}$, we see that $\mathbb{E}_0 \cong \mathbb{R}^n$ and for each $k \geq 1$, $\mathbb{E}_k$ is spanned by $\cos(kt)\varepsilon_j$ and $\sin(kt)\varepsilon_j$, $1 \leq j \leq n$, where $\{\varepsilon_1, \ldots, \varepsilon_n\}$ is the standard basis of $\mathbb{R}^n$. Therefore, $\mathbb{E}_k$, $k \geq 0$, are real $2n$-dimensional and so satisfy (SD1) of Sect. 6.5. To formulate the Hopf bifurcation problem as a fixed-point problem in the space of continuous functions of period 2π, we normalize the period of the $2\pi/\beta$-periodic solution (x, τ) in (6.74) by $(x(t), \tau(t)) = (y(\beta t), z(\beta t))$ and obtain

$$\dot{u}(t) = Q(u, \sigma, \beta)(t), \qquad (6.82)$$

where $u = (y, z)^T$ and $(\sigma, \beta) \in \mathscr{D}(\sigma_0, \beta_0)$, and

$$Q(u, \sigma, \beta)(t) = \frac{1}{\beta} \begin{bmatrix} f(y(t), y(t - \beta z(t)), \sigma) \\ g(y(t), z(t), \sigma) \end{bmatrix}.$$

Correspondingly, (6.76) is transformed into

$$\dot{u}(t) = \tilde{Q}(u, \sigma, \beta)(t), \qquad (6.83)$$

where $\tilde{Q} : \mathbb{E} \times \mathscr{D}(\sigma_0, \beta_0) \to \mathbb{E}$ is defined by

$$\tilde{Q}(u, \sigma, \beta)(t) = \frac{1}{\beta} \begin{bmatrix} \partial_1 f(\sigma)(y(t) - y_\sigma) + \partial_2 f(\sigma)(y(t - \beta z_\sigma) - y_\sigma) \\ \partial_1 g(\sigma)(y(t) - y_\sigma) + \partial_2 g(\sigma)(z(t) - z_\sigma) \end{bmatrix}$$

and $(y_\sigma, z_\sigma) = \eta(\sigma) = (x_\sigma, \tau_\sigma)$.

Before we state and prove the local Hopf bifurcation theorem, we need some technical preparations. We denote by $C_{2\pi}^1(\mathbb{R}; \mathbb{R}^{N+1})$ the Banach space of 2π-periodic and continuously differentiable functions equipped with the C^1 norm

$$\|x\|_{C^1} = \max\{ \sup_{t \in [0, 2\pi]} |x(t)|, \sup_{t \in [0, 2\pi]} |\dot{x}(t)| \}.$$

Lemma 6.14. *Let* $L\colon C^1_{2\pi}(\mathbb{R};\mathbb{R}^{N+1}) \to \mathbb{E}$ *and* $K\colon \mathbb{E} \to \mathbb{R}^{N+1}$ *be defined by*

$$Lu(t) = \dot{u}(t), \quad Ku(t) = \frac{1}{2\pi}\int_0^{2\pi} u(t)dt$$

for $t \in \mathbb{R}$. *Then* $L + K$ *has a compact inverse* $(L+K)^{-1}: \mathbb{E} \to \mathbb{E}$, *which is given by*

$$(L+K)^{-1}(v)(t) = \int_0^t v(s)ds + \frac{1}{2\pi}\int_0^{2\pi}(1 - \pi - t + s)v(s)ds.$$

This lemma can be found in [170] and is omitted here. It follows from Lemma 6.14 that $(L+K)^{-1}\circ[Q(\cdot,\alpha,\beta)+K]: \mathbb{E} \to \mathbb{E}$ and $(L+K)^{-1}\circ[\tilde{Q}(\cdot,\alpha,\beta)+K]: \mathbb{E} \to \mathbb{E}$ are completely continuous. That is, (SD2) and (SD4) are satisfied. Thus, finding a $2\pi/\beta$-periodic solution for the system (6.74) is equivalent to finding a solution of the following fixed-point problem:

$$u = (L+K)^{-1}[Q(u,\sigma,\beta)+K(u)], \qquad (6.84)$$

where $(u,\sigma,\beta) \in \mathbb{E} \times \mathbb{R} \times \mathbb{R}_+$. Define the maps $\mathscr{F}: \mathbb{E} \times \mathbb{R} \times \mathbb{R}_+ \to \mathbb{E}$ and $\tilde{\mathscr{F}}: \mathbb{E} \times \mathbb{R} \times \mathbb{R}_+ \to \mathbb{E}$ by

$$\mathscr{F}(u,\sigma,\beta) \stackrel{\text{def}}{=} u - (L+K)^{-1}[Q(u,\sigma,\beta)+K(u)],$$

$$\tilde{\mathscr{F}}(u,\sigma,\beta) \stackrel{\text{def}}{=} u - (L+K)^{-1}[\tilde{Q}(u,\sigma,\beta)+K(u)],$$

which are equivariant compact fields. Finding a $2\pi/\beta$-periodic solution of system (6.74) is equivalent to finding the solution of the problem

$$\mathscr{F}(u,\sigma,\beta) = 0, \quad (u,\sigma,\beta) \in \mathbb{E} \times \mathbb{R} \times \mathbb{R}_+.$$

It is an easy exercise to verify the following results.

Lemma 6.15. *For* $\sigma \in \mathbb{R}$ *and* $\beta > 0$, *the map* $Q(\cdot,\sigma,\beta): \mathbb{E} \to \mathbb{E}$ *defined by (6.82) is continuous.*

Lemma 6.16. *If system (6.76) has a nonconstant periodic solution with period* $T > 0$, *then there exists an integer* $m \geq 1$, $m \in \mathbb{N}$ *such that* $\pm im2\pi/T$ *are characteristic values of the stationary state* $(x_\sigma, \tau_\sigma, \sigma)$.

Lemma 6.17. *Assume (SHB1)–(SHB3) hold. If* $B_M(u_0,\sigma_0,\beta_0;r,\rho) \subseteq \mathbb{E} \times \mathbb{R}^2$ *is a special neighborhood of* $\tilde{\mathscr{F}}$, *where* $0 < \rho < \beta_0$, *then there exists* $r' \in (0,r]$ *such that the neighborhood*

$$B_M(u_0,\sigma_0,\beta_0;r',\rho) = \{(u,\sigma,\beta): \|u - \eta(\sigma)\| < r', |(\sigma,\beta) - (\sigma_0,\beta_0)| < \rho\}$$

satisfies $\dot{u}(t) \neq Q(u,\sigma,\beta)$ *for* $(u,\sigma,\beta) \in \overline{B_M(u_0,\sigma_0,\beta_0;r',\rho)}$ *with* $u = (y,z)^T \neq \eta(\sigma)$ *and* $|(\sigma,\beta) - (\sigma_0,\beta_0)| = \rho$.

Proof. Suppose not. Then for all $0 < r' \leq r$, there exists (u, σ, β) such that $0 < \|u - \eta(\sigma)\| < r'$, $|(\sigma, \beta) - (\sigma_0, \beta_0)| = \rho$ and $\dot{u}(t) = Q(u, \sigma, \beta)$ for $t \in \mathbb{R}$. Then there exists a sequence of nonconstant periodic solutions $\{(u_k, \sigma_k, \beta_k) = (y_k, z_k, \sigma_k, \beta_k)\}_{k=1}^{\infty}$ such that

$$\lim_{k \to +\infty} \|u_k - \eta(\sigma_k)\| = 0, \ |(\sigma_k, \beta_k) - (\sigma_0, \beta_0)| = \rho, \tag{6.85}$$

and

$$\dot{u}_k(t) = \frac{1}{\beta_k} \begin{pmatrix} f(y_k(t), y_k(t - \beta_k z_k(t)), \sigma_k) \\ g(y_k(t), z_k(t), \sigma_k) \end{pmatrix} \text{ for } t \in \mathbb{R}. \tag{6.86}$$

Note that $0 < \rho < \beta_0$ implies that $\beta_k \geq \beta_0 - \rho > 0$ for every $k \in \mathbb{N}$. Also, since the sequence $\{(\sigma_k, \beta_k)\}_{k=1}^{\infty}$ belongs to a bounded neighborhood of (σ_0, β_0) in $\mathbb{R}^2$, there exists a subsequence, denoted by $\{(\sigma_k, \beta_k)\}_{k=1}^{\infty}$, that converges to (σ^*, β^*), so that $|(\sigma^*, \beta^*) - (\sigma_0, \beta_0)| = \rho$ and $\beta^* > 0$. Without loss of generality, we denote this sequence by $\{(\sigma_k, \beta_k)\}_{k=1}^{\infty}$. Our strategy here is to show that the system

$$\dot{v}(t) = \frac{1}{\beta^*} \begin{bmatrix} \partial_1 f(\sigma^*) & 0 \\ \partial_1 g(\sigma^*) & \partial_2 g(\sigma^*) \end{bmatrix} v(t) + \frac{1}{\beta^*} \begin{bmatrix} \partial_2 f(\sigma^*) & 0 \\ 0 & 0 \end{bmatrix} v(t - \beta^* z_{\sigma^*}) \tag{6.87}$$

has a nonconstant periodic solution, which contradicts the assumption that $u_0 = (y_{\sigma_0}, z_{\sigma_0})^T$ is the only center of (6.83) in $\overline{B_M(u_0, \sigma_0, \beta_0; r, \rho)}$.

Put

$$v_k(t) = \frac{u_k(t) - \eta(\sigma_k)}{\|u_k - \eta(\sigma_k)\|}.$$

Then we have

$$\dot{v}_k(t) = \frac{1}{\beta_k} \int_0^1 \begin{bmatrix} \partial_1 f_k(\sigma_k, s)(t) & 0 \\ \partial_1 g_k(\sigma_k, s)(t) & \partial_2 g_k(\sigma_k, s)(t) \end{bmatrix} ds \, v_k(t)$$

$$+ \frac{1}{\beta_k} \int_0^1 \begin{bmatrix} \partial_2 f_k(\sigma_k, s)(t) & 0 \\ 0 & 0 \end{bmatrix} ds \, v_k(t - \beta_k z_k(t)), \tag{6.88}$$

where

$$\partial_j f_k(\sigma_k, s)(t) \stackrel{\text{def}}{=} \partial_j f(y_{\sigma_k} + s(y_k(t) - y_{\sigma_k}), y_{\sigma_k} + s(y_k(t - z_k(t)) - y_{\sigma_k}), \sigma_k)),$$

$$\partial_j g_k(\sigma_k, s)(t) \stackrel{\text{def}}{=} \partial_j g(y_{\sigma_k} + s(y_k(t) - y_{\sigma_k}), z_{\sigma_k} + s(z_k(t) - z_{\sigma_k}), \sigma_k)$$

for all $j = 1, 2$. We claim that there exists a convergent subsequence of $\{v_k\}_{k=1}^{+\infty}$. Indeed, by (6.85), we know that $\{(z_k, \beta_k)\}_{k=1}^{+\infty}$ is uniformly bounded in $C(\mathbb{R}; \mathbb{R}) \times \mathbb{R}$, and hence $\lim_{t \to +\infty} [t - \beta_k z_k(t)] = +\infty$. Then, we have

$$\|v_k\| = 1, \quad \|v_k(\cdot - \beta_k z_k(\cdot))\| = 1.$$

Recall that $\partial_i f(\sigma^*)$ and $\partial_i g(\sigma^*)$, $i = 1, 2$, are defined in (6.76). By (6.85), we know that $(y_{\sigma_k} + s(y_k(t) - y_{\sigma_k}), y_{\sigma_k} + s(y_k(t - z_k(t)) - y_{\sigma_k}), \sigma_k)$ converges to the stationary state $(x_{\sigma^*}, \tau_{\sigma^*}, \sigma^*)$ in $C(\mathbb{R}; \mathbb{R}) \times \mathbb{R}$ uniformly for all $s \in [0, 1]$. By (SHB1), we know that $f(\theta_1, \theta_2, \sigma)$ is C^2 in (θ_1, θ_2) and $\partial_1 f(\theta_1, \theta_2, \sigma)$ is C^1 in σ. Also, by (6.85), the sequence $\{(u_k, \beta_k, \sigma_k)\}_{k=1}^{+\infty}$ is uniformly bounded in $C(\mathbb{R}; \mathbb{R}^{N+1}) \times \mathbb{R}^2$. Then we obtain

$$\begin{cases} \lim_{k \to +\infty} \|\partial_j f_k(\sigma_k, s) - \partial_j f(\sigma^*)\| = 0, \\ \lim_{k \to +\infty} \|\partial_j g_k(\sigma_k, s) - \partial_j g(\sigma^*)\| = 0 \end{cases} \tag{6.89}$$

uniformly for $s \in [0, 1]$, $j = 1, 2$. It is clear from (6.89) that $\|\partial_j f_k(\sigma_k, s)\|$ and $\|\partial_j g_k(\sigma_k, s)\|$ $(j = 1, 2)$ are all uniformly bounded for all $k \in \mathbb{N}$ and $s \in [0, 1]$. Then it follows from (6.88) that there exists a constant $\check{L}_2 > 0$ such that $\|\dot{v}_k\| < \check{L}_2$ for every $k \in \mathbb{N}$. By the Arzelà–Ascoli theorem, there exists a convergent subsequence $\{v_{k_j}\}_{j=1}^{+\infty}$ of $\{v_k\}_{k=1}^{+\infty}$. That is, there exists $v^* \in \{v \in V : \|v\| = 1\}$ such that

$$\lim_{j \to +\infty} \|v_{k_j} - v^*\| = 0. \tag{6.90}$$

By the integral mean value theorem, we have

$$|v_{k_j}(t - \beta_{k_j} z_{k_j}(t)) - v_{k_j}(t - \beta^* z_{\sigma^*})|$$
$$= \left| \int_0^1 \dot{v}_{k_j}(t - \theta(\beta_{k_j} z_{k_j}(t) - \beta^* z_{\sigma^*})) d\theta (\beta_{k_j} z_{k_j}(t) - \beta^* z_{\sigma^*}) \right|$$
$$\leq \|\dot{v}_{k_j}\| \cdot |\beta_{k_j} z_{k_j}(t) - \beta^* z_{\sigma^*}|$$
$$\leq \check{L}_2(\beta_{k_j} |z_{k_j}(t) - z_{\sigma^*}| + |\beta_{k_j} - \beta^*| z_{\sigma^*}). \tag{6.91}$$

By (6.85) and (6.91), we have

$$\lim_{j \to +\infty} \|v_{k_j}(\cdot - \beta_{k_j} z_{k_j}(\cdot)) - v_{k_j}(\cdot - \beta^* z_{\sigma^*})\| = 0. \tag{6.92}$$

Therefore, it follows from (6.90) and (6.92) that

$$\lim_{j \to +\infty} \|v_{k_j}(\cdot - \beta_{k_j} z_{k_j}(\cdot)) - v^*(\cdot - \beta^* z_{\sigma^*})\| = 0. \tag{6.93}$$

It follows from (6.85), (6.89), (6.90), and (6.93) that the right-hand side of (6.88) converges uniformly to the right-hand side of (6.87). Therefore, v^* is differentiable and satisfies (6.87). Moreover, we have

$$\lim_{k \to +\infty} |\dot{v}_k(t) - \dot{v}^*(t)| = 0.$$

Since by (SHB3), the matrix

$$\begin{bmatrix} \partial_1 f(\sigma^*) + \partial_2 f(\sigma^*) & 0 \\ \partial_1 g(\sigma^*) & \partial_2 g(\sigma^*) \end{bmatrix}$$

is nonsingular, $v = 0$ is the only constant solution of (6.87). Also, we have $v^* \in \{v \in V : \|v\| = 1\}$, $\|v^*\| \neq 0$. Therefore, $(v^*(t), \sigma^*, \beta^*)$ is a nonconstant peri-odic solution of the linear equation (6.87). Then by Lemma 6.16, $(\eta(\sigma^*), \sigma^*, \beta^*)$ is also a center of (6.83) in $\overline{B_M(u_0, \sigma_0, \beta_0; r, \rho)}$. This contradicts the assumption that $B_M(u_0, \sigma_0, \beta_0; r, \rho)$ is a special neighborhood of (6.82). This completes the proof.   □

As preparation for the proof of the local Hopf bifurcation theorem, we need the following lemma.

Lemma 6.18. *Assume that (SHB1)–(SHB3) hold. If $\mathscr{U} = B_M(u_0, \sigma_0, \beta_0; r, \rho) \subseteq \mathbb{E} \times \mathbb{R}^2$ is a special neighborhood of $\tilde{\mathscr{F}}$ with $0 < \rho < \beta_0$, then there exists $r' \in (0, r]$ such that $\mathscr{F}_\theta = (\mathscr{F}, \theta)$ and $\tilde{\mathscr{F}}_\theta = (\tilde{\mathscr{F}}, \theta)$ are homotopic on $\overline{B_M(u_0, \sigma_0, \beta_0; r', \rho)}$, where θ is a completing function defined on $\overline{B_M(u_0, \sigma_0, \beta_0; r', \rho)}$.*

Proof. Since $\mathscr{U} = B_M(u_0, \sigma_0, \beta_0; r, \rho) \subseteq \mathbb{E} \times \mathbb{R}^2$ is a special neighborhood of $\tilde{\mathscr{F}}$ with $0 < \rho < \beta_0$, then by Lemma 6.17, both $\mathscr{F}_\theta = (\mathscr{F}, \theta)$ and $\tilde{\mathscr{F}}_\theta = (\tilde{\mathscr{F}}, \theta)$ are $\mathscr{U}$-admissible.

Suppose that the conclusion is not true. Then for every $r' \in (0, r]$, $\mathscr{F}_\theta = (\mathscr{F}, \theta)$ and $\tilde{\mathscr{F}}_\theta = (\tilde{\mathscr{F}}, \theta)$ are not homotopic on $\overline{B_M(u_0, \sigma_0, \beta_0; r', \rho)}$. That is, every homotopy map between $\mathscr{F}_\theta$ and $\tilde{\mathscr{F}}_\theta$ has a zero on the boundary of $\overline{B_M(u_0, \sigma_0, \beta_0; r', \rho)}$. In particular, the linear homotopy $h(\cdot, \alpha) \overset{\text{def}}{=} \alpha \mathscr{F}_\theta + (1 - \alpha)\tilde{\mathscr{F}}_\theta = (\alpha \mathscr{F} + (1 - \alpha)\tilde{\mathscr{F}}, \theta)$ has a zero on the boundary of $\overline{B_M(u_0, \sigma_0, \beta_0; r', \rho)}$, where $\alpha \in [0, 1]$.

Note that $\theta(u, \sigma, \beta) < 0$ if $\|u - \eta(\sigma)\| = r'$. Then there exist (u, σ, β) and $\alpha \in [0, 1]$ such that $\|u - \eta(\sigma)\| < r'$, $|(\sigma, \beta) - (\sigma_0, \beta_0)| = \rho$ and

$$H(u, \sigma, \beta, \alpha) \overset{\text{def}}{=} \alpha \mathscr{F} + (1 - \alpha)\tilde{\mathscr{F}} = 0. \tag{6.94}$$

Since $r' > 0$ is arbitrary in the interval $(0, r]$, there exists a nonconstant sequence $\{(y_k, z_k, \sigma_k, \beta_k, \alpha_k)\}_{k=1}^\infty$ of solutions of (6.94) such that

$$\lim_{k \to +\infty} \|u_k - \eta(\sigma_k)\| = 0, \; |(\sigma_k, \beta_k) - (\sigma_0, \beta_0)| = \rho, 0 \leq \alpha_k \leq 1, \tag{6.95}$$

and

$$\dot{u}_k = \alpha_k Q(u_k, \sigma_k, \beta_k) + (1 - \alpha_k)\tilde{Q}(u_k, \sigma_k, \beta_k), \text{ for all } k \in \mathbb{N}. \tag{6.96}$$

Note that $0 < \rho < \beta_0$ implies that $\beta_k \geq \beta_0 - \rho > 0$ for every $k \in \mathbb{N}$. From (6.95), we know that $\{(\sigma_k, \beta_k, \alpha_k)\}_{k=1}^\infty$ belongs to a compact subset of $\mathbb{R}^3$. Therefore, there exist a convergent subsequence, denoted still by $\{(\sigma_k, \beta_k, \alpha_k)\}_{k=1}^\infty$ without loss of generality, and $(\sigma^*, \beta^*, \alpha^*) \in \mathbb{R}^3$ such that $\beta^* \geq \beta_0 - \rho > 0$, $\alpha^* \in [0, 1]$ and

$$\lim_{k \to +\infty} |(\sigma_k, \beta_k, \alpha_k) - (\sigma^*, \beta^*, \alpha^*)| = 0.$$

Similarly to the proof of Lemma 6.17, we can show that the system

$$\dot{v}(t) = \frac{1}{\beta^*} \begin{bmatrix} \partial_1 f(\sigma^*) & 0 \\ \partial_1 g(\sigma^*) & \partial_2 g(\sigma^*) \end{bmatrix} v(t) + \frac{1}{\beta^*} \begin{bmatrix} \partial_2 f(\sigma^*) & 0 \\ 0 & 0 \end{bmatrix} v(t - \beta^* z_{\sigma^*})$$

with $\partial_i f(\sigma^*)$, $\partial_i g(\sigma^*)$, $i = 1, 2$, defined after (6.76), has a nonconstant periodic solution, which contradicts the assumption that $B_M(u_0, \sigma_0, \beta_0; r, \rho)$ is a special neighborhood that contains an isolated center of (6.83). This completes the proof. $\square$

Now we are able to state and prove the local Hopf bifurcation theorem.

Theorem 6.15. *Assume that* (SHB1)–(SHB3) *hold. Let* $(x_{\sigma_0}, \tau_{\sigma_0}, \sigma_0)$ *be an isolated center of system* (6.76). *If the crossing number defined by* (6.80) *satisfies*

$$\gamma(x_{\sigma_0}, \tau_{\sigma_0}, \sigma_0, \beta_0) \neq 0,$$

then there exists a bifurcation of nonconstant periodic solutions of (6.74) *near* $(x_{\sigma_0}, \tau_{\sigma_0}, \sigma_0)$. *More precisely, there exists a sequence* $\{(x_n, \tau_n, \sigma_n, \beta_n)\}$ *such that* $\sigma_n \to \sigma_0$, $\beta_n \to \beta_0$ *as* $n \to \infty$, *and* $\lim_{n \to \infty} \|x_n - x_{\sigma_0}\| = 0$, $\lim_{n \to \infty} \|\tau_n - \tau_{\sigma_0}\| = 0$, *where*

$$(x_n, \tau_n, \sigma_n) \in C(\mathbb{R}; \mathbb{R}^{N+1}) \times \mathbb{R}$$

is a nonconstant $2\pi/\beta_n$*-periodic solution of system* (6.74).

Proof. By (SHB1), we know that the linear operator $\tilde{Q}$ is continuous. By Lemma 6.15, we know that $Q(\cdot, \sigma, \beta): \mathbb{E} \to \mathbb{E}$ is continuous. Moreover, as stated before, by Lemma 6.14, (SD2) and (SD4) are satisfied. Since $(u_0, \sigma_0) \overset{\text{def}}{=} (x_{\sigma_0}, \tau_{\sigma_0}, \sigma_0)$ is an isolated center of system (6.76) with a purely imaginary characteristic value $i\beta_0$, $\beta_0 > 0$, $(u_0, \sigma_0, \beta_0) \in \mathbb{E} \times \mathbb{R} \times (0, +\infty)$ is an isolated $\mathbb{E}$-singular point of $\tilde{\mathscr{F}}$. One can define the following two-dimensional submanifold $M \subset \mathbb{E}^G \times \mathbb{R} \times (0, +\infty)$ by

$$M \overset{\text{def}}{=} \{(\eta(\sigma), \sigma, \beta) : \sigma \in (\sigma_0 - \delta, \sigma_0 + \delta), \beta \in (\beta_0 - \varepsilon, \beta_0 + \varepsilon)\}$$

such that the point $(\eta(\sigma_0), \sigma_0, \beta_0)) = (u_0, \sigma_0, \beta_0)$ is the only $\mathbb{E}$-singular point of $\tilde{\mathscr{F}}$ in M, which is the set of trivial solutions to the system (6.76); it satisfies assumption (SD3).

Moreover, $(u_0, \sigma_0, \beta_0) \in \mathbb{E} \times \mathbb{R} \times (0, +\infty)$ is an isolated $\mathbb{E}$-singular point of $\tilde{\mathscr{F}}$. That is, for $\rho > 0$ sufficiently small, the linear operator $D_u \tilde{\mathscr{F}}(\eta(\sigma), \sigma, \beta): \mathbb{E} \to \mathbb{E}$ with $|(\sigma, \beta) - (\sigma_0, \beta_0)| < \rho$ is not an isomorphism only if $(\sigma, \beta) = (\sigma_0, \beta_0)$. Then by the implicit function theorem, there exists $r > 0$ such that for all $(u, \sigma, \beta) \in \mathbb{E} \times \mathbb{R} \times (0, +\infty)$ with $|(\sigma, \beta) - (\sigma_0, \beta_0)| = \rho$ and $0 < \|u - \eta(\sigma)\| \leq r$, we have $\tilde{\mathscr{F}}(u, \sigma, \beta) \neq 0$. Then the set $B_M(u_0, \sigma_0, \beta_0; r, \rho)$ defined by

$$\{(u, \sigma, \beta) \in \mathbb{E} \times \mathbb{R} \times (0, +\infty) : |(\sigma, \beta) - (\sigma_0, \beta_0)| < \rho, \|u - \eta(\sigma)\| < r\}$$

is a special neighborhood for $\tilde{\mathscr{F}}$. By Lemma 6.17, there exists a special neighborhood $\mathscr{U} = B_M(u_0, \sigma_0, \beta_0; r', \rho)$ such that $\mathscr{F}$ and $\tilde{\mathscr{F}}$ are nonzero for $(u, \sigma, \beta) \in \overline{B_M(u_0, \sigma_0, \beta_0; r', \rho)}$ with $u \neq \eta(\sigma)$ and $|(\sigma, \beta) - (\sigma_0, \beta_0)| = \rho$. That is, (SD5) is satisfied.

Let θ be a completing function on $\mathscr{U}$. It follows from Lemma 6.18 that $(\mathscr{F}, \theta)$ is homotopic to $(\tilde{\mathscr{F}}, \theta)$ on $\mathscr{U}$.

For $(\sigma, \beta) \in \mathscr{D}(\sigma_0, \beta_0)$, we denote by $\Psi(\sigma, \beta)$ the map $D_u \tilde{\mathscr{F}}(u(\sigma), \sigma, \beta)$: $\mathbb{E} \to \mathbb{E}$. It is easy to see that $\Psi(\sigma, \beta)(\mathbb{E}_k) \subset \mathbb{E}_k$ for all $k = 0, 1, 2, \cdots$. Define $\Psi_k : \mathscr{D}(\sigma_0, \beta_0) \to L(\mathbb{E}_k, \mathbb{E}_k)$ by

$$\Psi_k(\sigma, \beta) \stackrel{\text{def}}{=} \Psi(\sigma, \beta)|_{\mathbb{E}_k}.$$

Thus, the matrix representation $[\Psi_k]$ of $\Psi_k(\sigma, \beta)$ $\{e^{ik \cdot} \varepsilon_j\}_{j=1}^{N+1}$ is given by

$$\frac{1}{ik\beta} \Delta_{(u(\sigma), \sigma)}(ik\beta).$$

For the application of Theorem 6.7, we now show that there exists some $k \in \mathbb{Z}, k \geq 1$, such that $\varepsilon_0 \mu_k(u(\sigma_0), \sigma_0, \beta_0) = \varepsilon_0 \deg_B(\det_{\mathbb{C}}[\Psi_k]) \neq 0$, where $\varepsilon_0 = \operatorname{sgn} \det \Psi_0(\sigma, \beta)$ for $(\sigma, \beta) \in \mathscr{D}(\sigma_0, \beta_0)$. For a constant map $v_0 \in \mathbb{E}_0$,

$$\Psi_0(\sigma, \beta) v_0 = -\frac{1}{\beta} \begin{bmatrix} \partial_1 f(\sigma) + \partial_2 f(\sigma) & 0 \\ \partial_1 g(\sigma) & \partial_2 g(\sigma) \end{bmatrix} v_0.$$

Then by (SHB3), we have $\varepsilon_0 \neq 0$, and therefore (SD6) is satisfied. In view of (6.81), we have

$$\mu_1(u(\sigma_0), \sigma_0, \beta_0) = \gamma(x_{\sigma_0}, \tau_{\sigma_0}, \sigma_0, \beta_0) \neq 0,$$

which by Theorem 6.7, implies that $(u(\sigma_0), \sigma_0, \beta_0)$ is a bifurcation point of the system (6.82). Consequently, there exists a sequence of nonconstant periodic solutions $(u_n, \sigma_n, \beta_n) = (x_n, \tau_n, \sigma_n, \beta_n)$ such that $\sigma_n \to \sigma_0$, $\beta_n \to \beta_0$ as $n \to \infty$, and $(x_n(t), \tau_n(t))$ is a $2\pi/\beta_n$-periodic solution of (6.74) such that $\lim_{n \to +\infty} \|(x_n, \tau_n) - (x_{\sigma_0}, \tau_{\sigma_0})\| = 0$. $\qquad \square$

Remark 6.2. A local Hopf bifurcation theory for FDEs with state-dependent delays was developed by Eichmann [86], where the existence of a local Hopf bifurcation is guaranteed by a transversality condition. This transversality implies that the crossing number defined by (6.80) is not zero, and hence the existence of a local Hopf bifurcation is also established in Theorem 6.15. Note that even in the case of a constant delay, one can have nontrivial crossing number while the transversality condition is not satisfied. Note also that the work of Eichmann gives more information about the local Hopf bifurcation such as smoothness of the bifurcation curve with respect to the parameter.

6.9.2 Global Bifurcation

To use Theorem 6.8 to describe the maximal continuation of bifurcated periodic solutions with large amplitudes when the bifurcation parameter σ is far away from the bifurcation value, we need to prove that there is a lower bound for the periods of periodic solutions of system (6.74).

Lemma 6.19 (Vidossich [287]). *Let X be a Banach space and $v : \mathbb{R} \to X$ a p-periodic function with the following properties:*

(i) $v \in L^1_{loc}(\mathbb{R}, X)$;
(ii) there exists $U \in L^1([0, \frac{p}{2}]; \mathbb{R}_+)$ such that $|v(t) - v(s)| \leq U(t - s)$ for almost every (in the sense of the Lebesgue measure) $s, t \in \mathbb{R}$ such that $s \leq t$, $t - s \leq \frac{p}{2}$;
(iii) $\int_0^p v(t) \, dt = 0$.

Then

$$\mathsf{p} \|v\|_{L^\infty} \leq 2 \int_0^{\frac{\mathsf{p}}{2}} U(t) \, dt.$$

We make the following assumption on system (6.74):

(SHB4)   There exist constants $L_f > 0$, $L_g > 0$ such that

$$|f(\theta_1, \theta_2, \sigma) - f(\overline{\theta}_1, \overline{\theta}_2, \sigma)| \leq L_f(|\theta_1 - \overline{\theta}_1| + |\theta_2 - \overline{\theta}_2|)$$
$$|g(\gamma_1, \gamma_2, \sigma) - g(\overline{\gamma}_1, \overline{\gamma}_2, \sigma)| \leq L_g(|\gamma_1 - \overline{\gamma}_1| + |\gamma_2 - \overline{\gamma}_2|)$$

for every $\theta_1, \theta_2, \overline{\theta}_1, \overline{\theta}_2, \gamma_1, \overline{\gamma}_1 \in \mathbb{R}^N$, $\gamma_2, \overline{\gamma}_2 \in \mathbb{R}$, $\sigma \in \mathbb{R}$.

Lemma 6.20. *Assume that system (6.74) satisfies the assumption (SHB4). If $u = (x, \tau)$ is a nonconstant periodic solution of (1.1), then the minimal period of u satisfies*

$$\mathsf{p} \geq \frac{4(|\dot{x}|_{L^\infty} + |\dot{\tau}|_{L^\infty})}{(2L_f + L_g)|\dot{x}|_{L^\infty} + L_g |\dot{\tau}|_{L^\infty} + L_f |\dot{x}|_{L^\infty} |\dot{\tau}|_{L^\infty}}.$$

Moreover, suppose $g(x, \tau, \sigma)$ satisfies

(SHB5)   *for every $\sigma \in \mathbb{R}$, there exists $L_0 > 0$ such that $-L_0 \leq g(x, \tau, \sigma) < 1$ for all $(x, \tau) \in \mathbb{R}^{N+1}$.*

Then the minimal period p of u satisfies

$$\mathsf{p} \geq \frac{4}{\max\{L_0, 1\} + 2(L_f + L_g)}.$$

Proof. Let $v(t) = \dot{u}(t)$. Then $\int_0^p v(t) dt = 0$, since $u(t)$ is a p-periodic solution. For $s \leq t$, by (SHB4) and the integral mean value theorem, we have

$$|v(t) - v(s)| \leq |\dot{x}(t) - \dot{x}(s)| + |\dot{\tau}(t) - \dot{\tau}(s)|$$
$$\leq L_f(|x(t) - x(s)| + |x(t - \tau(t)) - x(s - \tau(s))|)$$

$$+ L_g(|x(t) - x(s)| + |\tau(t) - \tau(s)|)$$
$$\leq L_f|\dot{x}|_{L^\infty}(t-s) + L_f|\dot{x}|_{L^\infty}(t-s+|\tau(t)-\tau(s)|)$$
$$+ L_g|\dot{x}|_{L^\infty}(t-s) + L_g|\dot{\tau}|_{L^\infty}(t-s)$$
$$\leq \left[(2L_f + L_g)|\dot{x}|_{L^\infty} + L_g|\dot{\tau}|_{L^\infty} + L_f|\dot{x}|_{L^\infty} \cdot |\dot{\tau}|_{L^\infty}\right](t-s).$$

Let

$$U(t) = \left[(2L_f + L_g)|\dot{x}|_{L^\infty} + L_g|\dot{\tau}|_{L^\infty} + |\dot{x}|_{L^\infty} \cdot |\dot{\tau}|_{L^\infty}\right]t.$$

Then by Lemma 6.19, we obtain

$$\mathsf{p}|(\dot{x}, \dot{\tau})|_{L^\infty} \leq 2\int_0^{\frac{\mathsf{p}}{2}} U(t)dt = \frac{\mathsf{p}^2}{4}\left[(2L_f + L_g)|\dot{x}|_{L^\infty} + L_g|\dot{\tau}|_{L^\infty} + |\dot{x}|_{L^\infty} \cdot |\dot{\tau}|_{L^\infty}\right].$$

Therefore,

$$\mathsf{p} \geq \frac{4|(\dot{x}, \dot{\tau})|_{L^\infty}}{(2L_f + L_g)|\dot{x}|_{L^\infty} + L_g|\dot{\tau}|_{L^\infty} + L_f|\dot{x}|_{L^\infty}|\dot{\tau}|_{L^\infty}}.$$

Moreover, if $-L_0 \leq g(x(t), \tau(t), \sigma) < 1$, then

$$|\dot{x}|_{L^\infty} \cdot |\dot{\tau}|_{L^\infty} \leq \max\{L_0, 1\}|\dot{x}|_{L^\infty},$$

and hence

$$\mathsf{p} \geq \frac{4|(\dot{x}, \dot{\tau})|_{L^\infty}}{(2L_f + L_g)|\dot{x}|_{L^\infty} + L_g|\dot{\tau}|_{L^\infty} + \max\{L_0, 1\}|\dot{x}|_{L^\infty}}$$
$$\geq \frac{4|(\dot{x}, \dot{\tau})|_{L^\infty}}{(2L_f + L_g)|(\dot{x}, \dot{\tau})|_{L^\infty} + L_g|(\dot{x}, \dot{\tau})|_{L^\infty} + \max\{L_0, 1\}|(\dot{x}, \dot{\tau})|_{L^\infty}}$$
$$= \frac{4}{\max\{L_0, 1\} + 2(L_f + L_g)}.$$

$\square$

The following result was first established by Mallet-Paret and Yorke [216] for ordinary differential equations and was extended to neutral equations by Wu [301].

Lemma 6.21. *Suppose that system (6.74) satisfies (SHB1)–(SHB2) and (SHB4)–(SHB5). Assume further that there exists a sequence of real numbers $\{\sigma_k\}_{k=1}^\infty$ such that:*

(i) For each k, system (6.74) with $\sigma = \sigma_k$ has a nonconstant periodic solution $u_k = (x_k, \tau_k) \in C(\mathbb{R}; \mathbb{R}^{N+1})$ with minimal period $T_k > 0$;
(ii) $\lim\limits_{k\to\infty} \sigma_k = \sigma_0 \in \mathbb{R}$, $\lim\limits_{k\to\infty} T_k = T_0 < \infty$, and $\lim\limits_{k\to\infty} \|u_k - u_0\| = 0$, where $u_0 : \mathbb{R} \to \mathbb{R}^{N+1}$ is a constant map with the value (x_0, τ_0).

Then (u_0, σ_0) is a stationary state of (6.74), and there exists $m \geq 1$, $m \in \mathbb{N}$ such that $\pm im2\pi/T_0$ are the roots of the characteristic equation (6.77) with $\sigma = \sigma_0$.

Proof. By Lemma 6.20, we conclude that $T_k \geq \frac{4}{\max\{L_0, 1\} + 2(L_f + L_g)}$ and therefore $T_0 \geq \frac{4}{\max\{L_0, 1\} + 2(L_f + L_g)} > 0$.

Now we show that (u_0, σ_0) is a stationary state of (6.74). Since by (ii), $\lim_{k \to \infty} \sigma_k = \sigma_0$ and $\lim_{k \to \infty} \|u_k - u_0\| = 0$, we have only to show that the derivatives $\{\dot{u}_k\}_{k=1}^{+\infty}$ converge uniformly to the right-hand side of system (6.74). That is,

$$\|f(x_k, x_k(\cdot - \tau_k), \sigma_k) - f(x_0, x_0, \sigma_0)\| + \|g(x_k, \tau_k, \sigma_k) - g(x_0, \tau_0, \sigma_0)\| \to 0 \quad (6.97)$$

as $k \to +\infty$. Note that we have used $f(x_k, x_k(\cdot - \tau_k), \sigma_k)$ to denote the function $f(x_k(\cdot), x_k(\cdot - \tau_k), \sigma_k)$. In fact, it follows from (SHB1) and assumption (ii) that $\lim_{k \to \infty} \|g(x_k, \tau_k, \sigma_k) - g(x_0, \tau_0, \sigma_0)\| = 0$. Moreover, by the integral mean value theorem, we have $\|f(x_k, x_k(\cdot - \tau_k), \sigma_k) - f(x_0, x_0, \sigma_0)\| \to 0$ as $k \to +\infty$. This completes the proof of (6.97). Therefore, $(u_0, \sigma_0) = (x_{\sigma_0}, \tau_{\sigma_0}, \sigma_0)$ is the stationary state of (6.74) with $\sigma = \sigma_0$.

Next, we show that the linear system

$$\dot{v}(t) = \begin{bmatrix} \partial_1 f(\sigma_0) & 0 \\ \partial_1 g(\sigma_0) & \partial_2 g(\sigma_0) \end{bmatrix} v(t) + \begin{bmatrix} \partial_2 f(\sigma_0) & 0 \\ 0 & 0 \end{bmatrix} v(t - \tau_0) \quad (6.98)$$

has a nonconstant periodic solution.

For $\rho \in (0, 1)$, define

$$\varepsilon_{k, \rho} = \max_{t \in \mathbb{R}} |u_k(t + \rho T_k) - u_k(t)|,$$

$$v_k(t) = \varepsilon_{k, \rho}^{-1} [u_k(t + \rho T_k) - u_k(t)].$$

Then $\|v_k\| = 1$, and $v_k(t) \stackrel{\text{def}}{=} (y_k(t), z_k(t))$ satisfies

$$\dot{v}_k(t) = \begin{bmatrix} \partial_1 f(\sigma_0) & 0 \\ \partial_1 g(\sigma_0) & \partial_2 g(\sigma_0) \end{bmatrix} v_k(t) + \begin{bmatrix} \partial_2 f(\sigma_0) & 0 \\ 0 & 0 \end{bmatrix} v_k(t - \tau_0) + \begin{pmatrix} \delta_{1k}(t) \\ \delta_{2k}(t) \end{pmatrix},$$

where

$$\begin{cases} \delta_{1k}(t) = \varepsilon_{k, \rho}^{-1} [f(x_k(t + \rho T_k), x_k(t + \rho T_k - \tau_k(t + \rho T_k)), \sigma_k) \\ \qquad - f(x_k(t), x_k(t - \tau_k(t)), \sigma_k) - \partial_1 f(\sigma_0)(x_k(t + \rho T_k) - x_k(t)) \\ \qquad - \partial_2 f(\sigma_0)(x_k(t + \rho T_k - \tau_0) - x_k(t - \tau_0))], \\ \delta_{2k}(t) = \varepsilon_{k, \rho}^{-1} [g(x_k(t + \rho T_k), \tau_k(t + \rho T_k), \sigma_k) - g(x_k(t), \tau_k(t), \sigma_k) \\ \qquad - \partial_1 g(\sigma_0)(x_k(t + \rho T_k) - x_k(t)) - \partial_2 g(\sigma_0)(\tau_k(t + \rho T_k) - \tau_k(t))]. \end{cases}$$

Using the integral mean value theorem, we can show that $|\delta_{1k}(t)| \to 0$, $|\delta_{2k}(t)| \to 0$ as $k \to +\infty$ uniformly for $t \in \mathbb{R}$. This, together with the fact that $\|v_k\| = 1$, implies that there exists $\tilde{L}_6 > 0$ such that $\|\dot{v}_k\| \leq \tilde{L}_6$ for all $k \in \mathbb{N}$. Also, by assumption (ii), the set of periods $\{T_k\}_{k=1}^{+\infty}$ is bounded. Then by the Arzelà–Ascoli theorem, $\{v_k\}_{k=1}^{+\infty}$ has a convergent subsequence, denoted by $\{v_{k_j}\}_{j=1}^{+\infty}$. Let

$$v_\rho(t) = \lim_{j \to +\infty} v_{k_j}(t).$$

Then v_ρ is a periodic solution of (6.98) with period T_0. Since $\|v_k\| = 1$ and the average value of each v_k is zero, the same is true for v_ρ. So v_ρ is a nonconstant T_0-periodic solution of (6.98). Then by Lemma 6.16, there exists $m \geq 1, m \in \mathbb{N}$, such that $\pm im2\pi/T_0$ are characteristic values of (6.77). This completes the proof. $\qquad\square$

Now we can describe the relation between $2\pi/\beta_k$ and the minimal period of u_k in Theorem 6.15.

Theorem 6.16. *Assume that* (SHB1)–(SHB5) *hold. In Theorem 6.15, every limit point of the minimal period of $u_k = (x_k, \tau_k)$ as $k \to +\infty$ is contained in the set*

$$\left\{ \frac{2\pi}{(n\beta_0)} : \pm imn\beta_0 \text{ are characteristic values of } (u_0, \sigma_0), m, n \geq 1, m, n \in \mathbb{N} \right\}.$$

Moreover, if $\pm imn\beta_0$ are not characteristic values of (u_0, σ_0) for any integers $m, n \in \mathbb{N}$ such that $mn > 1$, then $2\pi/\beta_k$ is the minimal period of $u_k(t)$ and $2\pi/\beta_k \to 2\pi/\beta_0$ as $k \to \infty$.

Proof. Let T_k denote the minimal period of $u_k(t)$. Then there exists a positive integer n_k such that $2\pi/\beta_k = n_k T_k$. Since $T_k \leq 2\pi/\beta_k \to 2\pi/\beta_0$ as $k \to \infty$, there exist a subsequence $\{T_{k_j}\}_{j=1}^\infty$ and T_0 such that $T_0 = \lim_{j \to \infty} T_{k_j}$. Since $2\pi/\beta_{k_j} \to 2\pi/\beta_0, T_{k_j} \to T_0$ as $j \to \infty$, n_{k_j} is identical to a constant n for k large enough. Therefore, $2\pi/\beta_0 = nT_0$. Thus $T_{k_j} \to 2\pi/(n\beta_0)$ as $j \to \infty$. By Lemma 6.21, $\pm im2\pi/T_0 = \pm imn\beta_0$ are characteristic values of (u_0, σ_0) for some $m \geq 1, m \in \mathbb{N}$.

Moreover, if $\pm imn\beta_0$ are not characteristic values of (u_0, σ_0) for any integers $m \in \mathbb{N}$ and $n \in \mathbb{N}$ with $mn > 1$, then $m = n = 1$. Therefore, for k large enough, $n_{k_j} = 1$ and $2\pi/\beta_k = T_k$ is the minimal period of $u_k(t)$ and $2\pi/\beta_k \to 2\pi/\beta_0$ as $k \to \infty$. This completes the proof. $\qquad\square$

The following lemma shows that we can locate all possible Hopf bifurcation points of system (6.74) with state-dependent delay at the centers of its corresponding formal linearization.

Lemma 6.22. *Assume that* (SHB1)–(SHB3) *hold. If (u_0, σ_0) is a Hopf bifurcation point of system (6.74), then it is a center of (6.76).*

Proof. If (u_0, σ_0) is a Hopf bifurcation point of system (6.74), then there exist a sequence $\{(u_k, \sigma_k, T_k)\}_{k=1}^{+\infty} \subseteq C(\mathbb{R}; \mathbb{R}^{N+1}) \times \mathbb{R}^2$ and $T_0 \geq 0$ such that $\lim_{k \to +\infty} \|(u_k, \sigma_k, T_k) - (u_0, \sigma_0, T_0)\| = 0$, where (u_k, σ_k) is a nonconstant T_k-periodic solution of system (6.74). Using a similar argument to that in the proof of Lemma 6.17, we see that the system

$$\dot{v}(t) = \begin{bmatrix} \partial_1 f(\sigma_0) & 0 \\ \partial_1 g(\sigma_0) & \partial_2 g(\sigma_0) \end{bmatrix} v(t) + \begin{bmatrix} \partial_2 f(\sigma_0) & 0 \\ 0 & 0 \end{bmatrix} v(t - \tau_{\sigma_0}) \qquad (6.99)$$

has a nonconstant periodic solution v^*. Therefore, $(v^* + u_0, \sigma_0)$ is a nonconstant periodic solution of (6.76). Then, by Lemma 6.16, (u_0, σ_0) is a center of (6.76). $\qquad\square$

Now we are able to consider the global Hopf bifurcation problem of system (6.74). Letting $(x(t), \tau(t)) = (y(\frac{2\pi}{\mathsf{p}}t), z(\frac{2\pi}{\mathsf{p}}t))$, we can reformulate the problem as a problem of finding 2π-periodic solutions to the following equation:

$$\dot{u}(t) = Q(u(t), \sigma, 2\pi/\mathsf{p}), \tag{6.100}$$

where $u(t) = (y(t), z(t))$. Accordingly, the formal linearization (6.76) becomes

$$\dot{u}(t) = \tilde{Q}(u(t), \sigma, 2\pi/\mathsf{p}). \tag{6.101}$$

Using the same notation as in the proof of Theorem 6.15, we can define $\mathcal{N}_0(u, \sigma, \mathsf{p}) = Q(u, \sigma, 2\pi/\mathsf{p})$, $\tilde{\mathcal{N}}_0(u, \sigma, \mathsf{p}) = \tilde{Q}(u, \sigma, 2\pi/\mathsf{p})$. Then the system

$$Lu = \mathcal{N}_0(u, \sigma, \mathsf{p}), \mathsf{p} > 0, \tag{6.102}$$

is equivalent to (6.100), and

$$Lu = \tilde{\mathcal{N}}_0(u, \sigma, \mathsf{p}), \mathsf{p} > 0, \tag{6.103}$$

is equivalent to (6.101). Let $\mathscr{S}$ denote the closure of the set of all nontrivial periodic solutions of system (6.102) in the space $\mathbb{E} \times \mathbb{R} \times \mathbb{R}_+$, where $\mathbb{R}_+$ is the set of all nonnegative real numbers. It follows from Lemma 6.20 that the constant solution $(u_0, \sigma_0, 0)$ does not belong to this set. Consequently, we can assume that problem (6.102) is well posed on the whole space $\mathbb{E} \times \mathbb{R}^2$, in the sense that if $\mathscr{S}$ exists in $\mathbb{E} \times \mathbb{R}^2$, then it must be contained in $\mathbb{E} \times \mathbb{R} \times \mathbb{R}_+$.

On the other hand, assume that (SHB3) holds at every center of (6.103). Then from the proof of Theorem 6.15, we know that the assumptions (SHB1–SHB3) are sufficient for the systems (6.102) and (6.103) to satisfy the conditions (SD1)–(SD6). Also, under the same assumptions, Lemma 6.22 implies (SD7), and Lemma 6.18 implies (SD8). Then by Theorem 6.8, we obtain the following global Hopf bifurcation theorem for system (6.102) with state-dependent delay.

Theorem 6.17. *Suppose that system (6.74) satisfies* (SHB1)–(SHB5) *and* (SHB3) *holds at every center of (6.103). Assume that all the centers of (6.103) are isolated. Let $\mathcal{M}$ be the set of trivial periodic solutions of (6.102) and suppose that $\mathcal{M}$ is complete. If $(u_0, \sigma_0, \mathsf{p}_0) \in \mathcal{M}$ is a bifurcation point, then either the connected component $C(u_0, \sigma_0, \mathsf{p}_0)$ of $(u_0, \sigma_0, \mathsf{p}_0)$ in $\mathscr{S}$ is unbounded, or*

$$C(u_0, \sigma_0, \mathsf{p}_0) \cap \mathcal{M} = \{(u_0, \sigma_0, \mathsf{p}_0), (u_1, \sigma_1, \mathsf{p}_1), \cdots, (u_q, \sigma_q, \mathsf{p}_q)\},$$

where $\mathsf{p}_i \in \mathbb{R}_+$, $(u_i, \sigma_i, \mathsf{p}_i) \in \mathcal{M}$, $i = 0, 1, 2, \cdots, q$. Moreover, in the latter case, we have

$$\sum_{i=0}^{q} \varepsilon_i \gamma(u_i, \sigma_i, 2\pi/\mathsf{p}_i) = 0,$$

where $\gamma(u_i, \sigma_i, 2\pi/\mathsf{p}_i)$ is the crossing number of $(u_i, \sigma_i, \mathsf{p}_i)$ defined by (6.80) and
$$\varepsilon_i = \operatorname{sgn} \det \begin{bmatrix} \partial_1 f(\sigma_i) + \partial_2 f(\sigma_i) & 0 \\ \partial_1 g(\sigma_i) & \partial_2 g(\sigma_i) \end{bmatrix}.$$

Definition 6.4. Let $\mathscr{C}$ be a connected component of the closure of all nonconstant periodic solutions of (6.74) in the Fuller space $C(\mathbb{R};\mathbb{R}^{N+1}) \times \mathbb{R}^2$. We call $\mathscr{C}$ a continuum of *slowly oscillating periodic solutions* if for every $(x, \tau, \sigma, p) \in \mathscr{C}$, there exists $t_0 \in \mathbb{R}$ such that $p > \tau(t_0) > 0$. Similarly, we call $\mathscr{C}$ a continuum of *rapidly oscillating periodic solutions* if for every $(x, \tau, \sigma, p) \in \mathscr{C}$, there exists $t_0 \in \mathbb{R}$ such that $0 < p < \tau(t_0)$.

Theorem 6.17 shows that for a given trivial solution (x^*, τ^*, σ^*) with virtual period p^*, either the connected component $C(x^*, \tau^*, \sigma^*, p^*)$ has finitely many bifurcation points with the sum of $\mathbb{S}^1$-equivariant degrees being zero or $C(x^*, \tau^*, \sigma^*, p^*)$ is unbounded in the Fuller space $C(\mathbb{R};\mathbb{R}^{N+1}) \times \mathbb{R}^2$. Therefore, if global persistence of periodic solutions when the parameter is far away from the local Hopf bifurcation value σ^* is desired, we should find conditions to ensure that the connected component $C(x^*, \tau^*, \sigma^*, p^*)$ of Hopf bifurcation is unbounded in the Fuller space $C(\mathbb{R};\mathbb{R}^{N+1}) \times \mathbb{R}^2$ and $C(x^*, \tau^*, \sigma^*, p^*)$ will not blow up to infinity at any given σ in the norm of the Fuller space $C(\mathbb{R}; \mathbb{R}^{N+1}) \times \mathbb{R}^2$. That is, there exists a continuous function $M : \mathbb{R} \ni \sigma \to M(\sigma) > 0$ such that for every $(x, \tau, \sigma, p) \in C(x^*, \tau^*, \sigma^*, p^*)$, we have

$$\|(x, \tau, p)\|_{C(\mathbb{R};\mathbb{R}^{N+1}) \times \mathbb{R}} \le M(\sigma). \tag{6.104}$$

To achieve this goal, we shall give some sufficient geometric conditions ensuring the uniform boundedness of all possible periodic solutions (x, τ, σ) of (6.74), that is, we show that there exists a continuous function $M_1 : \mathbb{R} \ni \sigma \to M_1(\sigma) > 0$ such that for every $(x, \tau, \sigma, p) \in C(x^*, \tau^*, \sigma^*, p^*)$, we have

$$\|(x, \tau)\|_{C(\mathbb{R};\mathbb{R}^{N+1})} \le M_1(\sigma). \tag{6.105}$$

Then we seek a continuous function $M_2 : \mathbb{R} \ni \sigma \to M_2(\sigma) > 0$ such that for every $(x, \tau, \sigma, p) \in C(x^*, \tau^*, \sigma^*, p^*)$, we have

$$|p| \le M_2(\sigma). \tag{6.106}$$

6.9.3 Uniform Bounds for Periods of Periodic Solutions in a Connected Component

In order, by way of contradiction, to exclude certain values of the period of the periodic solutions in a given connected component, we need some analytic properties of an interval map under the following assumptions:

(SHB6)   For every $(\sigma, \tau) \in \mathbb{R}^2$, $\frac{\partial g}{\partial \tau}(x_\sigma, \tau, \sigma) \ne 0$.

(SHB7) $\frac{\partial g}{\partial x}(x, \tau, \sigma)f(x, x, \sigma) \ne 0$ for $(x, \tau, \sigma) \in \mathbb{R}^{N+1} \times \mathbb{R}$ such that $x \ne x_\sigma$ and $g(x, \tau, \sigma) = 0$.

In this section, to avoid notational complications, we use superscripts to denote function compositions, e.g., $l^j(t)$ denotes the jth composition of l evaluated at time t.

The following result can be found in [171].

Lemma 6.23. *Suppose that (6.74) satisfies* (SHB1)–(SHB2) *and* (SHB6)–(SHB7) *and* (x, τ, σ_0) *is a nonconstant periodic solution of (6.74). If* (x, τ) *is* $\tau(t_0)$*-periodic and if* $\tau(t_0) \neq \tau_{\sigma_0}$, *then the function* $l(t) = t - \tau(t) + \tau(t_0)$ *defined on* $[t_0, t_0 + \tau(t_0)]$ *satisfies the following properties:*

(a) *$l(t)$ is a self-mapping on* $[t_0, t_0 + \tau(t_0)]$.
(b) *$l(t)$ has only finitely many fixed points* $\{t_i\}_{i=1}^n$ *in* $[t_0, t_0 + \tau(t_0)]$ *with* $t_i < t_{i+1}$ *for every* $i \in \{1, 2, \cdots, n-1\}$.
(c) *For every* $t \in (t_i, t_{i+1}) \subseteq [t_0, t_0 + \tau(t_0)]$,

$$\lim_{j \to +\infty} l^j(t) = \begin{cases} t_i, & \text{if there exists } \bar{t} \in [t_i, t_{i+1}] \text{ such that } \bar{t} > l(\bar{t}), \\ t_{i+1}, & \text{if there exists } \bar{t} \in [t_i, t_{i+1}] \text{ such that } \bar{t} < l(\bar{t}). \end{cases}$$

(d) *Let* $\{t_{i_k}\}_{k=1}^{k_0} \subseteq \{t_i\}_{i=1}^n$ *be all the fixed points such that* $\lim_{j \to +\infty} l^j(t) = t_{i_k}$ *for every* $t \in [t_{i_k}, t_{i_k+1})$. *Then for* $\delta > 0$ *small enough,*

$$\lim_{j \to +\infty} \sup_{t \in [t_{i_k}, t_{i_k+1}-\delta]} |l^j(t) - t_{i_k}| = 0,$$

$$\lim_{j \to +\infty} \sup_{t \in [t_i+\delta, t_{i+1}], t_i \in \{t_1, t_2, \cdots, t_n\} \setminus \{t_{i_k}\}_{k=1}^{k_0}} |l^j(t) - t_{i+1}| = 0.$$

(e) *Let* $h(t) = t - \tau(t)$. *Then* $l^j(t) = h^j(t) + j\tau(t_0)$ *for every* $t \in [t_0, t_0 + \tau(t_0)]$ *and* $j \in \mathbb{N}$;
(f) *$h^j(t + \tau(t_0)) = h^j(t) + \tau(t_0)$ for all* $t \in \mathbb{R}$ *and* $j \in \mathbb{N}$.

Recall that $C(x^*, \tau^*, \sigma^*, p^*)$ denotes the connected component of the closure of all the nonconstant periodic solutions of system (6.74) bifurcated at $(x^*, \tau^*, \sigma^*, p^*)$ in the Fuller space $C(\mathbb{R}; \mathbb{R}^{N+1}) \times \mathbb{R}^2$. We hope to exclude, for each periodic solution $(x_0, \tau_0, \sigma_0, p_0)$, certain values of the period. To be specific, we find an open interval I and a small open neighborhood $U \ni (x_0, \tau_0, \sigma_0, p_0)$ such that every $(x, \tau, \sigma, p) \in U \cap C(x^*, \tau^*, \sigma^*, p^*)$ satisfies $\tau(t) \neq mp$ for all $t \in I$ and $m \in \mathbb{N}$. Then we will glue up these local exclusions to a global upper bound for the period along the rescaled (by period normalization) connected component $C(y^*, z^*, \sigma^*, p^*)$.

We first consider the periods of the solutions in a neighborhood of a periodic solution that does not assume a certain period.

Lemma 6.24. *If a solution* $(x_0, \tau_0, \sigma_0, p_0) \in C(x^*, \tau^*, \sigma^*, p^*)$ *satisfies* $\tau_0(t_0) \neq mp_0$ *for some* $t_0 \in \mathbb{R}$ *and for all* $m \in \mathbb{N}$, *then there exist an open neighborhood* $I \ni t_0$ *and an open neighborhood* $U \ni (x_0, \tau_0, \sigma_0, p_0)$ *in* $C(\mathbb{R}; \mathbb{R}^{N+1}) \times \mathbb{R}^2$ *such that every solution* $(x, \tau, \sigma, p) \in U \cap C(x^*, \tau^*, \sigma^*, p^*)$ *satisfies* $\tau(t) \neq mp$ *for all* $m \in \mathbb{N}$ *and* $t \in I$.

Proof. By way of contradiction, we suppose that for every open interval $I \ni t_0$ and every open neighborhood $U \ni (x_0, \tau_0, \sigma_0, p_0)$ in $C(\mathbb{R}; \mathbb{R}^{N+1}) \times \mathbb{R}^2$, there exist $t \in I$, $m \in \mathbb{N}$, and a periodic solution $(x, \tau, \sigma, p) \in U \cap C(x^*, \tau^*, \sigma^*, p^*)$ such that $\tau(t) = mp$. Then there exist sequences $\{(x_k, \tau_k, \sigma_k, p_k, t_k)\}_{k=1}^{+\infty} \subseteq U \cap C(x^*, \tau^*, \sigma^*, p^*)$ and $\{m_k : m_k \in \mathbb{N}\}_{k=1}^{+\infty}$ such that

$$
\begin{cases}
\tau_k(t_k) = m_k \, p_k, \\
\lim\limits_{k \to +\infty} (x_k, \tau_k, \sigma_k, p_k, t_k) = (x_0, \tau_0, \sigma_0, p_0, t_0).
\end{cases}
\tag{6.107}
$$

Without loss of generality, we assume $m_k \to m_0 \in \mathbb{N}$ as $k \to +\infty$ (otherwise, we take a subsequence). Then it follows from (6.107), (SHB2), and (SHB5) that

$$
m_0 = \lim_{k \to +\infty} m_k = \lim_{k \to +\infty} \frac{\tau_k(t_k)}{p_k} = \frac{\tau_0(t_0)}{p_0}.
\tag{6.108}
$$

Therefore, we have $\tau_0(t_0) = m_0 p_0$, which is a contradiction to the assumption. $\quad\square$

We note that for a nonconstant periodic solution (x, τ, σ) of system (6.74), it is allowed that $\tau(t)$ assume its stationary value τ_σ, or even $\tau(t) = \tau_\sigma$ for all $t \in \mathbb{R}$. Ruling out these cases turns out to be crucial for us to exclude certain values of periods of the periodic solutions.

Now we consider the periods of the periodic solutions in a neighborhood of a given nonconstant periodic solution in the Fuller space for which the delay τ-component is not equal to the corresponding stationary value at some time t. We need the following condition:

(SHB8) (i) $f(0, 0, \sigma) = 0$ for all $\sigma \in \mathbb{R}$;

(ii) $x f(x, x, \sigma)$ is positive (or negative) if $f(x, x, \sigma) \neq 0$.

Theorem 6.18. *Suppose that system (6.74) satisfies (SHB6)–(SHB8). Let $(x_0, \tau_0, \sigma_0, p_0)$ be a nonconstant periodic solution in $C(x^*, \tau^*, \sigma^*, p^*)$. If $\tau_0(t_0) \neq \tau_{\sigma_0}$ for some t_0, then there exist an open interval I and an open neighborhood U of $(x_0, \tau_0, \sigma_0, p_0)$ in $C(\mathbb{R}; \mathbb{R}^{N+1}) \times \mathbb{R}^2$ such that every solution (x, τ, σ, p) in $U \cap C(x^*, \tau^*, \sigma^*, p^*)$ satisfies $\tau(t) \neq mp$ for all $m \in \mathbb{N}$ and $t \in I$.*

Proof. We first show that there exist an open neighborhood U of $(x_0, \tau_0, \sigma_0, p_0)$ and an open neighborhood I_0 of t_0 such that $\tau(t) \neq \tau_{\sigma_0}$ for every $(x, \tau, \sigma, p) \in U \cap C(x^*, \tau^*, \sigma^*, p^*)$ and $t \in I_0$.

By way of contradiction, suppose that for every neighborhood $\tilde{I}$ of t_0 and neighborhood U of $(x_0, \tau_0, \sigma_0, p_0)$, there exist $t \in \tilde{I}$ and a nonconstant solution $(x, \tau, \sigma, p) \in U \cap C(x^*, \tau^*, \sigma^*, p^*)$ such that $\tau(t) = \tau_{\sigma_0}$. Then there exist a sequence of periodic solutions $\{(x_k, \tau_k, \sigma_k, p_k)\}_{k=1}^{+\infty}$ and $\{t_k\}_{k=1}^{+\infty}$ such that

$$
\begin{cases}
\tau_k(t_k) = \tau_{\sigma_k}, \\
\lim\limits_{k \to +\infty} (x_k, \tau_k, \sigma_k, p_k, t_k) = (x_0, \tau_0, \sigma_0, p_0, t_0).
\end{cases}
$$

This, together with assumption (SHB2), implies that

$$|\tau_k(t_k) - \tau_0(t_0)| \le |\tau_k(t_k) - \tau_k(t_0)| + |\tau_k(t_0) - \tau_0(t_0)|$$
$$\le |t_k - t_0| + \sup_{t \in \mathbb{R}} \|\tau_k - \tau_0\|$$
$$\to 0 \text{ as } k \to +\infty.$$

Therefore, we have

$$\tau_0(t_0) = \lim_{k \to +\infty} \tau_k(t_k) = \lim_{k \to +\infty} \tau_{\sigma_k} = \tau_{\sigma_0}.$$

This is a contradiction to the assumption that $\tau_0(t_0) \ne \tau_{\sigma_0}$, and hence the claim is proved.

If $(x_0, \tau_0, \sigma_0, p_0)$ satisfies $\tau_0(t_0) \ne mp_0$ for all $m \in \mathbb{N}$, then the existence of I and U is followed from Lemma 6.24. Otherwise, $(x_0, \tau_0, \sigma_0, p_0)$ is $\tau_0(t_0)$-periodic. Let Γ_{σ_0} be the nonempty solution set of the equation $f(x, x, \sigma_0) = 0$ for $x \in \mathbb{R}^N$. Then by (SHB6), for every $x \in \Gamma_{\sigma_0}$, τ_{σ_0} is the unique solution of $g(x, \tau, \sigma_0) = 0$ for $\tau \in \mathbb{R}$. Now we distinguish two cases:

Case 1. $x_0(t_0) = x_{\sigma_0}$ for some $x_{\sigma_0} \in \Gamma_{\sigma_0}$. Since $\tau_0(t_0) \ne \tau_{\sigma_0}$, by system (6.74) and by (SHB6), we have

$$\begin{cases} \dot{x}_0(t_0) = f(x_{\sigma_0}, x_{\sigma_0}, \sigma_0) = 0, \\ \dot{\tau}_0(t_0) = g(x_{\sigma_0}, \tau_0(t_0), \sigma_0) \ne 0. \end{cases} \tag{6.109}$$

Without loss of generality, we suppose $\dot{\tau}_0(t) > 0$ for t in some open neighborhood of t_0. Then, by the continuity and local monotonicity of $\tau_0(t)$, there exists $\delta > 0$ small enough that

$$0 < \tau_0(t) - \tau_0(t_0) < p_{\min}, t \in (t_0, t_0 + \delta),$$

where $p_{\min} > 0$ is the minimal period of (x_0, τ_0). Then $\tau_0(t) \ne m p_{\min}$ for every $m \in \mathbb{N}$. Therefore, (x_0, τ_0) is not $\tau_0(t)$-periodic for all $t \in (t_0, t_0 + \delta)$. So we have $\tau_0(t) \ne mp_0$ for all $t \in (t_0, t_0 + \delta)$ and $m \in \mathbb{N}$.

By Lemma 6.24, for every $t^* \in (t_0, t_0 + \delta)$, there exist an open interval I of t^* and an open neighborhood U of $(x_0, \tau_0, \sigma_0, p_0)$ in $C(\mathbb{R}; \mathbb{R}^{N+1}) \times \mathbb{R}^2$ such that every solution (x, τ, σ, p) in $U \cap C(x^*, \tau^*, \sigma^*, p^*)$ satisfies $\tau(t) \ne mp$ for all $m \in \mathbb{N}$ and $t \in I$.

Case 2. $x_0(t_0) \ne x_\sigma$ for every $x_\sigma \in \Gamma_{\sigma_0}$. By Lemma 6.23 (c), there are finitely many fixed points $\{t_i\}_{i=1}^n$ of $l(t) = t - \tau_0(t) + \tau_0(t_0)$ in $[t_0, t_0 + \tau_0(t_0)]$ that are in ascending order (we assume in the proof that all the sequences of the fixed points of l are in ascending order). And we can let the subsequence $\{t_{i_k}\}_{k=1}^{k_0} \subseteq \{t_i\}_{i=1}^n$ be all the fixed points such that $\lim_{j \to +\infty} l^j(t) = t_{i_k}$ for every $t \in [t_{i_k}, t_{i_k+1})$. Note that $\tau_0(t_i) = \tau_0(t_0)$ and $\tau_0(t_0) \ne \tau_{\sigma_0}$ implies that $\tau_0(t_i) \ne \tau_{\sigma_0}$ for all $i \in \{1, 2, \cdots, n\}$. If $x_0(t_{i_0}) = x_{\sigma_0}$ for some $i_0 \in \{1, 2, \cdots, n\}$ and for some $x_{\sigma_0} \in \Gamma_{\sigma_0}$. Then the conclusion follows by Case 1 with t_0 replaced by t_{i_0}.

Now we exclude that $x_0(t_i) \ne x_\sigma$ for every $i \in \{0, 1, 2, \cdots, n\}$ and for every $x_\sigma \in \Gamma_{\sigma_0}$. Assume that the contrary is true. We want to obtain a contradiction under the assumption that (x_0, τ_0) is $\tau_0(t_0)$-periodic.

For $\delta > 0$ small enough, we consider the following compact subset I_δ of $[t_0, t_0 + \tau_0(t_0)]$:

$$I_\delta = \bigcup_{t_{i_k} \in \{t_{i_1}, t_{i_2}, \cdots, t_{i_{k_0}}\}} [t_{i_k}, t_{i_k+1} - \delta] \bigcup_{t_i \in \{t_1, t_2, \cdots, t_n\} \setminus \{t_{i_k}\}_{k=1}^{k_0}} [t_i + \delta, t_i].$$

Note that for each interval $[t_i, t_{i+1}]$, only one of the endpoints is the limit of $\lim_{j \to +\infty} l^j(t)$ for every $t \in (t_i, t_{i+1})$. Note also that when δ goes to zero, I_δ goes to $[t_0, t_0 + \tau_0(t_0)]$ in the sense of Lebesgue measure.

Now for $\delta > 0$ small enough, we introduce the following piecewise constant function $\chi(t)$ on the compact subset I_δ of $[t_0, t_0 + \tau_0(t_0)]$:

$$\chi(t) = \begin{cases} t_{i_k}, & \text{if } t \in [t_{i_k}, t_{i_k+1} - \delta], t_{i_k} \in \{t_{i_k}\}_{k=1}^{k_0}, \\ t_{i+1}, & \text{if } t \in [t_i + \delta, t_{i+1}], t_i \in \{t_1, t_2, \cdots, t_n\} \setminus \{t_{i_k}\}_{k=1}^{k_0}. \end{cases}$$

Since the number of intervals with endpoints the fixed points of $l(t)$ is finite, it is clear from Lemma 6.23 (d) that

$$\lim_{j \to +\infty} \sup_{t \in I_\delta} |l^j(t) - \chi(t)| = 0. \tag{6.110}$$

Note that $(x(t), \tau(t))$ is a periodic solution of system (6.74). There exists $\tilde{M} > 0$ such that $|\dot{x}(t)| \leq \tilde{M}$ for every $t \in [t_0, t_0 + \tau(t_0)]$. Let I_i with $i \in \{1, 2, \cdots, n\}$ be the subinterval of I_δ that is either $[t_{i-1}, t_i - \delta]$ or $[t_{i-1} + \delta, t_i]$. Then we have $\chi(t) = t_{i-1}$ or $\chi(t) = t_i$ for $t \in I_i$, and hence we have

$$x_0(\chi(t)) = x_0(t_{i-1}) \text{ or } x_0(\chi(t)) = x_0(t_i) \text{ for every } t \in I_i. \tag{6.111}$$

Since $x_0(t_i) \neq x_\sigma$ for every $i \in \{0, 1, 2, \cdots, n\}$ and for every $x_\sigma \in \Gamma_{\sigma_0}$, by (6.111), we have

$$x_0(\chi(t)) \notin \Gamma_{\sigma_0} \text{ for every } t \in I_\delta. \tag{6.112}$$

By (6.110), for every $\varepsilon > 0$, there exists $N_0 > 0$ large enough that

$$\sup_{t \in I_\delta} |l^j(t) - \chi(t)| \leq \varepsilon, \text{ for every } j > N_0. \tag{6.113}$$

Let $(x_j(t), \tau_j(t)) = (x_0(h^j(t)), \tau_0(h^j(t))$ for $j = 0, 1, 2, \cdots$, where we define $h^0(t) = t$. Then by Lemma 6.23 (e), we have $(x_j(t), \tau_j(t)) = (x_0(l^j(t)), \tau_0(l^j(t)))$. Note that I_δ is composed of finitely many subintervals. By applying the integral mean value theorem to each subinterval of I_δ and by (6.113), we have for every $j > N_0$ that

$$\sup_{t \in I_\delta} |x_0(l^j(t)) - x_0(\chi(t))| \leq \sup_{t \in I_\delta} |\dot{x}_0(t)| \sup_{t \in I_\delta} |l^j(t) - \chi(t)| \leq \tilde{M} \varepsilon. \tag{6.114}$$

Differentiating $x_j(t)$ for $j = 1, 2, \cdots$, we can obtain from system (6.74) that

$$\dot{x}_j(t) = \prod_{m=0}^{j-1} (1 - g(x_m(t), \tau_m(t)), \sigma_0) f(x_j(t), x_{j+1}(t), \sigma_0). \qquad (6.115)$$

Since $g(x, \tau, \sigma) < 1$, we have

$$\prod_{m=0}^{j-1} (1 - g(x_m(t), \tau_m(t)), \sigma_0) > 0, t \in \mathbb{R}. \qquad (6.116)$$

Also by (ii) of (SHB8), $xf(x, x, \sigma_0) > 0$ as long as $x \notin \Gamma_{\sigma_0}$. Then by (6.112) we have

$$x_0(\chi(t)) f(x_0(\chi(t)), x_0(\chi(t)), \sigma_0) > 0 \qquad (6.117)$$

for every $t \in I_\delta$. By (6.114), (6.117), and by the continuity of f, it follows that there exists $N_1 > N_0$ such that

$$x_j(t) f(x_j(t), x_{j+1}(t), \sigma_0) > 0 \text{ for } j > N_1 \text{ and } t \in I_\delta. \qquad (6.118)$$

Therefore, for every $t \in I_\delta$ and $j > N_1$, by (6.115), (6.116), and (6.118), we have

$$x_j(t) \cdot \dot{x}_j(t) = \prod_{m=0}^{j-1} (1 - g(x_m(t), \tau_m(t)), \sigma_0) x_j(t) f(x_j(t), x_{j+1}(t), \sigma_0) > 0. \quad (6.119)$$

Since $\delta > 0$ is arbitrary and I_δ goes to I in measure as $\delta \to 0$, by the continuity of $x_j \cdot \dot{x}_j$, we have $x_j(t) \cdot \dot{x}_j(t) \geq 0$ for every $t \in I$ and $j > N_1$. By (6.119), we know that $x_j \cdot \dot{x}_j \not\equiv 0$ on I with $j > N_1$. Therefore, $x_j \cdot x_j$ is a nonconstant increasing continuous function. But this is impossible, since $x_j \cdot x_j$ is continuous and periodic. This completes the proof. $\qquad \square$

We now consider the periods of nonconstant periodic solutions, where the delay coincides with the corresponding stationary value for every $t \in \mathbb{R}$.

Lemma 6.25. *Suppose system (6.74) satisfies (SHB7). Let (x, τ, σ, p) be a nonconstant p-periodic solution of system (6.74). If $\tau(t) = \tau_\sigma$ for every $t \in \mathbb{R}$, then (x, τ, σ, p) is not τ_σ-periodic.*

Proof. Suppose, by way of contradiction, that (x, τ, σ, p) is τ_σ-periodic. If $\tau(t) = \tau_\sigma$ for every $t \in \mathbb{R}$, then we have

$$\begin{cases} \dot{x}(t) = f(x(t), x(t), \sigma), \\ 0 = \dot{\tau}(t) = g(x(t), \tau_\sigma, \sigma). \end{cases} \qquad (6.120)$$

It follows from (6.120) that

$$\ddot{\tau}(t) = \frac{\partial g}{\partial x}(x(t), \tau_\sigma, \sigma) \cdot f(x(t), x(t), \sigma) = 0. \qquad (6.121)$$

Then by (SHB7) and (6.121), $x(t) = x_\sigma$ for every $t \in \mathbb{R}$. Thus, (x, τ, σ, p) is a constant periodic solution of (6.74). This is a contradiction. $\square$

We now formulate our next assumption:

(SHB9)   For every Hopf bifurcation point $(x, \tau, \sigma, p) \in C(x^*, \tau^*, \sigma^*, p^*)$, $mp \neq \tau$
for every $m \in \mathbb{N}$.

Theorem 6.19. *Assume that system (6.74) satisfies* (SHB6)–(SHB9). *Then for every solution* $(x_0, \tau_0, \sigma_0, p_0) \in C(x^*, \tau^*, \sigma^*, p^*)$, *there exist an open interval I and an open neighborhood* $U \ni (x_0, \tau_0, \sigma_0, p_0)$ *such that every solution*

$$(x, \tau, \sigma, p) \in U \cap C(x^*, \tau^*, \sigma^*, p^*)$$

satisfies $\tau(t) \neq mp$ *for all* $m \in \mathbb{N}$ *and* $t \in I$.

Proof. For a given $\sigma_0 \in \mathbb{R}$, if $(x_0, \tau_0, \sigma_0, p_0) \in C(x^*, \tau^*, \sigma^*, p^*)$ is a constant periodic solution, then it is a Hopf bifurcation point of system (6.74) (See Lemma 6.21). Thus the existence of an open interval I and an open neighborhood $U \ni (x_0, \tau_0, \sigma_0, p_0)$ follows immediately from (SHB9) and Lemma 6.24.

If $(x_0, \tau_0, \sigma_0, p_0) \in C(x^*, \tau^*, \sigma^*, p^*)$ is a nonconstant periodic solution and $\tau_0(t) = \tau_{\sigma_0}$ for all $t \in \mathbb{R}$, then by Lemma 6.25, $(x_0, \tau_0, \sigma_0, p_0)$ is not τ_{σ_0}-periodic. The conclusion is implied by Lemma 6.24.

If $(x_0, \tau_0, \sigma_0, p_0)$ is a nonconstant periodic solution and $\tau_0(t) \neq \tau_{\sigma_0}$ for some $t \in \mathbb{R}$, then the conclusion follows from Theorem 6.18. $\square$

We now start the process that uses the local exclusion of periods developed above to construct a uniform upper bound for periods of solutions in the Fuller space. To achieve this goal, we need to "glue" the local exclusion of periods along the connected component. Now we shall show that (6.106) is valid, provided that (6.105) holds.

Theorem 6.20. *Let* $C(y^*, z^*, \sigma^*, p^*)$ *be a connected component of the closure of all the nonconstant periodic solutions of system (6.100), bifurcated from* $(y^*, z^*, \sigma^*, p^*)$ *in the Fuller space* $C(\mathbb{R}/2\pi; \mathbb{R}^{N+1}) \times \mathbb{R}^2$. *Suppose that system (6.74) satisfies* (SHB6)–(SHB9). *Then for every* $(y_0, z_0, \sigma_0, p_0) \in C(y^*, z^*, \sigma^*, p^*)$, *there exist an open interval I and an open neighborhood* $U \ni (y_0, z_0, \sigma_0, p_0)$ *such that* $mp \neq z(t)$ *for every solution* $(y, z, \sigma, p) \in U \cap C(y^*, z^*, \sigma^*, p^*)$, $m \in \mathbb{N}$ *and* $t \in I$.

Proof. Note that $p > 0$ for every solution (y, z, σ, p) in $C(y^*, z^*, \sigma^*, p^*)$. We show that the mapping

$$\iota : C(y^*, z^*, \sigma^*, p^*) \to C(x^*, \tau^*, \sigma^*, p^*) \tag{6.122}$$

$$(y(\cdot), z(\cdot), \sigma, p) \to \left(y\left(\frac{2\pi}{p}\cdot\right), z\left(\frac{2\pi}{p}\cdot\right), \sigma, p \right)$$

is continuous, where $C(x^*, \tau^*, \sigma^*, p^*) \subseteq C(\mathbb{R}; \mathbb{R}^{N+1}) \times \mathbb{R}^2$. Indeed, if

$$\lim_{n \to +\infty} \|(y_n(\cdot), z_n(\cdot), \sigma_n, p_n) - (y_0(\cdot), z_0(\cdot), \sigma_0, p_0)\|_{C(\mathbb{R}/2\pi; \mathbb{R}^{N+1}) \times \mathbb{R}^2} = 0,$$

then we have

$$\|\iota(y_n(\cdot), z_n(\cdot), \sigma_n, p_n) - \iota(y_0(\cdot), z_0(\cdot), \sigma_0, p_0)\|_{C(\mathbb{R}; \mathbb{R}^{N+1}) \times \mathbb{R}}$$

$$= |y_n\left(\frac{2\pi}{p_n} \cdot\right) - y_0\left(\frac{2\pi}{p_0} \cdot\right)|_C + |z_n\left(\frac{2\pi}{p_n} \cdot\right) - z_0\left(\frac{2\pi}{p_0} \cdot\right)|_C$$

$$+ |\sigma_n - \sigma_0| + |p_n - p_0|$$

$$\leq |y_n - y_0|_C + 2\pi|\dot{y}_0|\left|\frac{1}{p_n} - \frac{1}{p_0}\right| + |z_n - z_0|_C + 2\pi|\dot{z}_0|\left|\frac{1}{p_n} - \frac{1}{p_0}\right|$$

$$+ |\sigma_n - \sigma_0| + |p_n - p_0|$$

$$\to 0 \text{ as } n \to +\infty,$$

where $|\cdot|_C$ denotes the supremum norm in either $C(\mathbb{R}/2\pi; \mathbb{R}^N)$ or $C(\mathbb{R}/2\pi; \mathbb{R})$. Therefore, $C(x^*, \tau^*, \sigma^*, p^*)$ is a connected component of periodic solutions of (6.74).

Let $(x_0, \tau_0, \sigma_0, p_0) = \iota(y_0, z_0, \sigma_0, p_0) \in C(x^*, \tau^*, \sigma^*, p^*)$. Then by Theorem 6.19, there exist an open interval I' and an open neighborhood $U' \ni (x_0, \tau_0, \sigma_0, p_0)$ such that every solution $(x, \tau, \sigma, p) \in U' \cap C(x^*, \tau^*, \sigma^*, p^*)$ satisfies $\tau(t) \neq mp$ for all $m \in \mathbb{N}$ and $t \in I'$.

Since ι is continuous, we can choose an open set $U \subseteq C(\mathbb{R}/2\pi; \mathbb{R}^{N+1}) \times \mathbb{R}^2$ small enough that $(y_0, z_0, \sigma_0, p_0) \in U \subseteq \iota^{-1}(U')$ and the open set

$$I \overset{\text{def}}{=} \bigcap_{\{p:(y,z,\sigma,p) \in U\}} \frac{p}{2\pi} \cdot I'$$

is nonempty. Then by the definition of ι, $mp \neq z(t)$ for every $(y, z, \sigma, p) \in U \cap C(y^*, z^*, \sigma^*, p^*)$, $m \in \mathbb{N}$, and $t \in I$. □

Lemma 6.26 (The generalized intermediate value theorem [227]). *Let $f : X \to Y$ be a continuous map from a connected space X to a linearly ordered set Y with order topology. If $a, b \in X$ and $y \in Y$ lies between $f(a)$ and $f(b)$, then there exists $x \in X$ such that $f(x) = y$.*

Definition 6.5. Let $C(y^*, z^*, \sigma^*, p^*)$ be a connected component of the closure of all the nonconstant periodic solutions of system (6.100), bifurcated from $(y^*, z^*, \sigma^*, p^*)$ in the Fuller space $C(\mathbb{R}/2\pi; \mathbb{R}^{N+1}) \times \mathbb{R}^2$. Let $I \subset \mathbb{R}$ be an interval and U a subset in $C(y^*, z^*, \sigma^*, p^*)$. We call $I \times (U \cap C(y^*, z^*, \sigma^*, p^*))$ a delay-period disparity set if every solution

$$(y, z, \sigma, p) \in U \cap C(y^*, z^*, \sigma^*, p^*)$$

satisfies $mp \neq z(t)$ for every $t \in I$ and $m \in \mathbb{N}$. We call $I \times (U \cap C(y^*, z^*, \sigma^*, p^*))$ a delay-period disparity set at $(t_0, y_0, z_0, \sigma_0, p_0)$ if $(t_0, y_0, z_0, \sigma_0, p_0) \in I \times (U \cap C(y^*, z^*, \sigma^*, p^*))$.

In the remainder of this subsection, the following assumption is sometimes needed:

(SHB10)   Every periodic solution (x, τ, σ) of (6.74) satisfies $\tau(t) > 0$ for every $t \in \mathbb{R}$.

Lemma 6.27. *Suppose that system* (6.74) *satisfies* (SHB6)–(SHB7) *and* (x, τ, σ) *is a nonconstant periodic solution. If*

(i) $\tau \not\equiv \tau_\sigma$ *and there exists* $t_0 \in \mathbb{R}$ *such that* $\tau(t_0) = \tau_\sigma$, *and*
(ii) (x, τ) *is* τ_σ-*periodic*,

then there exists $t_1 \in \mathbb{R}$ *such that* $\tau(t_1) > \tau_\sigma$.

Proof. We prove the result by contradiction. Suppose that

$$\tau(t) \le \tau_\sigma \text{ for every } t \in \mathbb{R}. \tag{6.123}$$

Then since $\tau \not\equiv \tau_\sigma$, there exists $t^* \in \mathbb{R}$ such that $\tau(t^*) < \tau_\sigma$. We can choose a maximal interval $[a, b] \subset \mathbb{R}$ that contains t^* in the sense that

$$\tau(t) < \tau_\sigma \text{ for any } t \in (a, b), \tag{6.124}$$
$$\tau(t) = \tau_\sigma \text{ for any } t = a \text{ and } t = b. \tag{6.125}$$

If $\dot{\tau}(a) \ne 0$ or $\dot{\tau}(b) \ne 0$, then it follows from the local monotonicity of $\tau(t)$ (at a or b) that there exists $t_1 \in \mathbb{R}$ in some neighborhood of a or b such that $\tau(t_1) > \tau_\sigma$. This is a contradiction to (6.123).

If $\dot{\tau}(a) = \dot{\tau}(b) = 0$, then we have

$$g(x(a), \tau_\sigma, \sigma) = g(x(b), \tau_\sigma, \sigma) = 0. \tag{6.126}$$

We distinguish the following two cases:

Case 1. $x(a) \ne x_\sigma$ or $x(b) \ne x_\sigma$. Without loss of generality, we suppose $x(a) \ne x_\sigma$. Then by (ii), we have

$$\ddot{\tau}(a) = \frac{\partial g}{\partial x}(x(a), \tau_\sigma, \sigma) f(x(a), x(a), \sigma). \tag{6.127}$$

It follows from (SHB7), (6.126), and (6.127) that $\ddot{\tau}(a) \ne 0$. Therefore, we have that $\dot{\tau}(t)$ is strictly monotonic in some neighborhood of a. Hence there exists $t_1 \in \mathbb{R}$ such that $\tau(t_1) > \tau_\sigma$. This is also a contradiction to (6.123).

Case 2. $x(a) = x(b) = x_\sigma$. By (S5), we have $\frac{\partial g}{\partial \tau}(x_\sigma, \tau_\sigma, \sigma) \ne 0$. Without loss of generality, we assume that

$$\frac{\partial g}{\partial \tau}(x_\sigma, \tau_\sigma, \sigma) < 0. \tag{6.128}$$

Then by (6.124), (6.126), (6.128), and the continuity of $x(t)$ and $\tau(t)$, we can choose $\varepsilon > 0$ small enough that

$$\dot{\tau}(t) = g(x(t), \tau(t), \sigma) > 0 \text{ for every } t \in (a, a+\varepsilon) \cup (b-\varepsilon, b). \tag{6.129}$$

Therefore, we have $\tau(a) < \tau(a+\varepsilon)$. That is, there exists $t_1 = a+\varepsilon$ such that $\tau(a) = \tau_\sigma < \tau(t_1)$. This is a contradiction to (6.123). The proof is complete. $\square$

Lemma 6.28. *Suppose that (6.74) satisfies* (SHB6)–(SHB10). *Let* $C(y^*, z^*, \sigma^*, p^*)$ *be a connected component of the closure of all the nonconstant periodic solutions of system (6.100), bifurcated from* $(y^*, z^*, \sigma^*, p^*)$ *in the Fuller space* $C(\mathbb{R}/2\pi; \mathbb{R}^{N+1}) \times \mathbb{R}^2$. *Let* $I \subset \mathbb{R}$ *be an open interval and* $\bar{v} \stackrel{\text{def}}{=} (\bar{y}, \bar{z}, \bar{\sigma}, \bar{p}) \in C(y^*, z^*, \sigma^*, p^*)$. *If there is no delay-period disparity set at* $(t, \bar{u})$ *for any* $t \in I$, *then*

   (i) *there exists* $m \in \mathbb{N}$ *such that* $m\bar{p} = \bar{z}(t) = z_{\bar{\sigma}}$ *for every* $t \in I$;
  (ii) $\bar{v}$ *is a nonconstant solution with* $\bar{z}(t) = z_{\bar{\sigma}}$ *for every* $t \in I$;
 (iii) *there exist an open interval* $I' \subseteq \mathbb{R}$ *and an open neighborhood* U' *of* $\bar{v}$ *such that* $I' \times (U' \cap C(y^*, z^*, \sigma^*, p^*))$ *is a delay-period disparity set with* $\bar{v} \in U' \cap C(y^*, z^*, \sigma^*, p^*)$, *and the inequality* $z_{\bar{\sigma}} < \bar{z}(t)$ *holds for every* $t \in I'$.

Proof. (i) By Definition 6.5, for every $t \in I$, there exists $m \in \mathbb{N}$ such that $\bar{z}(t) = m\bar{p}$. Note that $\bar{z}(t)$ is continuous, $\bar{z}(t) = m\bar{p}$ for every $t \in I$. Then for every $t \in I$, we have

$$\dot{\bar{y}}(t) = \frac{\bar{p}}{2\pi} f(\bar{y}(t), \bar{y}(t), \bar{\sigma}), \tag{6.130}$$

$$\dot{\bar{z}}(t) = \frac{\bar{p}}{2\pi} g(\bar{y}(t), m\bar{p}, \bar{\sigma}) = 0. \tag{6.131}$$

By (6.131), we have

$$\ddot{\bar{z}}(t) = \frac{\bar{p}^2}{4\pi^2} \frac{\partial g}{\partial x} (\bar{y}(t), m\bar{p}), \bar{\sigma}) \cdot f(\bar{y}(t), \bar{y}(t), \bar{\sigma}) = 0. \tag{6.132}$$

By (SHB7), (6.131), and (6.132), we have $\bar{y}(t) = y_{\bar{\sigma}}$ on I. Hence by (SHB6) and by (6.131), we have $\bar{z}(t) = z_{\bar{\sigma}} = m\bar{p}$ on I. This finishes the proof of (i).

 (ii) Note that the stationary solutions of (6.74) and (6.100) are equal. That is, $(x_\sigma, \tau_\sigma) = (y_\sigma, z_\sigma)$ for every $\sigma \in \mathbb{R}$.
If $\bar{v}$ is a constant solution, then by (i) we have $\bar{z}(t) = z_{\bar{\sigma}} = m\bar{p}$ and $\bar{y}(t) = y_{\bar{\sigma}}$ for all $t \in \mathbb{R}$. Then $(y_{\bar{\sigma}}, z_{\bar{\sigma}}, \bar{\sigma}, \bar{p})$ is a bifurcation point in $C(y^*, z^*, \sigma^*, p^*)$ that satisfies $z_{\bar{\sigma}} = m\bar{p}$ for some $m \in \mathbb{N}$. This contradicts assumption (SHB9). So $\bar{v}$ is a nonconstant solution with $\bar{z}(t) = z_{\bar{\sigma}}$ for all $t \in I$.

(iii) Now we show that there exists $t_0 \in \mathbb{R}$ such that $\bar{z}(t_0) \neq z_{\bar{\sigma}}$. If not, that is, if $\bar{z}(t) = z_{\bar{\sigma}}$ for all $t \in \mathbb{R}$, then

$$(\bar{x}(\cdot), \bar{\tau}(\cdot), \bar{\sigma}) \stackrel{\text{def}}{=} (\bar{y}(\frac{2\pi}{\bar{p}}\cdot), \bar{z}(\frac{2\pi}{\bar{p}}\cdot), \bar{\sigma}) = (\bar{y}(\frac{2\pi}{\bar{p}}\cdot), z_{\bar{\sigma}}, \bar{\sigma}) = (\bar{y}(\frac{2\pi}{\bar{p}}\cdot), \tau_{\bar{\sigma}}, \bar{\sigma})$$

is a solution of (6.74). Then by Lemma 6.25, $(\bar{x}, \bar{\tau})$ is not $\tau_{\bar{\sigma}}$-periodic. Then we have $m\bar{p} \neq z_{\bar{\sigma}}$ for every $m \in \mathbb{N}$. This is a contradiction to (i).

Therefore, there exists $t_0 \in \mathbb{R}$ such that $\bar{z}(t_0) \neq z_{\bar{\sigma}}$. That is, $\bar{\tau}(\frac{\bar{p}}{2\pi}t_0) \neq \tau_{\bar{\sigma}}$. Note that by (i), $(\bar{x}, \bar{\tau})$ is $\tau_{\bar{\sigma}}$-periodic and $\bar{\tau}(t) = \tau_{\bar{\sigma}}$ on $\frac{\bar{p}}{2\pi}I$. Then by Lemma 6.27, there exists $t_1 \in \mathbb{R}$ such that

$$\bar{\tau}(t_1) > \tau_{\bar{\sigma}}. \tag{6.133}$$

By the continuity of $\bar{\tau}$ and by (6.133), there exists a finite interval $(a, b) \ni t_1$ such that for every $t \in (a, b)$,

$$\bar{\tau}(t) > \tau_{\bar{\sigma}}. \tag{6.134}$$

We claim that there exists $t_0 \in (a, b)$ such that $\bar{v}$ is not $\bar{\tau}(t_0)$-periodic. Indeed, if not, then $\bar{v}$ would be $\bar{\tau}(t)$-periodic for every $t \in (a, b)$. Then by the continuity of $\bar{\tau}$ and by (6.134), there would exist $t_1, t_2 \in (a, b)$ and an interval $(\bar{\tau}(t_1), \bar{\tau}(t_2))$ with $\bar{\tau}(t_2) > \bar{\tau}(t_1)$, so that $\bar{\tau}$ would be p-periodic for all $p \in (\bar{\tau}(t_1), \bar{\tau}(t_2))$. Hence $\bar{v}$ would be a constant solution. This is a contradiction to (ii), and the claim is proved.

Then we have $\bar{\tau}(t_0) \neq m\bar{p}$ for all $m \in \mathbb{N}$. By Lemma 6.24, there exist an open interval $I_1 \ni t_0$ and an open neighborhood $U_1 \ni (\bar{x}, \bar{\tau}, \bar{\sigma}, \bar{p})$ such that every solution (x, τ, σ, p) of (6.74) in $U_1 \cap C(x^*, \tau^*, \sigma^*, p^*)$ satisfies $\tau(t) \neq mp$ for all $m \in \mathbb{N}$ and $t \in I_1$. Note that $\bar{\tau}$ is continuous at $t = t_0$. We can therefore choose I_1 small enough that (6.134) holds for all $t \in I_1$.

Let ι be the continuous mapping defined by (6.122). Then we can choose an open set $U' \subseteq C(\mathbb{R}/2\pi; \mathbb{R}^{N+1}) \times \mathbb{R}^2$ small enough that $\bar{v} \in U' \subseteq \iota^{-1}(U_1)$ and

$$I' \stackrel{\text{def}}{=} \bigcap_{\{p : (y, z, \sigma, p) \in U'\}} \frac{p}{2\pi} \cdot I_1$$

is nonempty. It follows from the definition of ι that $mp \neq z(t)$ for every solution $(y, z, \sigma, p) \in U' \cap C(y^*, z^*, \sigma^*, p^*)$, $m \in \mathbb{N}$, and $t \in I'$. In particular, noting that (6.134) holds for all $t \in I_1$ and $I' \subseteq \frac{\bar{p}}{2\pi}I_1$, we have

$$\bar{z}(t) > z_{\bar{\sigma}} \tag{6.135}$$

for every $t \in I'$. This completes the proof. $\qquad\qquad\square$

Now we are able to state our main result.

Theorem 6.21. *Let $C(y^*, z^*, \sigma^*, p^*)$ be a connected component of the closure of all the nonconstant periodic solutions of system (6.100), bifurcated from $(y^*, z^*, \sigma^*, p^*)$ in the Fuller space $C(\mathbb{R}/2\pi; \mathbb{R}^{N+1}) \times \mathbb{R}^2$. Suppose that (6.74) satisfies (SHB6)–(SHB10). If $p^* < z^*$, then for every $(y, z, \sigma, p) \in C(y^*, z^*, \sigma^*, p^*)$, $p < z(t)$ for some $t \in \mathbb{R}$.*

Proof. By Theorem 6.20 and (SHB9), there exist an open interval $I^* \subseteq \mathbb{R}$ and an open set U^* in $C(\mathbb{R}/2\pi; \mathbb{R}^{N+1}) \times \mathbb{R}^2$ such that $I^* \times (U^* \cap C(y^*, z^*, \sigma^*, p^*))$ is a delay-period disparity set with $(y^*, z^*, \sigma^*, p^*) \in U^*$.

Let $A^* \ni (y^*, z^*, \sigma^*, p^*)$ be a connected component of $(U^* \cap C(y^*, z^*, \sigma^*, p^*))$. Then $I^* \times A^*$ is connected in $\mathbb{R} \times C(\mathbb{R}/2\pi; \mathbb{R}^{N+1}) \times \mathbb{R}^2$. Define $S: \mathbb{R} \times C(\mathbb{R}/2\pi; \mathbb{R}^{N+1}) \times \mathbb{R}^2 \to \mathbb{R}$ by

$$S(t, y, z, \sigma, p) = p - z(t).$$

Note that we have $p^* < z^*$. Then it follows that $S(t, y^*, z^*, \sigma^*, p^*) = p^* - z^* < 0$. Note that S is continuous. By Lemma 6.26, we have

$$S(t, y, z, \sigma, p) = p - z(t) < 0 \qquad (6.136)$$

for every $(t, y, z, \sigma, p) \in I^* \times A^*$, for otherwise, there would exist $(t_0, y_0, z_0, \sigma_0, p_0) \in I^* \times A^*$ such that $p_0 = z_0(t_0)$, which contradicts the fact that $I^* \times A^*$ is a subset of the forbidden range of delay $I^* \times (U^* \cap C(y^*, z^*, \sigma^*, p^*))$.

Now we show that there exists a sequence of connected subsets of $C(y^*, z^*, \sigma^*, p^*)$, denoted by $\{A_n\}_{n=1}^{n_0}$, $n_0 \in \mathbb{N}$ or $n_0 = +\infty$, that satisfies

(i) $A^* \subseteq A_1 \subset A_2 \subset \cdots \subset A_{n_0}$ and $\cup_{n=1}^{n_0} A_n = C(y^*, z^*, \sigma^*, p^*)$;
(ii) for every $(y, z, \sigma, p) \in A_n$ with $n \in \{1, 2, \cdots, n_0\}$, $p < z(t)$ at some $t \in \mathbb{R}$.

Let $A_1 \overset{\text{def}}{=} A^*$. If $A_1 = C(y^*, z^*, \sigma^*, p^*))$, then we are done by (6.136). If not, since the only sets that are both closed and open in the connected topological space $C(y^*, z^*, \sigma^*, p^*)$ are the empty set and the connected component $C(y^*, z^*, \sigma^*, p^*)$ itself, $A_1 \ni (y^*, z^*, \sigma^*, p^*)$ is not both closed and open. Then the boundary of A_1 in the sense of the relative topology induced by $C(y^*, z^*, \sigma^*, p^*)$ is nonempty. That is,

$$\partial A_1 \neq \emptyset. \qquad (6.137)$$

Let $\bar{v} = (\bar{y}, \bar{z}, \bar{\sigma}, \bar{p}) \in \partial A_1$. If there exist $t_1 \in I_1 \overset{\text{def}}{=} I^*$ and a delay-period disparity set $I' \times (U' \cap C(y^*, z^*, \sigma^*, p^*))$ such that $(t_1, \bar{v}) \in \bar{I}' \times (U' \cap C(y^*, z^*, \sigma^*, p^*))$, and if $A_{\bar{v}} \ni \bar{v}$ is the connected component of $U' \cap C(y^*, z^*, \sigma^*, p^*)$, then it is clear that $A_1 \cup A_{\bar{v}}$ is connected. Since A_1 is closed, we have $\bar{p} < \bar{z}(t_1)$. Then by Lemma 6.26, we have

$$S(t, y, z, \sigma, p) = p - z(t) < 0 \text{ for every } (t, y, z, \sigma, p) \in I' \times A_{\bar{v}}. \qquad (6.138)$$

If for every $t \in I_1$, there is no delay-period disparity set at $(t, \bar{u})$, then by Lemma 6.28, there exists a delay-period disparity set $I'' \times (U'' \cap C(y^*, z^*, \sigma^*, p^*))$ with $\bar{v} \in U' \cap C(y^*, z^*, \sigma^*, p^*)$ and

$$m\bar{p} = z_{\bar{\sigma}} < \bar{z}(t) \text{ for every } t \in I'' \text{ and } m \in \mathbb{N}. \qquad (6.139)$$

Let $A_{\bar{v}} \ni \bar{v}$ be the connected component of $U'' \cap C(y^*, z^*, \sigma^*, p^*)$. It is clear that $A_1 \cup A_{\bar{v}}$ is connected. Then by (6.139) and Lemma 6.26,

$$S(t, y, z, \sigma, p) = p - z(t) < 0 \text{ for any } (t, y, z, \sigma, p) \in I'' \times A_{\bar{v}}. \qquad (6.140)$$

By (6.138) and (6.140), we know that if $\bar{v} \in \partial A_1$, then there exists a delay-period disparity set $\tilde{I} \times (\tilde{U} \cap C(y^*, z^*, \sigma^*, p^*))$ with $A_{\bar{v}} \ni \bar{v}$ the connected component of $\tilde{U} \cap C(y^*, z^*, \sigma^*, p^*)$ such that

$$S(t, y, z, \sigma, p) = p - z(t) < 0 \text{ for any } (t, y, z, \sigma, p) \in \tilde{I} \times A_{\bar{v}}. \tag{6.141}$$

For every $\bar{v} \in \partial A_1$, we find a $A_{\bar{v}}$ satisfying (6.141). Then we define

$$A_2 = A_1 \cup \bigcup_{\bar{v} \in \partial A_1} A_{\bar{v}}.$$

It follows from (6.136), (6.138), and (6.140) that for every $(y, z, \sigma, p) \in A_2$, $p < z(t)$ for some $t \in \mathbb{R}$. Note that for every $\bar{v} \in \partial A_1$, $A_1 \cup A_{\bar{v}}$ is connected. Therefore, A_2 is connected.

Note that the existence of A_2 depends only on the fact that $\partial A_1 \neq \emptyset$, in the sense of the relative topology induced by $C(y^*, z^*, \sigma^*, p^*)$. Beginning with $n = 1$, we can always recursively construct a connected subset for each $n \geq 1$, $n \in \mathbb{N}$, with $\partial A_n \neq \emptyset$,

$$A_{n+1} = A_n \cup \bigcup_{\bar{v} \in \partial A_n} A_{\bar{v}}, \tag{6.142}$$

satisfying that for every $(y, z, \sigma, p) \in A_{n+1}$,

$$p < z(t) \text{ for some } t \in \mathbb{R}, \tag{6.143}$$

where $I_n \times (U_n \cap C(y^*, z^*, \sigma^*, p^*))$ is a delay-period disparity set at $(t, \bar{v}) \in I_n \times \partial A_n$ and $A_{\bar{v}}$ is the connected component of U_n.

If the construction in (6.142) stops at some $n_0 \in \mathbb{N}$ with $\partial A_{n_0} = \emptyset$, then $A_{n_0} = C(y^*, z^*, \sigma^*, p^*)$, and we are done. If not, then $n_0 = +\infty$, and we obtain a sequence of sets $\{A_n\}_{n=1}^{+\infty}$ that is a totally ordered family of sets with respect to the set inclusion relation $\subseteq$. Note that $\cup_{n=1}^{+\infty} A_n$ is the upper bound of $\{A_n\}_{n=1}^{+\infty}$. Then by Zorn's lemma, there exists a maximal element A_∞ for the sequence $\{A_n\}_{n=1}^{+\infty}$.

Now we show that $\partial A_\infty = \emptyset$, in the sense of the relative topology induced by $C(y^*, z^*, \sigma^*, p^*)$. Suppose not. Then there exist $\bar{v} \in \partial A_\infty$ and $A_{\bar{v}}$, which is the connected component of U_∞, where $I_\infty \times (U_\infty \times C(y^*, z^*, \sigma^*, p^*))$ is a delay-period disparity set at $(t, \bar{v}) \in I_\infty \times \partial A_\infty$. We distinguish two cases:

Case 1. $A_{\bar{v}} \setminus A_\infty = \emptyset$ for all $\bar{v} \in \partial A_\infty$. Then A_∞ is a connected component of $C(y^*, z^*, \sigma^*, p^*)$. Recall that $C(y^*, z^*, \sigma^*, p^*)$ itself is a connected component of the closure of all the nonconstant periodic solutions of system (6.100). So we have $A_\infty = C(y^*, z^*, \sigma^*, p^*)$. That is, $\partial A_\infty = \emptyset$. This is a contradiction.

Case 2. $A_{\bar{v}} \setminus A_\infty \neq \emptyset$. But this means that $A_\infty \subset A_\infty \cup A_{\bar{v}}$, which contradicts the maximality of A_∞.

These contradictions show that $\partial A_\infty = \emptyset$, and hence $A_\infty = C(y^*, z^*, \sigma^*, p^*)$. Therefore, (6.143) holds for all $(y, z, \sigma, p) \in C(y^*, z^*, \sigma^*, p^*)$. This completes the proof. $\qquad\qquad\square$

Theorem 6.22. *Let $C(y^*, z^*, \sigma^*, p^*)$ be a connected component of the closure of all the nonconstant periodic solutions of system (6.100), bifurcated at $(y^*, z^*, \sigma^*, p^*)$ in the Fuller space $C(\mathbb{R}/2\pi; \mathbb{R}^{N+1}) \times \mathbb{R}^2$. Suppose that (6.74) satisfies (S5)–(S9). If there exists a continuous function $M_1 : \mathbb{R} \ni \sigma \to M_1(\sigma) > 0$ such that for every $(y, z, \sigma, p) \in C(y^*, z^*, \sigma^*, p^*)$, we have*

$$\|(y, z)\|_{C(\mathbb{R}; \mathbb{R}^{N+1})} \leq M_1(\sigma), \tag{6.144}$$

then $p^* < z^*$ implies that $p < M_1(\sigma)$ for every $(y, z, \sigma, p) \in C(y^*, z^*, \sigma^*, p^*)$.

Proof. By Theorem 6.21, we have, for every $(y, z, \sigma, p) \in C(y^*, z^*, \sigma^*, p^*)$, that $p < z(t)$ for some $t \in \mathbb{R}$. Then by (6.144), we have $p < M_1(\sigma)$. $\qquad\square$

6.9.4 Uniform Boundedness of Periodic Solutions

We refer to [254] for the concepts of balanced, convex, and absorbing subsets and the Minkowski functional.

Lemma 6.29. *Let G be a convex absorbing subset of a locally convex linear topological space X that defines a Minkowski functional $p_G : X \to \mathbb{R}$ with $p_G(x) = \inf\{\alpha > 0 : \alpha^{-1}x \overset{\text{def}}{=} x/\alpha \in G\}$. For each $\gamma > 0$, define*

$$G^\gamma = \{x : p_G(x) < \gamma\}. \tag{6.145}$$

Then $x \in \partial G^\gamma$ if and only if $p_G(x) = \gamma$.

Proof. It is clear that $G^\gamma = \gamma G$. By linearity, the Minkowski functional $p_{G^\gamma} : X \to \mathbb{R}$ determined by G^γ is well defined. By (6.145) and by the definition of Minkowski functional, we have

$$
\begin{aligned}
x \in \partial G^\gamma &\Longleftrightarrow p_{G^\gamma}(x) = 1 \\
&\Longleftrightarrow \inf\{\alpha > 0 : x/\alpha \in G^\gamma\} = 1 \\
&\Longleftrightarrow \inf\{\alpha > 0 : p_G(x/\alpha) < \gamma\} = 1 \\
&\Longleftrightarrow \inf\{\alpha > 0 : p_G(x)/\gamma < \alpha\} = 1 \\
&\Longleftrightarrow p_G(x) = \gamma.
\end{aligned}
$$

$\qquad\square$

Lemma 6.30. *Let G_1 and G_2 be convex absorbing subsets of locally convex linear topological spaces X_1 and X_2, respectively. Let the Minkowski functionals associated with G_1 and G_2 be $p_{G_1}(x)$ and $p_{G_2}(\tau)$, respectively. Then the Minkowski functional defined by $G = G_1 \times G_2$ exists and satisfies*

$$p_G(x, \tau) = \max\{p_{G_1}(x), p_{G_2}(\tau)\}.$$

Proof. The existence of $p_G(x, \tau)$ is clear from the definition of a Minkowski functional. Let $A = \{\alpha : x/\alpha \in G_1\}$, $B = \{\alpha : \tau/\alpha \in G_2\}$. Then it is clear that $\inf A \cap B \geq \inf A$ and $\inf A \cap B \geq \inf B$. It follows that $\inf A \cap B \geq \max\{\inf A, \inf B\}$, that is,

$$p_G(x, \tau) \geq \max\{p_{G_1}(x), p_{G_2}(\tau)\}. \tag{6.146}$$

On the other hand, if $\alpha_A = \inf A \geq \alpha_B = \inf B$, since G_1 and G_2 are absorbing, we have for every $\varepsilon > 0$ that $\alpha_A + \varepsilon \in A$, $\alpha_A + \varepsilon \in B$. Therefore, $\inf A \cap B \leq \alpha_A + \varepsilon$. Similarly, if $\alpha_A = \inf A \leq \alpha_B = \inf B$, we have $\inf A \cap B \leq \alpha_B + \varepsilon$. Hence we obtain $\inf A \cap B \leq \max\{\alpha_A, \alpha_B\} + \varepsilon$. By the arbitrariness of $\varepsilon > 0$, we get $\inf A \cap B \leq \max\{\alpha_A, \alpha_B\}$, that is,

$$p_G(x, \tau) \leq \max\{p_{G_1}(x), p_{G_2}(\tau)\}. \tag{6.147}$$

By (6.146) and (6.147), we have

$$p_G(x, \tau) = \max\{p_{G_1}(x), p_{G_2}(\tau)\}.$$

This completes the proof. $\square$

An immediate corollary of Lemmas 6.29 and 6.30 is the following.

Corollary 6.1. *Let G_1 and G_2 be convex absorbing subsets of locally convex linear topological spaces X_1 and X_2, respectively. Let $p_{G_1}(x)$ and $p_{G_2}(\tau)$ be the Minkowski functionals associated with G_1 and G_2, respectively. Let $G = G_1 \times G_2$, and for every $\gamma > 0$, define*

$$G^\gamma = \{(x, \tau) : p_G(x, \tau) < \gamma\},$$
$$G_1^\gamma = \{x : p_{G_1}(x) < \gamma\},$$
$$G_2^\gamma = \{\tau : p_{G_2}(\tau) < \gamma\}.$$

Then $G^\gamma = G_1^\gamma \times G_2^\gamma$ and $\bar{G}^\gamma = \bar{G}_1^\gamma \times \bar{G}_2^\gamma$.

In this section, we use "$\cdot$" to denote the usual inner product of a Euclidean space, and we use G^c and D^c to denote the complementary sets of G and D, respectively.

We can now state and prove the geometric conditions for uniform boundedness of the periodic solutions of (6.74) with $\sigma \in \Sigma$, where $\Sigma \subseteq \mathbb{R}$ is a given subset.

Theorem 6.23. *Suppose that $G_1 \subset \mathbb{R}^N$ and $G_2 \subset \mathbb{R}$ are bounded, balanced, convex, and absorbing open subsets with associated Minkowski functionals $p_{G_1}(x)$ and $p_{G_2}(\tau)$. Let $G = G_1 \times G_2$ and $(\bar{x}, \bar{\tau}) = \frac{1}{p_G(x, \tau)}(x, \tau) \in \partial G$ for $(x, \tau) \neq 0$. Assume that there exists a vector-valued function $N : \partial G \setminus (\partial G_1 \times \partial G_2) \to \mathbb{R}^{N+1} \setminus \{0\}$ satisfying*

$(i) : \bar{G} \subseteq U_1 \cup U_2$, where

$$U_1 = \bigcap_{(x, \tau) \in \partial G \setminus (\partial G_1 \times \partial G_2)} \{(u, v) : N(x, \tau) \cdot (u - x, v - \tau) \leq 0\};$$

$$U_2 = \bigcap_{(x,\tau)\in\partial G_1\times\partial G_2} \{(u,v) : x\cdot(u-x)\leq 0,\ \tau\cdot(v-\tau)\leq 0\};$$

(ii) : $N(\bar{x},\bar{\tau})\cdot(f(x,\tilde{x},\sigma),g(x,\tau,\sigma))$ *is positive (or negative) for all* $(x,\tau)\in G^c$
with $(\bar{x},\bar{\tau})\notin\partial G_1\times\partial G_2$, *and all* $(\tilde{x},\tau)\in\mathbb{R}^N\times\mathbb{R}$ *with* $p_G(\tilde{x},\tau)\leq p_G(x,\tau)$
and $\sigma\in\Sigma$;

(iii) : $x\cdot f(x,\tilde{x},\sigma)$ *and* $\tau\cdot g(x,\tau,\sigma)$ *are both positive (or negative) for all* $(x,\tau)\in$
G^c *with* $(\bar{x},\bar{\tau})\in\partial G_1\times\partial G_2$, *and all* $(\tilde{x},\tau)\in\mathbb{R}^N\times\mathbb{R}$ *with* $p_G(\tilde{x},\tau)\leq p_G(x,\tau)$
and $\sigma\in\Sigma$.

Then the range of all the periodic solutions of (6.74) with $\sigma\in\Sigma$ *is contained in G.*

Remark 6.3. The prototype of the vector-valued function $N(x,\tau)$ is the (outer or inner) normal of G, which is not defined on $\partial G_1\times\partial G_2$. If G is a rectangle in a planar space, $\partial G_1\times\partial G_2$ are the four corner points of G. Conditions (ii)–(iii) of Theorem 6.23 require that the vector field determined by the right-hand side of system (6.74) have positive (or negative) inner product with respect to the normal of a given rectangle G, where the vector field is evaluated at $(x,\tau)\in\mathbb{R}^{N+1}$, which satisfies $(x,\tau)\in G^c$ and $p_G(\tilde{x},\tau)\leq p_G(x,\tau)$.

Proof. Letting $(x,\tau)(t)=(y,z)(\beta t)$ with a normalization parameter $\beta>0$, we only need to consider the 2π-periodic solutions of the following system:

$$\begin{cases} \dot{y}(t) = \frac{1}{\beta}f(y(t),y(t-\beta z(t)),\sigma), \\ \dot{z}(t) = \frac{1}{\beta}g(y(t),z(t),\sigma), \end{cases} \tag{6.148}$$

where $x\in\mathbb{R}^N$ and $\tau\in\mathbb{R}$. It is clear that if $(x(t),\tau(t))$ and $(y(t),z(t))$ are solutions of (6.74) and (6.148), respectively, then $(x(t),\tau(t))\in G$ for all $t\in\mathbb{R}$ if and only if $(y(t),z(t))\in G$ for all $t\in\mathbb{R}$.

For simplicity, we denote $y(t-\beta z(t))$ by $\tilde{y}(t)$ for each solution $(y(t),z(t))$ of (6.148). Let $(\bar{y},\bar{z})$ be the positive constant multiple of (y,z) such that $(\bar{y},\bar{z})\in\partial G$. That is, for every $(y,z)\in\mathbb{R}^{N+1}\setminus\{0\}$, there exists $(\bar{y},\bar{z})\in\partial G$ such that $(y,z)=p_G(y,z)(\bar{y},\bar{z})$.

Suppose there exists a 2π-periodic solution of (6.148) such that $(y(t_0),z(t_0))\notin G$ for some $t_0\in[0,2\pi]$ and define the map $\gamma:\mathbb{R}\ni t\rightarrow p_G(y(t),z(t))\in\mathbb{R}$. Since $\mathbb{R}^{N+1}\ni(y,z)\mapsto p_G(y,z)\in\mathbb{R}$ and $\mathbb{R}\ni t\mapsto(y(t),z(t))\in\mathbb{R}^{N+1}$ are continuous, the map $\gamma:t\rightarrow p_G(y(t),z(t))$ is continuous and there exist $\gamma^*\geq 1$ and $t^*\in[0,2\pi]$ such that

$$\gamma^* = p_G(y(t^*),z(t^*)) = \max_{t\in[0,2\pi]} p_G(y(t),z(t)). \tag{6.149}$$

Then by Lemma 6.29 and (6.149), we have $(y(t^*),z(t^*))\in\partial G^{\gamma^*}$ and $G^{\gamma(t)}\subseteq G^{\gamma^*}$ for all $t\in\mathbb{R}$. Therefore, by Corollary 6.1, $(y(t),z(t))\in\bar{G}^{\gamma^*}=\bar{G}_1^{\gamma^*}\times\bar{G}_2^{\gamma^*}$ for all $t \in [0, 2\pi]$. In particular, by the periodicity of $(y(t), z(t))$, we obtain $(y(t - \beta z(t)), z(t)) \in \bar{G}^{\gamma^*}$ for all $t \in [0, 2\pi]$ and $\beta > 0$. Therefore, we have

$$p_G(y(t^* - \beta z(t^*)), z(t^*)) \leq p_G(y(t^*), z(t^*)). \qquad (6.150)$$

We first suppose that $(\bar{y}(t^*), \bar{z}(t^*)) = \frac{1}{p_G(y(t^*), z(t^*))}(y(t^*), z(t^*)) \in U_1$. Then by (6.149), (6.150), and assumption (ii), we have (we use the positivity assumption in the proof; the proof is similar if we use the negativity assumption; see Remark 6.4 for details)

$$N(\bar{y}(t^*), \bar{z}(t^*)) \cdot \left[\frac{1}{\beta} f(y(t^*), y(t^* - \beta z(t^*)), \sigma), \frac{1}{\beta} g(y(t^*), z(t^*), \sigma) \right] > 0. \quad (6.151)$$

Let us write

$$\begin{bmatrix} y(t^* + h) \\ z(t^* + h) \end{bmatrix} = \begin{bmatrix} y(t^*) \\ z(t^*) \end{bmatrix} + \begin{bmatrix} \int_0^1 \dot{y}(t^* + sh)ds\, h \\ \int_0^1 \dot{z}(t^* + sh)ds\, h \end{bmatrix}, \qquad (6.152)$$

and choose $h > 0$ small enough that

$$N(\bar{y}(t^*), \bar{z}(t^*)) \cdot \left[\frac{1}{\beta} f(y(t), y(t - \beta z(t)), \sigma), \frac{1}{\beta} g(y(t), z(t), \sigma) \right] > 0 \qquad (6.153)$$

for $t^* \leq t < t^* + h$. Then by (6.148), (6.152), and (6.153), we have

$$N(\bar{y}(t^*), \bar{z}(t^*)) \cdot (y(t^* + h) - y(t^*), z(t^* + h) - z(t^*)) > 0. \qquad (6.154)$$

Now we distinguish the following two cases in order to deduce contradictions:

Case 1. If $(y(t^* + h), z(t^* + h)) \in \bar{G}$, then $\gamma^{*-1}(y(t^* + h), z(t^* + h)) \in \bar{G}$, since $\gamma^* \geq 1$. Also, we have $(y(t^*), z(t^*)) = (\gamma^* \bar{y}(t^*), \gamma^* \bar{z}(t^*))$ with $(\bar{y}(t^*), \bar{z}(t^*)) \in \partial G$. Then by assumption (i), we have

$$N(\bar{y}(t^*), \bar{z}(t^*)) \cdot \left(\gamma^{*-1} y(t^* + h) - \bar{y}(t^*), \gamma^{*-1} z(t^* + h) - \bar{z}(t^*) \right) \leq 0. \qquad (6.155)$$

On the other hand, we have by (6.154),

$$\begin{aligned} 0 &< N(\bar{y}(t^*), \bar{z}(t^*)) \cdot (y(t^* + h) - y(t^*), z(t^* + h) - z(t^*)) \\ &= \gamma^* N(\bar{y}(t^*), \bar{z}(t^*)) \cdot \left(\gamma^{*-1} y(t^* + h) - \bar{y}(t^*), \gamma^{*-1} z(t^* + h) - \bar{z}(t^*) \right), \end{aligned} \qquad (6.156)$$

which contradicts (6.155).

Case 2. If $(y(t^* + h), z(t^* + h)) \notin \bar{G}$, then by (6.149), we have

$$1 \leq \gamma_h = p_G(y(t^* + h), z(t^* + h)) \leq p_G(y(t^*), z(t^*)) = \gamma^*. \qquad (6.157)$$

Also, we have $(y(t^* +h), z(t^* +h)) = \gamma_h(\bar{y}(t^* +h), \bar{z}(t^* +h))$ with $(\bar{y}(t^* +h), \bar{z}(t^* + h)) \in \partial G$. By the convexity of $\bar{G}$ and by the inequality $\gamma_h/\gamma^* \leq 1$, we have

$$\left(\frac{\gamma_h}{\gamma^*}\bar{y}(t^* +h), \frac{\gamma_h}{\gamma^*}\bar{z}(t^* +h)\right) \in \bar{G}.$$

Then by assumption (i), we have

$$N(\bar{y}(t^*), \bar{z}(t^*)) \cdot \left(\frac{\gamma_h}{\gamma^*}\bar{y}(t^* +h) - \bar{y}(t^*), \frac{\gamma_h}{\gamma^*}\bar{z}(t^* +h) - \bar{z}(t^*)\right) \leq 0. \qquad (6.158)$$

On the other hand, we have by (6.154),

$$0 < N(\bar{y}(t^*), \bar{z}(t^*)) \cdot (y(t^* +h) - y(t^*), z(t^* +h) - z(t^*))$$
$$= \gamma^* N(\bar{y}(t^*), \bar{z}(t^*)) \cdot \left(\frac{\gamma_h}{\gamma^*}\bar{y}(t^* +h) - \bar{y}(t^*), \frac{\gamma_h}{\gamma^*}\bar{z}(t^* +h) - \bar{z}(t^*)\right), \qquad (6.159)$$

which contradicts (6.158).

Second, we suppose that $(\bar{y}(t^*), \bar{z}(t^*)) = \frac{1}{p_G(y(t^*), z(t^*))}(y(t^*), z(t^*)) \in U_2$. By assumption (iii), we have

$$\begin{cases} \bar{y}(t^*) \cdot \frac{1}{\beta}f(y(t^*), y(t^* - \beta z(t^*)), \sigma) > 0, \\ \bar{z}(t^*) \cdot \frac{1}{\beta}g(y(t^*), z(t^*), \sigma) > 0. \end{cases} \qquad (6.160)$$

Therefore, we can choose $h > 0$ small enough that for $t^* \leq t < t^* +h$,

$$\begin{cases} \bar{y}(t^*) \cdot \frac{1}{\beta}f(y(t), y(t - \beta z(t)), \sigma) > 0, \\ \bar{z}(t^*) \cdot \frac{1}{\beta}g(y(t), z(t), \sigma) > 0. \end{cases} \qquad (6.161)$$

Then by (6.148), (6.152), and (6.161), we have

$$\begin{cases} \bar{y}(t^*) \cdot (y(t^* +h) - y(t^*)) > 0, \\ \bar{z}(t^*) \cdot (z(t^* +h) - z(t^*)) > 0. \end{cases} \qquad (6.162)$$

We distinguish the following two cases in order to deduce contradictions:

Case 1′. If $(y(t^* +h), z(t^* +h)) \in \bar{G}$, then $\gamma^{*-1}(y(t^* +h), z(t^* +h)) \in \bar{G}$, since $\gamma^* \geq 1$. Also, we have $(y(t^*), z(t^*)) = (\gamma^*\bar{y}(t^*), \gamma^*\bar{z}(t^*))$ with $(\bar{y}(t^*), \bar{z}(t^*)) \in \partial G$. Then by assumption (i), we have

$$\begin{cases} \bar{y}(t^*) \cdot (\gamma^{*-1}y(t^* +h) - \bar{y}(t^*)) \leq 0, \\ \bar{z}(t^*) \cdot (\gamma^{*-1}z(t^* +h) - \bar{z}(t^*)) \leq 0. \end{cases} \qquad (6.163)$$

On the other hand, we have by (6.162),

$$\begin{cases} \bar{y}(t^*) \cdot (y(t^*+h) - y(t^*)) = \gamma^* \bar{y}(t^*) \cdot (\gamma^{*-1} y(t^*+h) - \bar{y}(t^*)) > 0, \\ \bar{z}(t^*) \cdot (z(t^*+h) - z(t^*)) = \gamma^* \bar{z}(t^*) \cdot (\gamma^{*-1} z(t^*+h) - \bar{z}(t^*)) > 0, \end{cases} \tag{6.164}$$

which contradicts (6.163).

Case 2'. If $(y(t^*+h), z(t^*+h)) \notin \bar{G}$, then by (6.157) and the convexity of $\bar{G}$, we have

$$\left(\frac{\gamma_h}{\gamma^*} \bar{y}(t^*+h), \frac{\gamma_h}{\gamma^*} \bar{z}(t^*+h) \right) \in \bar{G},$$

where $\gamma_h = p_G(y(t^*+h), z(t^*+h))$. Then by assumption (i), we have

$$\begin{cases} \bar{y}(t^*) \cdot \left(\frac{\gamma_h}{\gamma^*} \bar{y}(t^*+h) - \bar{y}(t^*) \right) \le 0, \\ \bar{z}(t^*) \cdot \left(\frac{\gamma_h}{\gamma^*} \bar{z}(t^*+h) - \bar{z}(t^*) \right) \le 0. \end{cases} \tag{6.165}$$

On the other hand, we have by (6.162),

$$\begin{cases} \bar{y}(t^*) \cdot (y(t^*+h) - y(t^*)) = \gamma^* \bar{y}(t^*) \cdot \left(\frac{\gamma_h}{\gamma^*} \bar{y}(t^*+h) - \bar{y}(t^*) \right) > 0, \\ \bar{z}(t^*) \cdot (z(t^*+h) - z(t^*)) = \gamma^* \bar{z}(t^*) \cdot \left(\frac{\gamma_h}{\gamma^*} \bar{z}(t^*+h) - \bar{z}(t^*) \right) > 0, \end{cases} \tag{6.166}$$

which contradicts (6.165). Therefore, contradictions are obtained in all cases, and the proof is complete. □

Remark 6.4. If we use < 0 instead of > 0 in the inequality (6.151), we need to change (6.152) to be the difference between $(y(t^*), z(t^*))$ and $(y(t^*-h), z(t^*-h))$. That is,

$$\begin{bmatrix} y(t^*) \\ z(t^*) \end{bmatrix} = \begin{bmatrix} y(t^*-h) \\ z(t^*-h) \end{bmatrix} + \begin{bmatrix} \int_0^1 \dot{y}(t^*-sh)ds\,h \\ \int_0^1 \dot{z}(t^*-sh)ds\,h \end{bmatrix}.$$

Then the rest of the proof is similar.

Corollary 6.2. *Suppose that $G_1 \subset \mathbb{R}^N$ and $G_2 \subset \mathbb{R}$ are bounded, balanced, convex, and absorbing open subsets that define the Minkowski functionals $p_{G_1}(x)$ and $p_{G_2}(\tau)$. Suppose $N : \partial G \setminus (\partial G_1 \times \partial G_2) \to \mathbb{R}^{N+1} \setminus \{0\}$ is the outer normal of G. Fix $\sigma \in \Sigma$ and let $G = G_1 \times G_2$ and*

$$F_{\max}(x, \sigma) = \max_{\{\tilde{x}: p_{G_1}(\tilde{x}) \le p_{G_1}(x)\}} x \cdot f(x, \tilde{x}, \sigma),$$

$$F_{\min}(x, \sigma) = \min_{\{\tilde{x}: p_{G_1}(\tilde{x}) \le p_{G_1}(x)\}} x \cdot f(x, \tilde{x}, \sigma).$$

Then the range of all the periodic solutions of (6.74) are contained in G if either of the following conditions holds:

(H1) $F_{\max}(x, \sigma) < 0$ *for every* $x \in G_1^c$ *and* $\tau \cdot g(x, \tau) < 0$ *for every* $\tau \in G_2^c, x \in \mathbb{R}^N$.
(H2) $F_{\min}(x, \sigma) > 0$ *for every* $x \in G_1^c$ *and* $\tau \cdot g(x, \tau) > 0$ *for every* $\tau \in G_2^c, x \in \mathbb{R}^N$.

Proof. We prove the conclusions by applying Theorem 6.23. By Corollary 6.1, there exist Minkowski functionals $p_G(x, \tau)$, $p_{G_1}(x)$, and $p_{G_2}(\tau)$ defined on $\mathbb{R}^N \times \mathbb{R}$, $\mathbb{R}^N$, and $\mathbb{R}$, respectively. For every $(x, \tau) \in G^c$, let $(\bar{x}, \bar{\tau}) = (x, \tau)/p_G(x, \tau) \in \partial G$. Recall that $N : \partial G \setminus (\partial G_1 \times \partial G_2) \to \mathbb{R}^{N+1} \setminus \{0\}$ is the outer normal of the convex set G. Then condition (i) of Theorem 6.23 is satisfied.

Suppose (H1) holds. Then we have

$$x \cdot f(x, \bar{x}, \sigma) < 0, \text{for all } (x, \bar{x}) \in G_1^c \times \mathbb{R}^N \text{ with } p_{G_1}(\bar{x}) \leq p_{G_1}(x), \qquad (6.167)$$

$$\tau \cdot g(x, \tau, \sigma) < 0, \text{ for all } \tau \in G_2^c, x \in \mathbb{R}^N. \qquad (6.168)$$

For every $(x, \tau) \in G^c$ with $p_G(\bar{x}, \tau) \leq p_G(x, \tau)$, let $(\bar{x}, \bar{\tau}) = (x, \tau)/p_G(x, \tau) \in \partial G$. Note that $\partial G = (G_1 \times \partial G_2) \cup (\partial G_1 \times G_2) \cup (\partial G_1 \times \partial G_2)$. We distinguish the following three cases:

Case 1: If $(\bar{x}, \bar{\tau}) \in G_1 \times \partial G_2$, then $N(\bar{x}, \bar{\tau}) = (0, \tau)/p_G(x, \tau) \neq 0$ is an outer normal of G. We claim that $\tau \in G_2^c$ holds.

Indeed, since $\bar{x} \in G_1$, we have $p_{G_1}(\bar{x}) = p_{G_1}(x/p_G(x, \tau)) < 1$. Therefore, $p_{G_1}(x) < p_G(x, \tau)$. By Lemma 6.30, we know that $p_G(x, \tau) = \max\{p_{G_1}(x), p_{G_2}(\tau)\}$. Then we have $p_{G_1}(x) < p_{G_2}(\tau)$ and $p_G(x, \tau) = p_{G_2}(\tau) > 1$. Then by Lemma 6.29, we have $\tau \in G_2^c$.

Then by (6.168), we have

$$N(\bar{x}, \bar{\tau}) \cdot (f(x, \bar{x}, \sigma), g(x, \tau, \sigma)) = \tau \cdot g(x, \tau, \sigma)/p_G(x, \tau) < 0.$$

Case 2: If $(\bar{x}, \bar{\tau}) \in \partial G_1 \times G_2$, then $N(\bar{x}, \bar{\tau}) = (x, 0)/p_G(x, \tau) \neq 0$ is an outer normal of G. We claim that $x \in G_1^c$ and $p_{G_1}(\bar{x}) \leq p_{G_1}(x)$.

Indeed, since $\bar{\tau} \in G_2$, we have $p_{G_2}(\bar{\tau}) = p_{G_2}(\tau/p_G(x, \tau)) < 1$. Therefore, $p_{G_2}(\tau) < p_G(x, \tau)$. By Lemma 6.30, we know that $p_G(x, \tau) = \max\{p_{G_1}(x), p_{G_2}(\tau)\}$. Then we have $p_{G_2}(\tau) < p_{G_1}(x)$ and $p_G(x, \tau) = p_{G_1}(x) > 1$. Then by Lemma 6.29, we have $x \in G_1^c$. Moreover, it follows again by Lemma 6.30 that $p_G(\bar{x}, \tau) \leq p_G(x, \tau)$ implies $p_{G_1}(\bar{x}) \leq p_{G_1}(x)$. This proves the claim.

By (6.167), we have

$$N(\bar{x}, \bar{\tau}) \cdot (f(x, \bar{x}, \sigma), g(x, \tau, \sigma)) = x \cdot f(x, \bar{x}, \sigma)/p_G(x, \tau) < 0.$$

From Case 1 and Case 2, we know that $N(\bar{x}, \bar{\tau}) \cdot (f(x, \bar{x}, \sigma), g(x, \tau, \sigma))$ is negative definite for all $(x, \tau) \in G^c$ and $\sigma \in \Sigma$ with $(\bar{x}, \bar{\tau}) \notin \partial G_1 \times \partial G_2$, and all $(\bar{x}, \tau) \in \mathbb{R}^N \times \mathbb{R}$ with $p_G(\bar{x}, \tau) \leq p_G(x, \tau)$. That is, condition (ii) of Theorem 6.23 is satisfied.

Case 3: If $(\bar{x}, \bar{\tau}) \in \partial G_1 \times \partial G_2$, we claim that $(x, \tau) \in G_1^c \times G_2^c$ and $p_{G_1}(\bar{x}) = p_{G_1}(x)$ hold.

Indeed, since $(\bar{x}, \bar{\tau}) \in \partial G_1 \times \partial G_2$, we have $p_{G_1}(\bar{x}) = p_{G_1}(x/p_G(x, \tau)) = 1$ and $p_{G_2}(\bar{\tau}) = p_{G_2}(\tau/p_G(x, \tau)) = 1$. Therefore, $p_G(x, \tau) = p_{G_1}(x) = p_{G_2}(\tau)$. Since $(x, \tau) \in G^c$, we have $p_{G_1}(x) = p_{G_2}(\tau) = p_G(x, \tau) > 1$. Then by Lemma 6.29, we have $(x, \tau) \in G_1^c \times G_2^c$. Moreover, it follows again by Lemma 6.30 that $p_G(\bar{x}, \bar{\tau}) \le p_G(x, \tau)$ implies $p_{G_1}(\bar{x}) \le p_{G_1}(x)$. This proves the claim.

Then by (6.167) and (6.168), we have

$$x \cdot f(x, \bar{x}, \sigma) < 0 \text{ and } \tau \cdot g(x, \tau, \sigma) < 0.$$

From Case 3, we know that $x \cdot f(x, \bar{x}, \sigma)$ and $\tau \cdot g(x, \tau, \sigma)$ are both negative definite for all $(x, \tau) \in G^c$ and $\sigma \in \Sigma$ with $(\bar{x}, \bar{\tau}) \in \partial G_1 \times \partial G_2$, and all $(\bar{x}, \bar{\tau}) \in \mathbb{R}^N \times \mathbb{R}$ with $p_G(\bar{x}, \bar{\tau}) \le p_G(x, \tau)$. That is, condition (iii) of Theorem 6.23 is satisfied.

It follows from Theorem 6.23 that the range of all the periodic solutions of (6.74) with $\sigma \in \Sigma$ is contained in G. Similarly, if (H2) holds, we can obtain from Theorem 6.23 the same conclusion. This completes the proof. $\square$

6.9.5 Global Continuation of Rapidly Oscillating Periodic Solutions: An Example

In this section, we illustrate the general results in the previous subsections by applying them to the study of the global continua of rapidly oscillating periodic solutions for the following differential equations with state-dependent delay:

$$\begin{cases} \dot{x}(t) = -\mu x(t) + \sigma^2 b(x(t - \tau(t))), \\ \dot{\tau}(t) = 1 - h(x(t)) \cdot (1 + \tanh \tau(t)), \end{cases} \qquad (6.169)$$

where $\tanh(\tau) = (e^{2\tau} - 1)/(e^{2\tau} + 1)$ and $\mu > 0$ is a constant. We make the following assumptions:

(α_1) $b, h : \mathbb{R} \to \mathbb{R}$ are C^2 functions with $b'(0) = -1$;
(α_2)   There exist $h_0 < h_1$ in $(1/2, 1)$ such that $h_1 > h(x) > h_0$ for all $x \in \mathbb{R}$;
(α_3) b is decreasing on $\mathbb{R}$;
(α_4) $xb(x) < 0$ for $x \ne 0$, and there exists a continuous function $M : \mathbb{R} \ni \sigma \to M(\sigma) \in (0, +\infty)$ such that

$$\frac{b(x)}{x} > -\frac{\mu}{\sigma^2}$$

for every $x \in \mathbb{R}$ with $|x| \ge M(\sigma)$;
(α_5)   There exists $M_0 > 0$ such that $|b'(x)| < M_0$ for every $x \in \mathbb{R}$;
(α_6) $h'(x) = 0$ only if x satisfies $-\mu x + \sigma^2 b(x) = 0$.

Remark 6.5. We use $\tanh(\tau)$ just for the sake of simplicity. Other types of functions can be used with minor changes in our arguments below.

We start with the uniform boundedness of periodic solutions $(x(t), \tau(t))$ of (6.169).

Lemma 6.31. *Assume that $(\alpha_1)-(\alpha_4)$ hold. Then the range of every periodic solution (x, τ) of (6.169) with $\sigma \in \mathbb{R}$ is contained in*

$$\Omega_1 = (-M(\sigma), M(\sigma)) \times \left(0, -\frac{\ln(2h_0 - 1)}{2}\right).$$

Proof. If $\sigma = 0$, the only periodic solution is $\left(0, -\frac{\ln(2h(0)-1)}{2}\right)$, which is contained in Ω_1. Now we assume that $\sigma \neq 0$. If $x > 0$, then by assumptions (α_3) and (α_4), we have

$$\max_{y \in \{y: |y| \leq |x|\}} x \cdot (-\mu x + \sigma^2 b(y)) = -\sigma^2 x^2 \left(\frac{\mu}{\sigma^2} - \frac{b(-x)}{x}\right) < 0$$

for every $x \in \mathbb{R}$ with $x \geq M(\sigma)$. It follows that

$$\max_{y \in \{y: |y| \leq |x|\}} x \cdot (-\mu x + \sigma^2 b(y)) < 0 \text{ for } x \geq M(\sigma).$$

Similarly, we have

$$\max_{y \in \{y: |y| \leq |x|\}} x \cdot (-\mu x + \sigma^2 b(y)) < 0 \text{ for } x \leq -M(\sigma).$$

Thus,

$$\max_{y \in \{y: |y| \leq |x|\}} x \cdot (-\mu x + \sigma^2 b(y)) < 0 \text{ if } x \notin (-M(\sigma), M(\sigma)). \tag{6.170}$$

It is clear from (α_2) that for all $x \in \mathbb{R}$,

$$\lim_{\tau \to \pm \infty} \tau \cdot (1 - h(x)(1 + \tanh \tau)) < 0.$$

To obtain an upper bound for τ, where (x, τ) is a periodic solution of (6.169), we introduce the following change of variable:

$$z(t) = \tau(t) + \frac{\ln(2h_0 - 1)}{4}. \tag{6.171}$$

Then system (6.169) is transformed to

$$\begin{cases} \dot{x}(t) = -\mu x(t) + \sigma^2 b\left(x\left(t - z(t) + \frac{\ln(2h_0 - 1)}{4}\right)\right), \\ \dot{z}(t) = 1 - h(x(t))\left(1 + \tanh\left(z(t) - \frac{1}{4}\ln(2h_0 - 1)\right)\right). \end{cases} \tag{6.172}$$

By (α_2) and the monotonicity of $\tanh \tau$, we have, for every $z \notin \left(\frac{\ln(2h_0 - 1)}{4}, -\frac{\ln(2h_0 - 1)}{4}\right)$ and for all $x \in \mathbb{R}$,

$$z \cdot \left(1 - h(x)\left(1 + \tanh\left(z - \frac{1}{4}\ln(2h_0 - 1)\right)\right)\right) < 0. \tag{6.173}$$

Thus it follows from Corollary 6.2, (6.170), and (6.173) that the range of all the periodic solutions (x, z) of (6.172) is contained in $(-M(\sigma), M(\sigma)) \times \left(\frac{\ln(2h_0 - 1)}{4}, -\frac{\ln(2h_0 - 1)}{4} \right)$. Then by (6.171), all periodic solutions (x, τ) of (6.169) with $\sigma \neq 0$ are contained in Ω_1. The proof is complete.   □

Now we consider the global Hopf bifurcation problem of system (6.169) under the assumptions (α_1)–(α_6). By (α_4), $(x, \tau) = (0, \tau^*)$ is the only stationary solution of (6.169), where $\tau^* = -\frac{1}{2}\ln(2h(0) - 1) > 0$. Freezing the state-dependent delay $\tau(t)$ at τ^* for the term $x(t - \tau(t))$ of (6.169) and linearizing the resulting system with constant delay at the stationary solution $(0, \tau^*)$, we obtain the following formal linearization of system (6.169):

$$\begin{cases} \dot{X}(t) = -\mu X(t) - \sigma^2 X(t - \tau^*), \\ \dot{T}(t) = -\rho X(t) - qT(t), \end{cases} \tag{6.174}$$

where

$$\rho = \frac{h'(0)}{h(0)}, q = 2 - \frac{1}{h(0)} > 0. \tag{6.175}$$

In the following, we regard σ as the bifurcation parameter. We obtain the characteristic equation of the linear system corresponding to (6.174):

$$(\lambda + \mu + \sigma^2 e^{-\tau^* \lambda})(\lambda + q) = 0. \tag{6.176}$$

Since the zero of $\lambda + q = 0$ is $-q$, which is real, Hopf bifurcation points are related to zeros of only the first factor $(\lambda + \mu + \sigma^2 e^{-\tau^* \lambda})$. To locate local Hopf bifurcation points, we let $\lambda = i\beta$, $\beta > 0$, in $\lambda + \mu + \sigma^2 e^{-\tau^* \lambda} = 0$ and express the resulting equation in terms of its real and imaginary parts as

$$\begin{cases} \beta = \sigma^2 \sin(\tau^* \beta), \\ \mu = -\sigma^2 \cos(\tau^* \beta). \end{cases} \tag{6.177}$$

It is easy to verify the following lemma.

Lemma 6.32. *(i) All the positive solutions of (6.177) can be represented by an infinite sequence $\{\beta_n\}_{n=1}^{+\infty}$ that satisfies $0 < \beta_1 < \beta_2 < \cdots < \beta_n < \cdots$, $\lim_{n \to +\infty} \beta_n = +\infty$, and*

$$\beta_n \in \left(\frac{(4n-3)\pi}{2\tau^*}, \frac{(4n-2)\pi}{2\tau^*} \right) \text{ for } n \geq 1.$$

(ii) $\pm i\beta_n$ are characteristic values of the stationary solution $(0, \tau^*, \sigma_n)$, where

$$\sigma_n = \pm(\beta_n^2 + \mu^2)^{1/4}.$$

If $\sigma \neq \sigma_n$, then the stationary solution $(0, \tau^*, \sigma)$ has no purely imaginary characteristic value.

(iii) Let $\lambda_n(\sigma) = u_n(\sigma) + iv_n(\sigma)$ be the root of (6.176) for σ close to σ_n such that $u_n(\sigma_n) + iv_n(\sigma_n) = i\beta_n$. Then

$$u_n'(\sigma)\Big|_{\sigma=\sigma_n} = \frac{2}{\sigma_n} \frac{(\mu^2+\beta_n^2)\tau^* + \mu}{(1+\mu\tau^*)^2 + (\beta_n\tau^*)^2}.$$

Now we are able to state our main results.

Theorem 6.24. *Assume that (α_1)–(α_6) hold. Let $\beta_n \in \left(\frac{(4n-3)\pi}{2\tau^*}, \frac{(4n-2)\pi}{2\tau^*} \right)$, $n \geq 1$, be as given in (i) of Lemma 6.32. Let $\sigma_n = \pm(\mu^2 + \beta_n)^{1/4}$ for $n \geq 1$. Then:*

(a) There exists an unbounded connected component $C\left(0, \tau^*, \sigma_n, \frac{2\pi}{\beta_n}\right)$ of the closure of all the nonconstant periodic solutions of system (6.169), bifurcated from $(0, \tau^*, \sigma_n, \frac{2\pi}{\beta_n})$ in the Fuller space where σ satisfies $\mathrm{sgn}(\sigma_n)\sigma > 0$.

(b) $(0, \tau^*, \sigma_1, \frac{2\pi}{\beta_1}) \notin C\left(0, \tau^*, \sigma_n, \frac{2\pi}{\beta_n}\right)$ for every $n \geq 2$.

(c) For every $n \geq 2$, the projection of $C\left(0, \tau^*, \sigma_n, \frac{2\pi}{\beta_n}\right)$ onto the parameter space $\mathbb{R}$ is unbounded in $(0, +\infty)$ if $\sigma_n > 0$ and is unbounded in $(-\infty, 0)$ if $\sigma_n < 0$.

Proof. (a) We apply Theorem 6.15. We first verify assumptions (SHB1)–(SHB3) and (SHB5). It is clear that (α_2) and (α_1) imply (SHB1), (SHB2), and (SHB5). Let us check (SHB3). Indeed, noticing that $\sigma_n = \pm(\mu^2 + \beta_n^2)^{1/4}$, $b'(0) = -1$, and $\beta_n > 0$, we have

$$\left(\frac{\partial}{\partial\theta_1} + \frac{\partial}{\partial\theta_2} \right) \left[-\mu\theta_1 + \sigma^2 b(\theta_2) \right]_{\sigma=\sigma_n,\, \theta_1=\theta_2=0} = -\mu - \sigma_n^2 < 0. \qquad (6.178)$$

Also, it follows from $\tau^* = -\frac{\ln(2h(0)-1)}{2}$ that

$$\frac{\partial}{\partial\gamma_2}(1 - h(\gamma_1))(1 + \tanh(\gamma_2))\Big|_{\sigma=\sigma_n,\, \gamma_1=0,\, \gamma_2=\tau^*} = -h(0) \cdot \frac{4e^{2\tau^*}}{(e^{2\tau^*}+1)^2} < 0. \quad (6.179)$$

Therefore, condition (SHB3) is satisfied by system (6.169).

We note from Lemma 6.32 (i), (ii), and (iii) that every center (including those with $\sigma < 0$) of system (6.174) is isolated. We now calculate the crossing number of $(0, \tau^*, \sigma_n, \beta_n)$. Let $u_n(\sigma) + iv_n(\sigma)$ be the characteristic value of (6.174) such that $u_n(\sigma_n) + iv_n(\sigma_n) = i\beta_n$. By (iv) of Lemma 6.32, we have

$$\frac{d}{d\sigma} u_n(\sigma)\Big|_{\sigma=\sigma_n} = u_n'(\sigma_n)\big|_{\sigma=\sigma_n}$$

$$= \frac{2}{\sigma_n} \frac{(\mu^2+\beta_n^2)\tau^* + \mu}{(1+\mu\tau^*)^2 + (\beta_n\tau^*)^2}. \qquad (6.180)$$

That is, $\frac{d}{d\sigma} u_n(\sigma)\big|_{\sigma=\sigma_n}$ has the same sign as σ_n, since $\tau^* > 0$ and $\mu > 0$. We note from (6.80) that the crossing number $\gamma(0, \tau^*, \sigma_n, \frac{2\pi}{\beta_n})$ counts the difference, as σ varies from σ_n^- to σ_n^+, of the number of imaginary characteristic values with positive real parts in a small neighborhood of $i\beta_n$ in the complex plane, where $\sigma_n^- < \sigma_n < \sigma_n^+$ are numbers in a small neighborhood of σ_n. Then by (6.180), the crossing number of the isolated center $(0, \tau^*, \sigma_n, \frac{2\pi}{\beta_n})$ in the Fuller space $C(\mathbb{R}; \mathbb{R}^2) \times \mathbb{R}^2$ satisfies

$$\gamma(0, \tau^*, \sigma_n, \frac{2\pi}{\beta_n}) = -\mathrm{sgn}(\sigma_n) \text{ for every } n \in \mathbb{N}. \qquad (6.181)$$

Then by Theorem 6.15, there exists a connected component $C\left(0, \tau^*, \sigma_n, \frac{2\pi}{\beta_n}\right)$ of the closure of all the nonconstant periodic solutions of system (6.169), bifurcated from the stationary solution $(0, \tau^*, \sigma_n, \frac{2\pi}{\beta_n})$ in the Fuller space. Note that there is no nonconstant periodic solution for the system (6.169) if $\sigma = 0$, since in this case, x satisfies a scalar ordinary differential equation. Moreover, there is no bifurcation point at $\sigma = 0$. Therefore, $C\left(0, \tau^*, \sigma_n, \frac{2\pi}{\beta_n}\right)$ is located in the Fuller space where σ satisfies $\mathrm{sgn}(\sigma_n)\sigma > 0$.

To prove the unboundedness of $C\left(0, \tau^*, \sigma_n, \frac{2\pi}{\beta_n}\right)$ in the Fuller space, we apply the global Hopf bifurcation Theorem 6.17 to exclude the case that there are finitely many bifurcation points in $C\left(0, \tau^*, \sigma_n, \frac{2\pi}{\beta_n}\right)$.

Now we suppose there are finitely many bifurcation points $\{(0, \tau^*, \sigma_{n_j}, \frac{2\pi}{\beta_{n_j}})\}_{j=1}^q$, $q \in \mathbb{N}$, in $C\left(0, \tau^*, \sigma_n, \frac{2\pi}{\beta_n}\right)$. We know that $C\left(0, \tau^*, \sigma_n, \frac{2\pi}{\beta_n}\right)$ is located in the Fuller space where σ satisfies $\mathrm{sgn}(\sigma_n)\sigma > 0$. Then the bifurcation points $\{(0, \tau^*, \sigma_{n_j}, \frac{2\pi}{\beta_{n_j}})\}_{j=1}^q$ satisfy $\mathrm{sgn}(\sigma_n)\sigma_{n_j} > 0$ for all $j \in \{1, 2, \cdots, q\}$.

Let ε_{n_j} be the value of

$$\mathrm{sgn}\det \left[\begin{array}{cc} \left(\frac{\partial}{\partial\theta_1} + \frac{\partial}{\partial\theta_2}\right)\tilde{f}(\theta_1, \theta_2, \sigma) & 0 \\ \frac{\partial}{\partial\gamma_1}\tilde{g}(\gamma_1, \gamma_2, \sigma) & \frac{\partial}{\partial\gamma_2}\tilde{g}(\gamma_1, \gamma_2, \sigma) \end{array} \right]$$

evaluated at $(\theta_1, \theta_2, \sigma) = (0, 0, \sigma_{n_j})$ and $(\gamma_1, \gamma_2, \sigma) = (0, \tau^*, \sigma_{n_j})$, where

$$\tilde{f}(\theta_1, \theta_2, \sigma) = [-\mu\theta_1 + \sigma^2 b(\theta_2)], \quad \tilde{g}(\gamma_1, \gamma_2, \sigma) = (1 - h(\gamma_1))(1 + \tanh(\gamma_2)).$$

Then by (6.178) and (6.179), we have

$$\varepsilon_{n_j} = 1 \text{ for all } j = 1, 2, \cdots, q. \qquad (6.182)$$

By (6.181) and (6.182), we have

$$\sum_{j=1}^q \varepsilon_{n_j}\gamma((0, \tau^*, \sigma_{n_j}, \frac{2\pi}{\beta_{n_j}}) = -q\,\mathrm{sgn}(\sigma_n) \neq 0. \qquad (6.183)$$

Note that (α_5) and (α_6) implies (SHB4). Then by Theorem 6.17, (6.183) is a contradiction. The unboundedness of $C\left(0, \tau^*, \sigma_n, \frac{2\pi}{\beta_n}\right)$ follows.

(b) In order to verify assumption (SHB7), we claim that the virtual period p_n of every bifurcation point $(0, \tau^*, \sigma_n, 2\pi/\beta_n)$ satisfies

$$mp_n \neq \tau^* \text{ for every } m \in \mathbb{N}. \tag{6.184}$$

Suppose that there exist $m_0, n_0 \in \mathbb{N}$ such that $m_0 p_{n_0} = m_0 \cdot 2\pi/\beta_{n_0} = \tau^*$. We note that

$$\beta_n \in \left(\frac{(4n-3)\pi}{2\tau^*}, \frac{(4n-2)\pi}{2\tau^*} \right) \text{ for all } n \geq 1. \tag{6.185}$$

Then we have

$$4n_0 - 3 < 4m_0 < 4n_0 - 2.$$

This is a contradiction, and the claim is proved.

We note that by (6.185), a sufficient condition for $p_n = \frac{2\pi}{\beta_n} < \tau^*$, is that $\frac{2\pi}{\beta_n} < 4\tau^*/(4n-3) < \tau^*$, that is, $n \geq 7/4$. Therefore, every $(0, \tau^*, \sigma_n, p_n)$ with $n \geq 2$ is a bifurcation point of system (6.169) satisfying

$$p_n < \tau^* \text{ for all } n \geq 2. \tag{6.186}$$

For the bifurcation point $(0, \tau^*, \sigma_1, p_1)$, we can conclude from (6.185) that

$$2\tau^* < p_1 < 4\tau^*. \tag{6.187}$$

We want to obtain the uniform boundedness of the period in $C(0, \tau^*, \sigma_n, \frac{2\pi}{\beta_n})$ with $n \geq 2$. We only need to check the conditions (SHB6)–(SHB10) for applying Theorems 6.21 and 6.22.

It is clear that (α_4), (6.184), and (6.179) imply (SHB8), (SHB9), and (SHB6), respectively. Also we conclude from (SHB2), (SHB4), and Lemma 6.20 that

$$p > 0 \tag{6.188}$$

for every $(x, \tau, \sigma, p) \in C(0, \tau^*, \sigma_n, \frac{2\pi}{\beta_n})$. Also, by Lemma 6.31, we have

$$0 < \tau(t) < -\frac{1}{2}\ln(2h_0 - 1) \tag{6.189}$$

for every $t \in \mathbb{R}$, and hence (SHB10) is satisfied. To check (SHB7), we let

$$\begin{cases} 1 - h(x)(1 + \tanh \tau) = 0, \\ (1 + \tanh \tau)h'(x)\left(-\mu x + \sigma^2 b(x)\right) = 0. \end{cases} \tag{6.190}$$

Then by (α_1), (α_4), and (α_6), the solutions of (6.190) are stationary solutions of (6.169). This verifies (SHB7).

Therefore, we can use Theorems 6.21, 6.22, (6.186), (6.188), and (6.189) to conclude that there exists some $t \in \mathbb{R}$ such that

$$0 < p < \tau(t) < -\frac{1}{2} \ln(2h_0 - 1) \qquad (6.191)$$

for every $(x, \tau, \sigma, p) \in C(0, \tau^*, \sigma_n, \frac{2\pi}{\beta_n})$ with $n \geq 2$. Then by (6.187) and (6.191), we know that $(0, \tau^*, \sigma_1, \frac{2\pi}{\beta_1}) \notin C(0, \tau^*, \sigma_n, \frac{2\pi}{\beta_n})$ for every $n \geq 2$. This proves (b).

(c) Let Σ be the projection of $C(0, \tau^*, \sigma_n, \frac{2\pi}{\beta_n})$ on the σ-parameter space $\mathbb{R}$. By (a), we know that $\Sigma \subseteq (0, +\infty)$ if $\sigma_n > 0$ and $\Sigma \subseteq (-\infty, 0)$ if $\sigma_n < 0$. By Lemma 6.31, we know that for every $\sigma \in \Sigma$, there exists a constant $M_n(\sigma) > 0$ such that

$$\|(x, \tau)\|_{C(\mathbb{R}; \mathbb{R}^{N+1})} \leq M_n(\sigma), \qquad (6.192)$$

where (x, τ, σ, p) is the solution associated with σ in $C(0, \tau^*, \sigma_n, \frac{2\pi}{\beta_n})$ and $M_n : \mathbb{R} \ni \sigma \to M_n(\sigma) \in (0, +\infty)$ is a continuous function on $\mathbb{R}$.

We know from (6.191) that the projection of $C(0, \tau^*, \sigma_n, \frac{2\pi}{\beta_n})$ on the p-parameter space $\mathbb{R}$ is bounded. If Σ is bounded, then it follows from (a) that the projection of $C(0, \tau^*, \sigma_n, \frac{2\pi}{\beta_n})$ on the (x, τ)-space $C(\mathbb{R}; \mathbb{R}^{N+1})$ must be unbounded in the supremum norm. But by the continuity of M_n on $\mathbb{R}$ and by (6.192), the projection of $C(0, \tau^*, \sigma_n, \frac{2\pi}{\beta_n})$ on the (x, τ)-space $C(\mathbb{R}; \mathbb{R}^{N+1})$ is uniformly bounded with respect to $\sigma \in \Sigma$. This is a contradiction, and the proof is complete. $\qquad\square$

We conclude by noting that the global continuation of slowly oscillating periodic solutions is addressed in [172].

Chapter 7
Bifurcation in Symmetric FDEs

7.1 Introduction

In the local theory of one-parameter families of nonlinear dynamical systems with a loss of stability of an equilibrium, two types of bifurcation generically occur. These are fold (also referred to as steady-state) bifurcations, for which the linearization has a zero eigenvalue, and Hopf bifurcations, for which the eigenvalue is complex with zero real part. Typically, branches of solutions bifurcate from the original equilibrium and are approximated to leading order by the corresponding eigenfunctions at singularities; these branches are often referred to as modes. Generically we expect, in a one-parameter system, to have only one critical mode. Multiple critical modes are expected in systems with more than one parameter. A secondary bifurcation is thought of as resulting from an interaction of several critical modes, called mode interaction. Since there are two types of critical modes (steady-state and Hopf), there may exist four types of mode interactions in two-parameter systems: (a) Bogdanov–Takens bifurcations, (b) fold–fold bifurcation, (c) Hopf–fold, (d) Hopf–Hopf. For example, the interaction of a fold bifurcation with a Hopf bifurcation can lead to much richer dynamics than just the expected equilibria and periodic solutions, including the possibility of an invariant 2-torus on which the flow may be periodic or quasiperiodic; see Gavrilov [108], Langford [202], Guckenheimer [124], Broer et al. [36, 42, 43], Kielhöfer [190], Kuznetsov [200, 201], Iooss and Langford [176]. As this torus grows fatter, generic perturbations can also lead to chaotic dynamics; see Guckenheimer [126], Holmes [166], Langford [203–205].

If in addition the system is symmetric, that is, equivariant with respect to the action of some group Γ, then generically, the eigenspace corresponding to a single steady-state mode is irreducible under the action of the symmetry group, while the eigenspace corresponding to a single Hopf mode is Γ-simple. However, in the context of mode interactions, either the sum of zero eigenspaces can be assumed nonirreducible actions, or there may be a degeneracy in the imaginary eigenspace, which can split as the direct sum of two Γ-simple spaces.

S. Guo and J. Wu, *Bifurcation Theory of Functional Differential Equations*,
Applied Mathematical Sciences 184, DOI 10.1007/978-1-4614-6992-6_7,
© Springer Science+Business Media New York 2013

7.2 Fold Bifurcation

Consider the following one-parameter family of retarded functional differential equations (RFDEs):

$$\dot{x}(t) = f(\alpha, x_t), \quad \alpha \in \mathbb{R}, \tag{7.1}$$

where $f \in C^k(\mathbb{R} \times C_{n,\tau}, \mathbb{R}^n)$ for a large enough integer k, and $f(0,0) = 0$. We will make use of the invariant notation for higher-order derivatives of functions of several variables. If $v_1, v_2, \cdots, v_j \in C_{n,\tau}$, we define

$$\mathscr{F}^j(\alpha, v_1, v_2, \cdots, v_j) = \frac{\partial^j}{\partial t_1 \partial t_2 \cdots \partial t_j} f\left(\alpha, \sum_{s=1}^j t_s v_s\right)\Big|_{t_1 = t_2 = \cdots = t_j = 0}$$

for $j \in \mathbb{N}$. Then we have

$$f(\alpha, \varphi) = f(\alpha, 0) + \mathscr{F}^1(\alpha, \varphi) + \cdots + \frac{1}{k!} \mathscr{F}^k(\alpha, \varphi, \cdots, \varphi) + o(\|\varphi\|^k).$$

Obviously, $L = \mathscr{F}^1(0, \cdot)$ is the linearized operator of $f(\alpha, \varphi)$ with respect to φ at $(\alpha, \varphi) = (0,0)$. By the Riesz representation theorem, there exists an $n \times n$ matrix-valued function $\eta : [-\tau, 0] \to \mathbb{R}^{n^2}$ whose elements are of bounded variation such that

$$L\varphi = \int_{-\tau}^0 d\eta(\theta)\varphi(\theta), \quad \varphi \in C_{n,\tau}. \tag{7.2}$$

Let $\mathscr{A}$ be the infinitesimal generator associated with the linear equation $\dot{x}(t) = Lx_t$, and let $\Delta(\lambda)$ be the characteristic matrix of the operator $\mathscr{A}$. Recall that the bilinear form $\langle \cdot, \cdot \rangle$ is defined as

$$\langle \psi, \varphi \rangle = \overline{\psi}(0)\varphi(0) - \int_{-\tau}^0 \int_0^\theta \overline{\psi}(\xi - \theta)d\eta(\theta)\varphi(\xi)d\xi \tag{7.3}$$

for $\psi \in C_{n,\tau}^*$ and $\varphi \in C_{n,\tau}$.

If $0 \notin \sigma(\mathscr{A})$, that is, $x = 0$ is a hyperbolic equilibrium in the system for $\alpha = 0$, then under a small parameter variation, the equilibrium moves slightly but remains hyperbolic.

7.2.1 Standard Fold Bifurcation

In this subsection, we always assume that:

(FB)   The infinitesimal generator $\mathscr{A}$ has a simple eigenvalue 0.

Then $\det \Delta(0) = 0$, and hence there exist $p \in \mathbb{R}^{n*}$ and $q \in \mathbb{R}^n$ such that $p\Delta(0) = 0$, $\Delta(0)q = 0$, and $p\Delta_\lambda(0)q = 1$. Thus, the eigenspace of $\mathscr{A}$ associated with eigenvalue 0 is spanned by $\hat{q}$, with the adjoint space spanned by $\hat{p}$. Moreover, $\langle \hat{p}, \hat{q} \rangle = 1$.

By the center manifold theorem, the reduced equation on the center manifold $M_{\text{loc}}^{c,\alpha}$ is

$$\dot{x}(t) = G(\alpha,x), \quad x \in \mathbb{R}, \tag{7.4}$$

where $G(\alpha,x) = pF(\alpha, qx + W(\alpha,x))$, $F(\alpha,\varphi) = f(\alpha,\varphi) - L\varphi$ for $\alpha \in \mathbb{R}$ and $\varphi \in C_{n,\tau}$, and W satisfies

$$\frac{\mathrm{d}}{\mathrm{d}t}W = \mathscr{A}_Q W + H(\alpha,x), \quad W(0,0) = 0, \quad D_x W(0,0) = 0,$$

and $H(\alpha,x) = [X_0 - qp]F(\alpha, qx + W(\alpha,x))$. Let

$$G(\alpha,x) = \sum_{j+s=1}^{k} \frac{1}{j!s!} G_{js}\alpha^j x^s + o(|(\alpha,x)|^k),$$

$$W(\alpha,x) = \sum_{j+s=1}^{k} \frac{1}{j!s!} W_{js}\alpha^j x^s + o(|(\alpha,x)|^k),$$

$$H(\alpha,x) = \sum_{j+s=1}^{k} \frac{1}{j!s!} H_{js}\alpha^j x^s + o(|(\alpha,x)|^k).$$

Obviously, $G_{01} = W_{01} = H_{01} = 0$, and

$$\begin{aligned}
G_{10} &= pf_\alpha(0,0),\\
G_{11} &= p\mathscr{F}_\alpha^1(0,q),\\
G_{02} &= p\mathscr{F}^2(0,q,q),\\
G_{03} &= p\{3\mathscr{F}_2(0,q,W_{02}) + \mathscr{F}_3(0,q,q,q)\},\\
H_{10} &= [X_0 - qp]f_\alpha(0,0),\\
H_{11} &= [X_0 - qp]\mathscr{F}_\alpha^1(0,q),\\
H_{02} &= [X_0 - qp]\mathscr{F}^2(0,q,q).
\end{aligned}$$

We still need to compute W_{js}, $j+s = 1,2,\ldots$. Noticing that

$$\frac{\mathrm{d}}{\mathrm{d}t}W = D_x W(\alpha,x)\dot{x} = D_x W(\alpha,x)G(\alpha,x),$$

we have

$$\left[\sum_{j+s\le k} \frac{1}{j!(s-1)!}W_{js}\alpha^j x^{s-1}\right]\left[\sum_{j+s\le k} \frac{1}{j!s!}G_{js}\alpha^j x^s\right]$$

$$= \sum_{j+s\le k} \frac{1}{j!s!}[\mathscr{A}_Q W_{js} + H_{js}]\alpha^j x^s + o(|(\alpha,x)|^k).$$

Comparing coefficients, we obtain

$$\mathscr{A}_Q W_{js} + H_{js} = \sum_{(p+r,q+l)=(j,s+1)} \frac{j!l!}{p!(q-1)!r!s!} W_{pq}G_{rl} \tag{7.5}$$

for $j+s \le k$. In particular,

$$\mathscr{A}_Q W_{10} + H_{10} = 0,$$
$$\mathscr{A}_Q W_{02} + H_{02} = 0. \tag{7.6}$$

It follows from the first equation that

$$\dot{W}_{10} = qG_{10} \tag{7.7}$$

and

$$LW_{10} = qG_{10} - f_\alpha(0,0). \tag{7.8}$$

From (7.7), we have $W_{10}(\theta) = qG_{10}\theta + E_0$, $\theta \in [-\tau,0]$. Substituting this into (7.8) yields

$$\Delta(0)E_0 = f_\alpha(0,0) - qG_{10} + \int_{-\tau}^0 \theta d\eta(\theta)qG_{10}. \tag{7.9}$$

Because of Keller [188], we know that the unique solution E_0 to (7.9) satisfying $pE_0=0$ is

$$E_0 = [\Delta(0,0)]^{\mathrm{inv}}[f_\alpha(0,0) - qG_{10} + \int_{-\tau}^0 \theta d\eta(\theta)qG_{10}].$$

Namely,

$$W_{10} = qG_{10}\theta + [\Delta(0,0)]^{\mathrm{inv}}[f_\alpha(0,0) - qG_{10} + \int_{-\tau}^0 \theta d\eta(\theta)qG_{10}].$$

Similarly, it follows from the second equation that

$$W_{02} = qG_{02}\theta + [\Delta(0)]^{\mathrm{inv}}[\mathscr{F}^2(0,q,q) - qG_{02} + \int_{-\tau}^0 \theta d\eta(\theta)qG_{02}].$$

Thus, we can evaluate G_{03}.

Note that $G(0,0) = 0$ and $G_x(0,0) = 0$, the reduced equation (7.4) can undergo a saddle-node bifurcation near $(\alpha,x) = (0,0)$ under the following condition:

(SN) $G_{10}G_{02} \ne 0$.

Thus, we have the following results.

Theorem 7.1. *Under assumptions (FB) and (SN), system (7.1) undergoes a saddle-node bifurcation near $(\alpha,x) = (0,0)$. Moreover, if $G_{10}G_{02} < 0$ (respectively, > 0), then near the origin, only two equilibria exist for $\alpha > 0$ (respectively, < 0), only one equilibrium $x = 0$ exists for $\alpha = 0$, and no equilibria exist for $\alpha < 0$ (respectively, > 0).*

If $f(\alpha,0) = 0$ for all $\alpha \in \mathbb{R}$, then the function $G(\alpha,x)$ in the reduced equation (7.4) satisfies $G(\alpha,0) = 0$ for all $\alpha \in \mathbb{R}$. Thus, we have the following theorem.

Theorem 7.2. *In addition to condition (FB), assume that*

(TR) $f(\alpha,0) = 0$ *for all* $\alpha \in \mathbb{R}$, *and* $G_{11}G_{02} \neq 0$.

Then system (7.1) undergoes a transcritical bifurcation near $(\alpha,x) = (0,0)$. *Namely, besides the trivial solution, system (7.1) has a nonzero equilibrium, which continuously depends on* α *for all sufficiently small* $|\alpha|$. *Moreover, this nonzero equilibrium is stable if the remaining eigenvalues of* $\mathscr{A}$ *have negative real parts and* $\alpha G_{11} > 0$, *and is unstable otherwise.*

Furthermore, if $f(\alpha,-\varphi) = -f(\alpha,\varphi)$ for all $\alpha \in \mathbb{R}$ and $\varphi \in C_{n,\tau}$. Then the map W may be chosen such that $W(\alpha,-x) = -W(\alpha,x)$. As a result, $G(\alpha,-x) = -G(\alpha,x)$. Therefore, we have the following result.

Theorem 7.3. *In addition to condition (FB), assume that*

(PF) $f(\alpha,-\varphi) = -f(\alpha,\varphi)$ *for all* $\alpha \in \mathbb{R}$ *and* $\varphi \in C_{n,\tau}$, $G_{11}G_{03} \neq 0$.

Then system (7.1) undergoes a pitchfork bifurcation near $(\alpha,x) = (0,0)$. *Moreover, if* $G_{11}G_{03} < 0$ *(respectively,* > 0), *then two nontrivial equilibria exist for* $\alpha > 0$ *(respectively,* < 0), *and only the trivial equilibrium continues to exist for* $\alpha < 0$ *(respectively,* > 0). *Moreover, the two nontrivial equilibria coalesce into zero as* α *goes to 0.*

7.2.2 Fold Bifurcations with $\mathbb{Z}_2$-Symmetry

In this subsection, we consider system (7.1) under condition (FB) and the following assumption:

(FBZ)   System (7.1) is $\mathbb{Z}_2$-equivariant, where the group $\mathbb{Z}_2$ is equal to $\{\mathrm{Id}, \kappa\}$ with $\kappa \neq \mathrm{Id}$ and $\kappa^2 = \mathrm{Id}$.

Then we know that either $\kappa \cdot q = q$ or $\kappa \cdot q = -q$. If $\kappa \cdot q = q$, then $\mathbb{Z}_2$ induces only the identity map on $\mathbb{R}$, and hence the symmetry provides no additional condition on the reduced equation (7.4). Therefore, we can employ the same arguments as in the previous subsection and obtain Theorems 7.1–7.3. However, the bifurcated equilibria $u(\alpha)$ are invariant under κ, i.e., $\kappa \cdot u(\alpha) = u(\alpha)$.

Now we consider the case $\kappa \cdot q = -q$. Then the induced action of $\mathbb{Z}_2$ on $\mathbb{R}$ is given by $\kappa \cdot x = -x$ for all $x \in \mathbb{R}$. It follows from the symmetric center manifold theorem that the function $G(\alpha,x)$ in the reduced equation (7.4) satisfies $G(\alpha,-x) = -G(\alpha,x)$ for all α and $x \in \mathbb{R}$. In view of Theorem 7.3, we have the following result.

Theorem 7.4. *Under assumptions (FB), (FBZ), and* $G_{11}G_{03} \neq 0$, *if* $\kappa \cdot q = -q$, *then system (7.1) undergoes a pitchfork bifurcation near* $\alpha = 0$. *More precisely, we have the following statements:*

(i) *If $G_{11}G_{03} < 0$, then two nontrivial equilibria $u^*_{1,2}(\alpha)$ exist for $\alpha > 0$ (which are stable if $G_{11} > 0$ and all the remaining eigenvalues of $\mathscr{A}$ have negative real parts, and unstable otherwise), and only the trivial equilibrium continues to exist for $\alpha < 0$. Moreover, the two nontrivial equilibria $u^*_{1,2}(\alpha)$ satisfy $\kappa \cdot u^*_1(\alpha) = u^*_2(\alpha)$ and coalesce into the trivial equilibrium as $\alpha \to 0^+$.*

(ii) *If $G_{11}G_{03} > 0$, then two nontrivial equilibria $u^*_{1,2}(\alpha)$ exist for $\alpha < 0$ (which are stable if $G_{11} < 0$ and all the remaining eigenvalues of $\mathscr{A}$ have negative real parts, and unstable otherwise), and only the trivial equilibrium continues to exist for $\alpha > 0$. Moreover, the two nontrivial equilibria $u^*_{1,2}(\alpha)$ satisfy $\kappa \cdot u^*_1(\alpha) = u^*_2(\alpha)$ and coalesce into the trivial equilibrium as $\alpha \to 0^-$.*

7.2.3 Fold Bifurcations with $O(2)$-Symmetry

Assume that system (7.1) is $O(2)$-equivariant, where the representation of $O(2)$ on $\mathbb{R}^n$ is given by the linear maps κ (flip) and $\{R_\theta : \theta \in \mathbb{R}\} \cong \mathbb{S}^1$ (rotation) with $R_\theta \circ R_\upsilon = R_{\theta+\upsilon}$, $R_0 = \mathrm{Id}_n$, $\kappa^2 = \mathrm{I}$, and $\kappa R_\theta = R_{-\theta}\kappa$. Namely,

(FBO1) $f(\alpha, R_\theta\varphi) = R_\theta f(\alpha, \varphi)$ and $f(\alpha, \kappa\varphi) = \kappa f(\alpha, \varphi)$ for all $\alpha \in \mathbb{R}$, $\varphi \in C_{n,\tau}$, and $\theta \in \mathbb{S}^1$.

Moreover, assume that:

(FBO2)   The infinitesimal generator $\mathscr{A}$ has a double eigenvalue 0.

Assumption (FBO2) means that the eigenspace P is two-dimensional. There exist $\hat{\zeta}_1 \in P$ and $k \in \mathbb{Z}$ such that $R_\theta\zeta_1 = e^{ik\theta}\zeta_1$. Setting $\zeta_2 = \kappa\zeta_1$, we have $P = \mathrm{span}\{\hat{\zeta}_1, \hat{\zeta}_2\}$. Hence, the action of $O(2)$ on $P \cong \mathbb{R}^2$ is given by

$$R_\theta|_P = \begin{bmatrix} e^{ik\theta} & 0 \\ 0 & e^{-ik\theta} \end{bmatrix}, \quad \kappa|_P = \begin{bmatrix} 0 & 1 \\ 1 & 0 \end{bmatrix}.$$

Furthermore, it is possible to choose $\hat{\zeta} \in P$ such that $\kappa\zeta = \overline{\zeta}$. Indeed, $R_\theta\overline{\zeta} = \overline{R_\theta\zeta} = e^{-ik\theta}\overline{\zeta}$, so there is $c \in \mathbb{C}$ with $\kappa\zeta = c\overline{\zeta}$, which together with $\kappa^2 = \mathrm{Id}$, implies that $\zeta = |c|^2\zeta$. Thus, $\kappa\zeta = e^{2i\varepsilon}\zeta$ for some ε. Defining $\zeta' = e^{-i\varepsilon}\zeta$, we have $\zeta' \in P$ and $\kappa\zeta' = \overline{\zeta'}$. Therefore, we have $P = \{z\hat{\zeta} + z\hat{\overline{\zeta}} : z \in \mathbb{C}\}$, and the reduced action of $O(2)$ on $\mathbb{C}$ is given by

$$R_\theta z = e^{ik\theta}z \text{ and } \kappa z = \overline{z}. \tag{7.10}$$

Thus, the reduced equation takes the form

$$\dot{z} = G(\alpha, z, \overline{z}), \tag{7.11}$$

where G is a complex function satisfying

$$G(\alpha, e^{ik\theta}z, e^{-ik\theta}\overline{z}) = e^{ik\theta}G(\alpha, z, \overline{z}), \quad G(\alpha, \overline{z}, z) = \overline{G(\alpha, z, \overline{z})}$$

for all $\alpha \in \mathbb{R}$ and $z \in \mathbb{C}$. Now we assume that k in (7.10) is not equal to 0, i.e.,

(FBO3) $\{R_\theta : \theta \in \mathbb{R}\} \cong \mathbb{S}^1$ acts nontrivially on P.

Then the first equation yields $G(\alpha, z, \bar{z}) = zQ(\alpha, |z|^2)$, where Q is a complex function. The second function implies that the function Q is real. Therefore, in polar coordinates $z = re^{i\phi}$, we have

$$\dot{r} = rQ(\alpha, r^2), \quad \dot{\phi} = 0.$$

The equation for r is a standard *pitchfork bifurcation equation*. Thus, if

$$\frac{\partial Q}{\partial r^2}(0,0) \neq 0, \tag{7.12}$$

then there exists $r(\alpha)$ with $Q(\alpha, r^2(\alpha)) \equiv 0$ for all small α. Thus, for every α, system (7.11) has solutions: $z = z(\alpha, \phi) \overset{\text{def}}{=} r(\alpha)e^{i\phi}$. We have

$$R_{2\pi/k}z(\alpha, \phi) = z(\alpha, \phi), \quad \kappa z(\alpha, s\pi) = z(\alpha, s\pi)$$

for all $\alpha \in \mathbb{R}$, $\phi \in \mathbb{R}$, and $s \in \mathbb{Z}$. This, together with symmetric center manifold theorem, implies the following result.

Theorem 7.5. *Under assumptions (FBO1)–(FBO3), assume that (7.12) holds. Then system (7.1) undergoes a cyclic pitchfork bifurcation near $\alpha = 0$. Namely, there exists a circle of equilibria parameterized by ϕ, denoted by $x(\alpha, \phi)$, that satisfies $R_{2\pi/k}x(\alpha, \phi) = x(\alpha, \phi)$ and $\kappa x(\alpha, s\pi) = x(\alpha, s\pi)$ for all $\alpha \in \mathbb{R}$, $\phi \in \mathbb{R}$, and $s \in \mathbb{Z}$.*

7.3 Hopf Bifurcation

We now turn to the next simple case, in which the infinitesimal generator $\mathscr{A}$ associated with the linear system of (7.1) about the equilibrium has a pair of purely imaginary eigenvalues $\pm i\omega$ with $\omega > 0$, and all other eigenvalues of $\mathscr{A}$ are not integer multiples of $i\omega$. Thus, system (7.1) may undergo a Hopf bifurcation. The phenomenon of Hopf bifurcation concerns the birth of a periodic solution from an equilibrium solution through a local oscillatory instability. Since 0 is not an eigenvalue of the infinitesimal generator $\mathscr{A}$, the equilibrium in general moves as α varies but remains isolated and close to the origin for all sufficiently small $|\alpha|$. Thus, we can perform a coordinate shift, placing this equilibrium at the origin. Therefore, we may assume without loss of generality that $f(\alpha, 0) = 0$ for all $\alpha \in \mathbb{R}$, Thus, system (7.1) can be written as

$$\dot{u} = L(\alpha)u_t + \tilde{f}(\alpha, u_t), \quad \alpha \in \mathbb{R}, \tag{7.13}$$

where $L(\alpha)\varphi = \mathscr{F}^1(\alpha, \varphi)$ and $\tilde{f}(\alpha, \varphi) = f(\alpha, \varphi) - L(\alpha)\varphi$ for $(\alpha, \varphi) \in \mathbb{R} \times C_{n,\tau}$. Denote by $\mathscr{A}_\alpha$ the infinitesimal generator associated with the linear system

$\dot{x} = L(\alpha)x_t$. Obviously, $L = L(0)$ and $\mathscr{A} = \mathscr{A}_0$, where L and $\mathscr{A}$ are as defined in the previous section. Let $\Delta(\alpha, \lambda)$ be the characteristic matrix of the operator $\mathscr{A}_\alpha$, i.e.,

$$\Delta(\alpha, \lambda) = \lambda \operatorname{Id}_n - L(\alpha)e^{\lambda(\cdot)}.$$

In the subsequent subsections, we assume that $\tilde{f}(\alpha, \varphi)$ has the Taylor expansion

$$\tilde{f}(\alpha, \varphi) = \frac{1}{2}\mathscr{F}^2(\alpha, \varphi, \varphi) + \cdots + \frac{1}{k!}\mathscr{F}^k(\alpha, \varphi, \varphi, \ldots, \varphi) + o(\|\varphi\|^k), \qquad (7.14)$$

in φ and the Taylor expansion

$$\tilde{f}(\alpha, \varphi) = \frac{1}{2!}f_2(\alpha, \varphi) + \cdots + \frac{1}{k!}f_k(\alpha, \varphi) + O(|(\alpha, \varphi)|^k), \qquad (7.15)$$

in (α, φ), where $f_j(\alpha, u) = \mathscr{H}_j((\alpha, u), \ldots, (\alpha, u))$, and $\mathscr{H}_j$ is a continuous multilinear symmetric map from $(\mathbb{R} \times C_{n,\tau}) \times \cdots \times (\mathbb{R} \times C_{n,\tau})$ (j times) to $\mathbb{R}^n$. Finally, we also rewrite $L(\alpha)$ in the Taylor expansions

$$L(\alpha) = L(0) + \alpha L'(0) + \tfrac{1}{2}\alpha^2 L''(0) + O(\alpha^3).$$

7.3.1 A Little History

The first results on Hopf bifurcation for retarded FDEs date back to work by Chafee [54] in 1971. However, according to Hale [146], the first proof of the Hopf bifurcation theorem for RFDEs under analytically computable conditions was presented by Chow and Mallet-Paret [68] in 1977. Since then, a considerable number of studies have been done by many authors, treating many aspects related to bifurcation of periodic solutions. For existence, uniqueness and regularity of the bifurcating branch, several approaches have been undertaken: the averaging method was notably developed by Gumowski [127] and Chow and Mallet-Paret [68]. Another approach, based on integral manifolds, was developed by Hale [154] and was further extended to the case of infinite delay by Stech [271]. Arino [12] treated the same problem by formulating an *adapted* implicit function theorem. Stech [271] used the Lyapunov–Schmidt reduction method and generalized a proof given by De Oliveira and Hale [80] in the case of ODEs to infinite delay differential equations. He also gaves a computational scheme of bifurcation elements via an asymptotic expansion of the bifurcation function. Staffans [270] established the theorem in a case analogous to Stech's for neutral functional differential equations, using the Lyapunov–Schmidt reduction method. Adimy [3] proved a Hopf bifurcation theorem using integrated semigroup theory. Diekmann et al. [81] tackled the problem of a lack of regularity of the solution operator associated with a delay equation. Using the sun–star theory of dual semigroups, they reduced the problem of bifurcation on a center manifold to a planar ordinary differential equation. In [91, 92], Faria and Magalhães studied the Hopf bifurcation problem by developing a normal form theory for retarded FDEs.

If a system of FDEs is symmetric, i.e., it has a nontrivial group of symmetries, one expects that the system has symmetric orbits, symmetric fixed points, and
periodic orbits as well as symmetric attractors or repellers. Also, symmetric steady
states can generate symmetric patterns in the state space of the system. In 1998, Wu
[303] employed the same techniques as those in [146] to establish a Hopf bifurcation for RFDEs with symmetry under the condition that the imaginary eigenspace is
isomorphic to the direct sum of two copies of the same absolutely irreducible representation. Guo and Lamb [136] developed the theory of equivariant Lyapunov–
Schmidt procedures in NFDEs with symmetry to set up a more general equivariant
Hopf bifurcation theory and obtained some important explicit formulas giving the
relevant coefficients for the determinations of the monotonicity of the periods and
Hopf bifurcation direction of the bifurcating symmetric periodic solutions directly
in terms of the coefficients of the original equations.

The first results on bifurcation from periodic solutions in retarded FDEs dates
back to a work by Walther [289], who considered the bifurcation from slowly oscillating periodic solutions of the scalar retarded FDE

$$\frac{\mathrm{d}}{\mathrm{d}t}x(t) = -\alpha f(x(t-1)) \tag{7.16}$$

under some symmetry conditions on f. Dormayer [83] considered (7.16) with a
class of nonmonotone functions f, periodic solutions y with the more general symmetry (S) $y(\cdot + \tau) = -y$, for some $\tau > 0$, that bifurcate from the primary branch at
some critical parameter. Their initial values lie on a smooth curve, and $\tau \neq 2$ except
at the bifurcation point. An example is $f(x) = -x/(1+x^2)$.

7.3.2 Standard Hopf Bifurcation

We assume that

(HB1) $\mathscr{A}_\alpha$ has a pair of simple complex conjugate eigenvalues $\lambda(\alpha)$ and $\overline{\lambda(\alpha)}$,
       satisfying $\lambda(0) = \mathrm{i}\omega$ with $\omega > 0$ and crossing the imaginary axis transversely at
       $\alpha = 0$ (i.e., $\mathrm{Re}\lambda'(0) \neq 0$).
(HB2) All other eigenvalues of $\mathscr{A}_0$ are not integer multiples of $\mathrm{i}\omega$.

Assume that q is the eigenfunction of $\mathscr{A}_0$ associated with eigenvalue $\mathrm{i}\omega$. Then
the associated eigenspace P is spanned by q and $\overline{q}$. Assume that the adjoint space P^*
of P is spanned by p and $\overline{p}$, where $p \in C_{n,\tau}^*$ satisfies $\langle p,q \rangle = 1$ and $\langle p,\overline{q} \rangle = 0$. Let
$\Phi = (q,\overline{q})$ and $\Psi = (p,\overline{p})^T$. By the center manifold theorem, we obtain the reduced
equation

$$\dot{z} = \mathrm{i}\omega z + G(\alpha,z,\overline{z}), \tag{7.17}$$

where $z \in \mathbb{C}$, $G(\alpha,z,\overline{z}) = \overline{p}(0)F(\alpha,zq + \overline{z}\overline{q} + W(\alpha,z,\overline{z}))$, $F(\alpha,\varphi) = \tilde{f}(\alpha,\varphi) +$
$L(\alpha)\varphi - L\varphi$ for $(\alpha,\varphi) \in \mathbb{R} \times C_{n,\tau}$, and $W(\alpha,z,\overline{z})$ satisfies

$$\frac{d}{dt}W = \mathscr{A}W + H(\alpha,z,\bar{z}) \qquad (7.18)$$

with $H(\alpha,z,\bar{z}) = [X_0 - \Phi\Psi(0)]F(\alpha,zq + \bar{z}\bar{q} + W(\alpha,z,\bar{z}))$. Let

$$G(\alpha,z,\bar{z}) = \sum_{s+k\geq 1}\frac{1}{s!k!}G_{sk}^{\alpha}z^s\bar{z}^k \qquad \text{and} \qquad W(\alpha,z,\bar{z}) = \sum_{s+k\geq 1}\frac{1}{s!k!}W_{sk}^{\alpha}z^s\bar{z}^k.$$

Obviously, $G_{10}^{\alpha} = \lambda'(0)\alpha + O(|\alpha|^2)$ and $G_{01}^{\alpha} = O(|\alpha|^2)$. Moreover, in view of Sect. 3.4.1, we have

$$G_{20}^0 = \bar{p}(0)\mathscr{F}^2(0,q,q),$$
$$G_{11}^0 = \bar{p}(0)\mathscr{F}^2(0,q,\bar{q}),$$
$$G_{02}^0 = \bar{p}(0)\mathscr{F}^2(0,\bar{q},\bar{q}),$$
$$G_{21}^0 = \bar{p}(0)\mathscr{F}^3(0,q,q,\bar{q}) + \bar{p}(0)\mathscr{F}^2(0,W_{20}^0,\bar{q}) + 2\bar{p}(0)\mathscr{F}^2(0,W_{11}^0,q),$$

where

$$W_{20}^0(\theta) = \frac{i}{\omega}G_{20}^0q(\theta) + \frac{i}{3\omega}\overline{G_{02}^0}\bar{q}(\theta) + E_1 e^{2i\omega\theta},$$
$$W_{11}^0(\theta) = \frac{1}{i\omega}G_{11}^0q(\theta) + \frac{i}{\omega}\overline{G_{11}^0}\bar{q}(\theta) + E_2,$$

and E_1 and E_2 are both n-dimensional vectors given by

$$E_1 = [\Delta(0,2i\omega)]^{-1}\mathscr{F}_2(0,q,q), \quad E_2 = [\Delta(0,0)]^{-1}\mathscr{F}_2(0,q,\bar{q}).$$

Therefore,

$$G_{21}^0 = \bar{p}(0)\mathscr{F}^3(0,q,q,\bar{q}) + \bar{p}(0)\mathscr{F}^2(0,E_1 e^{2i\omega(\cdot)},\bar{q}) + 2\bar{p}(0)\mathscr{F}^2(0,E_2,q)$$
$$+ \tfrac{i}{3\omega}\left\{6|G_{11}^0|^2 + |G_{02}^0|^2 - 3G_{20}^0G_{11}^0\right\}.$$

Next, employing the normalization as in Sect. 4.2.1, we obtain the following normal form:

$$\dot{z} = (i\omega + G_{10}^{\alpha})z + \frac{1}{2}C_1(0)z|z|^2 + O(\alpha^2|z| + |(\alpha,z)|^4), \qquad (7.19)$$

where

$$C_1(0) = \frac{i}{\omega}\left[G_{20}^0G_{11}^0 - 2|G_{11}^0|^2 - \frac{1}{3}|G_{02}^0|^2\right] + G_{21}^0. \qquad (7.20)$$

Substituting the expressions for G_{20}^0, G_{11}^0, G_{02}^0, and G_{21}^0 into (7.20) gives

$$C_1(0) = \bar{p}(0)\mathscr{F}^3(0,q,q,\bar{q}) + \bar{p}(0)\mathscr{F}^2(0,E_1 e^{2i\omega(\cdot)},\bar{q}) + 2\bar{p}(0)\mathscr{F}^2(0,E_2,q).$$
$$(7.21)$$

In polar coordinates $z = r^{i\xi}$, we have

$$\dot{r} = \text{Re}\{\lambda'(0)\}\alpha r + \frac{1}{2}\text{Re}\{C_1(0)\}r^3 + O(\alpha^2 r + |(r,\alpha)|^4),$$
$$\dot{\xi} = \omega + \text{Im}\{\lambda'(0)\}\alpha + \frac{1}{2}\text{Im}\{C_1(0)\}r^2 + O(|(r,\alpha)|^3). \tag{7.22}$$

We consider the following truncated system of (7.22):

$$\dot{r} = \text{Re}\{\lambda'(0)\}\alpha r + \frac{1}{2}\text{Re}\{C_1(0)\}r^3,$$
$$\dot{\xi} = \omega + \text{Im}\{\lambda'(0)\}\alpha + \frac{1}{2}\text{Im}\{C_1(0)\}r^2. \tag{7.23}$$

System (7.23) exhibits the same local bifurcation in a small neighborhood of the origin with sufficiently small α. We first consider the amplitude equation,

$$\dot{r} = \text{Re}\{\lambda'(0)\}\alpha r + \frac{1}{2}\text{Re}\{C_1(0)\}r^3, \tag{7.24}$$

since it is decoupled from ξ. Equation (7.24) always has the trivial equilibrium $r_0 = 0$. Other equilibria r of (7.24) satisfy $2\text{Re}\{\lambda'(0)\}\alpha r + \text{Re}\{C_1(0)\}r^3 = 0$, which has exactly one positive solution

$$r_1 = \sqrt{\frac{-2\text{Re}\{\lambda'(0)\}\alpha}{\text{Re}\{C_1(0)\}}} \tag{7.25}$$

if and only if $\alpha\text{Re}\{\lambda'(0)\}\text{Re}\{C_1(0)\} < 0$. Obviously, $r_1 \to 0$ as $\alpha \to 0$. This implies that system (7.19) has a branch of periodic solutions bifurcated from the origin that exists for $\alpha > 0$ (respectively, $\alpha < 0$) if $\text{Re}\{\lambda'(0)\}\text{Re}\{C_1(0)\} < 0$ (respectively, $\text{Re}\{\lambda'(0)\}\text{Re}\{C_1(0)\} > 0$).

The stability of the bifurcated periodic solutions is the same as that of r_1. Note that the eigenvalue of the linearized operator of the right-hand side of (7.24) at $r = r_1$ is $r_1^2\text{Re}\{C_1(0)\}$. It follows that the bifurcated periodic solutions are stable if $\text{Re}\{C_1(0)\} < 0$ and unstable otherwise.

Finally, we consider the phase equation for the bifurcated periodic solution of (7.23) corresponding to the equilibrium r_1. Namely,

$$\dot{\xi} = \omega + \text{Im}\{\lambda'(0)\}\alpha + \frac{1}{2}\text{Im}\{C_1(0)\}r^2. \tag{7.26}$$

It follows from (7.25) that

$$\alpha = -\frac{r_1^2\text{Re}\{C_1(0)\}}{2\text{Re}\{\lambda'(0)\}}.$$

Substituting the above expression of α into (7.26) yields

$$\dot{\theta} = \omega + \frac{1}{2}r_1^2 T_2,$$

where

$$T_2 = \mathrm{Im}\{C_1(0)\} - \frac{\mathrm{Im}\{\lambda'(0)\}}{\mathrm{Re}\{\lambda'(0)\}}\mathrm{Re}\{C_1(0)\}.$$

Therefore, the period of the bifurcated periodic solutions is greater than $\frac{2\pi}{\omega}$ (respectively, $< \frac{2\pi}{\omega}$) if $T_2 < 0$ (respectively, > 0).

We summarize the above discussion as follows.

Theorem 7.6. *In addition to conditions (HB1)–(HB2), assume that* $\mathrm{Re}\{C_1(0)\} \neq 0$. *Then system (7.19) undergoes a Hopf bifurcation. Moreover,* $\mathrm{Re}\{\lambda'(0)\}\mathrm{Re}\{C_1(0)\}$ *determines the directions of the Hopf bifurcation: if* $\mathrm{Re}\{\lambda'(0)\}\mathrm{Re}\{C_1(0)\} < 0$ *(respectively, > 0), then the Hopf bifurcation is supercritical (subcritical) and the bifurcating periodic solutions exist for* $\alpha > 0$ *(respectively, < 0);* $\mathrm{Re}\{C_1(0)\}$ *determines the stability of the bifurcating periodic solutions: the periodic solutions are orbitally stable (unstable) if* $\mathrm{Re}\{C_1(0)\} < 0$ *(respectively, > 0); and* T_2 *determines the period of the bifurcating periodic solutions: the period increases (respectively, decreases) if* $T_2 < 0$ *(respectively, > 0).*

To apply this theorem to specific systems, we need to know $\mathrm{Re}\{\lambda'(0)\}$ and $\mathrm{Re}\{C_1(0)\}$. In principle, $\mathrm{Re}\{C_1(0)\}$ is relatively straightforward to calculate. We simply carefully keep track of the coefficients in the normal form transformation in terms of our original vector field. However, in practice, the algebraic manipulations are horrendous.

7.3.3 Equivariant Hopf Bifurcation

In this section we consider system (7.13) with $\alpha \in \mathbb{R}$ under the following assumptions:

(EHB1) $\mathscr{A}_\alpha$ has a pair of complex conjugate eigenvalues $\lambda(\alpha)$ and $\overline{\lambda}(\alpha)$, each of multiplicity m, satisfying $\lambda(0) = i\omega$ with $\omega > 0$ and crossing the imaginary axis transversely at $\alpha = 0$ (i.e., $\mathrm{Re}\lambda'(0) \neq 0$).

(EHB2)   All other eigenvalues of $\mathscr{A}_0$ are not integer multiples of $i\omega$.

(EHB3)   System (7.13) is Γ-equivariant, and the eigenspace P associated with eigenvalue $\pm i\omega$ is Γ-simple, where Γ is a compact Lie group.

Assume that $\varphi_j \in C_{n,\tau}$, $j = 1,2,\ldots,m$, are the eigenfunctions of $\mathscr{A}_0$ associated with eigenvalue $i\omega$. Then the associated eigenspace P is spanned by φ_j and $\overline{\varphi}_j$, $j = 1,2,\ldots,m$. Assume that the adjoint space P^* of P is spanned by ψ_j and $\overline{\psi}_j$, $j = 1,2,\ldots,m$, where $\psi_j \in C^*_{n,\tau}$ satisfies $\langle \psi_j, \varphi_s \rangle = \delta_{js}$ and $\langle \overline{\psi}_j, \varphi_s \rangle = 0$ for all $j, s \in \{1,2,\ldots,m\}$. Let $\Phi = (\varphi_1,\ldots,\varphi_m,\overline{\varphi}_1,\ldots,\overline{\varphi}_m)$ and $\Psi = (\psi_1,\ldots,\psi_m,\overline{\psi}_1,\ldots,\overline{\psi}_m)^T$. Then $\langle \Psi, \Phi \rangle = \mathrm{Id}_{2m}$. By the center manifold theorem, we obtain the reduced Γ-equivariant equation

$$\dot{z}_j = i\omega z_j + G_j(\alpha, z), \tag{7.27}$$

where $z = (z_1, z_2, \ldots, z_m)^T \in \mathbb{C}^m$, $G_j(\alpha, z) = \overline{\Psi}_j(0)F(\alpha, 2\sum_{j=1}^m \text{Re}\{z_j\varphi_j\}$
$+ W(\alpha, z))$, and $W(\alpha, z)$ satisfies

$$\frac{\mathrm{d}}{\mathrm{d}t}W = \mathscr{A}W + H(\alpha, z) \tag{7.28}$$

with $H(\alpha, z) = [X_0 - \Phi\Psi(0)]F(\alpha, 2\sum_{j=1}^m \text{Re}\{z_j\varphi_j\} + W(\alpha, z))$. In view of $F(\alpha, 0) = 0$ for all $\alpha = 0$, we have $W(\alpha, 0) = 0$ for all $\alpha \in \mathbb{R}$. Therefore,

$$\frac{\partial^2}{\partial\alpha\partial z_s}G_j(0,0) = \overline{\Psi}_j(0)L'(0)\varphi_s = \lambda'(0)\delta_{js}$$

and

$$\frac{\partial^2}{\partial\alpha\partial\overline{z}_s}G_j(0,0) = \overline{\Psi}_j(0)L'(0)\overline{\varphi}_s = 0.$$

Therefore, the eigenvalue of the Jacobian matrix of $G(\alpha, z) = (G_1(\alpha, z), \ldots, G_m(\alpha, z))^T$ at $z = 0$ is $\sigma(\alpha)$, of multiplicity m, where $\sigma(\alpha) = i\omega + \alpha\lambda'(0) + o(|\alpha|)$. It follows from assumption (EHB1) that $\sigma'(0) \neq 0$. Therefore, all the conditions of the equivariant Hopf bifurcation theorem for ODE (7.27) established by Golubitsky et al. [118] are satisfied. Thus, we have established the corresponding equivariant Hopf bifurcation theorem for the RFDE (7.13).

Theorem 7.7. *Under assumptions (EHB1)–(EHB3), let $\Sigma \leq \Gamma \times \mathbb{S}^1$ be such that Σ is a maximal isotropy subgroup of $\Gamma \times \mathbb{S}^1$ with respect to their representations on P. Then there exists a unique branch of small-amplitude periodic solutions to (7.13) with period near $\frac{2\pi}{\omega}$ having Σ as their group of symmetry.*

It is easy to see that the Hopf-bifurcating periodic solutions of (7.13) have the same bifurcation direction as those of the corresponding Hopf-bifurcating periodic solutions of (7.27) on the center manifold. However, there is a little difference in their stability. If $\mathscr{A}_0$ has one eigenvalue with positive real part, then there exists an unstable manifold containing the trivial solution of (7.13), and hence all Hopf-bifurcating periodic solutions of (7.13) near $\alpha = 0$ are unstable; even the corresponding Hopf-bifurcating periodic solutions of (7.27) on the center manifold are stable. If all eigenvalues but $\pm i\omega$ have strictly negative real parts, then the Hopf-bifurcating periodic solutions of (7.13) have the same stability as those of the corresponding Hopf-bifurcating periodic solutions of the reduced ODE (7.27).

In the following, we look for periodic solutions to (7.27) with period approximately $\frac{2\pi}{\omega}$ by rescaling time as

$$s = (1 + \varsigma)t \tag{7.29}$$

for a new period-scaling parameter ς near 0. This yields the system

$$(1 + \varsigma)\frac{\mathrm{d}}{\mathrm{d}s}v = i\omega v + G(\alpha, v), \tag{7.30}$$

where $v(s) = v((1 + \varsigma)t) = z(t)$. Then $\frac{2\pi}{\omega}$-periodic solutions to (7.30) correspond to $\frac{2\pi}{(1+\varsigma)\omega}$-periodic solutions to (7.27). It follows from the $\mathbb{S}^1$-equivariance of the above normal form that the $\mathbb{S}^1$-action on a solution is identified with the phase shift. Hence $v(t) = e^{i\omega t}v(0)$, where $v(0)$ satisfies the steady-state equation:

$$i\omega\varsigma v = G(\alpha, v). \qquad (7.31)$$

Therefore, the bifurcations of small-amplitude periodic solutions of (7.27) are completely determined by the solutions of (7.31), and their orbital stability is determined by the signs of the eigenvalues of

$$-i\omega\varsigma\mathrm{Id}_m + G_v(\alpha, v(0)). \qquad (7.32)$$

According to Theorem 6.5 on the Page 297 of Golubitsky et al. [118], we have the following results.

Theorem 7.8. *In addition to assumptions (EHB1)–(EHB3), assume that all eigenvalues but $\pm i\omega$ of $\mathscr{A}_0$ have strictly negative real parts. Let $u(t; \alpha)$ be a small-amplitude Σ-symmetric T-periodic solution of (7.13), which corresponds to the solution $v(\alpha, \varsigma)$ of (7.31). Then $u(t; \alpha)$ is orbitally asymptotically stable if the $(2m - 1 - \dim\Gamma + \dim\Sigma)$-eigenvalues of (7.32) that are not forced to zero by the group action have negative real parts, while it is unstable if one of these eigenvalues has positive real part.*

7.3.4 Application to $\mathbb{D}_n$-Equivariant Hopf Bifurcation

In this subsection, we consider a special case in which $m = 2$ and Γ is the $2n$-order dihedral group $\mathbb{D}_n$ with $n \geq 3$. Recall that $\mathbb{D}_n$ is generated by the permutation γ of order n and the flip κ of order 2. Thus, assumptions (EHB1)–(EHB3) reduce to the following assumptions:

(DHB1) $\mathscr{A}_\alpha$ has a pair of complex conjugate eigenvalues $\lambda(\alpha)$ and $\overline{\lambda}(\alpha)$, each of multiplicity 2, satisfying $\lambda(0) = i\omega$ with $\omega > 0$ and $\mathrm{Re}\{\lambda'(0)\} > 0$.

(DHB2) All other eigenvalues of $\mathscr{A}_0$ have strictly negative real parts.

(DHB3) System (7.13) is $\mathbb{D}_n$-equivariant ($n \geq 3$).

Assumptions (DHB1) and (DHB2) imply that the trivial solution of (7.13) is subcritically stable and loses stability as α passes through 0. Let P and P^* be the eigenspaces of $\mathscr{A}_0$ and $\mathscr{A}_0^*$ associated with $\pm i\omega$, respectively. Let $\frac{2\pi}{n}$ denote a generator of $\mathbb{Z}_n$. Then there exist $v_1 \in \mathbb{C}^n$ and $u_1 \in \mathbb{C}^{n*}$ such that $\frac{2\pi}{n} \cdot v_1 = e^{i\frac{2k\pi}{n}}v_1$ for some positive integer $k < n$, and

$$\Delta(0, i\omega)v_1 = 0, \quad \overline{u}_1\Delta(0, i\omega) = 0, \quad \overline{u}_1\Delta_\lambda(0, i\omega)v_1 = 1.$$

Setting $u_2 = \kappa \cdot u_1$ and $v_2 = \kappa \cdot v_1$, we have $P = \mathrm{span}\{\varphi_1, \overline{\varphi}_1, \varphi_2, \overline{\varphi}_2\}$ and $P^* = \mathrm{span}\{\psi_1, \overline{\psi}_1, \psi_2, \overline{\psi}_2\}$, where $\varphi_j(\theta) = v_j e^{i\omega\theta}$ and $\psi_j(\theta) = u_j e^{i\omega\theta}$ for $\theta \in [-\tau, 0]$, $j = 1, 2$. Then $\langle\psi_j, \varphi_s\rangle = \delta_{js}$ and $\langle\overline{\psi}_j, \varphi_s\rangle = 0$ for all $j, s \in \{1, 2\}$. Let $\Phi = (\varphi_1, \overline{\varphi}_1, \varphi_2, \overline{\varphi}_2)$ and $\Psi = (\psi_1, \overline{\psi}_1, \psi_2, \overline{\psi}_2)^T$. Then $\langle\Psi, \Phi\rangle = \mathrm{Id}_4$.

Clearly, P and P^* are $\mathbb{D}_n \times \mathbb{S}^1$-invariant. Let $\gamma = \frac{2k\pi}{n}$. Then γ can be regarded as a generator of $\mathbb{Z}_{n^*}$, where

$$n^* = \frac{n}{\gcd(k,n)}, \qquad (7.33)$$

and $\gcd(k,n)$ denotes the greatest common divisor of k and n. Then the action $\mathbb{D}_n \times \mathbb{S}^1$ on P induces the action of $\mathbb{D}_{n^*} \times \mathbb{S}^1$ on $\mathbb{C}^2$, which is given by

$$\begin{aligned}
\gamma \cdot (z_1, z_2) &= (e^{i\gamma} z_1, e^{-i\gamma} z_2), & \gamma \in \mathbb{Z}_{n^*}, \\
\kappa \cdot (z_1, z_2) &= (z_2, z_1), & \kappa \in \mathbb{Z}_2, \\
\theta \cdot (z_1, z_2) &= (z_1 e^{i\theta}, z_2 e^{i\theta}), & \theta \in \mathbb{S}^1.
\end{aligned} \qquad (7.34)$$

Obviously, $\mathbb{C}^2$ and so P are $\mathbb{D}_n$-simple.

In order to apply Theorem 7.7, we need to find all the maximal isotropy subgroups of $\mathbb{D}_{n^*} \times \mathbb{S}^1$. In fact, $\mathbb{D}_{n^*} \times \mathbb{S}^1$ always has three maximal isotropy subgroups, see Table 7.1, where $\tilde{\mathbb{Z}}_{n^*} = \{(\gamma, -\gamma) : \gamma \in \mathbb{Z}_{n^*}\}$, $\mathbb{Z}_2(\kappa) = \{(0,0), \kappa\}$, $\mathbb{Z}_2(\kappa, \pi) = \{(0,0), (\kappa, \pi)\}$, and $\mathbb{Z}_2^c = \{(0,0), (\pi, \pi)\}$, $\mathbb{Z}_2(\kappa) \oplus \mathbb{Z}_2^c = \{(0,0), (\pi, \pi), (\kappa, 0), (\kappa\pi, \pi)\}$, $\mathbb{Z}_2(\kappa, \pi) \oplus \mathbb{Z}_2^c = \{(0,0), (\kappa, \pi), (\kappa\pi, 0), (\pi, \pi)\}$. It follows from the equivariant Hopf theorem, Theorem 7.7, that there are (at least) three branches of periodic solutions occurring generically in Hopf bifurcation with $\mathbb{D}_{n^*}$ symmetry. In what follows, we discuss the stabilities along those branches.

Table 7.1 Isotropy subgroups of $\mathbb{D}_{n^*} \times \mathbb{S}^1$ acting on $\mathbb{C}^2$

n^*	Isotropy subgroups	Fixed-point subspace	Dimensions
	$\mathbb{D}_{n^*} \times \mathbb{S}^1$	$(0,0)$	0
	$\tilde{\mathbb{Z}}_{n^*}$	$\{(z_1, 0)\}$	2
n^* is odd	$\mathbb{Z}_2(\kappa)$	$\{(z_1, z_1)\}$	2
	$\mathbb{Z}_2(\kappa, \pi)$	$\{(z_1, -z_1)\}$	2
	$\mathbb{I}$	$\mathbb{C}^2$	4
	$\mathbb{D}_{n^*} \times \mathbb{S}^1$	$(0,0)$	0
	$\tilde{\mathbb{Z}}_{n^*}$	$\{(z_1, 0)\}$	2
$n^* = 2 \pmod 4$	$\mathbb{Z}_2(\kappa) \oplus \mathbb{Z}_2^c$	$\{(z_1, z_1)\}$	2
	$\mathbb{Z}_2(\kappa, \pi) \oplus \mathbb{Z}_2^c$	$\{(z_1, -z_1)\}$	2
	$\mathbb{Z}_2^c$	$\mathbb{C}^2$	4
	$\mathbb{D}_{n^*} \times \mathbb{S}^1$	$(0,0)$	0
	$\tilde{\mathbb{Z}}_{n^*}$	$\{(z_1, 0)\}$	2
$n^* = 0 \pmod 4$	$\mathbb{Z}_2(\kappa) \oplus \mathbb{Z}_2^c$	$\{(z_1, z_1)\}$	2
	$\mathbb{Z}_2(\kappa\gamma) \oplus \mathbb{Z}_2^c$	$\{(z_1, e^{i\gamma} z_1)\}$	2
	$\mathbb{Z}_2^c$	$\mathbb{C}^2$	4

The nonresonance conditions are obviously satisfied. Thus, we can obtain the following $\mathbb{D}_{n^*} \times \mathbb{S}^1$-equivariant normal form:

$$\dot{z}_j = i\omega z_j + G_j(\alpha, z_1, z_2), \quad j = 1, 2. \qquad (7.35)$$

It follows from the $\mathbb{D}_{n^*} \times \mathbb{S}^1$-equivariance that $G = (G_1, G_2)^T$ takes the form

$$G(\alpha, z_1, z_2) = B \begin{bmatrix} z_1 \\ z_2 \end{bmatrix} + C \begin{bmatrix} z_1^2 \bar{z}_1 \\ z_2^2 \bar{z}_2 \end{bmatrix} + D \begin{bmatrix} \bar{z}_1^{k^*-1} z_2^{k^*} \\ \bar{z}_1^{k^*} \bar{z}_2^{k^*-1} \end{bmatrix} + E \begin{bmatrix} z_1^{k^*+1} \bar{z}_2^{k^*} \\ \bar{z}_1^{k^*} z_2^{k^*+1} \end{bmatrix}, \quad (7.36)$$

where $k^* = n^*$ if n^* is odd and $\frac{n^*}{2}$ otherwise, and B, C, D, E are complex-valued $\mathbb{D}_{n^*} \times \mathbb{S}^1$-invariant functions. Namely, B, C, D, E are functions of α, ℓ_j, $j = 1, 2, 3, 4$, where $\ell_1 = |z_1|^2 + |z_2|^2$, $\ell_2 = |z_1|^2 |z_2|^2$, $\ell_3 = (z_1 \bar{z}_2)^{k^*} + (\bar{z}_1 z_2)^{k^*}$, $\ell_4 = i(|z_1|^2 - |z_2|^2)[(z_1 \bar{z}_2)^{k^*} - (\bar{z}_1 z_2)^{k^*}]$.

For the simplification of notation, we introduce the operator $\Theta : V_j^5(\mathbb{C}) \to V_j^5(\mathbb{C}^4)$ defined by

$$\Theta(c z_1^{q_1} \bar{z}_1^{q_2} z_2^{q_3} \bar{z}_2^{q_4} \alpha^l) = \begin{bmatrix} c z_1^{q_1} \bar{z}_1^{q_2} z_2^{q_3} \bar{z}_2^{q_4} \alpha^l \\ \bar{c} z_1^{q_2} \bar{z}_1^{q_1} z_2^{q_4} \bar{z}_2^{q_3} \alpha^l \\ c z_1^{q_3} \bar{z}_1^{q_4} z_2^{q_1} \bar{z}_2^{q_2} \alpha^l \\ \bar{c} z_1^{q_4} \bar{z}_1^{q_3} z_2^{q_2} \bar{z}_2^{q_1} \alpha^l \end{bmatrix}$$

for $c \in \mathbb{C}$, $q = (q_1, q_2, q_3, q_4) \in \mathbb{N}_0^4$, and $l \in \mathbb{N}_0$ with $|(q, l)| = j$. Obviously, $\Theta(z + y) = \Theta(z) + \Theta(y)$ for $z, y \in V_j^5(\mathbb{C})$. Let

$$\Theta(G_1(\alpha, z_1, z_2)) = \frac{1}{2} g_2^1(\alpha, z, 0) + \frac{1}{6} g_3^1(\alpha, z, 0) + \cdots .$$

Then $g_2^1(\alpha, z, 0) = \Theta(J_0 \alpha z_1)$, and

$$g_3^1(\alpha, z, 0) = \Theta\left(K_0 \alpha^2 z_1 + K_1 z_1 |z_1|^2 + K_2 z_1 |z_2|^2\right)$$

if $n^* \geq 3$ and $n^* \neq 4$, or

$$g_3^1(\alpha, z, 0) = \Theta\left(K_0 \alpha^2 z_1 + K_1 z_1 |z_1|^2 + K_2 z_1 |z_2|^2 + K_3 \bar{z}_1 z_2^2\right)$$

if $n^* = 4$. In the sequel, we will not calculate K_0, because it does not have much effect in our bifurcation analysis. Our main purpose is to obtain concrete formulas of the complex coefficients J_0, K_1, K_2, and K_3.

Let

$$\begin{aligned} f_j^1(\alpha, z, y) &= \overline{\Psi}(0) F_j(\alpha, \Phi z + y), \\ f_j^2(\alpha, z, y) &= [X_0 - \Phi \overline{\Psi}(0)] F_j(\alpha, \Phi z + y), \end{aligned}$$

for $j \geq 2$, $z \in \mathbb{C}^4$, $y \in Q$, where $\Phi z = 2\mathrm{Re}\{\varphi_1 z_1 + \varphi_2 z_2\}$, Q denotes the space complementary to P in BC_n, F_j is the jth Fréchet derivative of $F(\alpha, \varphi)$ with respect to $\alpha \in \mathbb{R}$ and $\varphi \in C_{n,\tau}$. As in Sect. 4.3.2, we write (7.13) as

$$\begin{aligned} \dot{z} &= \Omega z + \sum_{j \geq 2} \frac{1}{j!} f_j^1(\alpha, z, y), \\ \frac{dy}{dt} &= \mathscr{A}_Q y + \sum_{j \geq 2} \frac{1}{j!} f_j^2(\alpha, z, y), \end{aligned} \quad (7.37)$$

where $\Omega = \mathrm{diag}(i\omega, -i\omega, i\omega, -i\omega)$. With the change of variables

$$(z,y) = (\hat{z},\hat{y}) + \tfrac{1}{j!}(U_j^1(\alpha,\hat{z}), U_j^2(\alpha,\hat{z})), \tag{7.38}$$

and dropping the hats for simplicity of notation, (7.37) becomes

$$\dot{z} = \Omega z + \tfrac{1}{2}g_2^1(\alpha,z,y) + \tfrac{1}{3!}\bar{f}_3^1(\alpha,z,y) + \cdots,$$
$$\tfrac{dy}{dt} = \mathscr{A}_Q y + \tfrac{1}{2}g_2^2(\alpha,z,y) + \tfrac{1}{3!}\bar{f}_3^2(\alpha,z,y) + \cdots, \tag{7.39}$$

where $g_2 = (g_2^1, g_2^2)$, $\hat{f}_2 = (f_2^1, f_2^2)$, and $\bar{f}_3 = (f_3^1, f_3^2)$ satisfy

$$g_2 = f_2 - \mathbf{M}_2 U_2 \tag{7.40}$$

and

$$\bar{f}_3 = f_3 + \frac{3}{2}[(D_{z,y}f_2)U_2 - (D_{z,y}U_2)g_2]. \tag{7.41}$$

Here, the operator $\mathbf{M}_2 = (\mathbf{M}_2^1, \mathbf{M}_2^2)$ is defined in Chap. 4. Thus, we may have $g_2^1(\alpha,z,y) = \mathrm{Proj}_{\mathrm{Ker}\mathbf{M}_2^{1*}} f_2^1(\alpha,z,y)$.

Let

$$f_j(\alpha, \Phi z) = \sum_{|(q,l)|=j} A_{(q,l)} z^q \alpha^l. \tag{7.42}$$

Obviously,

$$A_{(q_1,q_2,q_3,q_4,l)} = \overline{A}_{(q_2,q_1,q_4,q_3,l)}$$

for all $q = (q_1,q_2,q_3,q_4) \in \mathbb{N}_0^4$ and $l \in \mathbb{N}_0$ with $|(q,l)| \geq 2$. It follows from $\tilde{f}(\alpha,0) = 0$ and $\tilde{f}_\varphi(\alpha,0) = 0$ for all $\alpha \in \mathbb{R}$ that $A_{(q,j-1)} = A_{(0,j)} = 0$ for all q with $|q| = 1$ and $j \geq 2$. Moreover, by the equivariance of f_j,

$$\kappa \cdot A_{(q_1,q_2,q_3,q_4,l)} = A_{(q_3,q_4,q_1,q_2,l)}$$

for all $q = (q_1,q_2,q_3,q_4) \in \mathbb{N}_0^4$ and $l \in \mathbb{N}_0$ with $|(q,l)| \geq 2$. Notice that $f_2^1(\alpha,z,0) = \Psi(0)f_2(\alpha, \Phi z) + 2\alpha\Psi(0)L'(0)\Phi z$. Then we have

$$f_2^1(\alpha,z,0) = \Theta\left(2\alpha\bar{u}_1 L'(0)\Phi z + \sum_{|q|=2}\bar{u}_1 A_{(q,0)} z^q\right).$$

Note that

$$\mathbf{M}_j^1(z^q \alpha^l e_k) = i\omega\left[q_1 - q_2 + q_3 - q_4 + (-1)^k\right] z^q \alpha^l e_k$$

for all $1 \leq k \leq 4$, $q \in \mathbb{N}_0^4$, and $l \in \mathbb{N}_0$ with $|(q,l)| = j$, where $\{e_1,e_2,e_3,e_4\}$ is the canonical basis for $\mathbb{C}^4$. Thus,

$$g_2^1(\alpha,z,0) = 2\alpha\Theta\left(\lambda'(0)z_1\right),$$

i.e., $J_0 = 2\lambda'(0)$. Moreover, we have $U_2(\alpha,z) = \mathbf{M}_2^{-1}[\hat{f}_2(\alpha,z,0) - g(\alpha,z,0)]$, that is,

$$U_2^1(\alpha,z) = \Theta\left(\frac{\bar{u}_1}{i\omega}\left[-\alpha L'(0)(\bar{z}_1\varphi_2 + \bar{z}_2\varphi_4) + \sum_{|q|=2}(q_1 - q_2 + q_3 - q_4 - 1)^{-1}A_{(q,0)}z^q\right]\right),$$

and $U_2^2(\alpha,z) = h(\alpha,z)$ is the unique solution in $V_2^5(Q)$ of the equation

$$(\mathbf{M}_2^2 h)(\alpha,z) = [X_0 - \Phi\overline{\Psi}(0)]\left[2\alpha L'(0)(\Phi z) + f_2(\alpha,\Phi z)\right]. \tag{7.43}$$

Namely,

$$i\omega[z_1 D_{z_1}h - \bar{z}_1 D_{\bar{z}_1}h + z_2 D_{z_2}h - \bar{z}_2 D_{\bar{z}_2}h] - \dot{h} + X_0[h(0) - L(0)h]$$
$$= [X_0 - \Phi\overline{\Psi}(0)][2\alpha L'(0)(\Phi z) + f_2(\alpha,\Phi z)].$$

We write $U_2^2(z,\alpha)$ in the form

$$U_2^2(\alpha,z) = \sum_{|(q,l)|=2} h_{(q,l)}z^q\alpha^l.$$

Then (7.43) is equivalent to the following equations:

$$\begin{aligned}
\dot{h}_{(2,0,0,0,0)}(\theta) - 2i\omega h_{(2,0,0,0,0)}(\theta) &= \Phi\overline{\Psi}(0)A_{(2,0,0,0,0)}, \\
\dot{h}_{(0,0,2,0,0)}(\theta) - 2i\omega h_{(0,0,2,0,0)}(\theta) &= \Phi\overline{\Psi}(0)A_{(0,0,2,0,0)}, \\
\dot{h}_{(1,0,1,0,0)}(\theta) - 2i\omega h_{(1,0,1,0,0)}(\theta) &= \Phi\overline{\Psi}(0)A_{(1,0,1,0,0)}, \\
\dot{h}_{(1,1,0,0,0)}(\theta) &= \Phi\overline{\Psi}(0)A_{(1,1,0,0,0)}, \\
\dot{h}_{(0,0,1,1,0)}(\theta) &= \Phi\overline{\Psi}(0)A_{(0,0,1,1,0)}, \\
\dot{h}_{(1,0,0,1,0)}(\theta) &= \Phi\overline{\Psi}(0)A_{(1,0,0,1,0)}, \\
\dot{h}_{(0,1,1,0,0)}(\theta) &= \Phi\overline{\Psi}(0)A_{(0,1,1,0,0)},
\end{aligned} \tag{7.44}$$

with the boundary conditions

$$\begin{aligned}
\dot{h}_{(2,0,0,0,0)}(0) - L(0)h_{(2,0,0,0,0)} &= A_{(2,0,0,0,0)}, \\
\dot{h}_{(0,0,2,0,0)}(0) - L(0)h_{(0,0,2,0,0)} &= A_{(0,0,2,0,0)}, \\
\dot{h}_{(1,0,1,0,0)}(0) - L(0)h_{(1,0,1,0,0)} &= A_{(1,0,1,0,0)}, \\
\dot{h}_{(1,1,0,0,0)}(0) - L(0)h_{(1,1,0,0,0)} &= A_{(1,1,0,0,0)}, \\
\dot{h}_{(0,0,1,1,0)}(0) - L(0)h_{(0,0,1,1,0)} &= A_{(0,0,1,1,0)}, \\
\dot{h}_{(1,0,0,1,0)}(0) - L(0)h_{(1,0,0,1,0)} &= A_{(1,0,0,1,0)}, \\
\dot{h}_{(0,1,1,0,0)}(0) - L(0)h_{(0,1,1,0,0)} &= A_{(0,1,1,0,0)}.
\end{aligned} \tag{7.45}$$

With U_2 at our disposal, we can find $g_3^1(\alpha,z,0)$, which is the projection of $\bar{f}_3(\alpha,z,0)$ on $\mathrm{Ker}\,\mathbf{M}_3^{1*}$. It follows that the complex coefficients K_1, K_2, and K_3 are given as follows:

$$\begin{aligned}
K_1 = &\,\bar{u}_1 A_{(2,1,0,0,0)} + \frac{3}{2i\omega}\left[-\bar{u}_1 A_{(2,0,0,0,0)}\bar{u}_1 A_{(1,1,0,0,0)} + |\bar{u}_1 A_{(1,1,0,0,0)}|^2\right. \\
&\left. + \frac{2}{3}|\bar{u}_1 A_{(0,2,0,0,0)}|^2 + \bar{u}_1 A_{(0,1,1,0,0)}\bar{u}_1 A_{(0,0,2,0,0)} - \bar{u}_1 A_{(1,0,1,0,0)}\bar{u}_1 A_{(0,0,1,1,0)}\right]
\end{aligned}$$

$$+\bar{u}_1 A_{(1,0,0,1,0)} u_1 A_{(0,0,1,1,0)} + \tfrac{1}{3}\bar{u}_1 A_{(0,1,0,1,0)} u_1 A_{0,0,2,0,0}\big]$$
$$+3\bar{u}_1 \mathscr{H}_2((0,\varphi_1),(0,h_{(1,1,0,0,0)})) + 3\bar{u}_1 \mathscr{H}_2((0,\overline{\varphi}_1),(0,h_{(2,0,0,0,0)})),$$

$$K_2 = \bar{u}_1 A_{(1,0,1,1,0)} + \tfrac{3}{2i\omega}\big[-2\bar{u}_1 A_{(2,0,0,0,0)}\bar{u}_1 A_{(0,0,1,1,0)}$$
$$+\bar{u}_1 A_{(1,1,0,0,0)} u_1 A_{(0,0,1,1,0)} + \tfrac{1}{3}|\bar{u}_1 A_{(0,1,0,1,0)}|^2$$
$$+|\bar{u}_1 A_{(0,1,1,0,0)}|^2 + \bar{u}_1 A_{(0,0,1,1,0)}\bar{u}_1 A_{(1,0,1,0,0)}$$
$$-2\bar{u}_1 A_{(0,0,2,0,0)}\bar{u}_1 A_{(0,1,1,0,0)} - \bar{u}_1 A_{(1,0,1,0,0)}\bar{u}_1 A_{(1,1,0,0,0)}$$
$$+\bar{u}_1 A_{(1,0,0,1,0)} u_1 A_{(1,1,0,0,0)} + \tfrac{2}{3}\bar{u}_1 A_{(0,0,0,2,0)} u_1 A_{(1,0,1,0,0)}$$
$$+\bar{u}_1 A_{(0,0,1,1,0)} u_1 A_{(0,1,1,0,0)}\big] + 3\bar{u}_1 \mathscr{H}_2((0,\varphi_1),(0,h_{(0,0,1,1,0)}))$$
$$+3\bar{u}_1 \mathscr{H}_2((0,\varphi_2),(0,h_{(1,0,0,1,0)})) + 3\bar{u}_1 \mathscr{H}_2((0,\overline{\varphi}_2),(0,h_{(1,0,1,0,0)})),$$

$$K_3 = \bar{u}_1 A_{(0,1,2,0,0)} + \tfrac{3}{2i\omega}\big[\bar{u}_1 A_{(1,1,0,0,0)}\bar{u}_1 A_{(0,0,2,0,0)} - \bar{u}_1 A_{(1,0,1,0,0)}\bar{u}_1 A_{(0,1,0,1,0)}$$
$$+\tfrac{2}{3}\bar{u}_1 A_{(0,2,0,0,0)} u_1 A_{(0,0,2,0,0)} + \bar{u}_1 A_{(0,1,1,0,0)}\bar{u}_1 A_{(2,0,0,0,0)}$$
$$-2\bar{u}_1 A_{(0,0,2,0,0)} u_1 A_{(1,0,0,1,0)} + \bar{u}_1 A_{(0,1,1,0,0)} u_1 A_{(0,1,1,0,0)}$$
$$+\tfrac{1}{3}\bar{u}_1 A_{(0,1,0,1,0)} u_1 A_{(2,0,0,0,0)} + \bar{u}_1 A_{(0,0,1,1,0)} u_1 A_{(1,0,0,1,0)}\big]$$
$$+3\bar{u}_1 \mathscr{H}_2((0,\overline{\varphi}_1),(0,h_{(0,0,2,0,0)})) + 3\bar{u}_1 \mathscr{H}_2((0,\varphi_2),(0,h_{(0,1,1,0,0)})).$$

Consequently, we can obtain the coefficients for up to third-order terms of G in (7.36) as follows:

$$B_\alpha(0) = \lambda'(0), \quad B_{\ell_1}(0) = \frac{1}{6}K_2, \quad C(0) = \frac{1}{6}(K_1 - K_2). \tag{7.46}$$

In particular, if $n^* = 4$, then $D(0) = K_3/6$.

It follows from Theorem 7.8 that bifurcations of small-amplitude Σ-symmetric periodic solutions of (7.13) correspond to the zeros of the function $\mathscr{H} : \mathbb{C}^2 \times \mathbb{R} \times \mathbb{R} \to \mathbb{C}^2$ given by

$$H(v,\alpha,\varsigma) = -i\varsigma\omega v + G(\alpha,v_1,v_2), \tag{7.47}$$

and their orbital stability is determined by the signs of the eigenvalues of

$$D_v H(v,\alpha,\varsigma) = -i\varsigma\omega \mathrm{Id}_2 + D_v G(\alpha,v_1,v_2). \tag{7.48}$$

By (7.46),

$$H(v,\alpha,\varsigma) = \Big[\alpha\lambda'(0) - i\varsigma\omega + \tfrac{1}{6}K_0\alpha^2 + \tfrac{1}{6}K_2(|v_1|^2 + |v_2|^2)\Big]\begin{bmatrix} v_1 \\ v_2 \end{bmatrix}$$
$$+\tfrac{1}{6}(K_1 - K_2)\begin{bmatrix} v_1^2\overline{v}_1 \\ v_2^2\overline{v}_2 \end{bmatrix} + O(|\alpha,v|^3) \tag{7.49}$$

if $n^* \geq 3$ and $n^* \neq 4$, and

$$
\begin{aligned}
H(v,\alpha,\varsigma) = {} & \left[\alpha\lambda'(0) - i\varsigma\omega + \tfrac{1}{6}K_0\alpha^2 + \tfrac{1}{6}K_2(|v_1|^2 + |v_2|^2)\right]\begin{bmatrix} v_1 \\ v_2 \end{bmatrix} \\
& + \tfrac{1}{6}(K_1 - K_2)\begin{bmatrix} v_1^2\overline{v}_1 \\ v_2^2\overline{v}_2 \end{bmatrix} + \tfrac{1}{6}K_3\begin{bmatrix} \overline{v}_1 v_1^2 \\ v_1^2\overline{v}_2 \end{bmatrix} + O(|\alpha,v|^3)
\end{aligned}
\tag{7.50}
$$

if $n^* = 4$. According to the properties of the $\mathbb{D}_{n^*}$-symmetric Hopf bifurcation equation (see [118, Theorem 3.1 on Page 382] for more details), we have the following results.

Theorem 7.9. *Under assumptions (DHB1)–(DHB3), assume that the following non-degeneracy condition holds:*

$$
\mathrm{Re}\{K_1\}\mathrm{Re}\{K_1 + K_2\}\mathrm{Re}\{K_1 - K_2\}\mathrm{Re}\{(K_1 - K_2)\overline{D(0)}\} \neq 0. \tag{7.51}
$$

If $n^* \geq 3$ and $n^* \neq 4$, then near $\alpha = 0$, system (7.13) undergoes Hopf bifurcations and has at least one branch of small amplitude, near-$\frac{2\pi}{\omega}$-periodic solutions, for each of the isotropy subgroups $\tilde{\mathbb{Z}}_{n^*}$, $\mathbb{Z}_2(\kappa)$ (when n^* is odd), or $\mathbb{Z}_2(\kappa) \oplus \mathbb{Z}_2^c$ (when n^* is even); and $\mathbb{Z}_2(\kappa,\pi)$ (when n^* is odd), $\mathbb{Z}_2(\kappa,\pi) \oplus \mathbb{Z}_2^c$ (when $n^* = 2\,(\mathrm{mod}\,4)$), or $\mathbb{Z}_2(\kappa\gamma) \oplus \mathbb{Z}_2^c$ (when $n^* = 0\,(\mathrm{mod}\,4)$). The bifurcation direction and stability of each branch of bifurcated periodic solutions are completely determined by K_1, K_2, and $D(0)$:

(i) *The $\tilde{\mathbb{Z}}_{n^*}$ branch is supercritical (respectively, subcritical) if $\mathrm{Re}(K_1) < 0$ (respectively, > 0). It is stable if $\mathrm{Re}(K_2) < \mathrm{Re}(K_1) < 0$.*

(ii) *The $\mathbb{Z}_2(\kappa)$ branch is supercritical (respectively, subcritical) if $\mathrm{Re}(K_1 + K_2) < 0$ (respectively, > 0). It is stable if $\mathrm{Re}(K_1) < \mathrm{Re}(K_2) < -\mathrm{Re}(K_1)$ and $\mathrm{Re}\{(K_1 - K_2)\overline{D(0)}\} < 0$.*

(iii) *The $\mathbb{Z}_2(\kappa,\pi)$, $\mathbb{Z}_2(\kappa,\pi) \oplus \mathbb{Z}_2^c$, or $\mathbb{Z}_2(\kappa\gamma) \oplus \mathbb{Z}_2^c$ branch is supercritical (respectively, subcritical) if $\mathrm{Re}(K_1 + K_2) < 0$ (respectively, > 0). It is stable if $\mathrm{Re}(K_1) < \mathrm{Re}(K_2) < -\mathrm{Re}(K_1)$ and $\mathrm{Re}\{(K_1 - K_2)\overline{D(0)}\} > 0$.*

Remark 7.1. As for $D(0)$, we have to resort to a more complicated calculation of $(2k^* - 1)$-order terms of the normal form (7.35). Nevertheless, it is not necessary to do so, because $D(0)$ has nothing to do with the bifurcation direction for each branch of bifurcated periodic solutions of (7.13) with $n^* \geq 3$ and $n^* \neq 4$. As for stability, the stability of the $\tilde{\mathbb{Z}}_{n^*}$ branch does not depend on $D(0)$. Even for other types of branches, $\mathrm{Re}(K_1) < \mathrm{Re}(K_2) < -\mathrm{Re}(K_1)$ is a necessary condition for stability. If this condition fails, we can immediately conclude that such branches are unstable without computing $D(0)$.

Remark 7.2. If the nondegeneracy condition (7.51) does not hold, then we must proceed with computing the coefficients of g_j^1 ($j \geq 4$) in the normal form, up to the first nonvanishing coefficients.

Theorem 7.10. *Under assumptions (DHB1)–(DHB3), assume that the following nondegeneracy conditions hold:*

$$\mathrm{Re}\{K_1\}\mathrm{Re}\{K_1 - K_2\} \neq 0,$$
$$\mathrm{Re}\{K_1 + K_2 + K_3\}\mathrm{Re}\{K_1 + K_2 - K_3\} \neq 0,$$
$$\mathrm{Re}\{K_1 - K_2 + 3K_3\}\mathrm{Re}\{K_1 - K_2 - 3K_3\} \neq 0,$$
$$|K_3|^2 + \mathrm{Re}\{(K_1 - K_2)K_3\} \neq 0,$$
$$|K_3|^2 - \mathrm{Re}\{(K_1 - K_2)K_3\} \neq 0,$$
$$|K_1 - K_2|^2 \neq |K_3|^2.$$

If $n^* = 4$, then near $\alpha = 0$, system (7.13) undergoes Hopf bifurcations and has at least one branch of small amplitude, near-$\frac{2\pi}{\omega}$-periodic solutions, for each of the isotropy subgroups $\tilde{\mathbb{Z}}_4$, $\mathbb{Z}_2(\kappa) \oplus \mathbb{Z}_2^c$, and $\mathbb{Z}_2(\kappa\gamma) \oplus \mathbb{Z}_2^c$. The bifurcation direction and stability of each branch of bifurcated periodic solutions are completely determined by K_1, K_2, and K_3:

(i) *The $\tilde{\mathbb{Z}}_4$ branch is supercritical (respectively, subcritical) if $\mathrm{Re}(K_1) < 0$ (respectively, > 0). It is stable if $\mathrm{Re}(K_2) < \mathrm{Re}(K_1) < 0$ and $|K_1 - K_2|^2 > |K_3|^2$.*
(ii) *The $\mathbb{Z}_2(\kappa) \oplus \mathbb{Z}_2^c$ branch is supercritical (respectively, subcritical) if $\mathrm{Re}(K_1 + K_2 + K_3) < 0$ (respectively, > 0). It is stable if $\mathrm{Re}(K_1 + K_2 + K_3) < 0$, $\mathrm{Re}(K_1 - K_2 - 3K_3) < 0$, and $|K_3|^2 - \mathrm{Re}\{(K_1 - K_2)K_3\} > 0$.*
(iii) *The $\mathbb{Z}_2(\kappa\gamma) \oplus \mathbb{Z}_2^c$ branch is supercritical (respectively, subcritical) if $\mathrm{Re}(K_1 + K_2 - K_3) < 0$ (respectively, > 0). It is stable if $\mathrm{Re}(K_1 + K_2 - K_3) < 0$, $\mathrm{Re}(K_1 - K_2 + 3K_3) < 0$, and $|K_3|^2 + \mathrm{Re}\{(K_1 - K_2)K_3\} > 0$.*

7.3.5 Hopf Bifurcation in a Ring Network

In this section, we consider the influence of the delay on the behavior of a ring network modeled by the following system of delay differential equations:

$$\dot{u}_i(t) = -u_i(t) + f(u_i(t - \tau)) - [g(u_{i-1}(t - \tau)) + g(u_{i+1}(t - \tau))], \quad (7.52)$$

where i (mod n), $f, g \in C^1(\mathbb{R}; \mathbb{R})$ with $f(0) = g(0) = 0$. It can be seen that system (7.52) is bidirectional in the sense that the growth rate of the ith neuron depends on the excitatory (positive) self-feedback and the inhibitory (negative) feedback from the $(i-1)$th and the $(i+1)$th neurons. Thus, if the transfer functions f and g are monotonically increasing, then the network modeled by (7.52) has on-center off-surround characteristics.

We note that with the transformation $x_i(t) = u_i(\tau t)$ for i(mod n) and $h = f - 2g$, we can rewrite (7.52) as the following system of delay differential equations:

$$\dot{x}_i = -\tau x_i(t) + \tau h(x_i(t - 1)) \\ - \tau[g(x_{i-1}(t - 1)) + g(x_{i+1}(t - 1)) - 2g(x_i(t - 1))], \quad (7.53)$$

where i(mod n).

We denote a symmetric circulant matrix by $J = \text{circ}(a_1, a_2, \cdots, a_n)$, where $J_{ij} = a_{j-i+1}$ and $a_i = a_{n-i+2}$, $i(\text{mod } n)$. Clearly, we have the following properties (for more details about the proof, we refer to [133]).

Lemma 7.1. *Denote by ρ the generator of the cyclic subgroup $\mathbb{Z}_n$ and κ the flip. Define the action of $\mathbb{D}_n$ on $\mathbb{R}^n$ by*

$$(\rho x)_i = x_{i+1}, \quad (\kappa x)_i = x_{n+2-i}$$

for all $i(\text{mod } n)$ and $x \in \mathbb{R}^n$. Then system (7.53) is $\mathbb{D}_n$-equivariant.

The linearization of (7.53) at the origin leads to

$$\dot{x}_i = -\tau x_i(t) + \tau \zeta x_i(t-1) - \tau \eta [x_{i-1}(t-1) + x_{i+1}(t-1) - 2x_i(t-1)], \quad (7.54)$$

where $i(\text{mod } n)$, $\zeta = h'(0)$, and $\eta = g'(0)$. The associated characteristic equation of (7.54) takes the form

$$\det \Delta(\tau, \lambda) = 0,$$

where the characteristic matrix $\Delta(\tau, \lambda)$ is given by

$$\Delta(\tau, \lambda) = (\lambda + \tau)\text{Id} - \tau M e^{-\lambda}, \quad \lambda \in \mathbb{C}$$

with Id denoting the identity matrix and $M = \text{circ}(\zeta + 2\eta, -\eta, 0, \cdots, -\eta)$. We put $\chi = e^{2\pi i/n}$ and $v_k = (1, \chi^k, \chi^{2k}, \cdots, \chi^{(n-1)k})^T$, $k \in \{0, 1, \ldots, n-1\}$. Clearly, $v_0 = (1, 1, \cdots, 1)^T$ and $v_k = \bar{v}_{n-k}$. It is easy to see that $Mv_k = (\zeta + 4\eta \sin^2 \frac{k\pi}{n})v_k$ for all k. Thus, we have

$$\det \Delta(\tau, \lambda) = \prod_{k=0}^{n-1} [\lambda + \tau - (\zeta + 4\eta \sin^2 \frac{k\pi}{n})\tau e^{-\lambda}].$$

Throughout this section, we always assume that there exists some $k \in \{0, 1, 2, \ldots, n-1\}$ such that

$$\left| \zeta + 4\eta \sin^2 \frac{k\pi}{n} \right| > 1. \quad (7.55)$$

Define

$$\beta_{k,s} \stackrel{\text{def}}{=} \begin{cases} 2s\pi + \arccos \frac{1}{\zeta + 4\eta \sin^2 \frac{k\pi}{n}} & \zeta + 4\eta \sin^2 \frac{k\pi}{n} < -1, \\ 2(s+1)\pi - \arccos \frac{1}{\zeta + 4\eta \sin^2 \frac{k\pi}{n}} & \zeta + 4\eta \sin^2 \frac{k\pi}{n} > 1; \end{cases}$$

$$\tau_{k,s} \stackrel{\text{def}}{=} \frac{\beta_{k,s}}{\sqrt{\{\zeta + 4\eta \sin^2 \frac{k\pi}{n}\}^2 - 1}},$$

for all $s \in \mathbb{N}_0 = \{0, 1, 2, \cdots\}$. In view of Guo and Huang [133], we have the following observations:

(a) At (and only at) $\tau = \tau_{k,s}$, $s \in \mathbb{N}_0$, $\det \Delta(\tau, \cdot)$ has purely imaginary eigenvalues. These eigenvalues are given by $\pm i\beta_{k,s}$.
(b) All other eigenvalues of $\det \Delta(\tau_{k,s}, \lambda)$ are not integer multiples of $\pm i\beta_{k,s}$.
(c) For each fixed k, $\tau_{k,s}$ is monotonically increasing in s.

Let $\alpha = \tau - \tau_{k,s}$ and $\mathscr{A}_\alpha$ be the infinitesimal generator associated with (7.54). Then we can rewrite (7.53) in the form of (7.13) and (7.14) with

$$L(\alpha)\varphi = -(\tau_{k,s} + \alpha)\varphi(0) + (\tau_{k,s} + \alpha)M\varphi(-1)$$

and

$$\mathscr{F}^2(\alpha, u, w) = (\tau_{k,s}+\alpha)\mathscr{M}''(0)(u_1(-1)w_1(-1), \cdots, u_n(-1)w_n(-1))^T,$$
$$\mathscr{F}^3(\alpha, u, w, v) = (\tau_{k,s}+\alpha)\mathscr{M}'''(0)(u_1(-1)w_1(-1)v_1(-1), \cdots, u_n(-1)w_n(-1)v_n(-1))^T$$

for $u, w, v \in C_{n,1}$, where $\mathscr{M}(u)$ is an $n \times n$ matrix function defined by

$$\mathscr{M}(u) = \operatorname{circ}(h(u) + 2g(u), -g(u), 0, \cdots, -g(u)).$$

In what follows, we distinguish three cases to discuss the exact form of the bifurcating periodic solutions.

Case I: $k = 0$ (mod n) in (7.55). Let $\varphi_j(\theta)$, $j = 1, 2$, be the eigenvector for $\mathscr{A}_0$ associated with $i\beta_{0,s}$ and $-i\beta_{0,s}$, respectively; namely,

$$\mathscr{A}_0\varphi_1(\theta) = i\beta_{0,s}\varphi_1(\theta), \quad \mathscr{A}_0\varphi_2(\theta) = -i\beta_{0,s}\varphi_2(\theta). \tag{7.56}$$

In view of $\Delta(\tau_{0,s}, i\beta_{0,s})v_0 = 0$, we can choose $\varphi_1(\theta) = \overline{\varphi}_2(\theta) = v_0 e^{i\beta_{0,s}\theta}$ for $\theta \in [-1, 0]$. So at $\tau = \tau_{0,s}$, the center space is $X = \operatorname{span}\{\varphi_1, \varphi_2\}$. Hence $\Phi = (\varphi_1, \varphi_2)$ is a basis for the center space X. Since $\pm i\beta_{0,s}$ are also eigenvalues for $\mathscr{A}_0^*$, there are two nonzero row-vector functions $\psi_j(\xi)$, $\xi \in [0, 1]$, $j = 1, 2$, such that

$$\mathscr{A}_0^*\psi_1(\xi) = -i\beta_{0,s}\psi_1(\xi), \quad \mathscr{A}_0^*\psi_2(\xi) = i\beta_{0,s}\psi_2(\xi).$$

Then, $\Psi = (\psi_1, \psi_2)^T$ is a basis for the adjoint space X^*. We normalize ψ_j ($j = 1, 2$) by the condition $\langle \psi_1, \varphi_1 \rangle = 0$ and $\langle \psi_2, \varphi_1 \rangle = 0$. We then obtain

$$\psi_1(\xi) = \overline{\psi}_2(\xi) = \overline{D}_0 v_0^T e^{i\beta_{0,s}\xi}, \quad \xi \in [0, 1],$$

where $D_0 = \frac{1}{n}(1 + \tau_{0,s} + i\beta_{0,s})^{-1}$.

It is obvious that the action of $\mathbb{D}_n$ on the center space X is trivial, and so the maximal isotropy subgroup is $\mathbb{D}_n$, which corresponds to a standard Hopf bifurcation in which $\mathbb{D}_n$ symmetry is preserved. Thus, all neurons are synchronous (i.e., have the same waveform and move in phase). Namely, the state $(x_1(t), x_2(t), \cdots, x_n(t))$ of system (7.53) satisfies $x_j(t) = u(t)$ for all j, where $u(t)$ is the periodic solution to the system

$$\dot{u}(t) = -\tau u(t) + \tau h(v(t-1)), \tag{7.57}$$

with period near $\frac{2\pi}{\beta_{0,s}}$.

In view of Sect. 7.2, we have

$$E_1 = [\Delta(\tau_{0,s}, 2i\beta_{0,s})]^{-1}\mathscr{F}^2(0, \varphi_1, \varphi_1)$$

and

$$E_2 = [\Delta(\tau_{0,s}, 0)]^{-1}\mathscr{F}^2(0, \varphi_1, \overline{\varphi}_1)$$

and hence

$$E_1 = \frac{\tau_{0,s}h''(0)v_0 e^{-2i\beta_{0,s}}}{2i\beta_{0,s} + \tau_{0,s} - \tau_{0,s}\zeta e^{-2i\beta_{0,s}}}, \qquad E_2 = \frac{h''(0)v_0}{1-\zeta}.$$

Thus, it follows from (7.21) that

$$C_1(0) = \frac{nD}{2}\left\{\tau_{0,s}h'''(0)e^{-i\beta_{0,s}} + (h''(0)\tau_{0,s})^2\left[\frac{2e^{-i\beta_{0,s}}}{1-\zeta} + \frac{e^{-3i\beta_{0,s}}}{2i\beta_{0,s}+\tau_{0,s}-\tau_{0,s}\zeta e^{-2i\beta_{0,s}}}\right]\right\}.$$

Notice that

$$\mathrm{Re}\{\lambda'(\tau_{0,s})\} = \frac{\beta_{0,s}^2}{\tau_{0,s}[(1+\tau_{0,s})^2 + \beta_{0,s}^2]} > 0.$$

It follows from Theorem 7.6 that $\mathrm{Re}\{C_1(0)\}$ determines the directions of the Hopf bifurcation: if $\mathrm{Re}\{C_1(0)\} < 0$ (respectively, > 0), then the Hopf bifurcation is supercritical (respectively, subcritical) and bifurcating periodic solutions exist for $\tau > \tau_{0,s}$ (respectively, $< \tau_{0,s}$). Of course, we can determine the stability of the bifurcating periodic solutions. By a direct computation, if $\zeta < -1$, then

$$\begin{aligned}\mathrm{Re}\{C_1(0)\} = &\frac{\tau_{0,s}h'''(0)}{2}\cdot\frac{1+\zeta^2\tau_{0,s}}{\zeta(1+2\tau_{0,s}+\zeta^2\tau_{0,s}^2)}\\&-\frac{\tau_{0,s}(h''(0))^2}{2}\cdot\frac{\zeta^2\tau_{0,s}(11\zeta^2+6\zeta-2)+(2\zeta^3+13\zeta^2+4\zeta-4)}{\zeta^2(5\zeta+4)(\zeta-1)(1+2\tau_{0,s}+\zeta^2\tau_{0,s}^2)};\end{aligned}$$

if $\zeta > 1$, then

$$\begin{aligned}\mathrm{Re}\{C_1(0)\} = &\frac{\tau_{0,s}h'''(0)}{2}\cdot\frac{1+\zeta^2\tau_{0,s}}{\zeta(1+2\tau_{0,s}+\zeta^2\tau_{0,s}^2)}\\&-\frac{\tau_{0,s}(h''(0))^2}{2}\cdot\frac{\zeta^2\tau_{0,s}(11\zeta^3+35\zeta^2+24\zeta-6)+(-2\zeta^4+11\zeta^3+43\zeta^2+24\zeta-12)}{\zeta^2(\zeta-1)(5\zeta^2+15\zeta+12)(1+2\tau_{0,s}+\zeta^2\tau_{0,s}^2)}.\end{aligned}$$

Therefore,

$$\begin{aligned}\mathrm{Re}\{C_1(0)\} < 0 &\Leftrightarrow h'''(0)h'(0) < [h''(0)]^2 C_s(\zeta);\\\mathrm{Re}\{C_1(0)\} > 0 &\Leftrightarrow h'''(0)h'(0) > [h''(0)]^2 C_s(\zeta);\\\mathrm{Re}\{C_1(0)\} = 0 &\Leftrightarrow h'''(0)h'(0) = [h''(0)]^2 C_s(\zeta),\end{aligned}$$

where

$$C_s(\zeta) = \begin{cases}\frac{11\zeta^2+6\zeta-2}{(5\zeta+4)(\zeta-1)} + \frac{2(\zeta+1)^2}{(5\zeta+4)(1+\zeta^2\tau_{0,s})}, & \text{if } \zeta < -1,\\[2mm]\frac{11\zeta^3+35\zeta^2+24\zeta-6}{(\zeta-1)(5\zeta^2+15\zeta+12)} - \frac{2(\zeta+1)(\zeta^2-3)}{(1+\zeta^2\tau_{0,s})(5\zeta^2+15\zeta+12)}, & \text{if } \zeta > 1.\end{cases} \tag{7.58}$$

Thus, we obtain the following results.

Theorem 7.11. *If (7.55) holds for $k = 0$, then for all $s \in \mathbb{N}_0$, near $\tau = \tau_{0,s}$, system (7.53) undergoes a Hopf bifurcation. The direction of Hopf bifurcation and stability of bifurcating synchronous periodic solutions satisfy the following properties:*

(i) *Assume that $h'''(0)h'(0) > [h''(0)]^2 C_s(\zeta)$. Then the bifurcating branch of periodic solutions exists for $\tau < \tau_{0,s}$ (subcritical bifurcation). Moreover, all periodic solutions on this branch are unstable.*

(ii) *Assume that $h'''(0)h'(0) < [h''(0)]^2 C_s(\zeta)$. Then the bifurcating branch of periodic solutions exists for $\tau > \tau_{0,s}$ (supercritical bifurcation). (i) If there exists some $j \in \{0, 1, 2, \dots, n-1\}$ such that $\zeta + 4\eta \sin^2 \frac{j\pi}{n} > 1$, or $\eta < 0$ and $\zeta < -1$, then all periodic solutions on this branch are unstable. (ii) If $\zeta < -1$ and $\eta > 0$ and $\zeta + 4\eta \sin^2 \frac{j\pi}{n} < 1$ for all $j \in \{1, 2, \cdots, n-1\}$, then only the slowly oscillating periodic solution arising at $\tau_{0,0}$ is stable; all the periodic solutions arising at $\tau = \tau_{0,s}$ ($s \geq 1$) are unstable.*

Notice that the sequence $\{C_s(\zeta)\}_{s \in \mathbb{N}_0}$ has the following properties:

(a) For $\zeta < -1$ or $\zeta > \sqrt{3}$, the sequence $\{C_s(\zeta)\}_{s \in \mathbb{N}_0}$ is positive, bounded, and strictly increasing.

(b) For $1 < \zeta < \sqrt{3}$, the sequence $\{C_s(\zeta)\}_{s \in \mathbb{N}_0}$ is positive, bounded, and strictly decreasing.

(c) For $\zeta = \sqrt{3}$, the sequence $\{C_s(\zeta)\}_{s \in \mathbb{N}_0}$ is a constant sequence.

(d) For $\zeta < -1$, $\frac{2(126+125\sqrt{7})}{(2+5\sqrt{7})(35+2\sqrt{7})} < C_s(\zeta) < 2.2$. For $\zeta > 1$, $C_s(\zeta) > 2.1$.

From the boundedness of $\{C_s(\zeta)\}_{s \in \mathbb{N}_0}$, we have the following.

Corollary 7.1. *If (7.55) holds for $k = 0$, then for all $s \in \mathbb{N}_0$, near $\tau = \tau_{0,s}$, system (7.53) undergoes a Hopf bifurcation. The direction of Hopf bifurcation satisfies the following:*

(i) *If $\zeta < -1$ and $h'''(0)h'(0) < \frac{2(126+125\sqrt{7})}{(2+5\sqrt{7})(35+2\sqrt{7})}[h''(0)]^2$, then all bifurcations are supercritical.*

(ii) *If $\zeta < -1$ and $h'''(0)h'(0) > 2.2[h''(0)]^2$, then all bifurcations are subcritical.*

(iii) *If $\zeta > 1$ and $h'''(0)h'(0) < 2.1[h''(0)]^2$, then all bifurcations are supercritical.*

Case II: $k = \frac{n}{2}$ (mod n) with even n in (7.55). Similar to the analysis in Case I, the purely imaginary eigenvalues associated with a Hopf bifurcation are simple, and the $\mathbb{D}_n \times \mathbb{S}^1$-action on $\mathbb{C}$ is given by $\rho \cdot z = -z$, $\kappa \cdot z = z$, and $\theta \cdot z = e^{i\beta_{k,s}\theta}z$ for $z \in \mathbb{C}$. Obviously, the maximal isotropy subgroup is $\tilde{\mathbb{Z}}_2$, generated by (ρ, π) and κ, which corresponds to an equivariant Hopf bifurcation involving a branch of symmetry-breaking oscillations with isotropy subgroup $\tilde{\mathbb{Z}}_2$. Every neuron has the same waveform but is half a period out of phase with (i.e., antisynchronous to) its nearest neurons. Namely, the state $(x_1(t), \cdots, x_n(t))$ of system (7.53) satisfies $x_{j-1}(t) = x_j(t + \frac{\omega}{2}) = u(t)$ for all j (mod n), where $u(t)$ is an ω-periodic function with period ω near $\frac{2\pi}{\beta_{k,s}}$. The analysis of bifurcation direction and stability of bifurcated periodic solution can be investigated similarly to Case I.

Case III: $n \geq 3$ and $2k \neq 0 \pmod{n}$ **in (7.55).** Let $\varphi_j(\theta)$, $j = 1,2,3,4$, be the eigenvectors of $\mathscr{A}_0$ associated with $i\beta_{k,s}$ and $-i\beta_{k,s}$, respectively. Since $\Delta(\tau_{k,s}, i\beta_{k,s})v_k = 0$ and $\Delta(\tau_{k,s}, i\beta_{k,s})\bar{v}_k = 0$, we can choose

$$\varphi_1(\theta) = v_k e^{i\beta_{k,s}\theta}, \quad \varphi_2(\theta) = \bar{v}_k e^{i\beta_{k,s}\theta}$$

for $\theta \in [-1,0]$. So the center space at $\mu = 0$ is $X = \text{span}\{\varphi_1, \bar{\varphi}_1, \varphi_2, \bar{\varphi}_2\}$. Hence, $\Phi = (\varphi_1, \bar{\varphi}_1, \varphi_2, \bar{\varphi}_2)$ is a basis for the center space X. Let $\Psi = (\psi_1, \bar{\psi}_1, \psi_2, \bar{\psi}_2)^T$ be a basis for the adjoint space X^*. Similarly, we normalize ψ_1 and ψ_2 to obtain

$$\psi_1(\theta) = \bar{D}_k v_k^T e^{i\beta_{k,s}\theta}, \quad \psi_2(\theta) = \bar{D}_k \bar{v}_k^T e^{i\beta_{k,s}\theta}$$

for $\xi \in [0,1]$, where $D_k = \frac{1}{n}(1 + \tau_{k,s} + i\beta_{k,s})^{-1}$. Similarly to the last subsection, define n^* as (7.33). Now we discuss the spatiotemporal patterns of these bifurcated periodic solutions according to their corresponding isotropy subgroups of $\mathbb{D}_{n^*} \times \mathbb{S}^1 \leq \mathbb{D}_n \times \mathbb{S}^1$.

First, the isotropy subgroup $\tilde{\mathbb{Z}}_{n^*}$ corresponds to discrete waves of (7.53), which take the form $x_j(t) = x_{j+1}(t - \frac{k\omega}{n})$, $t \in \mathbb{R}$, $j \pmod{n}$, where $\omega > 0$ is a period of $x(t)$. The discrete waves are also called synchronous oscillations (if $k = 0 \pmod{n}$) or phase-locked oscillations (if $k \neq 0 \pmod{n}$), since each neuron oscillates just like the others except not necessarily in phase with one another.

Next, the isotropy subgroups $\mathbb{Z}_2(\kappa)$ (when n^* is odd), $\mathbb{Z}_2(\kappa) \oplus \mathbb{Z}_2^c$ (when n^* is even) correspond to mirror-reflecting waves of (7.53), which take the form $x_j(t) = x_{n+2-j}(t)$, $t \in \mathbb{R}$, $j \pmod{n}$.

Finally, the isotropy subgroups $\mathbb{Z}_2(\kappa, \pi)$ (when n^* is odd), $\mathbb{Z}_2(\kappa, \pi) \oplus \mathbb{Z}_2^c$ (when $n^* = 2 \pmod{4}$), $\mathbb{Z}_2(\kappa\gamma) \oplus \mathbb{Z}_2^c$ (when $n^* = 0 \pmod{4}$) correspond to standing waves of (7.53), which take the form $x_j(t) = x_{2+n-j}(t - \frac{\omega}{2})$, $t \in \mathbb{R}$, $j \pmod{n}$, where $\omega > 0$ is a period of u.

Applying the equivariant Hopf bifurcation theorem to (7.53) yields the following theorem.

Theorem 7.12. *Assume that $n \geq 3$ and $2k \neq 0 \pmod{n}$ in (7.55). Then system (7.53) undergoes Hopf bifurcation. More precisely, near each $\tau = \tau_{k,s}$, $s \in \mathbb{N}_0$, system (7.53) has $2(n+1)$ distinct branches of asynchronous periodic solutions of period near $\frac{2\pi}{\beta_{k,s}}$, bifurcated from the trivial solution, including two phase-locked oscillations, n mirror-reflecting waves, and n standing waves.*

In order to discuss bifurcation direction and stability of bifurcated periodic solutions, we need to find a concrete expression for the coefficients K_j, $j = 1,2,3$. In view of (7.42), we have

$$A_{(1,1,0,0,0)} = A_{(0,0,1,1,0)} = 2\tau_{k,s}h''(0)v_0$$
$$A_{(1,0,1,0,0)} = \overline{A}_{(0,1,0,1,0)} = 2\tau_{k,s}h''(0)v_0 e^{-2i\beta_{k,s}}$$
$$A_{(1,0,0,1,0)} = \overline{A}_{(0,1,1,0,0)} = \tau_{k,s}H_2 v_{2k},$$
$$A_{(2,0,0,0,0)} = \overline{A}_{(0,2,0,0,0)} = \tau_{k,s}H_2 v_{2k}e^{-2i\beta_{k,s}},$$
$$A_{(0,0,2,0,0)} = \overline{A}_{(0,0,0,2,0)} = \tau_{k,s}H_2 \overline{v}_{2k}e^{-2i\beta_{k,s}},$$
$$A_{(1,0,1,1,0)} = 2A_{(2,1,0,0,0)} = 6\tau_{k,s}H_3 v_k e^{-i\beta_{k,s}},$$
$$A_{(0,1,2,0,0)} = 3\tau_{k,s}H_3 \overline{v}_{3k}e^{-i\beta_{k,s}},$$

where $H_j = h^{(j)}(0) + 4g^{(j)}(0)\sin^2\frac{jk\pi}{n}$, $(j \geq 1)$. Moreover, it follows from (7.44) and (7.45) that

$$h_{(q,0)}(\theta) = i\beta_{k,s}^{-1}\left[\overline{\psi}_1(0)A_{(q,0)}\varphi_1 + \tfrac{1}{3}\psi_1(0)A_{(q,0)}\overline{\varphi}_1 + \overline{\psi}_2(0)A_{(q,0)}\varphi_2 + \tfrac{1}{3}\psi_2(0)A_{(q,0)}\overline{\varphi}_2\right]$$
$$+[\Delta(\tau_{k,s},2i\beta_{k,s})]^{-1}A_{(q,0)},$$
$$h_{(p,0)}(\theta) = i\beta_{k,s}^{-1}\left[-\overline{\psi}_1(0)A_{(p,0)}\varphi_1 + \psi_1(0)A_{(p,0)}\overline{\varphi}_1 - \overline{\psi}_2(0)A_{(p,0)}\varphi_2 + \psi_2(0)A_{(p,0)}\overline{\varphi}_2\right]$$
$$+[\Delta(\tau_{k,s},0)]^{-1}A_{(p,0)}$$

for all $q \in \{(2,0,0,0),\ (0,0,2,0),\ (1,0,1,0)\}$ and $p \in \{(1,1,0,0),\ (0,0,1,1), (1,0,0,1),\ (0,1,1,0)\}$. Thus, K_j, $j = 1,2,3$, can be evaluated according to the formula given in the previous subsection.

In particular, we make a further assumption on functions h and g as follows:

$$H_2 = 0, \qquad H_1 H_3 \neq 0. \tag{7.59}$$

Then $K_2 = 2K_1 = nD\tau_{k,s}H_3 e^{-i\beta_{k,s}}$. Recall that

$$\mathrm{Re}(K_2) = \frac{H_3(\tau_{k,s} + \tau_{k,s}^2 + \beta_{k,s}^2)^2}{H_1[(1+\tau_{k,s})^2 + \beta_{k,s}^2]}.$$

We have $\mathrm{signRe}(K_1) = \mathrm{signRe}(K_2 - K_1) = \mathrm{signRe}(K_1 + K_2) = \mathrm{sign}(H_1 H_3)$. In view of Theorem 7.9, we have the following corollary.

Corollary 7.2. *Assume that $n \geq 3$ and $4k \neq 0 \pmod{n}$ in (7.55) and that (7.59) holds. Then near $\tau = \tau_{k,s}$, system (7.53) undergoes Hopf bifurcations. The bifurcation direction and stability of each branch of bifurcated periodic solutions are completely determined by the sign of $H_1 H_3$:*

(i) *Assume that $H_3 H_1 < 0$ (respectively, > 0). Then the bifurcating branch of periodic solutions exists for $\tau > \tau_{k,s}$ (respectively, $\tau < \tau_{k,s}$).*

(ii) *Each one of the following conditions ensures that all bifurcated periodic solutions near $\tau_{k,s}$ are unstable: (i) n is even; (ii) there exists some $j \in \{0,1,2,\cdots,n-1\}$ such that $\zeta + 4\eta\sin^2\frac{j\pi}{n} > 1$; (iii) n is odd, $H_1 < -1$, and $\eta > 0$; (iv) n is odd, $H_1 < -1$, $\eta < 0$, $k \neq \left[\frac{n}{2}\right]$; (v) n is odd, $H_1 < -1$, $H_3 < 0$, $\eta < 0$.*

(iii) *Assume that n is odd, $H_1 < -1$, $\zeta < 1$, $\eta < 0$, and $k = \left[\frac{n}{2}\right]$, and $H_3 > 0$. Then only near $\tau_{k,0}$ can the Hopf bifurcation provide two orbitally asymptotically stable phase-locked waves, n orbitally asymptotically unstable standing waves,*

and n orbitally asymptotically unstable mirror-reflecting waves. However, near
$\tau = \tau_{k,s}$ *(s ≥ 1), all the bifurcated periodic solutions are unstable.*

Proof. Conclusion (i) follows easily from Theorem 7.9. We prove only (ii) and (iii). As stated in Theorem 7.9, the orbital stability of the bifurcated periodic solutions is determined by eigenvalues of $\mathscr{A}_Q$, signs of $\mathrm{Re}(K_1)$, signs of $\mathrm{Re}(K_1 + K_2)$, and $\mathrm{Re}(K_1 - K_2)$. Each of conditions (i)–(v) of conclusion (ii) ensures that $\mathscr{A}_Q$ has at least one eigenvalue with positive real part. Here, we consider only the case that n is even. The other cases can be dealt with analogously. In what follows, we distinguish two cases to show that near $\tau_{k,s}(\geq \tau_{k,0})$, $\mathscr{A}_Q$ has at least one eigenvalue with positive real part.

Case 1: $\zeta + 4\eta \sin^2 \frac{k\pi}{n} > 1$. We have $\zeta > \zeta + 4\eta \sin^2 \frac{k\pi}{n} > 1$ (if $\eta < 0$) or $\zeta + 4\eta > \zeta + 4\eta \sin^2 \frac{k\pi}{n} > 1$ (if $\eta > 0$). If $\zeta > \zeta + 4\eta \sin^2 \frac{k\pi}{n} > 1$ (or $\zeta + 4\eta > \zeta + 4\eta \sin^2 \frac{k\pi}{n} > 1$), then $\lambda + \tau - \tau\zeta e^{-\lambda} = 0$ (respectively, $\lambda + \tau - (\zeta + 4\eta)\tau e^{-\lambda} = 0$) has at least one solution with positive real part if $\tau \in [0, \tau_{0,0})$(respectively, $\tau \in [0, \tau_{n/2,0})$). It follows from the fact that $\tau_{0,0} < \tau_{k,0}$ (respectively, $\tau_{n/2,0} < \tau_{k,0}$) that near $\tau_{k,s}(\geq \tau_{k,0})$, $\mathscr{A}_Q$ has at least one eigenvalue with positive real part.

Case 2: $\zeta + 4\eta \sin^2 \frac{k\pi}{n} < -1$. We have $\zeta < \zeta + 4\eta \sin^2 \frac{k\pi}{n} < -1$ (if $\eta > 0$) or $\zeta + 4\eta < \zeta + 4\eta \sin^2 \frac{k\pi}{n} < -1$ (if $\eta < 0$). If $\zeta < \zeta + 4\eta \sin^2 \frac{k\pi}{n} < -1$ (or $\zeta + 4\eta < \zeta + 4\eta \sin^2 \frac{k\pi}{n} < -1$), then $\lambda + \tau - \tau\zeta e^{-\lambda} = 0$ (respectively, $\lambda + \tau - (\zeta + 4\eta)\tau e^{-\lambda} = 0$) has at least two solutions with positive real part if $\tau \in [\tau_{0,0}, \tau_{0,1})$ (respectively, $\tau \in [\tau_{n/2,0}, \tau_{n/2,1})$). It follows form the fact that $\tau_{0,0} < \tau_{k,0}$ (respectively, $\tau_{n/2,0} < \tau_{k,0}$) that near $\tau_{k,s}(\geq \tau_{k,0})$, $\mathscr{A}_Q$ has at least one eigenvalue with positive real part.

To prove (ii), assume that n is odd, $H_1 < -1$, $\eta < 0$, $k = \left[\frac{n}{2}\right]$, and $H_3 > 0$. Then

$$\zeta + 4\eta \sin^2 \frac{k\pi}{n} < \zeta + 4\eta \sin^2 \frac{(k-1)\pi}{n} < \cdots < \zeta < 1.$$

Thus if $\zeta + 4\eta \sin^2 \frac{j\pi}{n} < -1$ for some $j \in \{0, 1, 2, \cdots, k-1\}$, then

$$\tau_{k,s} < \tau_{k-1,s} < \tau_{k-2,s} < \cdots < \tau_{j,s}.$$

Namely, $\tau_{k,0}$ is the first critical value of τ in $[0, +\infty)$. The characteristic equation of (7.54) has only solutions with negative real part if $0 \leq \tau < \tau_{k,0}$. Therefore, near $\tau_{k,0}$, all eigenvalues of $\mathscr{A}_Q$ have negative real part. This, together with $H_1 H_3 < 0$, implies that conclusion (iii) holds and completes the proof of Corollary 7.2. □

Remark 7.3. Assume that $n \geq 3$ and $2k \neq 0 \pmod{n}$ but $4k = 0 \pmod{n}$ in (7.55) and that (7.59) holds. Then we have $K_2 = 2K_1 = 2K_3 = nD\tau_{k,s}H_3 e^{-i\beta_{k,s}}$. Therefore, near $\tau = \tau_{k,s}$, system (7.53) undergoes Hopf bifurcations. If $H_3 H_1 < 0$ (respectively, > 0), then the entire bifurcating branch of periodic solutions exists for $\tau > \tau_{k,s}$ (respectively, $\tau < \tau_{k,s}$). We conclude that under the assumption that $H_1 < -1$, $H_3 > 0$, and $H_1 < \zeta + 4\eta \sin^2 \frac{j\pi}{n} < 1$, only at $\tau_{k,0}$ are the bifurcated mirror-reflecting waves stable. As for the stability of bifurcated phase-locked waves and standing waves, we need to calculate higher-order terms of the normal form, because $|K_1 - K_2|^2 = |K_3|^2$ and $|K_3|^2 + \mathrm{Re}\{(K_1 - K_2)K_3\} = 0$.

7.4 Bogdanov–Takens Bifurcation

Consider the following two-parameter family of RFDEs:

$$\dot{x}(t) = f(\alpha, x_t), \quad \alpha = (\alpha_1, \alpha_2) \in \mathbb{R}^2, \tag{7.60}$$

where $f \in C^k(\mathbb{R}^2 \times C_{n,\tau}, \mathbb{R}^n)$ for a large enough integer k, $f(0,0) = 0$. As in the previous sections, we expand the function f as

$$f(\alpha, \varphi) = f(\alpha, 0) + \mathscr{F}^1(\alpha, \varphi) + \cdots + \frac{1}{k!}\mathscr{F}^k(\alpha, \varphi, \cdots, \varphi) + o(\|\varphi\|^k)$$

for $(\alpha, \varphi) \in \mathbb{R}^2 \times C_{n,\tau}$. Obviously, $L = \mathscr{F}^1(0, \cdot)$ is the linearized operator of $f(\alpha, \varphi)$ with respect to φ at $(\alpha, \varphi) = (0,0)$. Let

$$F(\alpha, \varphi) = f(\alpha, \varphi) - L\varphi$$

for $(\alpha, \varphi) \in \mathbb{R}^2 \times C_{n,\tau}$, and let $\langle \cdot, \cdot \rangle \colon C_{n,\tau}^* \times C_{n,\tau} \to \mathbb{R}$ be the bilinear form associated with the operator L. Let

$$\Delta(\lambda) = \lambda \operatorname{Id}_n - L(e^{\lambda(\cdot)}),$$

and let $\mathscr{A}$ be the infinitesimal generator associated with the linear system $\dot{x}(t) = Lx_t$.

As we mentioned before, system (7.60) may undergo standard or equivariant fold bifurcation if $\mathscr{A}$ has a semisimple eigenvalue 0. What happens to the case of a nonsemisimple eigenvalue 0? In fact, in this case, system (7.60) will undergo multiple-zero bifurcation (see, for example, [11, 105, 130]).

7.4.1 Center Manifold Reduction

In this subsection, we always assume that:

(BT1)   The infinitesimal generator $\mathscr{A}$ has a nonsemisimple double eigenvalue 0.

At this bifurcation, there exist two real linearly independently (generalized) eigenvectors $q_1, q_2 \in C_{n,\tau}$ such that

$$\mathscr{A}q_1 = 0, \quad \mathscr{A}q_2 = q_1.$$

Obviously, q_1 is actually a constant column-vector function satisfying $Lq_1 = 0$. Moreover, there exist similar vectors $p_1, p_2 \in C_{n,\tau}^*$ of the adjoint operator $\mathscr{A}^*$:

$$\mathscr{A}^* p_2 = 0, \quad \mathscr{A}^* p_1 = p_2.$$

Obviously, p_1 is actually a constant row-vector function satisfying $p_1 \Delta(0) = 0$. One can select these vectors to satisfy

$$\langle p_j, q_s \rangle = \delta_{js}, \quad j, s = 1, 2.$$

Let $\Phi = (q_1, q_2)$ and $\Psi = (p_1, p_2)^T$. Then $\langle \Psi, \Phi \rangle = \mathrm{Id}_2$ and $\dot{\Phi} = \Phi B$, where

$$
B = \begin{bmatrix} 0 & 1 \\ 0 & 0 \end{bmatrix}.
$$

By the center manifold theorem, the reduced equation on the center manifold $M_{\mathrm{loc}}^{c,\alpha}$ is

$$
\dot{x}(t) = Bx + G(\alpha, x), \quad x = (x_1, x_2)^T \in \mathbb{R}^2, \tag{7.61}
$$

where $G(\alpha, x) = \Psi(0) F(\alpha, \Phi x + W(\alpha, x))$, W satisfies

$$
\frac{\mathrm{d}}{\mathrm{d}t} W = \mathscr{A}_Q W + H(\alpha, x), \quad W(0,0) = 0, \quad D_x W(0,0) = 0, \tag{7.62}
$$

and $H(\alpha, x) = [X_0 - \Phi \Psi(0)] F(\alpha, \Phi x + W(\alpha, x))$. Let

$$
G(\alpha, x) = \sum_{j+s=0}^{k} \frac{1}{j!s!} G_{js}(\alpha) x_1^j x_2^s + o(|x|^k).
$$

Then we have

$$
\begin{aligned}
G_{00}(\alpha) &= \Psi(0) f_{\alpha_1}(0,0)\alpha_1 + \Psi(0) f_{\alpha_2}(0,0)\alpha_2 + O(|\alpha|^2), \\
G_{10}(\alpha) &= \Psi(0) \mathscr{F}_{\alpha_1}^1(0, q_1)\alpha_1 + \Psi(0) \mathscr{F}_{\alpha_2}^1(0, q_1)\alpha_2 + O(|\alpha|^2), \\
G_{01}(\alpha) &= \Psi(0) \mathscr{F}_{\alpha_1}^1(0, q_2)\alpha_1 + \Psi(0) \mathscr{F}_{\alpha_2}^1(0, q_2)\alpha_2 + O(|\alpha|^2), \\
G_{20}(\alpha) &= \Psi(0) \mathscr{F}^2(0, q_1, q_1) + O(|\alpha|), \\
G_{11}(\alpha) &= \Psi(0) \mathscr{F}^2(0, q_1, q_2) + O(|\alpha|), \\
G_{02}(\alpha) &= \Psi(0) \mathscr{F}^2(0, q_2, q_2) + O(|\alpha|).
\end{aligned}
$$

Using the near-identity transformation

$$
u = x_1, \quad v = x_2 + p_1(0) F(\alpha, \Phi x + W(\alpha, x)), \tag{7.63}
$$

we transform (7.61) into

$$
\dot{u} = v,
$$
$$
\dot{v} = g_{00}(\alpha) + g_{10}(\alpha)u + g_{01}(\alpha)v + \frac{1}{2}g_{20}(\alpha)u^2 + g_{11}(\alpha)uv + \frac{1}{2}g_{02}(\alpha)v^2 + o(u^2 + v^2), \tag{7.64}
$$

where

$$
\begin{aligned}
g_{20}(0) &= p_2(0) \mathscr{F}^2(0, q_1, q_1), \\
g_{11}(0) &= p_1(0) \mathscr{F}^2(0, q_1, q_1) + p_2(0) \mathscr{F}^2(0, q_1, q_2), \\
g_{02}(0) &= p_2(0) \mathscr{F}^2(0, q_2, q_2) + 2p_1(0) \mathscr{F}^2(0, q_1, q_2),
\end{aligned}
$$

and

$$g_{00}(\alpha) = p_2(0)f_{\alpha_1}(0,0)\alpha_1 + p_2(0)f_{\alpha_2}(0,0)\alpha_2 + O(|\alpha|^2),$$

$$g_{10}(\alpha) = p_2(0)[\alpha_1 \mathscr{F}^1_{\alpha_1}(0,q_1) + \alpha_2 \mathscr{F}^1_{\alpha_2}(0,q_1)]$$
$$\quad + p_1(0)\mathscr{F}^2(0,q_1,q_2)p_2(0)[\alpha_1 f_{\alpha_1}(0,0) + \alpha_2 f_{\alpha_2}(0,0)]$$
$$\quad - p_2(0)\mathscr{F}^2(0,q_1,q_2)p_1(0)[\alpha_1 f_{\alpha_1}(0,0) + \alpha_2 f_{\alpha_2}(0,0)] + O(|\alpha|^2),$$

$$g_{01}(\alpha) = p_2(0)[\alpha_1 \mathscr{F}^1_{\alpha_1}(0,q_2) + \alpha_2 \mathscr{F}^1_{\alpha_2}(0,q_2)]$$
$$\quad + p_1(0)[\alpha_1 \mathscr{F}^1_{\alpha_1}(0,q_1) + \alpha_2 \mathscr{F}^1_{\alpha_2}(0,q_1)]$$
$$\quad + p_1(0)\mathscr{F}^2(0,q_2,q_2)p_2(0)[\alpha_1 f_{\alpha_1}(0,0) + \alpha_2 f_{\alpha_2}(0,0)]$$
$$\quad - p_2(0)\mathscr{F}^2(0,q_2,q_2)p_1(0)[\alpha_1 f_{\alpha_1}(0,0) + \alpha_2 f_{\alpha_2}(0,0)]$$
$$\quad - p_1(0)\mathscr{F}^2(0,q_1,q_2)p_1(0)[\alpha_1 f_{\alpha_1}(0,0) + \alpha_2 f_{\alpha_2}(0,0)] + O(|\alpha|^2).$$

7.4.2 Bogdanov Normal Form

According to Bogdanov (see Kuznetsov [200]), we can annihilate the term proportional to v_2 in the equation for v_2 under the following assumption:

(BT2) $g_{11}(0) \neq 0.$

Namely, then there exists a parameter-dependent shift of coordinates in the u-direction that transforms (7.64) into

$$\dot{u} = v,$$
$$\dot{v} = h_{00}(\alpha) + h_{10}(\alpha)u + \frac{1}{2}h_{20}(\alpha)u^2 + h_{11}(\alpha)uv + \frac{1}{2}h_{02}(\alpha)v^2 + o(u^2 + v^2),$$
$$(7.65)$$

where $h_{jk}(0) = g_{jk}(0)$ for all $j+k = 2$, and

$$h_{00}(\alpha) = g_{00}^{\text{lin}}(\alpha) + O(|\alpha|^2)$$
$$h_{10}(\alpha) = g_{10}^{\text{lin}}(\alpha) - \frac{g_{20}(0)}{g_{11}(0)}g_{01}^{\text{lin}}(\alpha) + O(|\alpha|^2).$$

Here $g_{jk}^{\text{lin}}(\alpha)$ denotes the linear parts of $g_{jk}(\alpha)$, $j,k = 0,1$.

By some time reparameterization and second-coordinate transformation (see Kuznetsov [200]), we can eliminate the v^2-term and obtain

$$\dot{u} = v,$$
$$\dot{v} = \mu_1(\alpha) + \mu_2(\alpha)u + A(\alpha)u^2 + B(\alpha)uv + o(u^2 + v^2),$$
$$(7.66)$$

where

$$\mu_1(\alpha) = h_{00}(\alpha), \quad \mu_2(\alpha) = h_{10}(\alpha) - \frac{1}{2}h_{00}(\alpha)h_{02}(\alpha),$$
$$(7.67)$$

and

$$A(\alpha) = \frac{1}{2}[h_{20}(\alpha) - h_{10}(\alpha)h_{02}(\alpha)], \quad B(\alpha) = h_{11}(\alpha). \tag{7.68}$$

By performing a scaling of time and variables, we may transform (7.66) into

$$\begin{aligned} \dot{u} &= v, \\ \dot{v} &= \beta_1 + \beta_2 u + u^2 + suv + o(u^2 + v^2), \end{aligned} \tag{7.69}$$

where

$$s = \text{sgn}\{A(0)B(0)\} = \text{sgn}\{G_{20}^2(0)[G_{20}^1(0) + G_{11}^2(0)]\} \tag{7.70}$$

and

$$\beta_1(\alpha) = \frac{B^4(\alpha)}{A^3(\alpha)}\mu_1(\alpha), \quad \beta_2(\alpha) = \frac{B^2(\alpha)}{A^2(\alpha)}\mu_2(\alpha). \tag{7.71}$$

Thus, we have proved the following result.

Theorem 7.13. *In addition to conditions (BT1) and (BT2), assume that the Jacobian matrix of (β_1, β_2) with respect to (α_1, α_2) is nonsingular at $\alpha = 0$. Then system (7.60) is locally topologically equivalent to the normal form (7.69).*

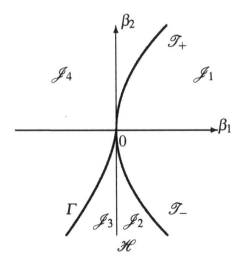

Fig. 7.1 Bifurcation sets for (7.72)

In what follows, we take $s = -1$ and consider system (7.69) without $o(u^2 + v^2)$ terms:

$$\begin{aligned} \dot{u} &= v, \\ \dot{v} &= \beta_1 + \beta_2 u + u^2 - uv. \end{aligned} \tag{7.72}$$

The bifurcation diagram of system (7.72) is presented in Fig. 7.1. Every equilibrium (u,v) of (7.72) is located on the line $v = 0$ and satisfies the equation $\beta_1 + \beta_2 u + u^2 = 0$. Hence, the fold bifurcation curve is

$$\mathscr{T} = \{(\beta_1,\beta_2) : \beta_2^2 = 4\beta_1\}.$$

Along this curve system, (7.72) has an equilibrium with a zero eigenvalue. If $\beta_2 \neq 0$, then crossing $\mathscr{T}$ from right to left implies the appearance of two equilibria. Let us denote the left one by E_1 and the right one by E_2. In particular, passing through $\mathscr{T}_-$ implies the coalescence of a stable node E_1 and a saddle point E_2, while crossing $\mathscr{T}_+$ generates an unstable node E_1 and a saddle point E_2. The Hopf bifurcation curve is

$$\mathscr{H} = \{(\beta_1,\beta_2) : \beta_1 = 0, \beta_2 < 0\}.$$

Along $\mathscr{H}$, the equilibrium E_1 has a pair of purely imaginary eigenvalues. The Hopf bifurcation gives rise to a stable limit cycle, which exists near $\mathscr{H}$ for $\beta_1 < 0$. Moreover, we have the following result (see Kuznetsov [200] for the proof)

Theorem 7.14. *There is a unique smooth curve Γ corresponding to a saddle homoclinic bifurcation in system (7.72) at $\beta = 0$. It has the following local representation:*

$$\Gamma = \{(\beta_1,\beta_2) : 25\beta_1 = -6\beta_2^2 + o(\beta_2^2)\}.$$

For $\|\beta\|$ small, system (7.72) has a unique and hyperbolic stable cycle for parameter values inside the region bounded by the Hopf bifurcation curve $\mathscr{H}$ and the homoclinic bifurcation curve Γ, and no cycles outside this region.

Thus, the stable limit cycle born via the Hopf bifurcation does not bifurcate in region $\mathscr{J}_3$. As we move clockwise, it grows and approaches the saddle, turning into a homoclinic orbit at Γ.

7.4.3 Normal Form of System (7.60) with a Fixed Equilibrium

In what follows, we investigate the normal form of Bogdanov–Takens bifurcation under condition (BT1) and the following assumption:

(BT3) $f(\alpha,0) = 0$ for all $\alpha \in \mathbb{R}^2$.

This implies that $x = 0$ is always an equilibrium of system (7.60). Then $g_{00}(\alpha) \equiv 0$ in (7.64). A truncated normal form up to second order for this equation is given by

$$\begin{aligned}
\dot{x}_1 &= x_2, \\
\dot{x}_2 &= c_{10}(\alpha)x_1 + c_{01}(\alpha)x_2 + c_{20}(\alpha)x_1^2 + c_{11}(\alpha)x_1 x_2,
\end{aligned} \tag{7.73}$$

where

$$c_{10}(\alpha) = p_2(0)[\alpha_1 \mathscr{F}_{\alpha_1}^1(0,q_1) + \alpha_2 \mathscr{F}_{\alpha_2}^1(0,q_1)] + O(|\alpha|^2),$$
$$c_{01}(\alpha) = p_2(0)[\alpha_1 \mathscr{F}_{\alpha_1}^1(0,q_2) + \alpha_2 \mathscr{F}_{\alpha_2}^1(0,q_2)]$$
$$\qquad + p_1(0)[\alpha_1 \mathscr{F}_{\alpha_1}^1(0,q_1) + \alpha_2 \mathscr{F}_{\alpha_2}^1(0,q_1)] + O(|\alpha|^2),$$
$$c_{20}(\alpha) = g_{20}(0) + O(|\alpha|),$$
$$c_{11}(\alpha) = g_{11}(0) + O(|\alpha|).$$

Now we assume that

(BT4) $g_{20}(0)g_{11}(0) \neq 0.$

Then, using the linear scalings

$$x = \frac{c_{11}^2(\alpha)}{c_{20}(\alpha)}x_1, \quad y = \frac{c_{11}^3(\alpha)}{c_{20}^2(\alpha)}x_2, \quad t^* = \frac{c_{20}(\alpha)}{c_{11}(\alpha)}t,$$

of the variables and time and then dropping $*$, we obtain the following system:

$$\begin{aligned}\dot{x} &= y, \\ \dot{y} &= \gamma_1 x + \gamma_2 y + x^2 + xy,\end{aligned} \tag{7.74}$$

where

$$\gamma_1 = \frac{c_{10}(\alpha)c_{11}^2(\alpha)}{c_{20}^3(\alpha)} \quad \gamma_2 = \frac{c_{01}(\alpha)c_{11}(\alpha)}{c_{20}(\alpha)}.$$

We assume that:

(BT5)   The mapping $(\alpha_1,\alpha_2) \mapsto (\gamma_1,\gamma_2)$ is regular at $\alpha = 0$, that is, the Jacobian matrix of (γ_1,γ_2) with respect to (α_1,α_2) is nonsingular at $(\alpha_1,\alpha_2) = (0,0)$.

Then, using the implicit function theorem, we can show that for sufficiently small $|\gamma|$, system (7.74) exhibits the same local bifurcations in a small neighborhood of the origin in the phase plane as (7.73) does with sufficiently small $|\alpha|$. In fact, an equilibrium (or limit cycle, homoclinic orbit, heteroclinic orbit) of (7.74) corresponds to an equilibrium (or limit cycle, homoclinic orbit, heteroclinic orbit) of (7.60). Moreover, they have the same stability if the trivial solution is stable before the bifurcation. Therefore, we have the following result.

Theorem 7.15. *Under assumptions (BT1), (BT3), (BT4), and (BT5), system (7.60) is locally topologically equivalent to the normal form (7.74).*

7.4.4 $\mathbb{D}_3$-Equivariant Bogdanov–Takens Bifurcation

In this section, we always assume that:

(DBT1) System (7.60) is $\mathbb{D}_3$-equivariant, this is, $f(\alpha, \rho \cdot \varphi) = \rho \cdot f(\alpha, \varphi)$ for all $(\alpha, \varphi) \in \mathbb{R}^2 \times C_{n,\tau}$ and $\rho \in \mathbb{D}_3$. Moreover, $f(\alpha, 0) = 0$ for all $\alpha \in \mathbb{R}^2$.

(DBT2) The infinitesimal generator $\mathscr{A}$ has a nonsemisimple eigenvalue 0, with geometric multiplicity two and algebraic multiplicity four.

Let γ denote a generator of $\mathbb{Z}_3$ and let κ be the flip. Assumption (DBT1) means that the geometric eigenspace $\mathscr{E}_0$ is two-dimensional. There exist $\zeta_1 \in \mathscr{E}_0$ and $k \in \mathbb{Z}$ such that $\gamma \cdot \zeta_1 = e^{ik\gamma}\zeta_1$. Setting $\zeta_2 = \kappa \cdot \zeta_1$, we have $\mathscr{E}_0 = \mathrm{span}\{\zeta_1, \zeta_2\}$. Furthermore, it is possible to choose $q_1 \in \mathscr{E}$ such that $\kappa \cdot q_1 = \bar{q}_1$. Indeed, $\gamma \cdot \bar{q}_1 = \overline{\gamma \cdot q_1} = e^{-ik\gamma}\bar{q}_1$, so there is $c \in \mathbb{C}$ with $\kappa \cdot q_1 = c\bar{q}_1$, which together with $\kappa^2 = \mathrm{Id}_n$ implies that $q_1 = |c|^2 q_1$. Thus, $\kappa \cdot q_1 = e^{2i\varepsilon}q_1$ for some ε. Defining $\zeta = e^{-i\varepsilon}q_1$, we have $\zeta \in \mathscr{E}_0$ and $\kappa \cdot \zeta = \bar{\zeta}$. Therefore, we have $\mathscr{E}_0 = \{zq_1 + \bar{z}\bar{q}_1 : z \in \mathbb{C}\}$. Since the zero eigenvalue of $\mathscr{A}$ has the algebraic multiplicity four, there exists $q_2 \in C_{n,\tau}$ such that $\gamma \cdot q_2 = e^{ik\gamma}q_2$ and $\mathscr{A}q_2 = q_1$. Thus, the generalized eigenspace P is spanned by $q_1, q_2, \bar{q}_1$, and $\bar{q}_2$, and the reduced action of $\mathbb{D}_3$ on $\mathbb{C}^2 \cong P$ is given by

$$\gamma \cdot (z_1, z_2) = (e^{ik\gamma}z_1, e^{ik\gamma}z_2), \quad \kappa \cdot (z_1, z_2) = (\bar{z}_1, \bar{z}_2). \tag{7.75}$$

Similarly, there exist eigenvectors p_1 and $p_2 \in C_{n,\tau}^*$ of the adjoint operator $\mathscr{A}^*$ such that $\gamma \cdot p_j = e^{ik\gamma}p_j$ and $\kappa \cdot p_j = \bar{p}_j$ for $j = 1, 2$, and

$$\mathscr{A}^* p_2 = 0, \quad \mathscr{A}^* p_1 = p_2.$$

One can select these vectors to satisfy $\langle p_j, q_k \rangle = \delta_{jk}$ and $\langle \bar{p}_j, q_k \rangle = 0$, $j, k = 1, 2$. Let $\Phi = (q_1, \bar{q}_1, q_2, \bar{q}_2)$ and $\Psi = (p_1, \bar{p}_1, p_2, \bar{p}_2)^T$. Then $\langle \Psi, \Phi \rangle = \mathrm{Id}_4$ and $\dot{\Phi} = \Phi B$, where

$$B = \begin{bmatrix} 0 & 1 & 0 & 0 \\ 0 & 0 & 0 & 0 \\ 0 & 0 & 0 & 1 \\ 0 & 0 & 0 & 0 \end{bmatrix}.$$

The reduced equation on the center manifold $M_{\mathrm{loc}}^{c,\alpha}$ is

$$\begin{aligned} \dot{z}_1 &= z_2 + G^1(\alpha, z), \\ \dot{z}_2 &= G^2(\alpha, z), \end{aligned} \tag{7.76}$$

where $z = (z_1, \bar{z}_1, z_2, \bar{z}_2) \in \mathbb{C}^4$ and $G^j(\alpha, z) = \bar{p}_j(0)F(\alpha, \Phi z + W(\alpha, z))$, W satisfies $W(0,0) = 0$, $D_z W(0,0) = 0$, $W(\alpha, \rho \cdot z) = \rho \cdot W(\alpha, z)$ for all $\rho \in \mathbb{D}_3$ and $z \in \mathbb{C}^4$,

$$\frac{d}{dt}W = \mathscr{A}_\varrho W + H(\alpha, z),$$

and $H(\alpha, z) = [X_0 - \Phi \Psi(0)]F(\alpha, \Phi z + W(\alpha, z))$.

Now we assume that in (7.75), $k \neq 0$, i.e.,

(DBT3) $\mathbb{Z}_3$ acts nontrivially on P.

Then by the symmetric center manifold theorem, the reduced equation (7.76) is $\mathbb{D}_3$-equivariant. So we need to consider the $\mathbb{D}_3$-invariants and -equivariants.

Lemma 7.2. *(i) The ring of all $\mathbb{D}_3$-invariant germs acting on $\mathbb{C}^2$ as in (7.75) is generated by $s_1 = z_1\bar{z}_1$, $s_2 = z_2\bar{z}_2$, $s_3 = z_1\bar{z}_2 + \bar{z}_1 z_2$, and $t_j = 2\mathrm{Re}(z_1^j z_2^{3-j})$, $j \in \{0,1,2,3\}$.*
(ii) The module of $\mathbb{D}_3$-equivariant smooth mappings of $\mathbb{C}^2 \to \mathbb{C}^2$ is generated by

$$\begin{bmatrix} 0 \\ z_1 \end{bmatrix}, \begin{bmatrix} 0 \\ z_2 \end{bmatrix}, \begin{bmatrix} z_1 \\ 0 \end{bmatrix}, \begin{bmatrix} z_2 \\ 0 \end{bmatrix}, \begin{bmatrix} 0 \\ \bar{z}_1^j \bar{z}_2^{2-j} \end{bmatrix}, \begin{bmatrix} \bar{z}_1^j \bar{z}_2^{2-j} \\ 0 \end{bmatrix}$$

for $j \in \{0,1,2\}$.

In view of the $\mathbb{D}_3$-equivariance of $G(\alpha,z) = (G^1(\alpha,z), G^2(\alpha,z))^T$, we see that the reduced equation (7.76) has the following Taylor expansion up to third order:

$$\begin{aligned}
\dot{z}_1 &= z_2 + \eta_1 z_1 + \zeta_1 z_2 \\
&\quad + a_1 \bar{z}_1^2 + b_1 \bar{z}_1 \bar{z}_2 + c_1 \bar{z}_2^2 \\
&\quad + z_1[d_1 z_1 \bar{z}_1 + e_1 z_2 \bar{z}_2 + f_1(z_1 \bar{z}_2 + \bar{z}_1 z_2)] \\
&\quad + z_2[g_1 z_1 \bar{z}_1 + h_1 z_2 \bar{z}_2 + i_1(z_1 \bar{z}_2 + \bar{z}_1 z_2)] + o(|z_1, z_2|^2), \\
\dot{z}_2 &= \eta_2 z_1 + \zeta_2 z_2 \\
&\quad + a_2 \bar{z}_1^2 + b_2 \bar{z}_1 \bar{z}_2 + c_2 \bar{z}_2^2 \\
&\quad + z_1[d_2 z_1 \bar{z}_1 + e_2 z_2 \bar{z}_2 + f_2(z_1 \bar{z}_2 + \bar{z}_1 z_2)] \\
&\quad + z_2[g_2 z_1 \bar{z}_1 + h_2 z_2 \bar{z}_2 + i_2(z_1 \bar{z}_2 + \bar{z}_1 z_2)] + o(|z_1, z_2|^2),
\end{aligned} \tag{7.77}$$

where the coefficients η_1, η_2, ζ_1, ζ_2, a_1, a_2, ..., depend on $\alpha \in \mathbb{R}^2$. In fact, we have for $j = 1, 2$,

$$\begin{aligned}
\eta_j &= \bar{p}_j(0)[\alpha_1 \mathscr{F}^1_{\alpha_1}(0, q_1) + \alpha_2 \mathscr{F}^1_{\alpha_2}(0, q_1)] + O(|\alpha|^2), \\
\zeta_j &= \bar{p}_j(0)[\alpha_1 \mathscr{F}^1_{\alpha_1}(0, q_2) + \alpha_2 \mathscr{F}^1_{\alpha_2}(0, q_2)] + O(|\alpha|^2), \\
a_j &= \bar{p}_j(0)\mathscr{F}^2(0, \bar{q}_1, \bar{q}_1) + O(|\alpha|), \\
b_j &= \bar{p}_j(0)\mathscr{F}^2(0, \bar{q}_1, \bar{q}_2) + O(|\alpha|), \\
c_j &= \bar{p}_j(0)\mathscr{F}^2(0, \bar{q}_2, \bar{q}_2) + O(|\alpha|).
\end{aligned}$$

Next, by some near-identity coordinate transformation, we can remove as many second-order terms as possible to obtain the following normal form to third order:

$$\begin{aligned}
\dot{u} &= v, \\
\dot{v} &= \mu_1 u + \mu_2 v + E\bar{u}^2 + F\bar{u}\bar{v} + [A|u|^2 + B|v|^2 + C(u\bar{v} + \bar{u}v)]u + D|u|^2 v,
\end{aligned} \tag{7.78}$$

where the coefficients μ_1, μ_2, E, F, A, B, and C depend on $\alpha \in \mathbb{R}^2$. In fact, we have

$$\begin{aligned}
\mu_1(\alpha) &= \bar{p}_2(0)[\alpha_1 \mathscr{F}^1_{\alpha_1}(0, q_1) + \alpha_2 \mathscr{F}^1_{\alpha_2}(0, q_1)] + O(|\alpha|^2), \\
\mu_2(\alpha) &= \bar{p}_2(0)[\alpha_1 \mathscr{F}^1_{\alpha_1}(0, q_1) + \alpha_2 \mathscr{F}^1_{\alpha_2}(0, q_1)] \\
&\quad + \bar{p}_2(0)[\alpha_1 \mathscr{F}^1_{\alpha_1}(0, q_2) + \alpha_2 \mathscr{F}^1_{\alpha_2}(0, q_2)] + O(|\alpha|^2), \\
E(\alpha) &= \frac{1}{2}\bar{p}_2(0)\mathscr{F}^2(0, \bar{q}_1, \bar{q}_1) + O(|\alpha|),
\end{aligned}$$

$$F(\alpha) = \frac{1}{2}\bar{p}_2(0)\mathscr{F}^2(0,\bar{q}_1,\bar{q}_2) + \bar{p}_1(0)\mathscr{F}^2(0,\bar{q}_1,\bar{q}_1) + O(|\alpha|).$$

Now we assume that

(DBT4) $E(0)F(0) \neq 0.$

Then using the linear scalings

$$z_1 = \frac{F^2}{E}u, \quad z_2 = \frac{F^3}{E^2}v, \quad t^* = \frac{E}{F}t$$

of the variables and time and then dropping $*$, we obtain

$$\begin{aligned}
\dot{z}_1 &= z_2, \\
\dot{z}_2 &= \delta_1 z_1 + \delta_2 z_2 + \bar{z}_1^2 + \bar{z}_1\bar{z}_2 \\
&\quad + [a|z_1|^2 + b|z_2|^2 + c(z_1\bar{z}_2 + \bar{z}_1 z_2)]u + d|z_1|^2 z_2,
\end{aligned} \tag{7.79}$$

where $\delta_1 = \mu_1 F^2/E^2$ and $\delta_2 = \mu_2 F/E$. Thus, we have proved the following result.

Theorem 7.16. *In addition to conditions (DBT1)–(DBT4), assume that the Jacobian matrix of (γ_1,γ_2) with respect to (α_1,α_2) is nonsingular at $\alpha = 0$. Then system (7.60) is locally topologically equivalent to the normal form (7.79).*

In what follows, we neglect the cubic terms in (7.79) and consider the truncated system

$$\begin{aligned}
\dot{z}_1 &= z_2, \\
\dot{z}_2 &= \delta_1 z_1 + \delta_2 z_2 + \bar{z}_1^2 + \bar{z}_1\bar{z}_2,
\end{aligned} \tag{7.80}$$

which always has four equilibria: $O = (0,0)$, $E = (-\delta_1,0)$, $\gamma \cdot E = (-\delta_1 e^{ik\gamma},0)$, and $\gamma^2 \cdot E = (-\delta_1 e^{-ik\gamma},0)$. Moreover, the characteristic polynomials of (7.80) at the equilibria O and E are $(\lambda^2 - \delta_2\lambda - \delta_1)^2 = 0$ and $[\lambda^2 - (\delta_2 - \delta_1)\lambda + \delta_1][\lambda^2 - (\delta_2 + \delta_1)\lambda - 3\delta_1] = 0$, respectively. This implies that $\delta_1 = 0$ is a bifurcation line of secondary steady state, along which system (7.80) undergoes a $\mathbb{D}_3$-equivariant transcritical bifurcation. We can check that the equilibrium O (or E) has a pair of double (respectively, simple) purely imaginary eigenvalues for (δ_1,δ_2) on the line l_1 (respectively, either l_2 or l_3), where

$$\begin{aligned}
l_1 &= \{(\delta_1,\delta_2) : \delta_1 < 0, \delta_2 = 0\}, \\
l_2 &= \{(\delta_1,\delta_2) : \delta_1 = \delta_2 \geq 0\}, \\
l_3 &= \{(\delta_1,\delta_2) : \delta_1 = -\delta_2 \leq 0\}.
\end{aligned}$$

Thus, we have the following results (Fig. 7.2).

Theorem 7.17. *(i) The equilibrium O undergoes a nondegenerate $\mathbb{D}_3$-equivariant Hopf bifurcation along the line l_1, giving rise to three different types of periodic solutions: isotropy types $\Sigma_\gamma^\pm = \langle(\gamma,\pm\frac{2\pi}{3})\rangle$ for $\delta_2 > 0$, isotropy type $\Sigma_\kappa^+ = \langle\kappa\rangle$, and isotropy type $\Sigma_\kappa^- = \langle(\kappa,\pi)\rangle$ both for $\delta_2 < 0$. All these solutions are of saddle type. In particular, the bifurcated periodic solutions of isotropy type Σ_κ^+ are inside $\mathrm{Fix}(\kappa)$.*

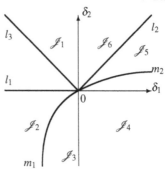

Fig. 7.2 Bifurcation sets for (7.80)

(ii) Along the line l_2 (or l_3), each of the equilibria E, $\gamma \cdot E$, and $\gamma^2 \cdot E$ undergoes a nondegenerate $\mathbb{Z}_2$-equivariant Hopf bifurcation, giving rise to an unstable periodic solution of isotropy type Σ_κ^+ (respectively, Σ_κ^-), which exists for $\delta_2 - \delta_1 < 0$ (respectively, $\delta_2 + \delta_1 < 0$).

In view of Sect. 1.8.1, inside the invariant fixed-point space $\mathrm{Fix}(\kappa)$, system (7.80) has an orbit homoclinic to the equilibrium O (or E) for (δ_1, δ_2) in m_1 (respectively, m_2), where m_1 and m_2 are homoclinic bifurcation curves defined by

$$m_1 = \left\{ (\delta_1, \delta_2) : \delta_2 = \frac{1}{7}\delta_1 + o(|\delta_1|), \, \delta_1 \le 0 \right\}$$

and

$$m_2 = \left\{ (\delta_1, \delta_2) : \delta_2 = \frac{6}{7}\delta_1 + o(|\delta_1|), \, \delta_1 \ge 0 \right\}.$$

By symmetry, we have the following results.

Theorem 7.18. *(i) For $(\delta_1, \delta_2) \in m_2$, there are three orbits homoclinic to the equilibrium O, and these three homoclinic orbits are inside the three invariant fixed-point spaces $\mathrm{Fix}(\kappa)$, $\gamma\mathrm{Fix}(\kappa)$, and $\gamma^2\mathrm{Fix}(\kappa)$, respectively.*
(ii) For $(\delta_1, \delta_2) \in m_1$, there are three orbits homoclinic to the equilibria E, $\gamma \cdot E$, and $\gamma^2 \cdot E$, respectively. Moreover, the orbit homoclinic to the equilibrium E (or $\gamma \cdot E$, $\gamma^2 \cdot E$) is inside $\mathrm{Fix}(\kappa)$ (respectively, $\gamma\mathrm{Fix}(\kappa)$, $\gamma^2\mathrm{Fix}(\kappa)$).

7.5 Double Hopf Bifurcation

Consider

$$\dot{u}(t) = L(\alpha)u_t + f(\alpha, u_t), \tag{7.81}$$

where $\alpha \in \mathbb{R}^2$, the linear operator $L(\alpha) : C_{n,\tau} \to \mathbb{R}^n$ is continuous with respect to $\alpha \in \mathbb{R}^2$, $f \in C^l(\mathbb{R}^2 \times C_{n,\tau}, \mathbb{R}^n)$ for a large enough integer l such that $f(\alpha, 0) = 0$, and $f_\varphi(\alpha, 0) = 0$ for all $\alpha \in \mathbb{R}^2$. The infinitesimal generator $\mathscr{A}_\alpha : C_{n,\tau} \to C_{n,\tau}$ is given by

$$(\mathscr{A}_\alpha \varphi)(\theta) = \begin{cases} d\varphi(\theta)/d\theta, & \text{if} \quad \theta \in [-\tau, 0), \\ L(\alpha)\varphi, & \text{if} \quad \theta = 0. \end{cases} \tag{7.82}$$

Moreover, $\lambda \in \sigma(\mathscr{A}_\alpha)$ if and only if $\Delta(\alpha, \lambda)v = 0$ for some $v \in \mathbb{C}^n \setminus \{0\}$, where

$$\Delta(\alpha, \lambda) = \lambda \operatorname{Id}_n - L(\alpha)(e^{\lambda(\cdot)}).$$

In this section, we consider a type of codimension-two bifurcation: double Hopf bifurcation, i.e., the infinitesimal generator $\mathscr{A}_0$ has at least two pairs of purely imaginary eigenvalues. The main dynamic feature of this bifurcation is the occurrence of an invariant 3-torus. Our purpose here is to obtain the normal form for the general system (7.81) at the nonresonant double Hopf bifurcation point $\alpha = 0$ and then express the coefficients of low-order terms in terms of the coefficients of (7.81). These expressions are expected to be of importance for a detailed discussion of the double Hopf interaction in ODEs. Throughout this section, we assume that:

(HHB1) $\mathscr{A}_\alpha$ has two pairs of simple complex conjugate eigenvalues, $\lambda_1(\alpha)$, $\overline{\lambda}_1(\alpha)$, $\lambda_2(\alpha)$, and $\overline{\lambda}_2(\alpha)$, satisfying $\lambda_j(0) = i\omega_j$ with $\omega_j > 0$, $j = 1, 2$.
(HHB2) $ki\omega_1 + si\omega_2 \notin \sigma(\mathscr{A}_0) \setminus \Lambda$ for all $k, s \in \mathbb{Z}$, where $\Lambda = \{i\omega_1, -i\omega_1, i\omega_2, -i\omega_2\}$.
(HHB3) ω_1/ω_2 is irrational.

Here, the nonresonance conditions are guaranteed by (HHB2). Assumption (HHB3) means that the double Hopf bifurcation discussed here is nonresonant. Assumption (HHB3) plays a key role in the form of the normal form. Assumption (HHB1) implies that there exist $v_j \in \mathbb{C}^n$ and $u_j \in \mathbb{C}^{n*}$ ($j = 1, 2$) such that

$$\Delta(0, i\omega_j)v_j = 0, \qquad \overline{u}_j \Delta(0, i\omega_j) = 0, \qquad \overline{u}_j \Delta_\lambda(0, i\omega_j)v_j = 1.$$

Let P (respectively, P^*) be the center space of $\mathscr{A}_0$ (respectively, $\mathscr{A}_0^*$) relative to the set Λ. Then we have $P = \operatorname{span}\{\varphi_1, \overline{\varphi}_1, \varphi_2, \overline{\varphi}_2\}$ and $P^* = \operatorname{span}\{\psi_1, \overline{\psi}_1, \psi_2, \overline{\psi}_2\}$, where $\varphi_j(\theta) = v_j e^{i\omega\theta}$ and $\psi_j(\theta) = u_j e^{i\omega\theta}$ for $\theta \in [-\tau, 0]$, $j = 1, 2$. Then $\langle \psi_j, \varphi_s \rangle = \delta_{js}$ and $\langle \overline{\psi}_j, \varphi_s \rangle = 0$ for all $j, s \in \{1, 2\}$. Let $\Phi = (\varphi_1, \overline{\varphi}_1, \varphi_2, \overline{\varphi}_2)$ and $\Psi = (\psi_1, \overline{\psi}_1, \psi_2, \overline{\psi}_2)^T$. Then $\langle \Psi, \Phi \rangle = \operatorname{Id}_4$. Obviously, P and P^* are $\mathbb{S}^1$-invariant. Assumption (HHB3) implies that the action of $\mathbb{S}^1$ on P induces the following action of $\mathbb{T}^2$ on $\mathbb{C}^2 \cong P$:

$$\theta_t \cdot (z_1, z_2) = (z_1 e^{i\omega_1 t}, z_2 e^{i\omega_2 t}) \quad \text{for} \quad \theta_t = (e^{i\omega_1 t}, e^{i\omega_2 t}) \in \mathbb{T}^2. \tag{7.83}$$

Thus, we can obtain the following $\mathbb{T}^2$-equivariant normal form:

$$\dot{z}_j = i\omega_j z_j + G_j(\alpha, z_1, z_2), \quad j = 1, 2, \tag{7.84}$$

where $G(\alpha, z_1, z_2) = (G_1(\alpha, z_1, z_2), G_2(\alpha, z_1, z_2))^T$ is $\mathbb{T}^2$-equivariant. Therefore, there exist complex-valued polynomials W_1 and W_2 such that

$$G_j(\alpha, z_1, z_2) = z_j W_j(|z_1|^2, |z_2|^2, \alpha) \text{ for } j = 1, 2.$$

Thus, we obtain the coefficients for up to third-order terms of $\mathcal{G}$ in (7.84) as follows:

$$
\begin{aligned}
\dot{z}_j = {} & i\omega_j z_j + \tfrac{1}{2}\sum_{l=1}^{2} C_{jl}\alpha_l z_j + \tfrac{1}{6}\sum_{k+s=2} E_{jks}\alpha_k\alpha_s z_j \\
& + \tfrac{1}{6}D_{j1}z_j|z_1|^2 + \tfrac{1}{6}D_{j2}z_j|z_2|^2 + O(|(\alpha, z_1, z_2, \bar{z}_1, \bar{z}_2)|^4)
\end{aligned}
\tag{7.85}
$$

for $j = 1, 2$. If we change to polar coordinates $z_1 = \rho_1 e^{i\xi_1}$ and $z_2 = \rho_2 e^{i\xi_2}$, system (7.85) can be rewritten as

$$
\begin{aligned}
\dot{\rho}_1 &= \gamma_1\rho_1 + a_{11}\rho_1^3 + a_{12}\rho_1\rho_2^2 + O(|(\alpha, \rho_1, \rho_2)|^4), \\
\dot{\rho}_2 &= \gamma_2\rho_2 + a_{21}\rho_2\rho_1^2 + a_{22}\rho_2^3 + O(|(\alpha, \rho_1, \rho_2)|^4), \\
\dot{\xi}_1 &= \omega_1 + \sigma_1 + b_{11}\rho_1^2 + b_{12}\rho_2^2 + O(|(\alpha, \rho_1, \rho_2)|^3), \\
\dot{\xi}_2 &= \omega_2 + \sigma_2 + b_{21}\rho_1^2 + b_{22}\rho_2^2 + O(|(\alpha, \rho_1, \rho_2)|^3),
\end{aligned}
\tag{7.86}
$$

where γ_j, σ_j, a_{js}, b_{js} ($j, s = 1, 2$) are all real such that $\gamma_j + i\sigma_j = \tfrac{1}{2}\sum_{l=1}^{2} C_{jl}\alpha_l + \tfrac{1}{6}\sum_{k+s=2} E_{jks}\alpha_k\alpha_s$ and $a_{js} + ib_{js} = \tfrac{1}{6}D_{js}$ for all $j, s = 1, 2$. System (7.86) allows us to decouple the amplitude equations and the phase equations. In particular, the dynamics of (7.86) is determined by the amplitude equations:

$$
\begin{aligned}
\dot{\rho}_1 &= \gamma_1\rho_1 + a_{11}\rho_1^3 + a_{12}\rho_1\rho_2^2 + O(|(\alpha, \rho_1, \rho_2)|^4), \\
\dot{\rho}_2 &= \gamma_2\rho_2 + a_{21}\rho_2\rho_1^2 + a_{22}\rho_2^3 + O(|(\alpha, \rho_1, \rho_2)|^4).
\end{aligned}
\tag{7.87}
$$

If $\alpha \to (\gamma_1, \gamma_2)$ is regular at $\alpha = 0$, i.e., if the rank of the Jacobian matrix

$$\frac{\partial(\gamma_1, \gamma_2)}{\partial(\alpha_1, \alpha_2)}$$

evaluated at $\alpha = 0$ is 2, then we can unfold this degenerate case by varying γ_1 and γ_2 in a full neighborhood of $(0,0)$. Usually, phase-plane analysis can be employed to discuss the steady-state solutions and bifurcation phenomena of (7.86) (see Guckenheimer and Holmes [125] and Kuznetsov [200] for more details).

Let us focus here on the symmetry of (7.87). In the entire (ρ_1, ρ_2)-plane, system (7.87) is $\mathbb{Z}_2 \oplus \mathbb{Z}_2$-equivariant, since the reflection $\rho_1 \to -\rho_1$ or $\rho_2 \to -\rho_2$ leaves it invariant. Hence, we can employ the equivariant Hopf theorem to seek periodic solutions for each maximal isotropy subgroup of $\mathbb{Z}_2 \oplus \mathbb{Z}_2$. Obviously, $\mathbb{Z}_2 \oplus \mathbb{Z}_2$ has only four isotropy subgroups: $\mathbb{Z}_2 \oplus \mathbb{Z}_2$, $\mathbb{Z}_2(-1, 1)$, $\mathbb{Z}_2(1, -1)$, and $\mathbb{I}$. Only the two isotropy subgroups $\mathbb{Z}_2(-1, 1)$ and $\mathbb{Z}_2(1, -1)$ are maximal, with fixed-point subspaces $\{(0, \rho_2)\}$ and $\{(\rho_1, 0)\}$, respectively. This means that periodic solutions with pure mode $(0, \rho_2)$ or pure mode $(\rho_1, 0)$ may be expected to bifurcate from the trivial solution of (7.81). This also implies that bifurcations of periodic solutions with periods near $\frac{2\pi}{\omega_1}$ and $\frac{2\pi}{\omega_2}$ are primary in system (7.81).

Since mixed mode $\{(\rho_1, \rho_2) : W_j(\rho_1^2, \rho_2^2, \alpha) = 0, j = 1, 2\}$ with trivial isotropy $\mathbb{I}$ can be expected, there probably exists an invariant 2-torus in system (7.81). Varying α may lead to an exchange of stability of mixed mode solutions of (7.87), and hence a limit cycle can appear in its neighborhood via Hopf bifurcation. This means that a branch of invariant 3-tori of (7.86) and hence of (7.81) may exist for α near 0. See Takens [274] for further details of Hopf–Hopf interaction.

In the remaining part of this section, we aim to figure out concrete expressions for the coefficients of (7.85). Again, for simplicity of notation, we introduce the operator $\Theta : V_j^5(\mathbb{C}^2) \to V_j^5(\mathbb{C}^4)$ defined by

$$\Theta \begin{bmatrix} c_1 z^q \alpha^l \\ c_2 z^p \alpha^k \end{bmatrix} = \left(c_1 z^q \alpha^l, \overline{c_1 z^q \alpha^l}, c_2 z^p \alpha^k, \overline{c_2 z^p \alpha^k} \right)^T$$

for $q, p \in \mathbb{N}_0^4$ and $l, k \in \mathbb{N}_0^2$ such that $|(q, l)| = |(p, k)| = j$. Let

$$\Theta \begin{bmatrix} G_1(\alpha, z_1, z_2) \\ G_2(\alpha, z_1, z_2) \end{bmatrix} = \frac{1}{2} g_2^1(\alpha, z, 0) + \frac{1}{6} g_3^1(\alpha, z, 0) + \cdots .$$

Then $g_2^1(z, 0, \alpha)$ and $g_3^1(z, 0, \alpha)$ in (7.84) take the form

$$g_2^1(\alpha, z, 0) = \Theta \begin{bmatrix} \sum_{j=1}^r C_{1j} \alpha_j z_1 \\ \sum_{j=1}^r C_{2j} \alpha_j z_2 \end{bmatrix},$$

$$g_3^1(\alpha, z, 0) = \Theta \begin{bmatrix} D_{11} z_1 |z_1|^2 + D_{12} z_1 |z_2|^2 + \sum_{j+k=2} E_{1jk} \alpha_j \alpha_k z_1 \\ D_{21} z_2 |z_1|^2 + D_{22} z_2 |z_2|^2 + \sum_{j+k=2} E_{2jk} \alpha_j \alpha_k z_2 \end{bmatrix}.$$

For $\Phi z = 2\text{Re}\{\varphi_1 z_1 + \varphi_2 z_2\}$, let

$$f_j(\alpha, \Phi z) = \sum_{|(q,l)|=j} A_{(q,l)} z^q \alpha^l \tag{7.88}$$

and

$$f_j^1(\alpha, z, y) = \overline{\Psi}(0) F_j(\alpha, \Phi z + y),$$
$$f_j^2(\alpha, z, y) = [X_0 - \Phi \overline{\Psi}(0)] F_j(\alpha, \Phi z + y),$$

for $j \geq 2$, $z \in \mathbb{C}^4$, $y \in Q$, where Q denotes the space complementary to P in BC_n, f_j and F_j are the jth Fréchet derivatives of $f(\alpha, \varphi)$ and $F(\alpha, \varphi) = L(\alpha)\varphi - L(0)\varphi + f(\alpha, \varphi)$ with respect to $(\alpha, \varphi) \in \mathbb{R}^2 \times C_{n, \tau}$, respectively. As in Sect. 4.3.2, we write (7.81) as

$$\dot{z} = \Omega z + \sum_{j \geq 2} \frac{1}{j!} f_j^1(\alpha, z, y),$$
$$\frac{dy}{dt} = \mathscr{A}_Q y + \sum_{j \geq 2} \frac{1}{j!} f_j^2(\alpha, z, y), \tag{7.89}$$

where $\Omega = \mathrm{diag}(i\omega_1, -i\omega_1, i\omega_2, -i\omega_2)$. With the change of variables

$$(z,y) = (\hat{z}, \hat{y}) + \tfrac{1}{j!}(U_j^1(\alpha, \hat{z}), U_j^2(\alpha, \hat{z})), \tag{7.90}$$

and dropping the hats for simplicity of notation, (7.37) becomes

$$\begin{aligned}
\dot{z} &= \Omega z + \tfrac{1}{2}g_2^1(\alpha, z, y) + \tfrac{1}{3!}\bar{f}_3^1(\alpha, z, y) + \cdots, \\
\tfrac{dy}{dt} &= \mathscr{A}_Q y + \tfrac{1}{2}g_2^2(\alpha, z, y) + \tfrac{1}{3!}\bar{f}_3^2(\alpha, z, y) + \cdots,
\end{aligned} \tag{7.91}$$

where $g_2 = (g_2^1, g_2^2)$, $\hat{f}_2 = (f_2^1, f_2^2)$, and $\bar{f}_3 = (f_3^1, f_3^2)$ satisfy

$$g_2 = f_2 - M_2 U_2 \tag{7.92}$$

and

$$\bar{f}_3 = f_3 + \frac{3}{2}[(D_{z,y}f_2)U_2 - (D_{z,y}U_2)g_2]. \tag{7.93}$$

Here, the operator $M_2 = (M_2^1, M_2^2)$ is as defined in Chap. 4. In fact,

$$M_j^1(z^q \alpha^l e_k) = \left\{ i\omega_1[q_1 - q_2 + (-1)^k] + i\omega_2[q_3 - q_4] \right\} z^q \alpha^l e_k$$

and

$$M_j^1(z^q \alpha^l e_{2+k}) = \left\{ i\omega_1[q_1 - q_2] + i\omega_2[q_3 - q_4 + (-1)^k] \right\} z^q \alpha^l e_{2+k}$$

for all $k = 1, 2$, $q \in \mathbb{N}_0^4$, and $l \in \mathbb{N}_0^2$ with $|(q,l)| = j$, where $\{e_1, e_2, e_3, e_4\}$ is the canonical basis for $\mathbb{C}^4$, $z = (z_1, \bar{z}_1, z_2, \bar{z}_2) \in \mathbb{C}^4$.

Because $f_2^1(\alpha, z, 0) = \overline{\Psi}(0)f_2(\alpha, \Phi z) + 2\overline{\Psi}(0)L^1(\alpha)\Phi z$, we have

$$f_2^1(\alpha, z, 0) = \Theta \begin{bmatrix} 2\bar{u}_1 L^1(\alpha)\Phi z + \sum_{|q|=2} \bar{u}_1 A_{(q,0)} z^q \\ 2\bar{u}_2 L^1(\alpha)\Phi z + \sum_{|q|=2} \bar{u}_2 A_{(q,0)} z^q \end{bmatrix},$$

where $L^1(\alpha) = \alpha_1 L_{\alpha_1}(0) + \alpha_2 L_{\alpha_2}(0)$ denotes the linear part of $L(\alpha) - L(0)$ with respect to $\alpha = (\alpha_1, \alpha_2)^T \in \mathbb{R}^2$. Then

$$C_{kj} = 2\frac{\partial \lambda_k(0)}{\partial \alpha_j}, \qquad j, k = 1, 2.$$

Moreover, $U_2(z, \alpha) = M_2^{-1} P_2 f_2(z, 0, \alpha)$, that is,

$$\begin{aligned}
U_2^1(z, \alpha) = {} &\Theta \begin{bmatrix} \bar{u}_1 L^1(\alpha)\left(i\omega_1^{-1}\overline{\varphi}_1 \bar{z}_1 - 2i\omega_2^{-1}\varphi_2 z_2 + 2i\omega_2^{-1}\overline{\varphi}_1 \bar{z}_1\right) \\ \bar{u}_2 L^1(\alpha)\left(-2i\omega_1^{-1}\varphi_1 z_1 + 2i\omega_1^{-1}\overline{\varphi}_1 \bar{z}_1 + i\omega_2^{-1}\overline{\varphi}_2 \bar{z}_2\right) \end{bmatrix} \\
&+ \Theta \begin{bmatrix} \sum_{|q|=2}[i\omega_1(q_1 - q_2 - 1) + i\omega_2(q_3 - q_4)]^{-1}\bar{u}_1 A_{(q,0)} z^q \\ \sum_{|q|=2}[i\omega_1(q_1 - q_2) + i\omega_2(q_3 - q_4 - 1)]^{-1}\bar{u}_2 A_{(q,0)} z^q \end{bmatrix},
\end{aligned}$$

and $U_2^2(z, \alpha) = h(z, \alpha)$ is the unique solution in $V_2^5(Q)$ of the equation

$$(M_2^2 h)(z, \alpha) = [X_0 - \Phi\overline{\Psi}(0)]\left[2L^1(\alpha)(\Phi z) + f_2(\alpha, \Phi z)\right]. \tag{7.94}$$

We write $U_2^2(z,\alpha)$ as

$$U_2^2(z,\alpha) = \sum_{|(q,l)|=2} h_{(q,l)} z^q \alpha^l.$$

After computing up to order 2, the term of order 3 of the normal form becomes $\tilde{f}_3 = f_3 + \frac{3}{2}[(D_{z,y}f_2)U_2 - (D_{z,y}U_2)g_2]$. Thus, the complex coefficients D_{jk} ($j,k = 1, 2$) are given as follows:

$$D_{11} = \bar{u}_1 A_{(2,1,0,0,0)} - \frac{3}{2i\omega_1}\bar{u}_1 A_{(2,0,0,0,0)}\bar{u}_1 A_{(1,1,0,0,0)} + \frac{3}{2i\omega_1}|\bar{u}_1 A_{(1,1,0,0,0)}|^2$$
$$+ \frac{1}{i\omega_1}|\bar{u}_1 A_{(0,2,0,0,0)}|^2 - \frac{3}{2i\omega_2}\bar{u}_1 A_{(1,0,1,0,0)}\bar{u}_2 A_{(1,1,0,0,0)}$$
$$+ \frac{3}{2i(2\omega_1-\omega_2)}\bar{u}_1 A_{(0,1,1,0,0)}\bar{u}_2 A_{(2,0,0,0,0)} + \frac{3}{2i\omega_2}\bar{u}_1 A_{(1,0,0,1,0)}u_2 A_{(1,1,0,0,0)}$$
$$+ \frac{3}{2i(2\omega_1+\omega_2)}\bar{u}_1 A_{(0,1,0,1,0)}u_2 A_{(0,2,0,0,0)} + 3\bar{u}_1 H_2((0,\varphi_1),(0,h_{(1,1,0,0,0)}))$$
$$+ 3\bar{u}_1 H_2((0,\bar{\varphi}_1),(0,h_{(2,0,0,0,0)})),$$

$$D_{12} = \bar{u}_1 A_{(1,0,1,1,0)} + \frac{3}{2i(2\omega_1-\omega_2)}|\bar{u}_1 A_{(0,1,1,0,0)}|^2 + \frac{3}{2i(2\omega_1+\omega_2)}|\bar{u}_1 A_{(0,1,0,1,0)}|^2$$
$$+ \frac{3}{i\omega_1-2i\omega_2}\bar{u}_1 A_{(0,0,2,0,0)}\bar{u}_2 A_{(1,0,0,1,0)} + \frac{3}{2i\omega_1}\bar{u}_1 A_{(0,0,1,1,0)}\bar{u}_2 A_{(1,0,1,0,0)}$$
$$+ \frac{3}{2i\omega_1}\bar{u}_1 A_{(0,0,1,1,0)}u_2 A_{(1,0,0,1,0)} + \frac{3}{i\omega_1+2i\omega_2}\bar{u}_1 A_{(0,0,0,2,0)}u_2 A_{(1,0,1,0,0)}$$
$$+ \frac{3}{2i\omega_1}\bar{u}_1 A_{(1,1,0,0,0)}u_1 A_{(0,0,1,1,0)} - \frac{3}{i\omega_1}\bar{u}_1 A_{(2,0,0,0,0)}\bar{u}_1 A_{(0,0,1,1,0)}$$
$$+ \frac{3}{2i\omega_2}\bar{u}_1 A_{(1,0,0,1,0)}u_2 A_{(0,0,1,1,0)} - \frac{3}{2i\omega_2}\bar{u}_1 A_{(1,0,1,0,0)}\bar{u}_2 A_{(0,0,1,1,0)}$$
$$+ 3\bar{u}_1 H_2((0,\varphi_1),(0,h_{(0,0,1,1,0)})) + 3\bar{u}_1 H_2((0,\varphi_2),(0,h_{(1,0,0,1,0)}))$$
$$+ 3\bar{u}_1 H_2((0,\bar{\varphi}_2),(0,h_{(1,0,1,0,0)})),$$

$$D_{21} = \bar{u}_2 A_{(1,1,1,0,0)} + \frac{3}{2i(2\omega_2-\omega_1)}|\bar{u}_2 A_{(1,0,0,1,0)}|^2 + \frac{3}{2i(2\omega_2+\omega_1)}|\bar{u}_2 A_{(0,1,0,1,0)}|^2$$
$$+ \frac{3}{i\omega_2-2i\omega_1}\bar{u}_2 A_{(2,0,0,0,0)}\bar{u}_1 A_{(0,1,1,0,0)} + \frac{3}{2i\omega_2}\bar{u}_2 A_{(1,1,0,0,0)}\bar{u}_1 A_{(1,0,1,0,0)}$$
$$+ \frac{3}{2i\omega_1}\bar{u}_2 A_{(1,1,0,0,0)}u_1 A_{(0,1,1,0,0)} + \frac{3}{i\omega_2+2i\omega_1}\bar{u}_2 A_{(0,2,0,0,0)}u_1 A_{(1,0,1,0,0)}$$
$$+ \frac{3}{2i\omega_1}\bar{u}_2 A_{(0,1,1,0,0)}u_1 A_{(1,1,0,0,0)} - \frac{3}{2i\omega_1}\bar{u}_2 A_{(1,0,1,0,0)}\bar{u}_1 A_{(1,1,0,0,0)}$$
$$+ \frac{3}{2i\omega_2}\bar{u}_2 A_{(0,0,1,1,0)}u_2 A_{(1,1,0,0,0)} - \frac{3}{i\omega_2}\bar{u}_2 A_{(0,0,2,0,0)}\bar{u}_2 A_{(1,1,0,0,0)}$$
$$+ 3\bar{u}_1 H_2((0,\varphi_1),(0,h_{(0,1,1,0,0)})) + 3\bar{u}_1 H_2((0,\bar{\varphi}_1),(0,h_{(1,0,1,0,0)}))$$
$$+ 3\bar{u}_1 H_2((0,\varphi_2),(0,h_{(1,1,0,0,0)})),$$

$$D_{22} = \bar{u}_1 A_{(0,0,2,1,0)} - \frac{3}{2i\omega_2}\bar{u}_2 A_{(0,0,2,0,0)}\bar{u}_2 A_{(0,0,1,1,0)} + \frac{3}{2i\omega_2}|\bar{u}_2 A_{(0,0,1,1,0)}|^2$$
$$+ \frac{1}{i\omega_2}|\bar{u}_2 A_{(0,0,0,2,0)}|^2 - \frac{3}{2i\omega_1}\bar{u}_2 A_{(1,0,1,0,0)}\bar{u}_1 A_{(0,0,1,1,0)}$$
$$+ \frac{3}{2i(2\omega_2-\omega_1)}\bar{u}_2 A_{(1,0,0,1,0)}\bar{u}_1 A_{(0,0,2,0,0)} + \frac{3}{2i\omega_1}\bar{u}_2 A_{(0,1,1,0,0)}u_1 A_{(0,0,1,1,0)}$$
$$+ \frac{3}{2i(\omega_1+2\omega_2)}\bar{u}_2 A_{(0,1,0,1,0)}u_1 A_{(0,0,2,0,0)} + 3\bar{u}_1 H_2((0,\varphi_2),(0,h_{(0,0,1,1,0)}))$$
$$+ 3\bar{u}_1 H_2((0,\bar{\varphi}_2),(0,h_{(0,0,2,0,0)})).$$

References

1. Abolinia, V.E., Mishkis, A.D.: A mixed problem for a linear hyperbolic system on the plane. Latvijas Valsts Univ. Zinatn. Raksti **20**, 87–104 (1958)
2. Abolinia, V.E., Mishkis, A.D.: Mixed problems for quasi-linear hyperbolic systems in the plane. Mat. Sb. (N.S.) **50**, 423–442 (1960)
3. Adimy, M.: Integrated semigroups and delay differential equations. J. Math. Anal. Appl. **177**, 125–134 (1993)
4. Afraimovich, V., Shil'nikov, L.: On singular trajectories of dynamical systems. Usp. Mat. Nauk **5**, 189–190 (1972) (in Russian)
5. Ait Babram, M.: An algorithmic scheme for approximating center manifolds and normal forms for functional differential equations. In: Arino, O., Hbid, M.L., Ait Dads, E. (eds.) Delay Differential Equations and Applications. NATO Sci. Ser. II Math. Phys. Chem., vol. 205, pp. 193–226. Springer, Dordrecht (2006)
6. Ait Babram, M., Arino, O., Hbid, M.L.: Computational scheme of a center manifold for neutral functional differential equations. J. Math. Anal. Appl. **258**(2), 396–414 (2001)
7. Ait Babram, M., Hbid, M.L., Arino, O.: Approximation scheme of a center manifold for functional-differential equations. J. Math. Anal. Appl. **213**(2), 554–572 (1997)
8. Alexander, J.C.: Bifurcation of zeros of parametrized functions. J. Funct. Anal. **29**, 37–53 (1978)
9. Alexander, J.C., Fitzpatrick, P.M.: The homotopy of a certain spaces of nonlinear equations, and its relation to global bifurcation of the fixed points of parametrized condensing operators. J. Funct. Anal. **34**, 87–106 (1979)
10. Alexander, J.C., Yorke, J.A.: Global bifurcations of periodic orbits. Am. J. Math. **100**, 263–292 (1978)
11. Algaba, A., Merino, M., Freire, E., Gamero, E., Rodrguez-Luis, A.J.: Some results on Chua's equation near a triple-zero linear degeneracy. Int. J. Bifurcat. Chaos Appl. Sci. Eng. **13**(3), 583–608 (2003)
12. Arino, O.: Contribution à l'étude des comportements des solutions d'équation différentielle à retard par des méthodes de monotonie et de bifurcation. Thése d'état, Université de Bordeaux 1 (1980)
13. an der Heiden, U.: Periodic solutions of a nonlinear second-order differential equations with delay. J. Math. Anal. Appl. **70**, 599–609 (1979)
14. Andronov, A.A.: Application of Poincaré's theorem on "bifurcation points" and "change in stability" to simple auto-oscillatory systems. C. R. Acad. Sci. Paris **189**(15), 559–561 (1929)
15. Andronov, A.A., Leontovich, E.: Some cases of dependence of limit cycles on a parameter. J. State Univ. Gorki **6**, 3–24 (1937) (in Russian)
16. Andronov, A.A., Pontryagin, L.: Systémes grossiéres. Dokl. Akad. Nauk SSSR **14**, 247–251 (1937) (in Russian).

S. Guo and J. Wu, *Bifurcation Theory of Functional Differential Equations*,
Applied Mathematical Sciences 184, DOI 10.1007/978-1-4614-6992-6,
© Springer Science+Business Media New York 2013

17. Arino, O., Hbid, M.L.: Existence of periodic solutions for a delay differential equation via the Poincaré procedure. Differ. Equat. Dyn. Syst. **4**(2), 125–148 (1996)

18. Arino, O., Sánchez, E.: A variation of constants formula for an abstract functional-differential equation of retarded type. Differ. Integr. Equat. **9**(6), 1305–1320 (1996)

19. Arnold, V.I.: Geometrical Methods in the Theory of Ordinary Differential Equations. Springer, New York (1983)

20. Arnold, V.I.: Lectures on bifurcations in versal families. Russ. Math. Surv. **27**, 54–123 (1972)

21. Arrowsmith, D.K., Place, C.M.: An Introduction to Dynamical Systems. Cambridge University Press, Cambridge (1990)

22. Ashkenazi, M., Chow, S.N.: Normal forms near critical points for differential equations and maps. IEEE Trans. Circuits Syst. **35**, 850–862 (1988)

23. Aubin, J.P.: Applied Functional Analysis. Wiley, New York (1979)

24. Balanov, Z., Krawcewicz, W.: Remarks on the equivariant degree theory. Topol. Methods Nonlinear Anal. **13**, 91–103 (1999)

25. Balanov, Z., Krawcewicz, W., Steinlein, H.: Reduced $SO(3) \times S^1$-equivariant degree with applications to symmetric bifurcations problems. Nonlinear Anal. **47**, 1617–1628 (2001)

26. Bélair, J.: Population models with state-dependent delays. Lect. Notes Pure Appl. Math. **131**, 165–176 (1991)

27. Bélair, J., Campbell, S.A.: Stability and bifurcations of equilibria in a multiple-delayed differential equation. SIAM J. Appl. Math. **54**, 1402–1424 (1994)

28. Bélair, J., Campbell, S.A., van den Driessche, P.: Frustration, stability, and delay-induced oscillations in a neural network model. SIAM J. Appl. Math. **56**, 245–255 (1996)

29. Bélair, J., Dufour, S.: Stability in a three-dimensional system of delay-differential equations. Can. Appl. Math. Q. **4**(2), 135–156 (1996)

30. Bellman, R., Cooke, K.L.: Differential Difference Equations. Academic Press, New York (1963)

31. Bernfeld, S.R., Negrini, P., Salvadori, L.: Generalized Hopf bifurcation and h-asymptotic stability. J. Nonlinear Anal. Theor. Meth. Appl. **4**, 109–1107 (1980)

32. Bernfeld, S.R., Negrini, P., Salvadori, L.: Quasi-invariant manifolds stability and generalized Hopf bifurcation. Ann. Math. Pura Appl. **4**, 105–119 (1982)

33. Birkhoff, G.D.: Dynamical Systems. AMS, Providence (1927)

34. Birkhoff, G.D.: Nouvelles recherches sur les systèmes dynamiques. Memoriae Pont. Acad. Sci. Novi. Lincaei Ser. 3 **1**, 85–216 (1935)

35. Bogdanov, R.: Versal deformations of a singular point on the plane in the case of zero eigenvalues. In: Proceedings of Petrovskii Seminar, Moscow State University, vol. 2, pp. 37–65 (1976) (in Russian) (English translation: Selecta Math. Soviet. **1**(4), 389–421, 1981)

36. Braaksma, B.L.J., Broer, H.W.: Quasiperiodic flow near a codimension one singularity of a divergence free vector field in dimension four. In: Bifurcation, Ergodic Theory and Applications (Dijon, 1981). Astérisque, vol. 98–99, pp. 74–142. Soc. Math. France, Paris (1982)

37. Brayton, R.K.: Bifurcation of periodic solutions in a nonlinear difference-differential equation of neutral type. Q. Appl. Math. **24**, 215–224 (1966)

38. Brayton, R.K.: Nonlinear oscillations in a distributed network. Q. Appl. Math. **24**, 289–301 (1967)

39. Brayton, R.K., Miranker, W.L.: A stability theory for nonlinear mixed initial boundary value problems. Arch. Ration. Mech. Anal. **17**, 358–376 (1964)

40. Brayton, R.K., Moser, J.K.: A theory of nonlinear networks. I. Q. Appl. Math. **22**, 1–33 (1964)

41. Bredon, G.E.: Introduction to Compact Transformation Groups. Academic, New York (1972)

42. Broer, H.W.: Coupled Hopf-bifurcations: persistent examples of n-quasiperiodicity determined by families of 3-jets. Geometric methods in dynamics. I. Astérisque **286**, xix, 223–229 (2003)

43. Broer, H.W.: Quasiperiodicity in local bifurcation theory. In: Bruter, C.P., Aragnol, A., Lichnérowicz, A. (eds.) Bifurcation Theory, Mechanics and Physics. Mathematics and Its Applications, pp. 177–208. Reidel, Dordrecht (1983)

44. Broer, H.W., Vegter, G.: Subordinate Sil'nikov bifurcations near some singularities of vector fields having low codimension. Ergod. Theor. Dyn. Syst. **4**, 509–525 (1984)
45. Brokate, M., Colonius, F.: Linearizing equations with state-dependent delays. Appl. Math. Optim. **21**, 45–52 (1990)
46. Brouwder, F.E.: Fixed point theory and nonlinear problems. Bull. Am. Math. Soc. **1**, 1–39 (1983)
47. Brouwer, L.E.J.: Über Abbildung der Mannigfaltigkeiten. Math. Ann. **70**, 97–115 (1912)
48. Bruno, A.D.: Local Method of Nonlinear Analysis of Differential Equations (in Russian). Izdatel'stvo Nauka, Moscow (1979)
49. Buono, P.L., Bélair, J.: Restrictions and unfolding of double Hopf bifurcation in functional differential equations. J. Differ. Equat. **189**, 234–266 (2003)
50. Busenberg, S., Huang, W.: Stability and Hopf bifurcation for a population delay model with diffusion effects. J. Differ. Equat. **124**(1), 80–107 (1996)
51. Busenberg, S., Travis, C.C.: On the use of reducible functional differential equations. J. Math. Anal. Appl. **89**, 46–66 (1982)
52. Campbell, S.A.: Time delays in neural systems. In: McIntosh, R., Jirsa, V.K. (eds.) Handbook of Brain Connectivity. Springer, New York (2007)
53. Carr, J.: Applications of Centre Manifold Theory. Applied Mathematical Sciences, vol. 35. Springer, New York (1981)
54. Chafee, N.: A bifurcation problem for a functional differential equation of finitely retarded type. J. Math. Anal. Appl. **35**, 312–348 (1971)
55. Chafee, N.: Generalized Hopf bifurcation and perturbation in a full neighborhood of a given vector field. Indiana Univ. Math. J. **27**, 173–194 (1978)
56. Chen, G., Della Dora, J.: Rational normal form for dynamical systems via Carleman linearization. In: Proceeding of ISSAC-99, pp. 165–172. ACM Press–Addison Wesley, Vancouver (1999)
57. Chen, G., Della Dora, J.: Further reduction of normal forms for dynamical systems. J. Differ. Equat. **166**, 79–106 (2000)
58. Chen, Y.: Existence and unstable sets of oscillating periodic orbits for delayed excitatory networks of two neurons. Differ. Equat. Dyn. Syst. **9**, 169–185 (2001)
59. Chen, Y., Wu, J.: Existence and attraction of a phase-locked oscillation in a delayed network of two neurons. Differ. Integr. Equat. **14**, 1181–1236 (2001)
60. Chen, Y., Wu, J.: Slowly oscillating periodic solutions for a delayed frustrated network of two neurons. J. Math. Anal. Appl. **259**, 188–208 (2001)
61. Chen, Y., Wu, J., Krisztin, T.: Connecting orbits from synchronous periodic solutions to phase-locked periodic solutions in a delay differential system. J. Differ. Equat. **163**, 130–173 (2000)
62. Chossat, P., Lauterbach, R.: Methods in Equivariant Bifurcations and Dynamical Systems. World Scientific, Singapore (2000)
63. Chow, S.N.: Existence of periodic solutions of autonomous functional differential equations. J. Differ. Equat. **15**, 350–378 (1974)
64. Chow, S.-N., Diekmann, O., Mallet-Paret, J.: Multiplicity of symmetric periodic solutions of a nonlinear Volterra integral equation. Jpn. J. Appl. Math. **2**, 433–469 (1985)
65. Chow, S.N., Hale, J.: Methods of Bifurcation Theory. Springer, New York (1982)
66. Chow, S.-N., Li, C., Wang, D.: Normal Forms and Bifurcations of Planar Vector Fields. Cambridge University Press, Cambridge (1994)
67. Chow, S.-N., Lin, X.-L., Mallet-Paret, J.: Transition layers for singularly perturbed delay differential equations with monotone nonlinearities. J. Dynam. Differ. Equat. **1**, 3–43 (1989)
68. Chow, S.N., Mallet-Paret, J.: Integral averaging and bifurcation. J. Differ. Equat. **26**, 112–159 (1977)
69. Chow, S.-N., Mallet-Paret, J.: The Fuller index and global Hopf bifurcation. J. Differ. Equat. **29**, 66–85 (1978)
70. Chow, S.-N., Mallet-Paret, J.: Singularly perturbed delay differential equations. In: Chandra, J., Scott, A. (eds.) Coupled Oscillators, pp. 7–12. North-Holland, Amsterdam (1983)

71. Chow, S.N., Mallet-Paret, J., Yorke, J.A.: Global Hopf bifurcation from a multiple eigenvalue. Nonlinear Anal. **2**, 753–763 (1978)

72. Cooke, K.L., Huang, W.Z.: On the problem of linearization for state-dependent delay differential equations. Proc. Am. Math. Soc. **124**, 1417–1426 (1996)

73. Coppel, W.A.: Stability and Asymptotic Behavior of Differential Equations. Health, Boston (1965)

74. Crandall, M.G., Rabinowitz, P.H.: The Hopf bifurcation theorem in infinite dimension. Arch. Ration. Mech. Anal. **67**, 53–72 (1977/78)

75. Cicogna, G.: Symmetry breakdown from bifurcation. Lettere al Nuovo Cimento **31**, 600–602 (1981)

76. Cushing, J.M.: Integrodifferential Equations and Delay Models in Population Dynamics. Lecture Notes in Biomathematics, vol. 20. Springer, New York (1977)

77. Cushman, R., Sanders, J.A.: Nilpotent normal forms and representation theory of $sl(2, R)$. In: Golubitsky, M., Guckenheimer, J. (eds.) Multiparameter Bifurcation Theory. Contemporary Mathematics, vol. 56, pp. 31–51. AMS, Providence (1986)

78. Cushman, R., Sanders, J.A.: Splitting algorithm for nilpotent normal forms. Dynam. Stabil. Syst. **2**(3–4), 235–246 (1988)

79. Cushman, R., Sanders, J.A.: A survey of invariant theory applied to normal forms of vector fields with nilpotent linear part. In: Proceedings of Invariant Theory, pp. 82–106. Springer, New York (1990)

80. de Oliveira, J.C., Hale, J.K.: Dynamic behavior from the bifurcation function. Tôhoku Math. J. **32**, 577–592 (1980)

81. Diekmann, O., van Gils, S.A.: The center manifold for delay equations in the light of suns and stars. In: Roberts, M., Stewart, I.N. (eds.) Singularity Theory and Its Application, Warwick, 1989, Part II, Springer LMN 1463, pp. 122–141. Springer, New York (1991)

82. Diekmann, O., van Gils, S.A., Verduyn Lunel, S.M., Walther, H.O.: Delay Equations, Functional-, Complex-, and Nonlinear Analysis. Springer, New York (1995)

83. Dormayer, P.: Smooth bifurcation of symmetric periodic solutions of functional-differential equations. J. Differ. Equat. **82**, 109–155 (1989)

84. Dumortier, F., Ibáñez, S.: Singularities of vector fields on $\mathbb{R}^3$. Nonlinearity **11**, 1037–1047 (1998)

85. Dylawerski, G., Gęba, K., Jodel, J., Marzantowicz, W.: S^1-equivalent degree and the Fuller index. Ann. Polon. Math. **52**, 243–280 (1991)

86. Eichmann, M.: A local Hopf bifurcation theorem for differential equations with state-dependent delays. Ph.D. Dissertation, Justus-Liebig University in Giessen (2006)

87. Elphick, C., Tirapegui, E., Brachet, M.E., Coullet, P., Iooss, G.: A simple global characterization for normal forms of singular vector fields. Phys. D **29**, 95–127 (1987)

88. Erbe, L.H., Krawcewicz, W., Geba, K., Wu, J.: S^1-degree and global Hopf bifurcation theory of functional differential equations. J. Differ. Equat. **98**, 227–298 (1992)

89. Erbe, L.H., Krawcewicz, W., Peschke, G.: Bifurcations of a parametrized family of boundary value problems for second order differential inclusions. Ann. Math. Pura Appl. **165**, 169–195 (1993)

90. Erbe, L.H., Krawcewicz, W., Wu, J.: Leray-Schauder degree for semilinear Fredholm maps and periodic boundary value problems of neutral equations. Nonlinear Anal. **15**, 747–764 (1990)

91. Faria, T., Magalhães, L.T.: Normal forms for retarded functional-differential equations with parameters and applications to Hopf bifurcation. J. Differ. Equat. **122**(2), 181–200 (1995)

92. Faria, T., Magalhães, L.T.: Normal forms for retarded functional-differential equations and applications to Bogdanov-Takens singularity. J. Differ. Equat. **122**(2), 201–224 (1995)

93. Faria, T., Magalhães, L.T.: Restrictions on the possible flows of scalar retarded functional differential equations in neighborhoods of singularities. J. Dynam. Differ. Equat. **8**, 35–70 (1996)

94. Feigenbaum, M.J.: Quantitative universality for a class of nonlinear transformations. J. Stat. Phys. **19**(1), 25–52 (1978)

95. Fenichel, N.: Persistence and smoothness of invariant manifolds for flows. Indiana Univ. Math. J. **21**, 193–226 (1971)
96. Fenichel, N.: Geometric singular perturbation theory for ordinary differential equations. J. Differ. Equat. **31**, 53–98 (1979)
97. Fermi, E., Pasta, J., Ulam, S.: Los Alamos Report LA-1940 (E. Fermi, Collected Papers II (1955)), pp. 977–988. University of Chicago Press, Chicago (1965)
98. Field, M.J.: Lectures on Bifurcations, Dynamics and Symmetry. Pitman Research Notes in Mathematics, vol. 356. Longman, Harlow (1996)
99. Field, M.J., Melbourne, I., Nicol, M.: Symmetric attractors for diffeomorphisms and flows. Proc. Lond. Math. Soc. **72**, 657–696 (1996)
100. Fiedler, B.: Global Hopf bifurcation in porous catalysts. In: Knobloch, H.W., Schmidt, K. (eds.) Proceedings Equadiff 82. Lecture Notes in Mathematics 1017, pp. 177–184. Springer, New York (1983)
101. Fiedler, B.: An index for global Hopf bifurcation in parabolic systems. J. Reine Angew. Math. **359**, 1–36 (1985)
102. Fiedler, B.: Global Bifurcation of Periodic Solutions with Symmetry. Lecture Notes in Mathematics, vol. 1309. Springer, New York (1988)
103. Fiedler, M.: Additive compound matrices and inequality for eigenvalues of stochastic matrices. Czech. Math. J. **99**, 392–402 (1974)
104. Filip, A.M., Venakides, S.: Existence and modulation of traveling waves in particle chains. Comm. Pure Appl. Math. **52**, 693–735 (1999)
105. Freire, E., Gamero, E., Rodríguez-Luis, A.J., Algaba, A.: A note on the triplezero linear degeneracy: normal forms, dynamical and bifurcation behaviors of an unfolding. Int. J. Bifurcat. Chaos Appl. Sci. Eng. **12**, 2799–820 (2002)
106. Gamero, E., Freire, E., Rodríguez-Luis, A.J.: Hopf-zero bifurcation: normal form calculation and application to an electronic oscillator. In: International Conference on Differential Equations, vol. 1, 2 (Barcelona, 1991), pp. 517–524. World Scientific, River Edge, NJ (1993)
107. Gaspard, P.: Local birth of homoclinic chaos. Phys. D **62**, 94–122 (1993)
108. Gavrilov, N.: On some bifurcations of an equilibrium with one zero and a pair of pure imaginary roots. In: Methods of Qualitative Theory of Differential Equations (in Russian). GGU, Gorkii (1978)
109. Gavrilov, N.: Bifurcations of an equilibrium with two pairs of pure imaginary roots. In: Methods of Qualitative Theory of Differential Equations (in Russian). GGU, Gorkii (1980)
110. Gavrilov, N.K., Shil'nikov, L.P.: On three-dimensional systems close to systems with a structurally unstable homoclinic curve: II. Math. USSR-Sb. **19**, 139–156 (1973)
111. Geba, K., Marzantowicz, W.: Global bifurcation of periodic solutions. Topol. Methods Nonlinear Anal. **1**, 67–93 (1993)
112. Geba, K., Krawcewicz, W., Wu, J.: An equivariant degree with applications to symmetric bifurcation problems 1: construction of the degree. Bull. Lond. Math. Soc. **69**, 377–398 (1994)
113. Giannakopoulos, F., Zapp, A.: Local and global Hopf bifurcation in a scalar delay differential equation. J. Math. Anal. Appl. **237**(2), 425–450 (1999)
114. Giannakopoulos, F., Zapp, A.: Bifurcations in a planar system of differential delay equations modeling neural activity. Phys. D **159**, 215–232 (2001)
115. Golubitsky, M., Guillemin, V.: Stable Mappings and Their Singularities. Graduate Texts in Mathematics, vol. 14. Springer, New York (1973)
116. Golubitsky, M., Marsden, J., Stewart, I., Dellnitz, M.: The constrained Lyapunov-Schmidt procedure and periodic orbits. Field. Inst. Comm. **4**, 81–127 (1995)
117. Golubitsky, M., Schaeffer, D.G.: Singularities and Groups in Bifurcation Theory, vol. 1. Springer, New York (1985)
118. Golubitsky, M., Stewart, I., Schaeffer, D.G.: Singularities and Groups in Bifurcation Theory, vol. 2. Springer, New York (1988)
119. Govaerts, W., Pryce, J.: Mixed block elimination for linear systems with wider borders. IMA J. Numer. Anal. **13**, 161–180 (1993)

120. Grabosch, A., Moustakas, U.: A semigroup approach to retarded differential equations. In: Nagel, R. (ed.) One-parameter Semigroups of Positive Operators. Lecture Notes in Mathematics, vol. 1184, pp. 219–232. Springer, Berlin (1986)

121. Grafton, R.B.: A periodicity theorem for autonomous functional differential equations. J. Differ. Equat. **6**, 87–109 (1969)

122. Grimmer, R.: Existence of periodic solutions of functional differential equations. J Math. Anal. Appl. **72**(2), 666–673 (1979)

123. Grobman, D.: Homeomorphisms of systems of differential equations. Dokl. Akad. Nauk SSSR **128**, 880 (1959)

124. Guckenheimer, J.: On a codimension two bifurcation. In: Rand, D.A., Young, L.-S. (eds.) Dynamical Systems and Turbulence. Warwick 1980 (Coventry, 1979/1980), vol. 898 of Lecture Notes in Mathematics, pp. 99–142. Springer, Berlin (1981)

125. Guckenheimer, J., Holmes, P.J.: Nonlinear Oscillations: Dynamical System and Bifurcations of Vector Fields. Springer, New York (1983)

126. Guckenheimer, J.: Multiple bifurcation problems of codimension two. SIAM J. Math. Anal. **15**, 1–49 (1984)

127. Gumowski, I.: Sur le calcul des solutions périodiques de l'équation de Cherwell-Wright. C.R. Acad. Sci. Paris Ser. A-B **268**, 157–159 (1969)

128. Guo, S.: Equivariant normal forms for neutral functional differential equations. Nonlinear Dyn. **61**(1), 311–329 (2010)

129. Guo, S.: Spatio-temporal patterns of nonlinear oscillations in an excitatory ring network with delay. Nonlinearity **18**, 2391–2407 (2005)

130. Guo, S.: Zero singularities in a ring network with two delays. Z. Angew. Math. Phys. 64(2), 201–222 (2013)

131. Guo, S., Chen, Y., Wu, J.: Equivariant normal forms for parameterized delay differential equations with applications to bifurcation theory. Acta Math. Sin. Engl. Ser. **28**(4), 825–856 (2012)

132. Guo, S., Chen, Y., Wu, J.: Two-parameter bifurcations in a network of two neurons with multiple delays. J. Differ. Equat. **244**, 444–486 (2008)

133. Guo, S., Huang, L.: Hopf bifurcating periodic orbits in a ring of neurons with delays. Phys. D **183**(1–2), 19–44 (2003)

134. Guo, S., Huang, L.: Global continuation of nonlinear waves in a ring of neurons. Proc. Math. Roy. Soc. Edinb. **135A**, 999–1015 (2005)

135. Guo, S., Huang, L.: Stability of nonlinear waves in a ring of neurons with delays. J. Differ. Equat. **236**, 343–374 (2007)

136. Guo, S., Lamb, J.S.W.: Equivariant Hopf bifurcation for neutral functional differential equations. Proc. Am. Math. Soc. **136**, 2031–2041 (2008)

137. Guo, S., Lamb, J.S.W., Rink, B.W.: Branching patterns of wave trains in the FPU lattice. Nonlinearity **22**, 283–299 (2009)

138. Guo, S., Man, J.: Center manifolds theorem for parameterized delay differential equations with applications to zero singularities. Nonlinear Anal. Theor. Meth. Appl. **74**(13), 4418–4432 (2011)

139. Guo, S., Man, J.: Patterns in hierarchical networks of neuronal oscillators with D3 × Z3 symmetry. J. Differ. Equat. **254**, 3501–3529 (2013)

140. Guo, S., Yuan, Y.: Pattern formation in a ring network with delay. Math. Model. Meth. Appl. Sci. **19**(10), 1797–1852 (2009)

141. Guo, S., Wu, J.: Generalized Hopf bifurcation in delay differential equations (in Chinese). Sci. Sin. Math. **42**, 91–105 (2012)

142. Gurney, W.S.C., Blythe, S.P., Nisbee, R.M.: Nicholson's blowflies revisited. Nature **287**, 17–21 (1980)

143. Hadeler, K.P., Tomiuk, J.: Periodic solutions of difference differential equations. Arch. Ration. Anal. **1**, 87–95 (1977)

144. Hale, J.K.: Linear Functional-Differential Equations with Constant Coefficients. Contributions to Differential Equations II, pp. 291–317. Research Institute for Advanced Studies, Baltimore (1963)

145. Hale, J.K.: Critical cases for neutral functional differential equations. J. Differ. Equat. **10**, 59–82 (1971)
146. Hale, J.K.: Theory of Functional Differential Equations. Springer, New York (1977)
147. Hale, J.K.: Flows on centre manifolds for scalar functional differential equations. Proc. Math. Roy. Soc. Edinb. **101A**, 193–201 (1985)
148. Hale, J.K.: Large diffusivity and asymptotic behavior in parabolic systems. J. Differ. Equat. **118**, 455–466 (1986)
149. Hale, J.K.: Partial neutral functional-differential equations. Rev. Roum. Math. Pure. Appl. **39**, 339–344 (1994)
150. Hale, J.K.: Diffusive coupling, dissipation, and synchronization. J. Dynam. Differ. Equat. **9**(1), 1–52 (1997)
151. Hale, J.K., Huang, W.: Period doubling in singularly perturbed delay equations. J. Differ. Equat. **114**, 1–23 (1994)
152. Hale, J.K., Kocak, H.: Dynamics and Bifurcations. Springer, New York (1991)
153. Hale, J.K., Tanaka, S.M.: Square and pulse waves with two delays. J. Dynam. Differ. Equat. **12**, 1–30 (2000)
154. Hale, J.K., Verduyn Lunel, S.M.: Introduction to Functional Differential Equations. Springer, New York (1993)
155. Hale, J.K., Weedermann, M.: On perturbations of delay differential equations with periodic orbits. J. Differ. Equat. **197**, 219–246 (2004)
156. Hartung, F.: Linearized stability in periodic functional differential equations with state-dependent delays. J. Comput. Appl. Math. **174**, 201–211 (2005)
157. Hartung, F., Turi, J.: On differentiability of solutions with respect to parameters in state-dependent delay equations. J. Differ. Equat. **135**, 192–237 (1997)
158. Hartung, F., Krisztin, T., Walther, H.-O., Wu, J.: Functional differential equations with state-dependent delays: theory and applications. In: Canada, A. (ed.) Handbook of Differential Equations: Ordinary Differential Equations, vol. 3. Elsevier, North Holland (2006)
159. Hassard, B.D., Wan, Y.H.: Bifurcation formulae derived from center manifold theory. J. Math. Appl. Math. **42**, 297–260 (1978)
160. Hassard, B.D., Kazarinoff, N.D., Wan, Y.H.: Theory and Applications of Hopf Bifurcation. Cambridge University Press, Cambridge (1981)
161. Hartman, P.: A lemma in the theory of structural stability of differential equations. Proc. Am. Math. Soc. **11**, 610–620 (1960)
162. Hartman, P.: Ordinary Differential Equations. Wiley, New York (1964)
163. Hirsch, M.W., Smale, S.: Differential Equations, Dynamical Systems and Linear Algebra. Academic, New York (1974)
164. Hirsch, M.W., Push, C.C., Shub, M.: Invariant Manifolds. Springer Lecture Notes in Mathematics, vol. 583. Springer, New York (1977)
165. Hirschberg, P., Knobloch, E: Silnikov-Hopf bifurcation. Phys. D **62**, 202–216 (1993)
166. Holmes, P.J.: Unfolding a degenerate nonlinear oscillators: a codimension two bifurcation. In: Helleman, R.H.G. (ed.) Nonlinear Dynamics, pp. 473–488. New York Academy of Science, New York (1980)
167. Hopf, E.: Abzweigung einer periodischen lösung eines Differential Systems. Berichen Math. Phys. Kl. Säch. Akad. Wiss. Leipzig **94**, 1–22 (1942)
168. Hopfield, J.J.: Neurons with graded response have collective computational properties like those of two-state neurons. Proc. Nat. Acad. Sci. U.S.A. **81**, 3088–3092 (1984)
169. Hsu, I.D., Kazarinoff, N.D.: An applicable Hopf bifurcation formula and instability of small periodic solutions of the Field-Noyes model. J. Math. Anal. Appl. **55**, 61–89 (1976)
170. Hu, Q., Wu, J.: Global Hopf bifurcation for differential equations with state-dependent delay. J. Differ. Equat. **248**, 2801–2840 (2010)
171. Hu, Q., Wu, J.: Global continua of rapidly oscillating periodic solutions of state-dependent delay differential equations. J. Dynam. Differ. Equat. **22**, 253–284 (2010)
172. Hu, Q., Wu, J., Zou, X.: Estimates of periods and global continua of periodic solutions of differential equations with state-dependent delay. SIAM J. Math. Anal. **44**, 2401–2427 (2012)

173. Humphreys, J.E.: Introduction to Lie Algebras and Representation Theory. Springer, New York (1978)
174. Iooss, G.: Travelling waves in the Fermi-Pasta-Ulam lattice. Nonlinearity **13**, 849–866 (2000)
175. Iooss, G., Adelmeyer, M.: Topics in Bifurcation Theory and Applications. World Scientific, Singapore (1992)
176. Iooss, G., Langford, W.F.: Conjectures on the routes to turbulence via bifurcation. In: Helleman, R.H.G. (ed.) Nonlinear Dynamics, pp. 489–505. New York Academy of Science, New York (1980)
177. Ize, J., Bifurcation Theory for Fredholm Operators, vol. 174. Memoirs of the American Mathematical Society, Providence (1976)
178. Ize, J.: Obstruction theory and multiparameter Hopf bifurcation. Trans. Am. Math. Soc. **289**, 757–792 (1985)
179. Ize, J., Massabó, I., Vignoli, V.: Degree theory for equivariant maps, I. Trans. Am. Math. Soc. **315**, 433–510 (1989)
180. Ize, J., Massabó, I., Vignoli, V.: Degree theory for equivariant maps, the $\mathbb{S}^1$-action. Memoirs of the American Mathematical Society, vol. 418. American Mathematical Society, Providence (1992)
181. Ize, J., Vignoli, A.: Equivariant degree for abelian actions, Part I; equivariant homotopy groups. Topol. Methods Nonlinear Anal. **2**, 367–413 (1993)
182. Ize, J., Vignoli, A.: Equivariant degree for abelian actions, Part II; Index computations. Topol. Methods Nonlinear Anal. **7**, 369–430 (1996)
183. Jolly, M.S., Rosa, R.: Computation of non-smooth local centre manifolds. IMA J. Numer. Anal. **25**(4), 698–725 (2005)
184. Joseph, D.D., Sattinger, D.H.: Bifurcating time periodic solutions and their stability. Arch. Ration. Mech. Anal. **45**, 79–109 (1972)
185. Kaplan, L., Yorke, J.A.: Ordinary differential equations which yield periodic solutions of differential delay equations. J. Math. Anal. Appl. **48**(2), 317–324 (1974)
186. Kato, T.: Perturbation Theory for Linear Operators, Grundlehren der mathematischen Wissenschaften, vol. 132. Springer, New York (1976)
187. Keener, J.: Infinite period bifurcation and global bifurcation branches. SIAM J. Appl. Math. **41**, 127–144 (1981)
188. Keller, H.: Numerical solution of bifurcation and nonlinear eigenvalue problems. In: Rabinowitz, P. (ed.) Applications of Bifurcation Theory, pp. 359–384. Academic, New York (1977)
189. Kelley, A.: The stable, center-stable, center, center-unstable and unstable manifolds. J. Differ. Equat. **3**, 546–570 (1967)
190. Kielhöfer, H.: Hopf bifurcation at multiple eigenvalues. Arch. Ration. Mech. Anal. **69**, 53–83 (1979)
191. Krasnosel'skii, M.A.: Topological Methods in the Theory of Nonlinear Integral Equations. Pergamon, New York (1964)
192. Krawcewicz, W., Vivi, P.: Normal bifurcation and equivariant degree. Indian J. Math. **42**, 55–68 (2000)
193. Krawcewicz, W., Wu, J.: Theory of Degrees with Applications to Bifurcations and Differential Equations. Canadian Mathematical Society Series of Monographs and Advanced Texts. Wiley, New York (1997)
194. Krawcewicz, W., Wu, J.: Theory and applications of Hopf bifurcations in symmetric functional-differential equations. Nonlinear Anal. Theor. Meth. Appl. **35**(7), 845–870 (1999)
195. Krawcewicz, W., Wu, J., Xia, H.: Global Hopf bifurcation theory for condensing fields and neutral equations with applications to lossless transmission problems. Can. Appl. Math. Q. **1**, 167–220 (1993)
196. Krisztin, T.: A local unstable manifold for differential equations with state-dependent delay. Discrete Contin. Dyn. Syst. **9**, 993–1028 (2003)
197. Krisztin, T., Walther, H.-O., Wu, J.: Shape, Smoothness and Invariant Stratification of an Attracting Set for Delayed Monotone Positive Feedback. The Fields Institute Monograph Series. American Mathematical Society, Providence (1999)

198. Kuang, K.: Global attractivity and periodic solutions in delay-differential equations related to models in physiology and population biology. Jpn. J. Ind. Appl. Math. **9**, 205–238 (1992)
199. Kulenovic, M.R.S., Ladas, G.: Linearized oscillations in population dynamics. Bull. Math. Biol. **49**, 615–627 (1987)
200. Kuznetsov, Y.A.: Elements of Applied Bifurcation Theory. Applied Mathematical Sciences, vol. 112, 2nd edn. Springer, Berlin (1998)
201. Kuznetsov, Y.A.: Numerical normalization techniques for all codim 2 bifurcations of equilibria in ODEs. SIAM J. Numer. Anal. **36**, 1104–1124 (1999)
202. Langford, W.F.: Periodic and steady-state mode interactions lead to tori. SIAM J. Appl. Math. **37**, 649–686 (1979)
203. Langford, W.F.: Chaotic dynamics in the unfoldings of degenerate bifurcations. In: Proceedings of the International Symposium on Applied Mathematics and Information Science, Kyoto University, Japan, pp. 241–247 (1982)
204. Langford, W.F.: A review of interactions of Hopf and steady-state bifurcations. In: Barenblatt, G.I., Iooss, G., Joseph, D.D. (eds.) Nonlinear Dynamics and Turbulence, pp. 215–237. Pitman Advanced Publishing Program, Boston (1983)
205. Langford, W.F.: Hopf bifurcation at a hysteresis point. In: Szőkefalvi-Nagy, B., Hatvani, L. (eds.) Differential Equations: Qualitative Theory, Colloq. Math. Soc. János Bolyai, vol. 47, pp. 649–686. North Holland, Amsterdam (1987)
206. Lenhart, S.N., Travis, C.C.: Stability of functional partial differential equations. J. Differ. Equat. **58**, 212–227 (1985)
207. Leray, J., Schauder, J.: Topologie et équations fonctionnelles. Ann. Sci. Ecole. Norm. Sup. **51**, 45–78 (1934)
208. Levinger, B.W.: A Folk theorem in functional differential equations. J. Differ. Equat. **4**, 612–619 (1968)
209. Li, S., Liao, X., Li, C., Wong, K.-W.: Hopf bifurcation of a two-neuron network with different discrete time delays. Int. J. Bifurcat. Chaos Appl. Sci. Eng. **15**, 1589–1601 (2005)
210. Li, M.Y., Muldowney, J.S.: On Bendixson's criterion. J. Differ. Equat. **106**, 27–39 (1993)
211. Ma, T., Wang, S.: Bifurcation theory and applications. World Scientific Series on Nonlinear Science. Series A: Monographs and Treatises, vol. 53. World Scientific, Hackensack, NJ (2005)
212. Mallet-Paret, J.: Generic periodic solutions of functional differential equation. J. Differ. Equat. **25**, 163–183 (1977)
213. Mallet-Paret, J.: Morse decomposition for delay differential equations. J. Differ. Equat. **72**, 270–315 (1988)
214. Mallet-Paret, J., Nussbaum, R.: Global continuation and asymptotic behavior for periodic solutions of a delay differential equation. Ann. Math. Pura Appl. **145**, 33–128 (1986)
215. Mallet-Paret, J., Nussbaum, R.D., Paraskevopoulos, P.: Periodic solutions for functional-differential equations with multiple state-dependent time lags. Topol. Methods Nonlinear Anal. **3**, 101–162 (1994)
216. Mallet-Paret, J., Yorke, J.A.: Snakes: oriented families of periodic orbits, their sources, sinks and continuation. J. Differ. Equat. **43**, 419–450 (1982)
217. Marsden, J., McCracken, M.: The Hopf Bifurcation and Its Applications. Applied Mathematical Sciences, vol. 19. Springer, New York (1976)
218. Medvedev, V.: On a new type of bifurcations on manifolds. Mat. Sbornik **113**, 487–492 (1980) (in Russian)
219. Memory, M.C.: Bifurcation and asymptotic behaviour of solutions of a delay-differential equation with diffusion. SIAM J. Math. Anal. **20**, 533–546 (1989)
220. Memory, M.C.: Stable and unstable manifolds for partial functional differential equations. Nonlinear Anal. **16**, 131–142 (1991)
221. Memory, M.C.: Invariant manifolds for partial functional differential equations. In: Arino, O., Axelrod, D.E., Kimmel, M. (eds.) Mathematical Population Dynamics, pp. 223–232. Marcel Dekker, New York (1991)
222. Metz, J.Z., Diekmann, O.: The Dynamics of Physiologically Structured Populations. Springer, New York (1986)

223. Michel, L.: Points critiques des fonctions G-invariantes. Note aux Comptes-Rendus Acad. Sci. Paris sér. A-B **272**, A433–A436 (1971)

224. Milton, J.: Dynamics of Small Neural Populations. American Mathematical Society, Providence, RI (1996)

225. Morita, Y.: Destablization of periodic solutions arising in delay-diffusion systems in several space dimensions. Jpn. J. Appl. Math. **1**, 39–65 (1984)

226. Muldowney, J.S.: Compound matrices and ordinary differential equations. Rocky Mt. J. Math. **20**, 857–871 (1990)

227. Munkres, J.: Topology, 2nd edn. Prentice Hall, Englewood Cliffs (1975)

228. Negrini, P., Salvadori, L.: Attractivity and Hopf bifurcation. Nonlinear Anal. **3**, 87–99 (1979)

229. Neimark, J.I.: Motions close to doubly-asymptotic motion. Soviet Math. Dokl. **8**, 228–231 (1967)

230. Newhouse, S., Palis, J., Takens, F.: Bifurcations and stability of families of diffeomorphisms. Publ. Math. Inst. Hautes Etud. Sci. **57**, 5–71 (1983)

231. Nicholson, A.J.: An outline of the dynamics of animal populations. Aust. J. Zool. **2**, 9–65 (1954)

232. Nussbaum, R.D.: Periodic solutions of some nonlinear functional differential equations. Ann. Math. Pura Appl. **101**, 263–338 (1974)

233. Nussbaum, R.D.: A global bifurcation theory with application to functional differential equations. J. Funct. Anal. **19**, 319–338 (1975)

234. Nussbaum, R.D.: Global bifurcation of periodic solutions of some autonomous functional differential equations. J. Math. Anal. Appl. **55**, 699–725 (1976)

235. Nussbaum, R.D.: The range of periods of periodic solutions of $x'(t) = -\alpha f(x(t-1))$. J. Math. Anal. Appl. **58**, 280–292 (1977)

236. Nussbaum, R.D.: A global Hopf bifurcation theorem of functional differential systems. Trans. Am. Math. Soc. **238**, 139–164 (1978)

237. Nussbaum, R.D.: Circulant matrices and differential-delay equations. J. Differ. Equat. **60**, 201–217 (1985)

238. Olien, L., Bélair, J.: Bifurcations, stability, and monotonicity properties of a delayed neural network model. Phys. D **102**, 349–363 (1997)

239. Oster, G., Ipaktchi, A.: Population cycles. In: Eyring, H. (ed.) Periodicities in Chemistry and Biology, pp. 111–132. Academic, New York (1978)

240. Palais, R.: Homotopy theory of infinite dimensional manifolds. Topology **5**, 1–16 (1966)

241. Palis, J., de Melo, W.: Geometric Theory of Dynamical Systems. Springer, New York (1982)

242. Palis, J., Pugh, C.: Fifty problems in dynamical systems. In: Manning, A. (ed.) Dynamical Systems – Warwick 1974, vol. 468 of Lecture Notes in Mathematics, pp. 345–353. Springer, Berlin (1975)

243. Palis, J., Takens, F.: Hyperbolicity and Sensitive Chaotic Dynamics at Homoclinic Bifurcations: Fractal Dimensions and Infinitely Many Attractors, vol. 35 of Cambridge Studies in Advanced Mathematics. Cambridge University Press, Cambridge (1993)

244. Peixoto, M.M.: Structural stability on two-dimensional manifolds. Topology **1**, 101–120 (1962)

245. Pliss, V.: Principal reduction in the theory of stability of motion. Izv. Akad. Nauk. SSSR Math. Ser. **28**, 1297–1324 (1964) (in Russian)

246. Poincaré, H.: Sur les propriétés des fonctions définies par les équations aux différences partielles. Thése. Gauthier-Villars, Paris (1879)

247. Poincaré, H.: Mémoire sur les courbes définis par une equation différentielle IV. J. Math. Pures Appl. **1**, 167–244 (1885)

248. Poincaré, H.: Les Méthodes Nouvelles de la Mécanique Céleste, vol. I. Cauthier-Villars, Paris (1892)

249. Pontryagin, L.: On the dynamical systems close to Hamiltonian systems. J. Exp. Theor. Phys. **4**, 234–238 (1934) (in Russian)

250. Poore, A.B.: On the theory and application of the Hopf-Friedrichs bifurcation theory. Arch. Ration. Mech. Anal. **60**, 371–393 (1976)

251. Rabinowitz, P.: Some global results for nonlinear eigenvalue problems. J. Funct. Anal. **7**, 487–513 (1971)
252. Ruan, S., Filfil, R.F.: Dynamics of a two-neuron system with discrete and distributed delays. Phys. D **191**, 323–342 (2004)
253. Ruan, S., Wei, J.: Periodic solutions of planar systems with two delays. Proc. Math. Roy. Soc. Edinb. **129**, 1017–1032 (1999)
254. Rudin, W.: Functional Analysis. McGraw-Hill Science, New York (1991)
255. Ruelle, D.: Bifurcations in the presence of a symmetry group. Arch. Ration. Mech. Anal. **51**, 136–152 (1973)
256. Ruelle, D., Takens, F.: On the nature of turbulence. Comm. Math. Phys. **20**, 167–192, and **23**, 343–344 (1971)
257. Rustichini, A.: Hopf bifurcation for functional differential equations of mixed type. J. Dynam. Differ. Equat. **1**, 145–177 (1989)
258. Sacker, R.: On invariant surfaces and bifurcations of periodic solutions of ordinary differential equations. Report IMM-NYU 333, New York University (1964)
259. Sanders, J.: On the computation of normal forms. Computational aspects of Lie group representations and related topics. In: Cohen, A.M. (ed.) Proceedings of the 1990 Computational Algebra Seminar, CWI Tracts 84, Amsterdam, pp. 129–142 (1991)
260. Sattinger, D.H.: Bifurcation of periodic solutions of the Navier-Stokes equations. Arch. Ration. Mech. Anal. **41**, 66–80 (1971)
261. Sattinger, D.H.: Bifurcation and symmetry breaking in applied mathematics. Bull. Am. Math. Soc. **3**, 779–819 (1980)
262. Shayer, L.P., Campbell, S.A.: Stability, bifurcation, and multistability in a system of two coupled neurons with multiple time delays. SIAM J. Appl. Math. **61**, 673–700 (2000)
263. Shil'nikov, L.P.: On a Poincaré-Birkhoff problem. Math. USSR-Sb. **3**, 353–371 (1967)
264. Shu, Y., Wang, L., Wu, J.: Global dynamics of the Nicholson's blowflies equation revisited: onset and termination of nonlinear oscillations (preprint)
265. Sieberg, H.W.: Some historical remarks concerning degree theory. Am. Math. Mon. **88**, 125–139 (1981)
266. Sijbrand, J.: Properties of center manifolds. Trans. Am. Math. Soc. **289**, 431–469 (1985)
267. Skinner, F.K., Bazzazi, H., Campbell, S.A.: Two-cell to N-cell heterogeneous, inhibitory networks: precise linking of multistable and coherent properties. J. Comput. Neurosci. **18**, 343–352 (2005)
268. Smale, S.: Diffeomorphisms with many periodic points. In: Cairns, S. (ed.) Differential and Combinatorial Topology, pp. 63–80. Princeton University Press, Princeton, NJ (1963)
269. Smith, H.L.: Hopf bifurcation in a system of functional equations modelling the spread of infectious disease. SIAM J. Appl. Math. **43**, 370–385 (1983)
270. Staffans, O.J.: Hopf bifurcation for functional and functional differential equations with infinite delay. J. Differ. Equat. **70**, 114–151 (1987)
271. Stech, H.: Hopf bifurcation calculations for functional differential equations. J. Math. Anal. Appl. **1109**, 472–491 (1985)
272. Takens, F.: A nonstabilizable jet of a singularity of a vector field. In: Dynamical Systems (Proceedings Symposium, University of Bahia, Salvador, 1971), pp. 583–597. Academic, New York (1973)
273. Takens, F.: Normal forms for certain singularities of vector fields. Ann. Inst. Fourier (Grenoble) **23**, 163–195 (1973)
274. Takens, F.: Singularities of vector fields. Publ. Math. IHES **43**, 47–100 (1974)
275. Thom, R.: Topological models in biology. Topology **8**, 313–335 (1969)
276. Thom, R.: Stabilité structurelle et morphogénése. Benjamin, New York (1972)
277. Tsiligiannis, C.A., Lyberatos, G.: Normal forms, resonance and bifurcation analysis via the Carleman linearization. J. Math. Anal. Appl. **139**, 123–138 (1989)
278. Tu, F., Liao, X., Zhang, W.: Delay-dependent asymptotic stability of a two-neuron system with different time delays. Chaos Solitons Fractals **28**, 437–447 (2006)
279. Turaev, D., Shil'nikov, L.: Blue sky catastrophes. Dokl. Math. **51**, 404–407 (1995)

280. Ushiki, S.: Normal forms for singularities of vector fields. Jpn. J. Appl. Math. **1**, 1–34 (1984)
281. van der Pol, B.: A theory of the amplitude of free and forced triode vibrations. Radio Rev. **1**, 701–710, 754–762 (1920)
282. van Gils, S.A., Valkering, T.: Hopf bifurcation and symmetry: standing and traveling waves in a circular–chain. Jpn. J. Appl. Math. **3**, 207–222 (1986)
283. Vanderbauwhede, A.: Symmetry and bifurcation near families of solutions. J. Diff. Equat. **36**, 173–178 (1980)
284. Vanderbauwhede, A.: Local Bifurcation and Symmetry. Research Notes in Mathematics, vol. 75. Pitman, London (1982)
285. Vanderbauwhede, A.: Center manifolds, normal forms and elementary bifurcations. Dynamics Reported, vol. 2. Wiley, New York (1989)
286. Vanderbauwhede, A., Iooss, G.: Center manifold theory in infinite dimension. Dynam. Report. Exposition Dynam. Syst. (N.S.) **1**, 125–163 (1992)
287. Vidossich, G.: On the structure of periodic solutions of differential equations. J. Differ. Equat. **21**, 263–278 (1976)
288. Walther, H.-O.: A theorem on the amplitudes of periodic solutions of differential delay equations with application to bifurcation. J. Differ. Equat. **29**, 396–404 (1978)
289. Walther, H.-O.: Bifurcation from periodic solutions in functional differential equations. Math. Z. **182**, 269–290 (1983)
290. Walther, H.-O.: The solution manifold and C^1 smoothness of solution operators for differential equations with state-dependent delay. J. Differ. Equat. **195**, 46–65 (2003)
291. Walther, H.-O.: Bifurcation of periodic solutions with large periods for a delay differential equation. Ann. Math. Pura Appl. **185**(4), 577–611 (2006)
292. Weedermann, M.: Normal forms for neutral functional differential equations. Field. Inst. Comm. **29**, 361–368 (2001)
293. Weedermann, M.: Hopf bifurcation calculations for scalar delay differential equations. Nonlinearity **19**, 2091–2102 (2006)
294. Wei, J., Li, M.Y.: Hopf bifurcation analysis in a delayed Nicholson blowflies equation. Nonlinear Anal. **60**, 1351–1367 (2005)
295. Wei, J., Ruan, S.: Stability and bifurcation in a neural network model with two delays. Phys. D **130**, 255–272 (1999)
296. Wei, J.J., Velarde, M.G.: Bifurcation analysis and existence of periodic solutions in a simple neural network with delays. Chaos **14**, 940–953 (2004)
297. Wiggins, S.: Normally Hyperbolic Invariant Manifolds in Dynamical Systems. Springer, New York (1994)
298. Wiggins, S.: Introduction to Applied Nonlinear Dynamical Systems and Chaos, 2nd edn. Springer, New York (2003)
299. Wittenberg, R.W., Holmes, P.: The limited effectiveness of normal forms: a critical review and extension of local bifurcation studies of the Brusselator PDE. Phys. D **100**, 1–40 (1997)
300. Wright, E.M.: A nonlinear differential difference equation. J. Reine Angew. Math. **194**, 66–87 (1955)
301. Wu, J.: Global continua of periodic solutions to some differential equations of neutral type. Tôhoku Math J. **45**, 67–88 (1993)
302. Wu, J.: Theory and Applications of Partial Functional Differential Equations. Springer, New York (1996)
303. Wu, J.: Symmetric functional differential equations and neural networks with memory. Trans. Am. Math. Soc. **350**, 4799–4838 (1998)
304. Wu, J.: Introduction to Neural Dynamics and Signal Transmission Delay. Walter de Gruyter, Berlin (2001)
305. Wu, J., Xia, H.: Self-sustained oscillations in a ring array of coupled lossless transmission lines. J. Differ. Equat. **124** 247–278 (1996)
306. Wu, J., Xia, H.: Rotating waves in neutral partial functional-differential equations. J. Dynam. Differ. Equat. **11**, 209–238 (1999)

Index

additive compound matrix, 175
adjoint
  equation, 48, 78
  operator, 48, 94, 126, 136, 259, 265
algebraic multiplicity, 22, 125, 265
Arzelà–Ascoli theorem, 201
Arzelà-Ascoli theorem, 195
auxiliary function, 159, 161, 166

Bautin bifurcation, 33
Bendixson's criterion, 172, 175
bifurcation, 1
  codimension, 6, 15, 17, 19, 21, 269
  diagram, 1, 10, 13, 14, 25, 34, 263
  point, 10, 19, 25, 159, 161, 163, 165, 166,
    198, 203, 204, 213, 228, 229, 239
  value, 1, 13, 14, 17, 120, 145, 150, 172
bilinear form, 50, 53, 59, 77, 78, 111, 126, 232,
  259
Birkhoff normal form
  truncated, 89
Bogdanov normal form, 21, 261
Bogdanov–Takens bifurcation, 21, 259
Bogdanov-Takens bifurcation
  equivariant, 264

center, 168, 170
  isolated, 170, 191
center manifold, 43, 61, 63, 65, 67–69, 72, 74,
  76, 77, 86, 103, 110, 111, 117, 233, 243,
  265
center manifold theorem, 7, 25, 61, 63, 81,
  233, 239, 260
  symmetric, 83, 235, 237
center subspace, 65, 116
characteristic

equation, 22, 43, 47, 50, 54, 69, 110, 116,
  141, 173, 180, 191, 200, 226, 252, 258
matrix, 47, 54, 77, 134, 142, 167, 170, 232,
  238, 252
value, 167, 168, 170, 177, 191–193, 202,
  226
circulant matrix, 180, 184
  symmetric, 252
closed orbit, 10, 17, 37–39
completing function, 161–163, 165, 166, 196,
  198
connected component, 169, 171, 177, 181,
  203–205, 210, 211, 213–217, 227, 228
critical value, 1
crossing number, 162, 168, 169, 171, 181, 192,
  197, 198, 203, 227
cusp bifurcation, 25, 26
cyclic system, 179, 180

degree
  $\mathbb{S}^1$-equivariant, 160–164, 166
  Brouwer, 154, 156, 157
  Leray-Schauder, 156–158
delay-period disparity set, 211, 213–216
diffeomorphic, 2, 3
dihedral group ($\mathbb{D}_n$), 145, 244, 252, 253
  action, 253
double Hopf bifurcation (*see also* Hopf–Hopf
  bifurcation), 269
double Hopf interaction, 269
Dugundji's extension theorem, 161, 164

eigenspace, 49, 231, 244
  generalized, 49, 57, 58, 65, 79
eigenvalues, 3–7, 10, 11, 15, 20, 27, 28, 30,
  31, 47, 48, 50, 57, 59, 61, 65, 69, 72,
  74, 79, 86, 89, 90, 94, 111, 116, 117,

S. Guo and J. Wu, *Bifurcation Theory of Functional Differential Equations*,
Applied Mathematical Sciences 184, DOI 10.1007/978-1-4614-6992-6,
© Springer Science+Business Media New York 2013

Printed in the United States
By Bookmasters